STRESS

STRESS
NEURAL, ENDOCRINE AND MOLECULAR STUDIES

Edited by

Richard McCarty

American Psychological Association
Washington, DC, USA

Greti Aguilera

National Institute of Child Health and Human Development
Bethesda, MD, USA

Esther L. Sabban

Department of Biochemistry and Molecular Biology
New York Medical College, Valhalla, NY, USA

and

Richard Kvetňanský

Institute of Experimental Endocrinology
Bratislava, Slovak Republic

London and New York

First published 2002
by Taylor & Francis
11 New Fetter Lane, London EC4P 4EE

Simultaneously published in the USA and Canada
by Taylor & Francis Inc
29 West 35th Street, New York, NY 10001

Taylor & Francis is an imprint of the Taylor & Francis Group

© 2002 Taylor & Francis

Typeset in Malaysia by Expo Holdings
Printed and bound in Great Britain by
St Edmundsbury Press, Bury St Edmunds, Suffolk

Every effort has been made to ensure that the advice and information in this book is true and accurate at the time of going to press. However, neither the publisher nor the authors can accept any legal responsibility or liability for any errors or omissions that may be made. In the case of drug administration, any medical procedure or the use of technical equipment mentioned within this book, you are strongly advised to consult the manufacturer's guidelines.

British Library Cataloguing in Publication Data
A catalogue record for this book is available from the British Library

Library of Congress Cataloging in Publication Data
A catalog record for this book has been requested

ISBN 0-415-27220-3

CONTENTS

PART 4   MOLECULAR REGULATION DURING STRESS

PART 5   NEURAL-ENDOCRINE-IMMUNE INTERACTIONS

PART 9   PROLOGUE

PREFACE

The Seventh Symposium on Catecholamines and Other Neurotransmitters in Stress once again presented an unbeatable combination – Smolenice Castle in the beautiful Slovak countryside, a distinguished collection of stress researchers from many different countries, and a carefully planned scientific and social program. The seventh symposium continued the strong tradition of scientific excellence that began in 1975 when Dr Richard Kvetňanský assembled a small but impressive group of stress researchers in Bratislava to discuss the most recent advances in basic and clinical research on catecholamines and stress. Now, nearly a quarter century after that first gathering, these symposia have provided a showcase for the best researchers in this interdisciplinary field to come together and discuss their recent findings as well as look to future research opportunities.

A strong tradition of careful planning and Slovak hospitality have infused each of these gatherings. Dr Kvetňanský and his colleagues at the Institute of Experimental Endocrinology have planned carefully such that each symposium is marked by innovative research presentations by leaders in the field. In addition, the social program enhances the scientific program and creates an atmosphere of collegiality that facilitates an active exchange of ideas.

A significant measure of the credit for the success of this symposium must go to the members of the Local Organizing Committee, headed by Dr I. Jaroscakova. This committee stood ready to assist any meeting attendee from the opening ceremony in Bratislava Castle to the closing ceremony in Smolenice Castle five days later. Many accompanying members are forever indebted to Mrs Darina Kvetňanský, who organized a varied and stimulating social program during the days when the scientific sessions were held.

Plans are already underway for the Eighth Symposium on Catecholamines and Other Neurotransmitters in Stress in 2003 that will once again bring the world's best researchers in these fields of inquiry back to Smolenice Castle for another scientific and cultural tour de force. We hope to see you there!

Richard McCarty

CORRESPONDING AUTHORS

Greti Aguilera
Section on Endocrine Physiology
National Institute of Child Health and Human
    Development
National Institutes of Health
Bethesda, Maryland 20892, USA

Hymie Anisman
Institute of Neuroscience
Life Science Research Building
Carleton University
Ottawa, Ontario KlS 5B6 Canada

Julius Brtko
Institute of Experimental Endocrinology
Slovak Academy of Sciences
Vlarska 3, 833 06 Bratislava, Slovak Republic

Errol B. De Souza
Hoechst AG-Hoechst Marion Roussel
Building H811
D-65926 Frankfurt am Main, Germany

Sladjana Dronjak
Institute of Nuclear Sciences ''Vinca''
Belgrade, Yugoslavia

Adrian J. Dunn
Department of Pharmacology and Therapeutics
Louisiana State University Medical Center
P.O. Box 33932
Shreveport, Louisiana 71130, USA

Nikolai N. Dygalo
Institute of Cytology and Genetics
Siberian Department of the Russian Academy of
    Sciences
Pr. Lavrentjeva 10
Novosibirsk 630090, Russia

Maxim L. Filipenko
Institute of Bioorganic Chemistry
Institute of Cytology and Genetics
Siberian Department of the Russian Academy of
    Sciences
Pr. Lavrentjeva 8
Novosibirsk 630090, Russia

Allison J. Fulford
Division of Medicine
University of Bristol
Marlborough Street
Bristol BS2 8HW, UK

Rachel Gitau
Fetal and Neonatal Stress Research Centre
Division of Paediatrics
Obstetrics and Gynaecology
Imperial College School of Medicine
Queen Charlotte's and Chelsea Hospital
Goldhawk Road
London W6 OXG, UK

David S. Goldstein
Clinical Neurocardiology Section
National Institute of Neurological Disorders and
    Stroke
National Institutes of Health
Bethesda, Maryland 20892, USA

Dirk H. Hellhammer
Center for Psychobiological and Psychosomatic Research
University of Trier
D-54286 Trier, Germany

James P. Herman
Department of Psychiatry
University of Cincinnati Medical Center
Cincinnati, Ohio, USA

N.B. Igosheva
Department of Biology
State University
Astrakhanskaya str. 83
Saratov 410026, Russia

Mark M. Knuepfer
Department of Pharmacological and Physiological
    Science
St. Louis University School of Medicine
1402 South Grand Boulevard
St. Louis, Missouri 63104, USA

Juraj Koška
Institute of Experimental Endocrinology
Slovak Academy of Sciences
Vlarska 3, 833 06 Bratislava, Slovak Republic

Ol'ga Križanová
Institute of Molecular Physiology and Genetics
Slovak Academy of Sciences
Vlarska 5, 833 34 Bratislava, Slovak Republic

Lucia Kšinantová
Institute of Experimental Endocrinology
Slovak Academy of Sciences
Vlarska 3, 833 06 Bratislava, Slovak Republic

Richard Kvetňanský
Institute of Experimental Endocrinology
Slovak Academy of Sciences
Vlarska 3, 833 06 Bratislava, Slovak Republic

Seymour Levine
Department of Psychology
University of Delaware
Newark, Delaware 19716, USA

Christopher A. Lowry
Division of Medicine
University of Bristol
Marlborough Street
Bristol BS2 8HW, UK

Richard McCarty
Science Directorate
American Psychological Association
750 First Street NE
Washington, DC 20002, USA

Toshiharu Nagatsu
Institute of Comprehensive Medical Science
Graduate School of Medicine
Fujita Health University
Toyoake
Aichi 470-1192, Japan

Winfried Otten
Research Institute for the Biology of Farm Animals
Department of Physiological Principles in Animal
    Behaviour
D-18196 Dummerstorf, Germany

Karel Pacak
Clinical Neurocardiology Section
National Institute of Neurological Disorders and
    Stroke
National Institutes of Health
Bethesda, Maryland 20892, USA

Miklós Palkovits
Laboratory of Neuromorphology
Semmelweis University
1094 Budapest
Tuzolto-utca 58, Hungary

Patricia E. Patterson-Buckendahl
808 East Somers Landing Road
Absecon, New Jersey 08201, USA

Franciszek Przekop
The Kielanowski Institute of Animal Physiology and
    Nutrition
05–110 Jabloma, Poland

Cristina Rabadan-Diehl
Section on Endocrine Physiology
National Institute of Child Health and Human
    Development
National Institutes of Health
Bethesda, Maryland 20892, USA

Milan Rusnák
Institute of Experimental Endocrinology
Slovak Academy of Sciences
Vlarska 3, 833 06 Bratislava, Slovak Republic

Esther L. Sabban
Department of Biochemistry and Molecular Biology
New York Medical College
Valhalla, New York 10595, USA

Hiroshi Saito
College of Pharmacy
Nihon University
7-7-1 Narashino-dai
Funabashi-shi
Ciba, 274-8555, Japan

Audrey F. Seasholtz
Mental Health Research Institute
University of Michigan
205 Zona Pitcher Place
Ann Arbor, Michigan 48109, USA

Miro Smriga
Central Research Laboratories
Ajinomoto Company Inc.
1-1 Suzuki-cho
Kawasaki 210-8681, Japan

S. Clare Stanford
Department of Pharmacology
University College London
Gower Street
London WC1E 6BT, UK

A. William Tank
Department of Pharmacology and Physiology
Box 711, University of Rochester Medical Center
601 Elmwood Avenue
Rochester, New York 14642, USA

Steven A. Thomas
Department of Pharmacology
University of Pennsylvania
Philadelphia, Pennsylvania 19104, USA

Nihal Tümer
Geriatric Research
Education and Clinical Center
Department of Veterans Affairs Medical Center
1601 S.W. Archer Road
Gainesville, Florida 32608, USA

Takashi Ueyama
Department of Anatomy and Neurobiology
Wakayama Medical College
811-1 Kimiidera
Wakayama 641-0012, Japan

Marta Weinstock
Department of Pharmacology
School of Pharmacy
Hebrew University Hadassah Medical Center
Ein Kerem
Jerusalem 91120, Israel

Dona Lee Wong
Laboratory of Molecular and Developmental
    Neurobiology
Department of Psychiatry
Harvard Medical School
McLean Hospital
115 Mill Street, MRC 116
Belmont, Massachusetts 02478, USA

Yoshio Yamakami
Department of Physiology
Showa University School of Dentistry
1-5-8 Hatanodai
Shinagawa-ku
Tokyo 142-8555, Japan

L'ubomíra Žačiková
Institute of Molecular Physiology and Genetics
Slovak Academy of Sciences
Vlarska 5, 833 34 Bratislava, Slovak Republic

ACKNOWLEDGEMENTS

The Organizing Committee of the Seventh Symposium on Catecholamines and Other Neurotransmitters in Stress acknowledges with gratitude the generous financial support provided by the following organizations:

National Institute of Child Health and Human Development, USA
The Catecholamine Foundation, USA
Slovak Academy of Sciences, Slovak Republic
Nippon Zoki Pharmaceutical Company, Ltd., Japan
The Wellcome Trust, UK
Wakunaga Pharmaceutical Company, Ltd., Japan
Hoechst Marion Roussel AG, Germany
Ajinomoto Company, Inc., Japan
Bristol-Myers Squibb Pharmaceutical Research Institute, USA
Novartis Pharma Services, Slovak Republic
Immunotech, s.r.o., Slovak Republic
Amersham Pharmacia Biotech, M.G.P. s.r.o., Czech Republic
Clinalfa AG, Switzerland
Baxter Export Corporation, Slovak Republic

1. Stress-Related Central Neuronal Regulatory Circuits

M. Palkovits

Introduction

Recent studies on the actions of various stressful stimuli have provided two basic conclusions: 1. Different stressors may elicit different stress responses, and 2. different stressors may act on different brain areas, and stressful stimuli may reach these areas through different neuronal pathways (Pacak et al., 1998).

We have utilized five different stressors: 1. Pain evoked by subsutaneous administration of 4% formalin (0.2 ml/100 g body weight) into the hind limb, 2. two hours immobilization, 3. cold exposure (5 hours on either on +4°C or –3°C), 4. Insulin-evoked (1.0 or 3.0 IU/kg) hypoglycemia, 5. hemorrhage (acute removal of 10% or 25% of the estimated blood volume) (Pacak et al., 1998). Plasma levels of ACTH, corticosterone, norepinephrine (NE) and epinephrine (EPI) were measured. Using an in vivo microdialysis technique, catecholamines (and their metabolites) were determined before, during and after stress (Pacak et al., 1998).

Painful stimuli can induce immediate hormonal, behavioral and autonomic responses by activating sensory pathways that innervate hypothalamic, brainstem and spinal cord neurons. It is generally accepted that the hypothalamic paraventricular nucleus (PVN) plays a central role in most circuits directing stimuli both to the hypothalamo-pituitary-adrenal (HPA) axis and to sympathetic and parasympathetic preganglionic neurons (ref. see Palkovits, 1999).

Several experiments confirmed that different stressors elicit marked heterogeneity of neuroendocrine and behavioral responses (Gaillet et al., 1991; Pacak et al., 1995a; Pacak et al., 1998). Both immobilization stress and formalin-pain resulted in intense c-fos activation in the parvocellular subdivisions of the PVN, whereas cold stress and insulin failed to have such effects. In contrast, hemorrhage induced c-fos expression in the magnocellular portion of the PVN, as well as in the subfornical organ (SFO), the nucleus of the solitary tract (NTS) and the area postrema (Palkovits et al., 1995). Insulin markedly elevated plasma EPI levels, cold produced high plasma NE levels, while hemorrhage elicited a high ACTH response. Formalin and immobilization increased plasma levels of all three compounds (Pacak et al., 1998).

Central Organization of Responses to Various Stressful Stimuli: Stress-Related Neuronal Circuits

The heterogeneity of stress responses clearly indicates that different stressors may use different afferent and efferent pathways. The suggested stressor-specific neuroanatomical circuits are briefly summarized below.

The afferent loop of the stress pathways can be neuronal and humoral. The neuronal inputs are carried by either somatosensory or viscerosensory nerve fibers into the dorsal horn of the spinal cord, or to brainstem sensory nuclei. From here, stressful signals diverge in different directions and activate different brain areas. Stress responses, organized by different brain regions also can be divided into neuronal and humoral responses. The neuronal responses include short (spinal) and long (supraspinal) circuits. The spinal reflexes (autonomic spinal and withdrawal reflexes) represent the short routes of immediate stress responses. The spinal autonomic reflex is based on either monosynaptic sensory-autonomic connections, or short interneuronal transfer between these two systems. The withdrawal reflex is bilateral (flexor/extensor innervation) and both inhibitory and excitatory interneurons are involved in the signal transduction.

The efferent (descending) leg of the long-loop system activates either the hypothalamo-pituitary system (neuroendocrine stress response), or terminates on medullary parasympathetic or spinal sympathetic preganglionic neurons (autonomic stress response). Descending pathways may activate spinal or brainstem motoneurons (motor stress responses). Descending pathways may arise from the hypothalamus. The majority of the descending fibers arise in the PVN, but a substantial number of axons descend from the arcuate, medial preoptic, and dorsomedial nuclei and from neurons of the lateral hypothalamus (Figure 1). The other significant sources of descending fibers are of amygdaloid origin. Among these fibers, only a small percentage consists of monosynaptic descending projections, the majority of the amygdaloid efferents to preganglionic neurons may be relayed by cells in the bed nucleus of the stria terminalis, the parabrachial nucleus, and some hypothalamic neurons (Figure 2).

The fine anatomical localization of long-loop circuit neurons for different stressors is rather difficult since several brain areas and pathways are involved in the

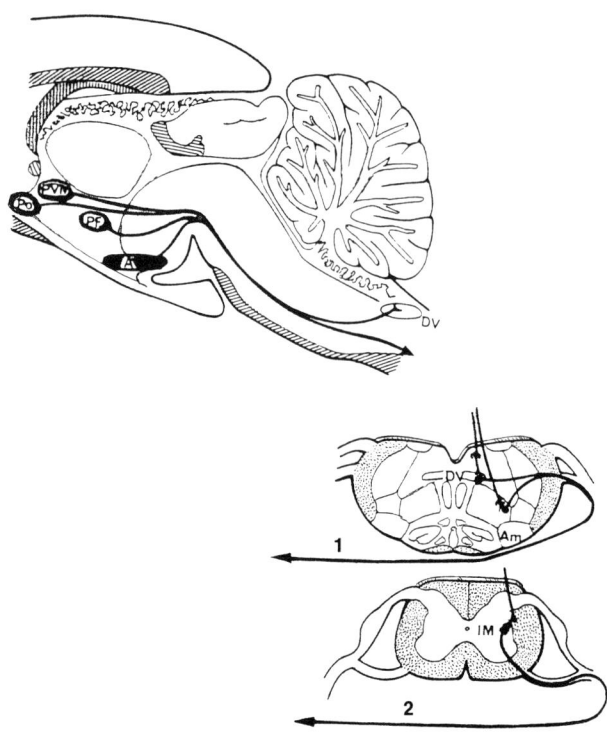

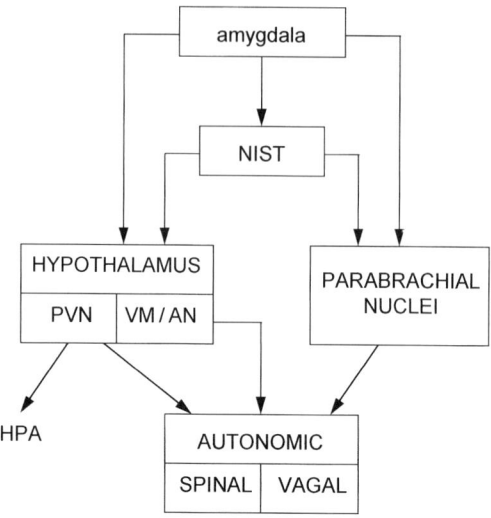

Fig. 2 Neuronal projections from the amygdala. *Abbreviations*: AN – arcuate nucleus, HPA – hypothalamo-pituitary-adrenal axis, NIST – bed nucleus of the stria terminalis, PVN – paraventricular nucleus, VM – hypothalamic ventromedial nucleus.

Fig. 1 Descending neuronal projections from the hypothalamus to medullary and spinal autonomic preganglionic neurons. *Abbreviations*: A – arcuate nucleus, Am – ambiguus nucleus, DV – dorsal motor nucleus of the vagus, IM – intermediolateral cell column, PF – perifornical nucleus, Po – medial preoptic nucleus, PVN – paraventricular nucleus, 1 – parasympathetic preganglionic fibers from the dorsal motor vagal and ambiguus nuclei, 2 – sympathetic preganglionic fibers from the thoracic intermediolateral cell column.

organization of stress responses, and almost all types of stressors use their own individual neuronal avenues.

Formalin-evoked Painful Stimuli

Formalin elicits large corticosterone, ACTH, NE and EPI responses with in a few minutes after an injection (Makara *et al.*, 1969; Pacak *et al.*, 1998; Palkovits *et al.*, 1999). Formalin induces neurons in the dorsal horn of the spinal cord to express *c-fos* (Presley *et al.*, 1990; Abbadie *et al.*, 1992; Senba *et al.*, 1993; Palkovits *et al.*, 1995; Baffi *et al.*, 1996; Palkovits *et al.*, 1997). Fos-immunopositive cell nuclei were detected only ipsilateral to the formalin injections (Figure 3A). Through different pathways (see below), pain-related signals lead to the hypothalamus, mainly to the PVN, where formalin evokes a rapid and marked expression of *c-fos* on both sides (Bullitt, 1990; Senba *et al.*, 1993; Palkovits *et al.*, 1995; Palkovits *et al.*, 1999). Pain induces corticotropin-releasing hormone (CRH) synthesis and release from this nucleus (Swanson and Simmons, 1989; Makino *et al.*, 1995). Using a retrograde transneuronal tracing technique (pseudorabies virus, Bartha strain), the route of the virus could be followed from the skin to the dorsal horn of the spinal cord (Figure 3B) and after a short time the virus appears in PVN neurons (Figure 4).

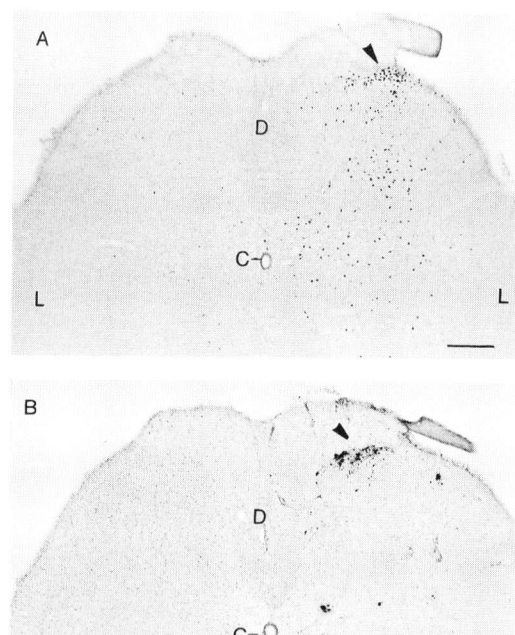

Fig. 3 A) Fos-immunoreactive cells in ipsilateral dorsal horn neurons (lumbar 4 segment) 60 min after a 4% formalin injection (0.1 (1/100 g) into the right hindlimb. B) Virus-infected neurons in the ipsilateral dorsal horn (lumbar 4 segment) 3 days after viral (pseudorabies, Bartha strain) injection into the right hindlimb. No labeling is seen in the spinal cord contralateral to the site of the injections. *Abbreviations*: C – central canal, D – dorsal funiculi, L – lateral funiculus. The superficial layers (laminae I and IIo) of the dorsal horn are pointed by arrowheads. Bar scales: 200 μm.

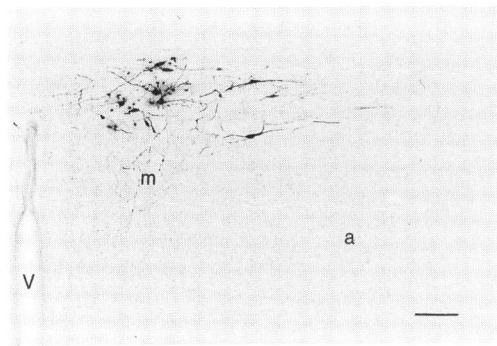

Fig. 4 Virus-labeled neurons in the dorsal subdivision of the paraventricular nucleus 3 days after viral (pseudorabies, Bartha strain) injection into the hindlimb. *Abbreviations*: a – anterior hypothalamic nucleus, m – medial parvocellular PVN, V – third ventricle. Bar scale: 100 μm.

Neuronal Circuits Activated by Formalin-Induced Pain

Somatosensory signals through primary sensory neurons reach the dorsal horn of the spinal cord and the sensory trigeminal nucleus. From these areas, pain-induced signals transfer through at least four major pathways to the hypothalamus and the limbic system (Figure 5):

1. Spinoreticulothalamic tract. This is a multisynaptic pathway with axon collaterals to the brainstem reticular

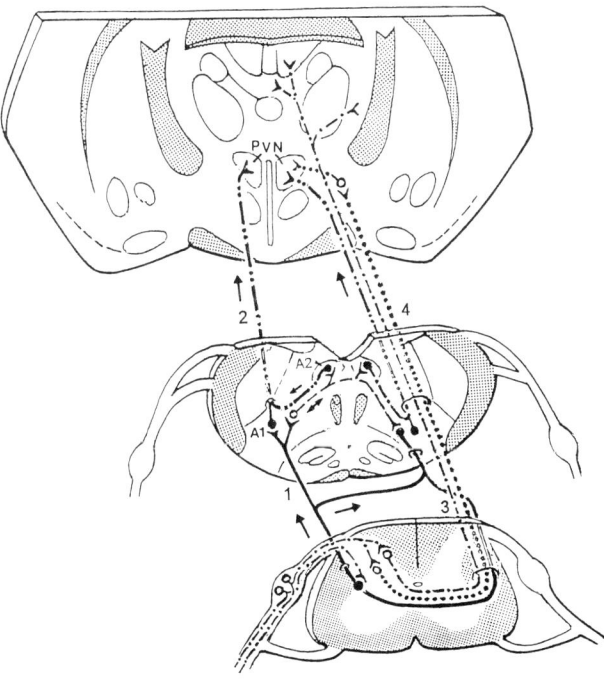

Fig. 5 Ascending pain-related pathways from the spinal cord to the hypothalamus. 1. spinoreticular tract (from the spinal cord to medullary catecholaminergic neurons), 2. ventral noradrenergic bundle, 3. spinoreticulothalamic tract, 4. spinohypothalamic tract. *Abbreviations*: A1 and A2–A1 and A2 catecholaminergic cell groups, PVN – paraventricular nucleus.

formation (Chaouch *et al.*, 1983), the NTS (Otake *et al.*, 1994), the periaqueductal central gray (Hylden *et al.*, 1989) and parabrachial nuclei (Liu, 1986). The pathways finally reach the midline and intralaminar thalamic nuclei from where the pain signals are relayed to limbic cortical areas which are involved in the organization of behavioral responses to painful stimuli. Pain elicits acute and strong *c-fos* expression in the midline thalamic nuclei and in the cingulate and piriform cortical neurons (Senba *et al.*, 1993; Palkovits *et al.*, 1995).

2. Direct spinohypothalamic pathway (Burstein *et al.*, 1987; Burstein *et al.*, 1990; Cliffer *et al.*, 1991; Burstein *et al.*, 1996). Axons arise in the dorsal horn (from neurons located in lamina I, IV and X) and ascend in the contralateral spinal cord and brainstem to the lateral hypothalamus. Marked *c-fos* activation was seen in lateral hypothalamic relay neurons 60 min after formalin injections into the hind paw (Palkovits *et al.*, 1995). From the lateral hypothalamus fibers project to the other side of the hypothalamus and to the limbic system (Giesler *et al.*, 1994; Giesler, 1995; Burstein *et al.*, 1996; Li *et al.*, 1997).

3. The nociceptive signals may elicit stress responses from hypothalamic and amygdaloid nuclei through the activation of brainstem catecholaminergic neurons. Noradrenergic neurons in the caudal ventrolateral medulla (A1 cell group), the dorsomedial medulla (A2 cell group) and the locus coeruleus, as well as adrenergic neurons in the rostral ventrolateral (C1 cell group) and dorsomedial medulla (C2 cell group) are the major sources of hypothalamic and limbic catecholaminergic fibers (Palkovits *et al.*, 1980; Sawchenko and Swanson, 1982; Cunningham and Sawchenko, 1988; Cunningham *et al.*, 1990). These catecholaminergic neurons receive monosynaptic sensory inputs from the spinal cord and sensory brainstem nuclei through the spinoreticular, spinosolitary and trigeminosolitary pathways (Menétrey *et al.*, 1983; Shokunbi *et al.*, 1985; Menétrey and Basbaum, 1987; Lima *et al.*, 1991; Esteves *et al.*, 1993; Tavares *et al.*, 1993). The catecholaminergic fibers from the brainstem ascend in the ventral and to a lesser extent in the dorsal noradrenergic bundle to the hypothalamus (Figure 5). After formalin injections, 5-8 fold increases in NE levels were measured in the PVN by *in vivo* microdialysis technique (Pacak *et al.*, 1995b; Pacak *et al.*, 1998; Palkovits *et al.*, 1999). The positive correlation between NE release in the PVN and plasma ACTH levels in response to a single formalin injection (Pacak *et al.*, 1995b) indicates that PVN NE is stimulatory to PVN CRH neurons.

4. Spinal and trigeminal sensory fibers terminate in peptidergic neurons of the NTS. They, in turn, project to the hypothalamus and the limbic system (Riche *et al.*, 1990; Sawchenko *et al.*, 1990). Painful stimuli from the NTS may reach the hypothalamus and the amygdala through the parabrachial nuclei (Saper and Loewy, 1980).

The *descending* (output) loop of the pain-induced stress response from the hypothalamus and the amygdala may be humoral (activation of the HPA axis) and/or neuronal (Figure 1 and 2). Long descending mono- and multisynaptic pathways from the PVN and the central nucleus of the amygdala terminate on vagal preganglionic neurons in the medulla and on intermediolateral neurons in the thoracic spinal cord (Swanson and Kuypers, 1980; ter Horst *et al.*, 1984; Luiten *et al.*, 1985; Tucker and Saper, 1985; Cechetto and Saper, 1988; Wallace *et al.*, 1989; Hosoya *et al.*, 1991; Portillo *et al.*, 1998; Tóth *et al.*, 1999).

Immobilization Stress

Immobilization stress is considered one of the strongest experimental stress situations. It evokes large increases in plasma levels of corticosterone, ACTH, NE and EPI (Pacak *et al.*, 1998). Immobilization stress should be viewed as a combination of physical (pain, depleted body temperature) and psychological stressors. Thus, neuronal circuits of various input and ouput systems are activated during immobilization.

Induction of *c-fos* expression in response to immobilization stress was found in several brain areas (Ceccatelli *et al.*,1989; Kononen *et al.*,1992; Pezzone *et al.*, 1992; Herbert and Howes, 1993; Senba *et al.*, 1993; Senba *et al.*, 1994; Chen and Herbert, 1995; Cullinan *et al.*,1995; Palkovits *et al.*,1995; Baffi *et al.*, 1996; Palkovits *et al.*,1997). Immobilization induced prominent *c-fos* activation in the paraventicular and supraoptic nuclei in the hypothalamus, in the midline thalamic nuclei (Fig. 6), in the central nucleus of the amygdala (Figure 7A), and in the brainstem catecholaminergic system, especially in the locus coeruleus and the A1 noradrenergic cell group (Baffi *et al.*, 1996; Palkovits *et al.*, 1997).

It is well demonstrated that immobilization increases CRH and vasopressin mRNA expression in the parvocellular PVN (Bartanusz *et al.*, 1993a; Bartanusz *et al.*, 1993b; Ceccatelli and Orazzo, 1993; Harbuz *et al.*, 1994; Kalin *et al.*, 1994; Pacak *et al.*, 1996). Using an *in vivo* microdialysis technique, Pacak *et al.* (Pacak *et al.*, 1992) were the first to demonstrate immobilization stress-induced NE increases in the PVN. Furthermore, a similar effect of immobilization stress has been reported on NE release from the central nucleus of the amygdala

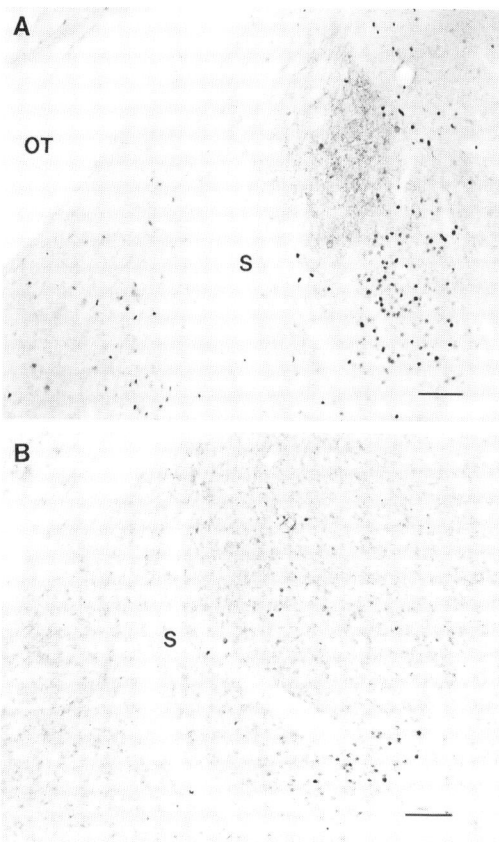

Fig. 7 Fos-immunoreactive cells in the central amygdaloid nucleus. A) Immobilization stress (3 hours): labeled cells in the lateral subdivision of the nucleus. (CRH positive neurons [arrow] did not show Fos-immunopositivity.) B) Insulin hyperglycemia (1 h after 3 IU. insulin): labeled cells at the lateral part of the central nucleus. *Abbreviations*: OT – optic tract, S – stria terminalis. Bar scales: 100 μm.

(Tanaka *et al.*, 1991; Pacak *et al.*, 1993). These two brain areas appear to occupy critical positions in the organization of responses to immobilization stress.

Neuronal Circuits Activated by Immobilization Stress

1. *Inputs* to the PVN and the central nucleus of the amygdala. The ascending loop of the neuronal circuit involves several neuronal pathways. Stressful stimuli, as in pain-evoked stress (Figure 5), may reach these nuclei through ascending catecholaminergic and non-catecholaminergic pathways from the lower brainstem and the spinal cord (Figure 8). In addition to these, cortical areas, like the somatosensory and insular cortex, and hippocampus (representing behavioral components of immobilization stress in animals, and psychological components in humans) are also activated during immobilization. These brain areas are neuronally connected to the PVN and the amygdala.

2. The *output routes*, i.e. response to immobilization stress, as to other stressors could be neurohumoral (activation of the HPA axis) and neuronal (activating the autonomic system). Both the PVN and the central nucleus of the amygdala have mono- and multisynaptic

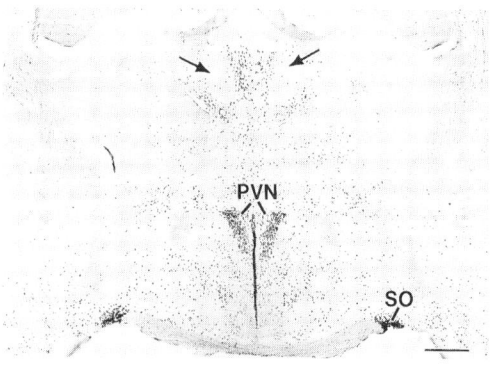

Fig. 6 Fos-immunoreactive cells in the hypothalamic and midline/intralaminar thalamic (arrows) nuclei after 3 hours of immobilization stress. *Abbreviations*: PVN – paraventricular nucleus, SO – supraoptic nucleus. Bar scale: 400 μm.

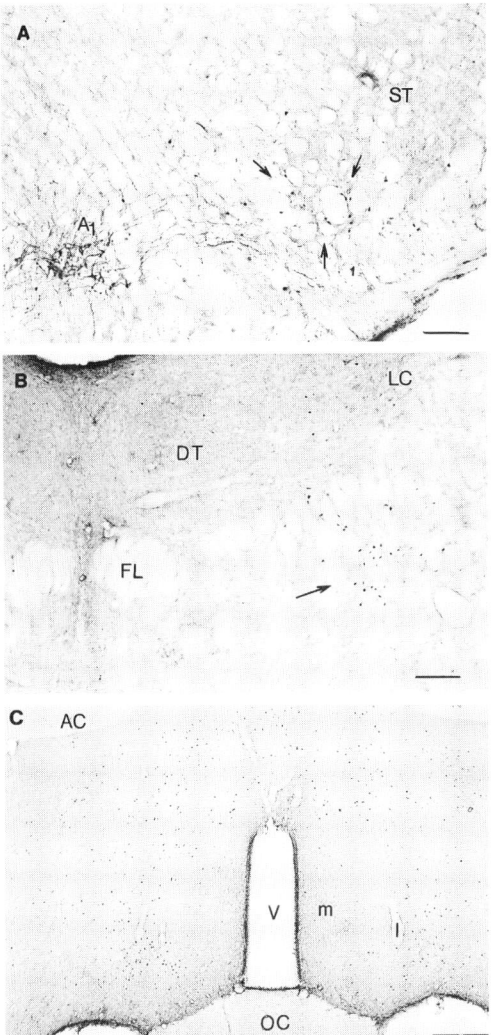

Fig. 8 Neuronal connections between the hypothalamus and vagal and spinal autonomic preganglionic neurons. (Neuronal circuits involved in the organization of responses to immobilization stress.) Ascending inputs from spinal and trigeminal viscero- and somatosensory neurons. Ascending (from A1 and C1) and descending (to A5 and C1) connections with brainstem catecholaminergic (CA) neurons. Descending fibers to medullary (dorsal vagal nucleus) and spinal (intermediolateral) preganglionic neurons.

projections to medullary and spinal autonomic neurons (Figure 1 and 2). Some descending signals from the PVN to the spinal cord are relayed by A5 noradrenergic neurons (Hosoya et al., 1990; Palkovits, 1999). Immobilization stress also activates the somatomotor system as indicated by strong c-fos activation of the cortico-ponto-cerebellar system (Palkovits et al., 1995).

Cold Stress

Plasma levels of NE increase markedly during cold stress, whereas plasma corticosterone and ACTH levels are elevated to a lesser extent (Pacak et al., 1998).

It is well demonstrated and generally accepted that the preoptic region of the hypothalamus is the major organizor for thermoregulation. Neurons in the lateral and median parts of the medial preoptic region receive cold- or heat-evoked signals from medullary and pontine neurons (Boulant, 1980; Lipton and Clark, 1986; Boulant, 1998).

Three hours exposure of rats to –3°C resulted in c-fos activation in several brain regions (Kiyohara et al., 1995; Miyata et al., 1995; Palkovits et al., 1995; Baffi et al., 1996; Palkovits et al., 1997). Especially strong Fos-like immunoreactivity was exhibited in the peritrigeminal nucleus in the medulla oblongata (Figure 9A), in the pontine tegmentum (Figure 9B) and in the medial preoptic nucleus (Figure 9C).

Acute or chronic cold stress increases CRH and thyrotropin-releasing hormone (TRH) mRNA expression in the PVN (Zoeller et al., 1990) and induces TRH release from the median eminence (Arancibia et al., 1983).

Fig. 9 Fos-immunoreactive cells of thermosensitive brain areas in rats following cold stress (+4°C for 3 h). A) Peritrigeminal nucleus (arrows). B) "Pontine thermosensitive area" (arrow). C) Lateral subdivision of the medial preoptic nucleus. *Abbreviations*: AC – anterior commissure, A1–A1 NE cell group, DT – dorsal tegmental nucleus, FL – fasciculus longitudinalis medialis, I – lateral part of the medial preoptic nucleus, LC – locus coeruleus, m – medial part of the medial preoptic nucleus, OC – optic chiasm, ST – nucleus of the spinal trigeminal tract, V – third ventricle. Bar scales: 100 μm (A and B), 200 μm (C).

Neuronal Circuits Activated by Cold Stress

Cold stress affects metabolic, endocrine, autonomic and behavioral systems. Consequently, several brain areas and pathways are involved in the response to thermal stimuli. Thermal stimulation from the environment activates thermoreceptors from which signals are carried by primary sensory fibers to neurons in the dorsal horn of the spinal cord and to the sensory trigeminal nucleus (Figure 10). From these neurons, stimuli reach medullary and pontine thermosensitive nuclei and are further relayed to the medial preoptic

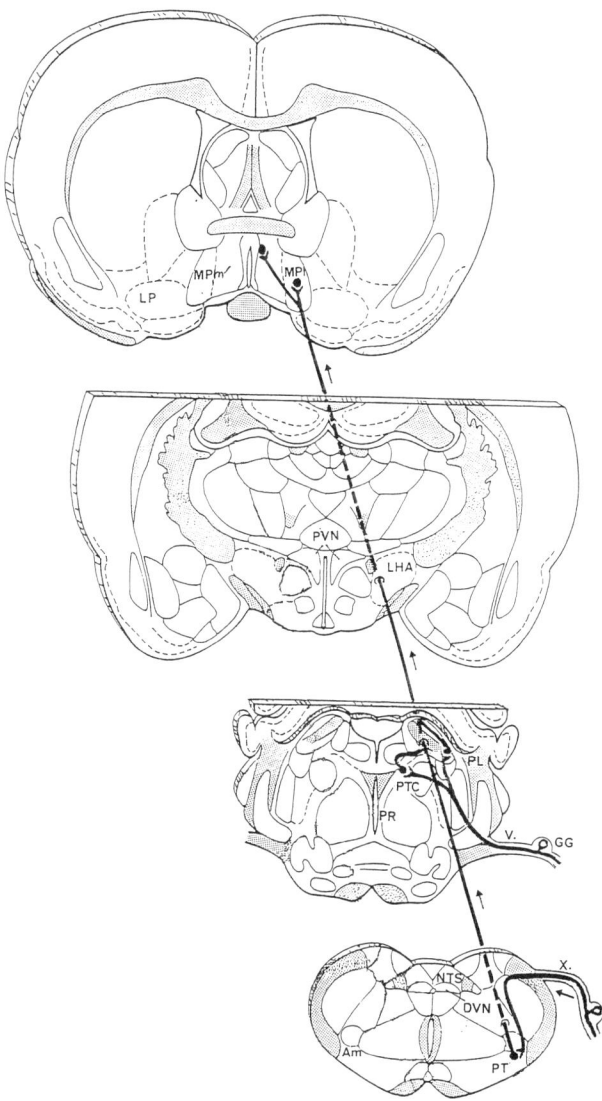

Fig. 10 Ascending thermoregulatory pathway from medullary and pontine thermosensitive areas to the lateral subdivision of the medial preoptic nucleus. *Abbreviations*: Am – ambiguus nucleus, DVN – dorsal vagal nucleus, GG – Gasserian ganglion, LHA – lateral hypothalamic area, LP – lateral preoptic area, MPI – lateral subdivision of the medial preoptic nucleus, MPM – medial subdivision of the medial preoptic nucleus, NTS – nucleus of the solitary tract, PL – lateral parabrachial nucleus, PR – pontine raphe nucleus, PT – peritrigeminal nucleus, PTC – "pontine thermoregulatory center", PVN – paraventricular nucleus, V. – trigeminal nerve, X. – vagal nerve.

nucleus (Figure 10). Connections between brainstem thermosensitive and preoptic neurons are monosynaptic and almost completely ipsilateral (Bratincsák and Palkovits, in preparation).

The *descending* loop of the cold stress response has not yet been topographically localized. Two proposed pathways arising from the medial preoptic nucleus should be considered (Figure 10):

1. Preoptic neurons project to the PVN and increase TRH expression. The activation of the hypothalamo-pituitary-thyroid axis is one of the possible routes for the response to cold-evoked stressful stimulation.

2. Preoptic neurons also project to the lower brainstem and, through the serotoninergic raphe nuclei, to the spinal cord (Smith *et al.*, 1998). In the brainstem, preoptic axons innervate superior salivatory preganglionic neurons. Electrical stimulation of the medial preoptic neurons produces a marked increase in secretion from the salivary glands. Warming of the preoptic area induces salivation which is suppressed by cooling of the skin (Kanosue *et al.*, 1990). Preoptic fibers also terminate in the dorsomedial medulla, in the vacinity of parasympathetic preganglionic neurons (Simerly and Swanson, 1986). In conclusion, medial preotic neurons may control body temperature by influencing the activity of the thyroid and salivary glands and circulation in cutaneous blood vessels (Figure 10).

Insulin-Evoked Hypoglycemia

In our studies, the largest increase in plasma EPI levels was measured after administration of insulin. It is likely, however, that insulin-induced hypoglycemia did not activate brain catecholaminergic neurons (Pacak *et al.*, 1998).

Data based on *c-fos* immunohistochemistry after insulin administration are rather contradictory (Bahjaoui-Bouhaddi *et al.*, 1994; Niimi *et al.*, 1995; Palkovits *et al.*, 1995; Porter and Bokil, 1997). It was repeatedly found that neurons in the lateral hypothalamus, the dorsomedial nucleus and the zona incerta responded to insulin-evoked hypoglycemia. In the PVN, Fos-like immunoreactivity was found only in a few dorsal and caudal parvocellular neurons, mainly in those which project to the brainstem and the spinal cord. In the dorsomedial medulla, *c-fos* expression were seen only in non-catecholaminergic neurons in the NTS (Baffi *et al.*, 1996; Palkovits *et al.*, 1997). In our experiments, no *c-fos* expression was found in the limbic system, except a well-outlined area in the ventrolateral part of the central amygdaloid nucleus (Figure 7B).

Hypoglycemia increases CRH mRNA expression in the PVN, CRH turnover in the median eminence, and CRH levels in the hypophyseal portal and peripheral blood (Suda *et al.*, 1988; Berkenbosch *et al.*, 1989). In contrast to these findings, Plotsky found unchanged CRH concentrations but elevated vasopressin levels in portal blood during hypoglycemia (Plotsky *et al.*, 1985).

Neuronal Circuits Activated During Insulin-Evoked Hypoglycemia

Insulin may induce stress responses through activation of the HPA axis and/or the central autonomic system. Descending projections arise in the hypothalamus, mainly from the PVN and the dorsomedial and the arcuate nuclei (Figure 11). Axons from PVN neurons project to the dorsal motor vagal nucleus and terminate on cells which innervate the pancreas (Loewy and Haxhiu, 1993; Hopkins *et al.*, 1996; Portillo *et al.*, 1996). The role of the dorsomedial and arcuate nuclei in the

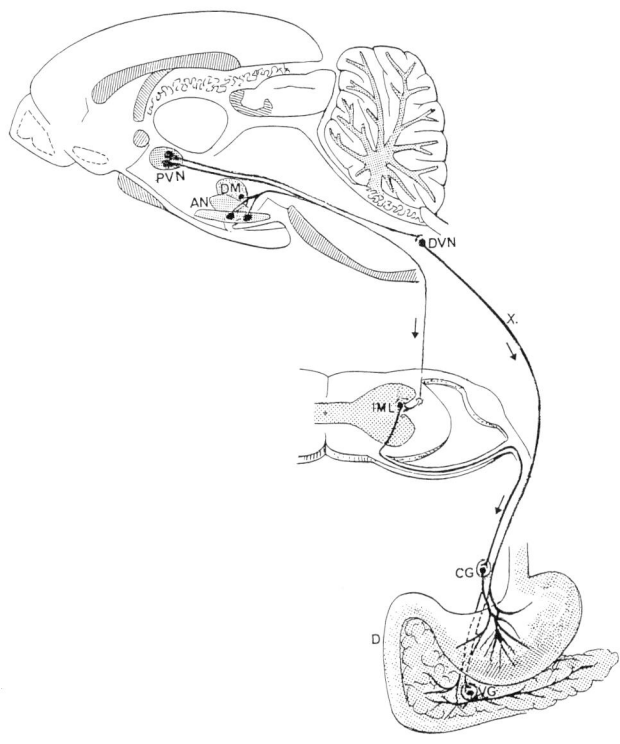

Fig. 11 Hypothalamic neuronal inputs to autonomic preganglionic neurons innervating the stomach and the pancreas. *Abbreviations*: AN – arcuate nucleus, CG – celiac ganglion, D – duodenum, DM – dorsomedial nucleus, DVN – dorsal vagal nucleus, IML – intermediolateral cell column, PVN – paraventricular nucleus, VG – vagal parasympathetic ganglionic cells, X. – vagal nerve.

response to insulin stress needs further verification. Immunohistochemistry revealed that insulin evokes *c-fos* activation in the dorsomedial but not the arcuate nucleus (Palkovits *et al.*, 1995; Porter and Bokil, 1997). Autonomic preganglionic neurons receive neuronal input from the central nucleus of the amygdala (Loewy and Haxhiu, 1993). Insulin-induced Fos immuno-reactivity in this nucleus (Figure 7B) suggests that the amygdala may contribute to the organization of the response to acute hypoglycemic stress.

Hemorrhage

Hemorrhage evoked small NE and EPI responses relative to elevations in plasma ACTH and corticosterone levels (Darlington *et al.*, 1986; Pacak *et al.*, 1998). In the ventrolateral medulla, Fos-like immunostaining was reported in catecholaminergic neurons which project to the intermediolateral cell column of the thoracic spinal cord and to the PVN (Badoer and Merolli, 1993; Dun *et al.*, 1993; Chan and Sawchenko, 1995; Krukoff *et al.*, 1997). Hypotensive hemorrhage also activated *c-fos* in non-catecholaminergic neurons of the NTS (Figure 12). These neurons in the dorsomedial medulla may represent the baroreceptor neuronal input of the hypothalamic circuit neurons (Chan and Sawchenko, 1995).

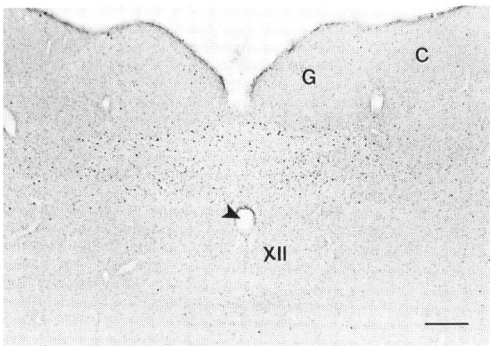

Fig. 12 Fos-immunoreactive neurons in the nucleus of the solitary tract 60 min after hypovolemic hemorrhage. *Abbreviations*: C – cuneate nucleus, G – gracile nucleus, XII. – motor hypoglossal nucleus. The central canal is indicated by an arrowhead. Bar scale: 200 μm.

In the hypothalamus, hypotensive hemorrhage induced marked *c-fos* expression in the magnocellular PVN and the supraoptic nucleus, as well as in parvicellular PVN neurons which project to the medulla and the spinal cord. Activated neurons were observed in several other forebrain regions such as the preoptic area, the central amygdaloid nucleus, the bed nucleus of the stria terminalis and the piriform cortex (Badoer *et al.*, 1993; Chan *et al.*, 1993; Palkovits *et al.*, 1995; Krukoff *et al.*, 1997; Badoer *et al.*, 1998).

Increased plasma corticosterone and ACTH levels, activate CRH mRNA expression, and elevated CRH levels in the portal blood indicate that the HPA axis is activated upon exposure to acute hemorrhage (Plotsky and Vale, 1984; Darlington *et al.*, 1986; Darlington *et al.*, 1992). Hypotensive hemorrhage has also been shown to increase vasopressin and oxytocin release in the PVN (Plotsky *et al.*, 1985; Ota *et al.*, 1994).

Neuronal Circuits for Activating the Central Organization of Body Electrolyte and Water Homeostasis

Hypovolemic hemorrhage-induced stress responses activate the HPA axis, the sympathoadrenal system, the renin-angiotensin system, and hypothalamic peptidergic circuits. This angiotensin-atrial natriuretic polypeptide-vasopressin circuit (Figure 13) is responsible for central nervous system control of body water and electrolyte homeostasis (ref. see Palkovits, 1995). The hemorrhage-induced increase in circulating angiotensin II acts on neuronal cells having angiotensin II receptors in the subfornical organ and the area postrema. Subfornical neurons project to the PVN and the supraoptic nucleus and excite vasopressin- and oxytocin-containing neurons. Subfornical neurons also project to atrial natriuretic polypeptide-containing neurons in the organum vasculosum laminae terminalis and the periventricular preoptic nucleus. These neurons, in turn, innervate vasopressinergic neurons in the PVN and the supraoptic nucleus. Each component of this hypothalamic circuit receives ascending inputs from brainstem catecholaminergic neurons and peptidergic

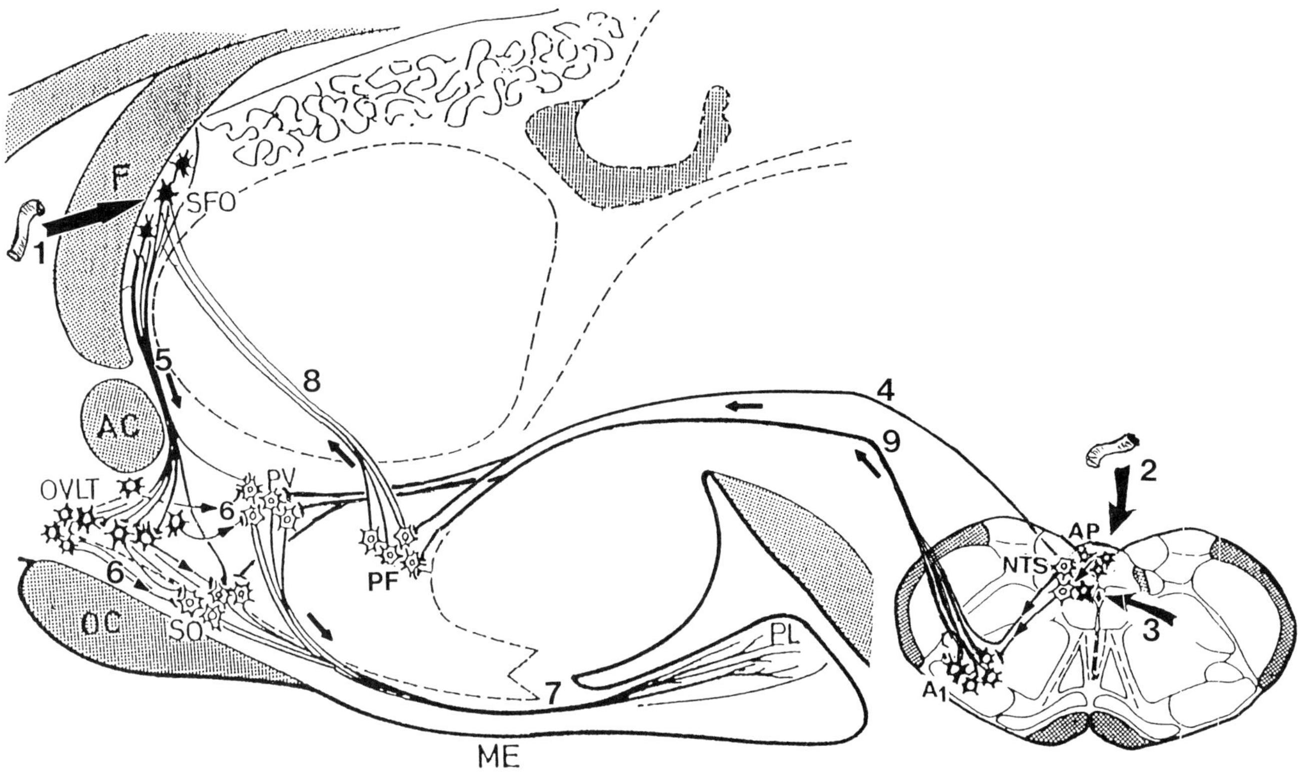

Fig. 13 Hypothalamic neuropeptidergic circuit in the organization of body fluid and mineralocorticoid homeostasis. Humoral and neuronal inputs to hypothalamic nuclei. 1 – Humoral inputs through the subfornical organ, 2 – Humoral inputs through the area postrema, 3 – viscerosensory inputs to the nucleus of the solitary tract, 4 – ascending peptidergic projections from the nucleus of the solitary tract to the paraventricular, supraoptic and perifornical nuclei, 5 – angiotensin II-containing neurons in the subfornical organ with their projections to the medial preoptic area and the supraoptic nucleus, 6 – atrial natriuretic hormone-containing neurons in the organum vasculosum laminae terminalis and the median preotic nucleus with supraoptic and paraventricular projections, 7 – supraoptic and paraventricular vasopressinergic projections to the posterior pituitary, 8 – angiotensin II-containing fibers from the perifornical nucleus to the subfornical organ, 9 – ventral noradrenergic bundle from the A1 and A2 catecholaminergic neurons. *Abbreviations*: AC – anterior commissure, AP – area postrema, F – fornix, ME – median eminence, NTS – nucleus of the solitary tract, OC – optic chiasm, OVLT – organum vasculosum laminae terminalis, PF – perifornical nucleus, PP – posterior pituitary lobe, PVN – paraventricular nucleus, SFO – subfornical organ, SO – supraoptic nucleus.

neurons of the NTS which may relay baroreceptor and viscerosensory signals from the periphery to the hypothalamus (Figure 13).

Concluding Remarks

In this brief review, I have summarized stressor-specific responses of the HPA axis and the central autonomic system and proposed neuronal circuits that are involved in responses. Stressful stimuli evoke characteristic endocrine, autonomic and behavioral responses. Responses to various stressors are extremely variable depending on the type and nature of the stressors. The data presented here indicate that brain regions and pathways are activated differently following exposure to different stressful stimuli. Increased knowledge about the existence of stressor-specific pathways and neuronal circuits may help us to understand the mechanism of action of various stressors and to elucidate the specific target areas in the brain which participate in orchestrating stress responses.

References

Abbadie, C., Lombard, M.-C., Morain, F., and Besson, J.M. (1992). Fos-like immunoreactivity in the rat superficial dorsal horn induced by formalin injection in the forepaw: effects of dorsal rhizotomies. *Brain Research* **578**, 17–25.

Arancibia, S., Tapia-Arancibia, L., Assenmacher, I., and Astier, H. (1983). Direct evidence of short-term cold-induced TRH release in the median eminence of unanesthetized rats. *Neuroendocrinology* **37**, 225–228.

Badoer, E., McKinley, M.J., Oldfield, B.J., and McAllen, R.M. (1993). A comparison of hypotensive and non-hypotensive hemorrhage on Fos expression in spinally projecting neurons of the paraventricular nucleus and rostral ventrolateral medulla. *Brain Research* **610**, 216–223.

Badoer, E., and Merolli, J. (1998). Neurons in the hypothalamic paraventricular nucleus that project to the rostral ventrolateral medulla are activated by haemorrhage. *Brain Research* **791**, 317–320.

Baffi, J., Pacak, K., Szabó, S., and Palkovits, M. (1996). Different stressors have different effects on c-fos activity of catecholaminergic neurons. In R. McCarty, G. Aguilera, E. Sabban, & R. Kvetňanský (Eds.), *Stress: Molecular Genetic and Neurological Advances* (pp. 143–155). New York: Gordon and Breach.

Bahjaoui-Bouhaddi, M., Fellmann, D., and Bugnon, C. (1994). Induction of fos-immunoreactivity in prolactin-like containing neurons of the rat lateral hypothalamus after insulin injection. *Neuroscience Letters* **168**, 11–15.

Bartanusz, V., Aubry, J.M., Jezova, D., Baffi, J., and Kiss, J.Z. (1993a). Up-regulation of vasopressin mRNA in paraventricular hypophysiotrophic neurons after acute immobilization stress. *Neuroendocrinology* **58**, 625–629.

Bartanusz, V., Jezova, D., Bertini, L.T., Tilders, F.J., Aubry, J.M., and Kiss, J.Z. (1993b). Stress-induced increase in vasopressin and corticotropin-releasing factor expression in hypophysiotrophic paraventricular neurons. *Endocrinology* **132**, 895–902.

Berkenbosch, F., De Goeji, D., and Tilders, F. (1989). Hypoglycemia enhances turnover of corticotropin-releasing factor and vasopressin in the zona externa of the rat median eminence. *Endocrinology* **125**, 28–34.

Boulant, J.A. (1980). Hypothalamic control of thermoregulation. In P.J. Morgane and J. Panksepp (Eds.), *Handbook of the Hypothalamus* (Vol. 2, pp. 1–82). New York: Marcell Decker.

Boulant, J.A. (1998). Cellular mechanisms of temperature sensitivity in hypothalamic neurons. *Progress in Brain Research* **115**, 3–8.

Bullitt, E. (1990). Expression of c-fos-like protein as a marker for neuronal activity following noxious stimulation in the rat. *Journal of Comparative Neurology* **296**, 517–530.

Burstein, R., Cliffer, K.D., and Giesler, G.J.J. (1987). Direct somatosensory projections from the spinal cord to the hypothalamus and telencephalon. *Journal of Neuroscience* **7**, 4159–4164.

Burstein, R., Cliffer, K.D., and Giesler, G.J.J. (1990). Cells of origin of the spinohypothalamic tract in the rat. *Journal of Comparative Neurology* **291**, 329–344.

Burstein, R., Falkowsky, O., Borsook, D., and Strassman, A. (1996). Distinct lateral and medial projections of the spinohypothalamic tract of the rat. *Journal of Comparative Neurology* **373**, 549–574.

Ceccatelli, S., and Orazzo, C. (1993). Effect of different types of stressors on peptide messenger ribonucleic acids in the hypothalamic paraventricular nucleus. *Acta Endocrinologia* (Copenh) **128**, 485–492.

Ceccatelli, S., Villar, M.J., Goldstein, M., and Hökfelt, T. (1989). Expression of c-fos immunoreactivity in transmitter-characterized neurons after stress. *Proceedings of the National Academy of Sciences U S A* **86**, 9569–9573.

Cechetto, D.F., and Saper, C.B. (1988). Neurochemical organization of the hypothalamic projection to the spinal cord in the rat. *Journal of Comparative Neurology* **272**, 579–604.

Chan, R.K.W., Brown, E.R., Ericsson, A., Kovács, K.J., and Sawchenko, P.E. (1993). A comparison of two immediate-early genes, c-fos and NGFI-B, as markers for functional activation in stress-related neuroendocrine circuitry. *Journal of Neuroscience* **13**, 5126–5138.

Chan, R.K.W., and Sawchenko, P.E. (1995). Hemodynamic regulation of tyrosine hydroxylase messenger RNA in medullary catecholamine neurons: A c-fos-guided hybridization histochemical study. *Neuroscience* **66**, 377–390.

Chaouch, A., Ménétrey, D., Binder, R., and Besson, J.M. (1983). Neurons at the origin of the medial spinoreticular tract in the rat: An anatomical study using horseradish peroxidase retrograde transport. *Journal of Comparative Neurology* **214**, 309–320.

Chen, X., and Herbert, J. (1995). Regional changes in *c-fos* expression in the basal forebrain and brainstem during adaptation to repeated stress: correlations with cardiovascular, hypothermic and endocrine responses. *Neuroscience* **64**, 675–685.

Cliffer, K.D., Burstein, R., and Giesler, G.J.J. (1991). Distributions of spinothalamic, spinohypothalamic, and spinotelencephalic fibers revealed by anterograde transport of PHA-L in rats. *Journal of Neuroscience* **11**, 852–868.

Cullinan, W.E., Herman, J.P., Battaglia, D.F., Akil, H., and Watson, S.J. (1995). Pattern and time course of immediate early gene expression in rat brain following acute stress. *Neuroscience* **64**, 477–505.

Cunningham, E.T., and Sawchenko, P.E. (1988). Anatomical specificity of noradrenergic inputs to the paraventricular and supraoptic nuclei of the rat hypothalamus. *Journal of Comparative Neurology* **274**, 60–76.

Cunningham, E.T.J., Bohn, M.C., and Sawchenko, P.E. (1990). Organization of adrenergic inputs to the paraventricular and supraoptic nuclei of the hypothalamus in the rat. *Journal of Comparative Neurology* **292**, 651–667.

Darlington, D.N., Barraclough, C.A., and Gann, D.S. (1992). Hypotensive hemorrhage elevates corticotropin-releasing hormone messenger ribonucleic acid (mRNA) but not vasopressin mRNA in the rat hypothalamus. *Endocrinology* **130**, 1281–1288.

Darlington, D.N., Shinsako, J., and Dallman, M.F. (1986). Responses of ACTH, epinephrine, norepinephrine, and cardiovascular system to hemorrhage. *American Journal of Physiology* **251**, H612–H618.

Dun, N.J., Dun, S.L., and Chiaia, N.L. (1993). Hemorrhage induces Fos immunoreactivity in rat medullary catecholaminergic neurons. *Brain Research* **608**, 223–232.

Esteves, F., Lima, D., and Coimbra, A. (1993). Structural types of spinal cord marginal (lamina-1) neurons projecting to the nucleus of the tractus solitarius in the rat. *Somatosensory and Motor Research* **10**, 203–216.

Gaillet, S., Lachuer, J., Malaval, F., Assenmacher, I., and Szafarczyk, A. (1991). The involvement of noradrenergic ascending pathways in the stress-induced activation of ACTH and corticosterone secretions is dependent on the nature of stressors. *Experimental Brain Research* **87**, 173–180.

Giesler, G.J. (1995). Evidence of direct nociceptive projections from the spinal cord to the hypothalamus and telencephalon. *Neuroscience*, **253–261**.

Giesler, G.J., Katter, J.T., and Dado, R.J. (1994). Direct spinal pathways to the limbic system for nociceptive information. *Trends in Neuroscience* **17**, 244–250.

Harbuz, M.S., Jessop, D.S., Lightman, S.L., and Chowdrey, H.S. (1994). The effects of restraint or hypertonic saline stress on corticotropin-releasing factor, arginine vasopressin, and proenkephalin A mRNAs in the CFY, Sprague-Dawley and Wistar strains of rat. *Brain Research* **667**, 6–12.

Herbert, J., and Howes, S.R. (1993). Interactions between corticotropin-releasing factor and endogenous opiates on the cardioaccelerator, hypothermic, and corticoid responses to restraint in the rat. *Peptides* **14**, 145–152.

Hopkins, D.A., Bieger, D., De Vente, J., and Steinbusch, H.W.M. (1996). Vagal efferent projections: viscerotopy, neurochemistry, and effects of vagotomy. *Progress in Brain Research* **107**, 79–96.

Hosoya, Y., Sugiura, Y., Ito, R., and Kohno, K. (1990). Descending projections from the hypothalamic paraventricular nucleus to the A5 area, including the superior salivatory nucleus, in the rat. *Experimental Brain Research* **82**, 513–518.

Hosoya, Y., Sugiura, Y., Okado, N., Loewy, A.D., and Kohno, K. (1991). Descending input from the hypothalamic paraventricular nucleus to sympathetic preganglionic neurons in the rat. *Experimental Brain Research* **85**, 10–20.

Hylden, J.L.K., Anton, F., and Nahin, R.L. (1989). Spinal lamina I projection neurons in the rat: Collateral innervation of parabrachial area and thalamus. *Neuroscience* **28**, 27–37.

Kalin, N.H., Takahashi, L.K., and Chen, F.-L. (1994). Restraint stress increases corticotropin-releasing hormone mRNA content in the amygdala and paraventricular nucleus. *Brain Research* **656**, 182–186.

Kanosue, K., Nakayama, T., Tanaka, H., Yanase, M., and Yasuda, H. (1990). Modes of action of local hypothalamic and skin thermal stimulation on salivary secretion in rats. *Journal of Physiology (London)* **424**, 459–471.

Kiyohara, T., Miyata, S., Nakamura, T., Shido, O., Nakashima, T., and Shibata, M. (1995). Differences in Fos expression in the rat brains between cold and warm ambient exposures. *Brain Research Bulletin* **38**, 193–201.

Kononen, J., Honkaniemi, J., Alho, H., Koistinaho, J., Iadarola, M., and Pelto-Huikko, M. (1992). Fos-like immunoreactivity in the rat hypothalamic-pituitary axis after immobilization stress. *Endocrinology* **130**, 3041–3047.

Krukoff, T.L., MacTavish, D., and Jhamandas, J.H. (1997). Activation by hypotension of neurons in the hypothalamic paraventricular nucleus that project to the brainstem. *Journal of Comparative Neurology* **385**, 285–296.

Li, J.L., Kaneko, T., Shigemoto, R., and Mizuno, N. (1997). Distribution of trigeminohypothalamic and spinohypothalamic tract neurons displaying substance P receptor-like immunoreactivity in the rat. *Journal of Comparative Neurology* **378**, 508–521.

Lima, D., Mendes-Ribeiro, J.A., and Coimbra, A. (1991). The spino-latero-reticular system of the rat: projections from the superficial dorsal horn and structural characterization of marginal neurons involved. *Neuroscience* **45**, 137–152.

Lipton, J.M., and Clark, W.G. (1986). Neurotransmitters in temperature control. *Annual Review of Physiology* **48**, 613–623.

Liu, R.P.C. (1986). Spinal neuronal collaterals to the intralaminar thalamic nuclei and periaqueductal gray. *Brain Research* **365**, 145–150.

Loewy, A.D., and Haxhiu, M.A. (1993). CNS cell groups projecting to pancreatic parasympathetic preganglionic neurons. *Brain Research* **620**, 323–330.

Luiten, P.G., ter Horst, G.J., Karst, H., and Steffens, A.B. (1985). The course of paraventricular hypothalamic efferents to autonomic structures in medulla and spinal cord. *Brain Research* **329**, 374–378.

Makara, G.B., Stark, E., and Mihály, K. (1969). Corticotrophin release induced by injection of formalin in rats with hemisection of the spinal cord. *Acta Physiologia Academy of Science Hungary* **35**, 331–333.

Makino, S., Schulkin, J., Smith, M.A., Pacák, K., Palkovits, M., and Gold, P.W. (1995). Regulation of corticotropin-releasing hormone receptor messenger ribonucleic acid in the rat brain and pituitary by glucocorticoids and stress. *Endocrinology* **136**, 4517–4525.

Menétrey, D., and Basbaum, A.I. (1987). Spinal and trigeminal projections to the nucleus of the solitary tract. A possible substrate for somatovisceral and visceroviseral reflex activation. *Journal of Comparative Neurology* **255**, 439–450.

Menétrey, D., Roudier, F., and Besson, J.M. (1983). Spinal neurons reaching the lateral reticular nucleus as studied in the rat by retrograde transport of horseradish peroxidase. *Journal of Comparative Neurology* **220**, 439–452.

Miyata, S., Ishiyama, M., Shido, O., Nakashima, T., Shibata, M., and Kiyohara, T. (1995). Central mechanism of neural activation with cold acclimation of rats using fos immunohistochemistry. *Neuroscience Research* **22**, 209–218.

Niimi, M., Sato, M., Tamaki, Y., Wada, Y., Takahara, J., and Kawanishi, K. (1995). Induction of Fos protein in the rat hypothalamus elicited by insulin-induced hypoglycemia. *Neuroscience Research* **23**, 361–364.

Ota, M., Crofton, J., and Share, L. (1994). Hemorrhage-induced vasopressin release in paraventricular nucleus measured by *in vivo* microdialysis. *Brain Research* **658**, 49–54.

Otake, K., Reis, D.J., and Ruggiero, D.A. (1994). Afferents to the midline thalamus issue collaterals to the nucleus tractus solitarii: An anatomical basis for thalamic and visceral reflex integration. *Journal of Neuroscience* **14**, 5694–5707.

Pacak, K., Armando, I., Fukuhara, K., Kvetňanský, R., Palkovits, M., Kopin, I.J., and Goldstein, D.S. (1992). Noradrenergic activation in the paraventricular nucleus during acute and chronic immobilization stress in rats: An *in vivo* microdialysis study. *Brain Research* **589**, 91–96.

Pacak, K., Palkovits, M., I.J., K., and Goldstein, D.S. (1995a). Stress-induced norepinephrine release in the hypothalamic paraventricular nucleus and pituitary-adrenocortical and sympathoadrenal activity: *In vivo* microdialysis studies. *Frontiers in Neuroendocrinology* **16**, 89–150.

Pacak, K., Palkovits, M., Kvetňanský, R., Fukuhara, K., Armando, I., Kopin, I.J., and Goldstein, D.S. (1993). Effects of single or repeated immobilization on release of norepinephrine and its metabolites in the central nucleus of the amygdala in conscious rats. *Neuroendocrinology* **57**, 626–633.

Pacak, K., Palkovits, M., Kvetňanský, R., Yadid, G., Kopin, I.J., and Goldstein, D.S. (1995b). Effects of various stressors on in vivo norepinephrine release in the hypothalamic paraventricular nucleus and on the pituitary-adrenocortical axis. *Annals of the New York Academy of Sciences* **771**, 115–130.

Pacak, K., Palkovits, M., Makino, S., Kopin, I.J., and Goldstein, D.S. (1996). Brainstem hemisection decreases corticotropin-releasing hormone mRNA in the paraventricular nucleus but not in the central amygdaloid nucleus. *Journal of Neuroendocrinology* **8**, 543–551.

Pacak, K., Palkovits, M., Yadid, G., Kvetňanský, R., Kopin, I.J., and Goldstein, D.S. (1998). Heterogeneous neurochemical responses to different stressors: a test of Selye's doctrine of nonspecificity. *American Journal of Physiology* **275**, R1247–R1255.

Palkovits, M. (1999). Interconnections between the neuroendocrine hypothalamus and the central autonomic system. *Frontiers in Neuroendocrinology* **20**, 270–298.

Palkovits, M., Baffi, J.S., and Dvori, S. (1995). Neuronal organization of stress response. Pain-induced c-fos expression in brain stem catecholaminergic cell groups. *Annals of the New York Academy of Sciences* **771**, 313–326.

Palkovits, M., Baffi, J.S., and Pacak, K. (1997). Stress-induced fos-like immunoreactivity in the pons and the medulla oblongata of rats. *Stress* **1**, 155–168.

Palkovits, M., Baffi, J.S., and Pacak, K. (1999). The role of ascending neuronal pathways in stress-induced release of noradrenaline in the hypothalamic paraventricular nucleus of rats. *Journal of Neuroendocrinology* **11**, 529–539.

Palkovits, M., Záborszky, L., Feminger, A., Mezey, É., Fekete, M.I.K., Herman, J.P., Kanyicska, B., and Szabó, D. (1980). Noradrenergic innervation of the rat hypothalamus: Experimental, biochemical and electron microscopic studies. *Brain Research* **191**, 161–171.

Pezzone, M.A., Lee, W.-S., Hoffman, G.E., and Rabin, B.S. (1992). Induction of c-Fos immunoreactivity in the rat forebrain by conditioned and unconditioned aversive stimuli. *Brain Research* **597**, 41–50.

Plotsky, P.M., Bruhn, T.O., and Vale, W. (1985). Hypophysiotrophic regulation of adrenocorticotropin secretion in response to insulin-induced hypoglycemia. *Endocrinology* **117**, 323–329.

Plotsky, P.M., and Vale, W. (1984). Hemorrhage-induced secretion of corticotropin-releasing factor-like immunoreactivity into the rat hypophysial portal circulation and its inhibition by glucocorticoids. *Endocrinology* **114**, 164–169.

Porter, J.P., and Bokil, H.S. (1997). Effect of intracerebroventricular and intravenous insulin on Fos-immunoreactivity in the rat brain. *Neuroscience Letters* **224**, 161–164.

Portillo, F., Carrasco, M., and Vallo, J.J. (1996). Hypothalamic neuron projection to autonomic preganglionic levels related with glucose metabolism: A fluorescent labelling study in the rat. *Neuroscience Letters* **210**, 197–200.

Portillo, F., Carrasco, M., and Vallo, J.J. (1998). Separate populations of neurons within the paraventricular hypothalamic nucleus of the rat project to vagal and thoracic autonomic preganglionic levels and express c-Fos protein induced by lithium chloride. *Journal of Chemical Neuroanatomy* **14**, 95–102.

Presley, R.M., Menétrey, D., Levine, J.D., and Basbaum, A.I. (1990). Systemic morphine suppresses noxious stimulus-evoked fos protein-like immunoreactivity in the rat spinal cord. *Journal of Neuroscience* **10**, 323–335.

Riche, D., Pommery, J.D., and Menétrey, D. (1990). Neuropeptides and catecholamines in efferent projections of the nuclei of the solitary tract in the rat. *Neuroscience* **293**, 399–424.

Saper, C.B., and Loewy, A.D. (1980). Efferent connections of the parabrachial nucleus in the rat. *Brain Research* **197**, 291–317.

Sawchenko, P.E., Arias, C., and Bittencourt, J.C. (1990). Inhibin β, somatostatin, and enkephalin immunoreactivities coexist in caudal medullary neurons that project to the paraventricular nucleus of the hypothalamus. *Journal of Comparative Neurology* **291**, 269–280.

Sawchenko, P.E., and Swanson, L.W. (1982). The organization of noradrenergic pathways from the brainstem to the paraventricular and supraoptic nuclei in the rat. *Brain Research Reviews* **4**, 275–325.

Senba, E., Matsunaga, K., Tohyama, M., and Noguchi, K. (1993). Stress-induced c-fos expression in the rat brain: Activation mechanism of sympathetic pathway. *Brain Research Bulletin* **31**, 329–341.

Senba, E., Umemoto, S., Kawai, Y., and Noguchi, K. (1994). Differential expression of fos family and jun family in mRNAs in the rat hypothalamo-pituitary-adrenal axis after immobilization stress. *Molecular Brain Research* **24**, 283–294.

Shokunbi, M.T., Hrycyshin, A.W., and Flumerfelt, B.A. (1985). Spinal projections to the lateral reticular nucleus in the rat: A retrograde labeling study using horseradish peroxidase. *Journal of Comparative Neurology* **239**, 216–226.

Simerly, R.B., and Swanson, L.W. (1986). The organization of neuronal inputs to the medial preoptic nucleus of the rat. *Journal of Comparative Neurology* **246**, 312–342.

Smith, J.E., Jansen, A.S., Gilbey, M.P., and Loewy, A.D. (1998). CNS cell groups projecting to sympathetic outflow of tail artery: Neural circuits involved in heat loss in the rat. *Brain Research* **786**, 153–164.

Suda, T., Tozawa, F., Yamada, M., Ushiyama, T., Tomori, N., Sumimoto, T., Nakagami, Y., Demura, H., and Shizume, K. (1988). Insulin-induced hypglycemia increases corticotropin-releasing factor messenger ribonucleic acid levels in rat hypothalamus. *Endocrinology* **123**, 1371–1375.

Swanson, L.W., and Kuypers, H.G.J.M. (1980). The paraventricular nucleus of the hypothalamus: cytoarchitectonic subdivisions and organization of projections to the pituitary, dorsal vagal complex and spinal cord as demonstrated by retrograde fluorescence double labeling methods. *Journal of Comparative Neurology* **194**, 555–570.

Swanson, L.W., and Simmons, D.M. (1989). Differential steroid hormone and neural influences on peptide mRNA levels in CRH cells of the paraventricular nucleus: A hybridization histochemical study in the rat. *Journal of Comparative Neurology* **285**, 413–435.

Tanaka, T., Yokoo, H., Mizoguchi, K., Yoshida, M., Tsuda, A., and Tanaka, M. (1991). Noradrenaline release in the rat amygdala is increased by stress: studies with intracerebral microdialysis. *Brain Research* **544**, 174–176.

Tavares, I., Lima, D., and Coimbra, A. (1993). Neurons in the superficial dorsal horn of the rat spinal cord projecting to the medullary ventrolateral reticular formation express c-fos after noxious stimulation of the skin. *Brain Research* **623**, 278–286.

ter Horst, G.J., Luiten, P.G., and Kuipers, F. (1984). Descending pathways from hypothalamus to dorsal motor vagus and ambiguus nuclei in the rat. *Journal of the Autonomic Nervous System* **11**, 59–75.

Tóth, Z.E., Gallatz, K., Fodor, M., and Palkovits, M. (1999). Decussations of the descending paraventricular pathways to the brainstem and spinal cord autonomic centers. *Journal of Comparative Neurology* **414**, 255–266.

Tucker, D.C., and Saper, C.B. (1985). Specificity of spinal projections from hypothalamic and brainstem areas which innervate sympathetic preganglionic neurons. *Brain Research* **360**, 159–164.

Wallace, D.M., Magnuson, D.J., and Gray, T.S. (1989). The amygdalo-brainstem pathway: selective innervation of dopaminergic, noradre-nergic and adrenergic cells in the rat. *Neuroscience Letters* **97**, 252–258.

Zoeller, R.T., Kabeer, N., and Albers, H.E. (1990). Cold exposure elevates cellular levels of messenger ribonucleic acid encoding thyrotropin-releasing hormone in paraventricular nucleus despite elevated levels of thyroid hormones. *Endocrinology* **127**, 2955–2962.

2. Re-evaluation of the Role of Serotonin in Stress – A "Systems-Level" Approach

C.A. Lowry

Introduction

Serotonergic systems may play an important role in regulating neuroendocrine, immunological, behavioral, and autonomic responses to stress. However, a multitude of studies measuring changes in serotonergic metabolism or serotonergic activity following exposure to stress-related stimuli have yielded seemingly contradictory results. For example, multiple studies can be cited indicating that exposure to stressful stimuli results in increases, decreases, or no effect on, serotonergic synthesis, metabolism, or release (reviewed in Anisman, 1978; Dunn and Kramarcy, 1984; Anisman *et al.*, 1985; Rueter *et al.*, 1997; Lucki, 1998). Furthermore, electrophysiological studies in behaving animals suggest that the firing rates of serotonergic neurons are unaltered in response to a variety of physical and psychological stressors, even when associated with strong increases in sympathetic activation. These stressors include thermoregulatory and glucoregulatory challenges, as well as loud noise, restraint, and exposure to a natural predator (reviewed by Jacobs and Azmitia, 1982). A re-evaluation of previous studies in light of recent advances reveals that this variability and complexity arises in part from the unique topographical and functional properties of different serotonergic systems. This review provides a "systems level" analysis of the role of serotonin (5-hydroxytryptamine, 5-HT) in stress responses, focusing on defining the roles of topographically and functionally distinct subpopulations of serotonergic neurons. The heuristic value of this "systems level" approach is illustrated by evidence from our laboratory and others suggesting that stress or stress-related neurochemicals (including corticotropin-releasing factor, CRF) can selectively activate a topographically organized subpopulation of mesolimbocortical serotonergic neurons. Evidence suggests that these neurons are predominantly located in the median raphe nucleus and the midline, caudal portion of the dorsal raphe nucleus. Selective effects of stress and stress-related neurochemicals on a topographically organized mesolimbocortical serotonergic system support the hypothesis that specific subpopulations of serotonergic neurons play a *dedicated* role in mediating stress responses, including behavioral anxiety. In addition to stress-induced activation of a mesolimbocortical serotonergic system with neuronal cell bodies in the brainstem raphe, recent functional and anatomical studies support the hypothesis that "inducible" serotonergic systems – dependent upon increased tryptophan availability – may exist in some stress-related brain regions, such as the dorsomedial hypothalamic nucleus and the nucleus of the solitary tract. The overall hypothesis indicating "stress-specific" functions for subpopulations of serotonergic neurons is compared and contrasted with the alternative (or complementary) hypothesis that the increase in 5-HT release under a variety of conditions simply reflects, or is largely due to, an increase in behavioral arousal or motor activity associated with the manipulation.

Subpopulations of Serotonergic Neurons may have Stress-related Functions

"Although it is still possible that some facet of serotonin release does promote the actual feeling of anxiety, that would have to be a very small component of overall brain serotonergic activity." (Panksepp, 1990)

Serotonin and Behavioral State – an Overriding Influence

Strong evidence has been accumulated from electrophysiological studies in behaving animals indicating that serotonergic neuronal activity in a variety of brainstem raphe structures is correlated with level of arousal or behavioral state (reviewed in Jacobs and Azmitia, 1992; Jacobs and Fornal, 1995). Based on these studies, three separate models have been proposed to account for the role of serotonin in diverse physiological and behavioral processes: 1) various subpopulations of serotonergic neurons might be separately dedicated to specific functions, such as feeding, mood regulation, etc., 2) the entire, diffuse serotonergic system might be activated as a whole to exert its diverse effects, or 3) the serotonergic system might be activated in association with motor activity. In this case the physiological and behavioral functions ascribed to serotonin would be secondary to the primary association of serotonin with motor activity or arousal (Jacobs and Fornal, 1995). Strong arguments have been made previously for the third model (Jacobs and Fornal, 1995; Rueter *et al.*, 1997). Although it seems certain that the activity of a vast majority of brainstem serotonergic neurons is correlated with behavioral state, it is equally certain that there are small subpopulations of serotonergic neurons that are not. In the current review we explore the possibility that these smaller subpopulations of serotonergic neurons may have more specific physiological or behavioral functions – with a special emphasis on the possibility that they may play an important role in mediating physiological and behavioral responses associated with stress.

"Classical" Serotonergic Neuronal Cell Groups and Stress-related Functions

Overview

"The monoamine-containing nerve cells have a widespread distribution within the lower brain stem. It is at present difficult to systematize them on account of our scanty knowledge of what functional or anatomical systems they belong to. ... Where possible, however, the cells have been divided into groups. This has been done with consideration not only to the topographical conditions and to the morphology of the cells but also with a view to the degree to which the cells seem to form more or less continuous systems, separated from other cell systems. It is possible – and perhaps even probable – that the division is partly artificiel [sic]. This is true especially of the 5-HT type cells." (Dahlström and Fuxe, 1964)

Thirty-five years following the description of serotonergic neuronal cell groups in the brainstem by Dahlström and Fuxe (1964) we are still faced with the task of accurately defining functional properties of groups of serotonergic neurons. There are several excellent reviews describing topographically organized serotonergic neuronal systems (Törk, 1985; Azmitia and Gannon, 1986; Imai *et al.*, 1986; Törk, 1990; Jacobs and Azmitia, 1992; Kazakov *et al.*, 1993; Halliday *et al.*, 1995). However, none of these have discussed serotonergic systems in the context that different topographically organized serotonergic systems may be differentially responsive to stress. For reference, the topographical distribution of serotonergic neuronal cell groups described by Dahlström and Fuxe (1964) is presented diagrammatically in Figure 1. For abbreviations, see Table 1.

A Mesolimbocortical Serotonergic System – a Definition

A pattern is emerging – based on neurochemical studies – suggesting that stress-related stimuli result in a selective activation of mesolimbocortical serotonergic systems. Increases in serotonin metabolism or release are frequently observed in the medial prefrontal cortex (Dunn, 1988; Inoue *et al.*, 1993, 1994; Kawahara *et al.*, 1993; Yoshioka *et al.*, 1995; Adell *et al.*, 1997), nucleus accumbens (Inoue *et al.*, 1993, 1994), amygdala (Inoue *et al.*, 1993, 1994; Kawahara *et al.*, 1993; Adell *et al.*, 1997; Amat *et al.*, 1998), and hippocampus (Joseph and Kennett, 1983b). Meanwhile, retrograde tracing studies have determined that these terminal regions receive input from serotonergic neurons located in the median raphe nucleus and the caudal portion of the dorsal raphe nucleus, particularly the intrafascicular part (Wyss *et al.*, 1979; Mehler *et al.*, 1980; Ottersen, 1981; Köhler and Steinbusch, 1982; Porrino and Goldman-Rakic, 1982; Imai *et al.*, 1986; Van Bockstaele *et al.*, 1993; Petrov *et al.*, 1994). In some cases, serotonergic neurons with collateral projections to multiple limbic forebrain regions have been identified (Köhler and Steinbusch, 1982; Van Bockstaele *et al.*, 1993).

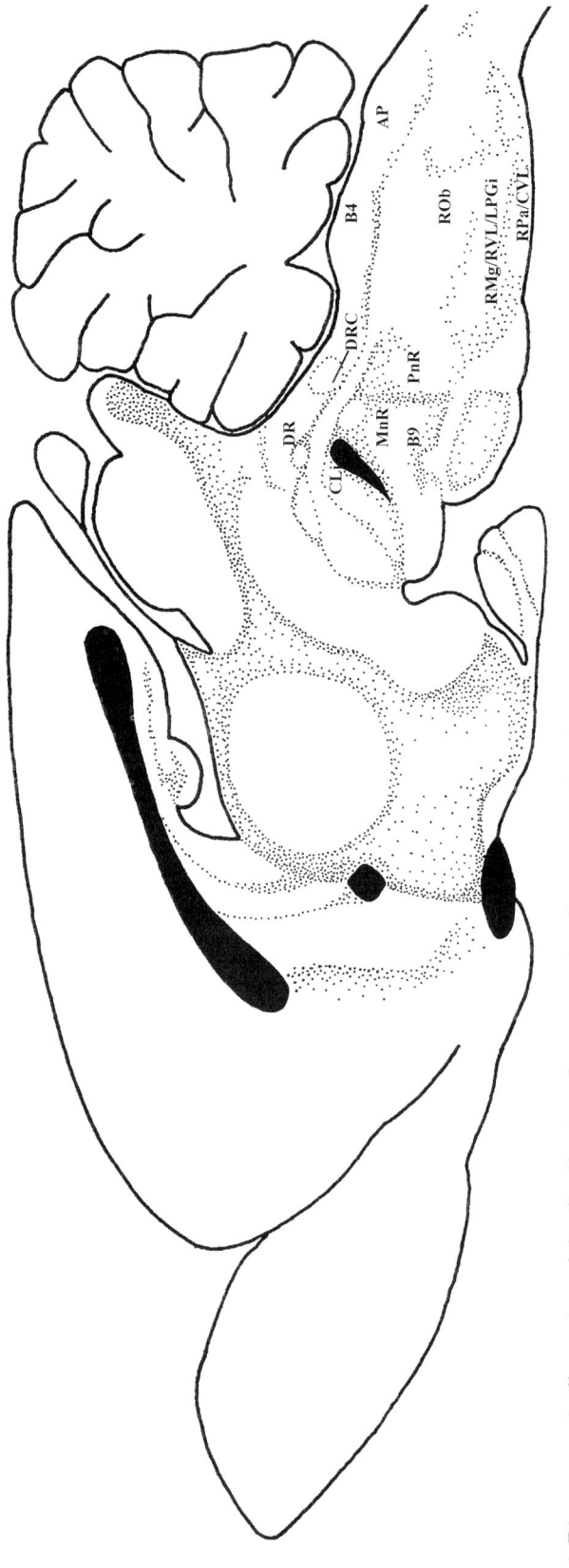

Fig. 1 Diagrammatic illustration of the distribution of serotonergic cell groups as described by Dahlström and Fuxe (1964). For abbreviations, see Table 1.

Table 1 Nomenclature assigned to "classical" serotonergic neurons (according to Dahlström and Fuxe, 1964).

Alphanumeric designation	Cytoarchitectonic regions
B1	Raphe pallidus nucleus (RPa) Caudal ventrolateral medulla (CVL)
B2	Raphe obscurus nucleus (ROb)
B3	Raphe magnus nucleus (RMg) Rostral ventrolateral medulla (RVL) Lateral paragigantocellular reticular nucleus (LPGi)
B4	Central gray of the medulla oblongata
B5	Pontine raphe nucleus (PnR)
B6	Dorsal raphe nucleus, caudal part (DRC)
B7	Dorsal raphe nucleus (DR)
B8	Median raphe nucleus (MnR); (Nucleus centralis superior in primates) Pontomesencephalic reticular formation (PMRF); (Nucleus pontis oralis in humans) Caudal linear nucleus (CLi)
B9	Medial lemniscus

Together these observations suggest that stress may reliably and selectively activate a subpopulation of serotonergic neurons located in the median raphe nucleus and caudal portions of the dorsal raphe nucleus, particularly its midline and intrafascicular components. In support of this hypothesis, we have found that the stress-related neuropeptide, corticotropin-releasing factor (CRF), selectively and dose-dependently increases the *in vitro* firing rates of a small subpopulation of serotonergic neurons in the intrafascicular and midline regions of the caudal dorsal raphe nucleus (Lowry *et al.*, 1999). The location of these responsive neurons coincides with the location of mesolimbocortical serotonergic neurons as defined by retrograde tracing studies (see above) and the location of a small subpopulation of serotonergic neurons with unique behavioral correlates (Rasmussen *et al.*, 1984, see Table 2). These results are exciting because they represent the first study to identify a topographically organized subpopulation of serotonergic neuronal cell bodies that may selectively contribute to stress-induced increases in serotonergic activity.

The Unique Dependence of Stress-related Serotonergic Activity on Rapid Increases in Tryptophan Availability

Stress-induced increases in serotonin metabolism and release are dependent on increased tryptophan availability (Kennett and Joseph, 1981; Joseph and Kennett, 1983a,b; Dunn and Welch, 1991). This is most likely due to increases in the free tryptophan in plasma (unbound to plasma albumin; Knott and Curzon, 1972). Global increases in free tryptophan in plasma may be due in part to sympathetic nervous system induced increases in plasma concentrations of unesterified free fatty acids

Table 2 Comparison of Type I and Type II serotonergic cells[a].

Type I	Type II
Decreased firing rate during the inactive period of the sleep wake cycle	No change in firing rate during the inactive period of the sleep wake cycle
Virtually silent during REM sleep	Active during REM sleep
Short latency, short duration excitation following phasic auditory and visual stimuli	Short latency, long-duration inhibition following phasic auditory and visual stimuli
Present throughout the dorsal raphe nucleus and median raphe nucleus	Restricted to a highly confined region between the medial longitudinal fasciculi at the caudal interface of the dorsal raphe nucleus and the median raphe nucleus (from P 2.5 to P 1.5 in the cat)

[a] From Rasmussen *et al.*, 1984.

(which displace tryptophan from albumin; Taggart and Carruthers, 1971; Knott and Curzon, 1972), but how do we account for regional changes in tryptophan uptake into the brain, based on regional demand? One possibility is that regional vasodilation increases local concentrations of free tryptophan in the plasma (Pardride and Fierer, 1990; Pardridge, 1998). Candidate vasoactive neuromodulators associated with serotonergic systems include nitric oxide (Dun *et al.*, 1994), and corticotropin-releasing factor (Ruggiero *et al.*, 1999). A second possibility is direct regulation of sodium-independent long chain neutral amino acid carriers on the luminal and abluminal surfaces of endothelial cells at the blood-brain barrier (Pardridge, 1998).

Vascular Networks Around Mesolimbocortical Serotonergic Systems – a Clue to Functional Specialization?

Several serotonergic cell groups are remarkable for their close association with blood vessels, including the median raphe nucleus (nucleus centralis superior medialis in primates, Azmitia and Gannon, 1986; Törk, 1985), the caudal portion of the interpeduncular nucleus (Groenewegen *et al.*, 1986), and the nucleus paragigantocellularis lateralis (Azmitia and Gannon, 1986). Likewise, the evolutionarily conserved group of serotonergic neurons in the dorsomedial hypothalamic nucleus (DMN, Lowry *et al.*, 1996) is closely associated with blood vessels in a broad range of vertebrate species (reviewed in Vigh *et al.*, 1967). This association with cerebrovasculature in the dorsomedial hypothalamus of nonmammalian vertebrates is so striking that this region has been referred to as the nucleus organi paraventricularis, or the "paraventricular organ", and likened to other circumventricular organs like the subfornical organ, organum vasculosum of the lamina terminalis, and area postrema that have reduced blood-brain barriers. As in mammals (Frankfurt *et al.*, 1981), the serotonin-containing neurons in the dorsomedial hypothalamus of non-mammalian vertebrates are bipolar with a primitive neuronal structure; in nonmammalian vertebrates the apical process extends through the ependymal lining of the third ventricle where the process forms a "club-like" ending in the cerebrospinal fluid. These cerebrospinal fluid-contacting processes arising from DMN neurons have not been observed in adult mammalian brain (possibly due to extensive neuronal migration away from the ventricular wall), but have been visualized in neonatal rats (Tantaoui, 1986). The basal processes of the serotonin-accumulating neurons often make direct contact with the laterally displaced cerebrovasculature (reviewed in Vigh *et al.*, 1967). Although a "paraventricular organ" has not been historically recognized in mammals, the DMN is highly vascularized and receives vascular input from three separate cerebral arteries (Ambach and Palkovits, 1977; reviewed by Bernardis and Bellinger, 1987). Thus, in summary, the median raphe nucleus, the interpeduncular nucleus, the nucleus paragigantocellularis lateralis in the reticular formation of the ventrolateral medulla, and the DMN are remarkable for their associations with extensive vascular networks. These networks may contribute to rapid stress-induced activation of serotonergic activity via rapid, local changes in tryptophan transport across the blood-brain-barrier.

Integrated Roles for the Sympathetic Nervous System and the Hypothalamo-Pituitary-Adrenal Axis in Regulating Tryptophan Availability

Stress-induced increases in brain tryptophan (due to electric footshock, restraint, or interleukin-(IL)-1α or lipopolysaccharide (LPS) injection are blocked by prior treatment with the ganglionic blocker chlorisondamine (Dunn and Welch, 1991), indicating that sympathetic nervous system activation is a necessary prerequisite. It was suggested that this effect may be mediated by β-adrenergic receptors. Stress-related activation of the hypothalamo-pituitary-adrenal axis is not required for increases in brain tryptophan because it occurs in adrenalectomized animals (Curzon *et al.*, 1972; Dunn 1988). However, corticosterone may be necessary for enhancement of high-affinity uptake of tryptophan from the extracellular compartment within the brain to the intraneuronal compartment of serotonergic neurons, possibly via a rapid membrane bound corticosterone receptor-mediated mechanism (Towle and Sze, 1983). Further evidence that corticosterone may enhance brain serotonergic activity during stress is provided by the observations that glucocorticoid administration can increase brain tryptophan (Neckers and Sze, 1975), tryptophan hydroxylase activity (Azmitia and McEwen, 1969), and serotonin turnover (Neckers and Sze, 1975; Telegdy and Vermes, 1975; Van Loon *et al.*, 1981). Together, these observations suggest that the sympathetic nervous system and the hypothalamo-pituitary-adrenal axis may act together to increase 5-HT function during stress.

Summary and Future Directions

Evidence suggests that a variety of stress-related stimuli may selectively activate a subpopulation of mesolimbocortical serotonergic neurons located primarily in the median raphe nucleus and the caudal midline portion of the dorsal raphe nucleus, particularly the intrafascicular portion. Serotonergic neurons in these regions innervate primarily limbic forebrain regions, many of which previously have been associated with behavioral anxiety. Future studies should provide additional insights into the behavioral correlates of these neurons, as well as the significance of the unique associations between these cells and the extensive cerebrovascular networks that exist within corresponding regions of the brainstem raphe.

Acknowledgements

I gratefully acknowledge the financial support of the Wellcome Trust (Project Grant 045843/Z/95/Z), and of the Neuroendocrinology Charitable Trust during the preparation of this chapter. I also acknowledge Professor Adrian J. Dunn for rewarding academic discussions.

References

Adell, A., Casanovas, J.M., and Artigas, F. (1997). Comparative study in the rat of the actions of different types of stress on the release of 5-HT in raphe nuclei and forebrain areas. *Neuropharmacology* **36**, 735–741.

Amat, J., Matus-Amat, P., Watkins, L.R., and Maier, S.F. (1998). Escapable and inescapable stress differentially alter extracellular levels of 5-HT in the basolateral amygdala of the rat. *Brain Research* **812**, 113–120.

Ambach, G., and Palkovits, M. (1977). Blood supply of the rat hypothalamus. V. The medial hypothalamus (nucleus ventromedialis, nucleus dorsomedialis, nucleus perifornicalis). *Acta Morphologica Acad. Sci. Hung.* **25**, 259–278.

Anisman, H. (1978). Neurochemical changes elicited by stress. In: H. Anisman and G. Bignami (Eds.), *Psychopharmacology of Aversively Motivated Behavior*, pp. 119–172. New York: Plenum Press.

Anisman, H.L. Kokkinidis, and L.S. Sklar (1985). Neurochemical consequences of stress. In: S.R. Burchfield (Ed.), *Stress: Psychological and Physiological Interactions* pp. 67–98. Washington: Hemisphere Publishing.

Azmitia, E.C., and Gannon, P.J. (1986). The primate serotonergic system: A review of human and animal studies and a report on *Macaca fascicularis*. In: S. Fahn *et al.*, (Eds.), *Myoclonus* pp. 407–468. New York: Raven Press.

Azmitia, E.C., and McEwen, B.S. (1969). Corticosterone regulation of tryptophan hydroxylase in midbrain of the rat. *Science* **166**, 1274–1276.

Bernardis, L.L., and Bellinger, L.L. (1987). The dorsomedial hypothalamic nucleus revisited: 1986 update. *Brain Research Reviews* **12**, 321–381.

Curzon, G., Joseph, M.H., and Knott, P.J. (1972). Effects of immobilization and food deprivation on rat brain tryptophan metabolism. *Journal of Neurochemistry* **19**, 1967–1974.

Dahlström, A., and Fuxe, K. (1964). Evidence for the existence of monoamine-containing neurons in the central nervous system. I. Demonstration of monoamines in the cell bodies of brainstem neurons. *Acta Physiologica Scandinavica* **62** Suppl. 232, 1–55.

Dun, N.J., Dun, S.L., and Förstermann, U. (1994). Nitric oxide synthase immunoreactivity in rat pontine medullary neurons. *Neuroscience* **59**, 429–445.

Dunn, A.J. (1988). Stress-related changes in cerebral catecholamine and indoleamine metabolism: lack of effect of adrenalectomy and corticosterone. *Journal of Neurochemistry* **51**, 406–412.

Dunn, A.J., and Kramarcy, N.R. (1984). Neurochemical responses in stress: relationships between the hypothalamic-pituitary-adrenal and catecholamine systems. In: L.L. Iversen, S.D. Iversen, and S.H. Snyder (Eds.), *Handbook of Psychophamacology* Vol. 18, pp. 455–515. Plenum Press.

Dunn, A.J., and Welch, J. (1991). Stress- and endotoxin-induced increases in brain tryptophan and serotonin metabolism depend on sympathetic nervous system activity. *Journal of Neurochemistry* **57**, 1615–1622.

Frankfurt, M., Lauder, J.M., and Azmitia, E.C. (1981). The immunocytochemical localization of serotonergic neurons in the rat hypothalamus. *Neuroscience Letters* **24**, 227–232.

Groenewegen, H.J., Ahlenius, S., Haber, S.N., Kowall, N.W., and Nauta, W.J.H. (1986). Cytoarchitecture, fiber connections, and some histochemical aspects of the interpeduncular nucleus in the rat. *Journal of Comparative Neurology* **249**, 65–102.

Halliday, G., Harding, A., and Paxinos, G. (1995). Serotonin and tachykinin systems. In: G. Paxinos (Ed.), *The Rat Nervous System*, Second Edition, pp. 929–974. San Diego: Academic Press.

Imai, H., Steindler, D. A., and Kitai, S. T. (1986). The organization of divergent axonal projections from the midbrain raphe nuclei in the rat. *Journal of Comparative Neurology* **243**, 363–380.

Inoue, T., Koyama, T., and Yamashita, I. (1993). Effect of conditioned fear stress on serotonin metabolism in the rat brain. *Pharmacology Biochemistry and Behavior* **44**, 371–374.

Inoue, T., Tsuchiya, K., and Koyama, T. (1994). Regional changes in dopamine and serotonin activation with various intensity of physical and psychological stress in the rat brain. *Pharmacology Biochemistry and Behavior* **49**, 911–920.

Jacobs, B.L., and Azmitia, E.C. (1992). Structure and function of the brain serotonin system. *Physiological Reviews* **72**, 165–229.

Jacobs, B.L., and Fornal, C.A. (1995). Activation of 5-HT neuronal activity during motor behavior. *Seminars in the Neurosciences* **7**, 401–408.

Joseph, M.H., and Kennett, G.A. (1983a). Corticosteroid response to stress depends upon increased tryptophan availability. *Psychopharmacology* **79**, 79–81.

Joseph, M.H., and Kennett, G.A. (1983b). Stress-induced release of 5-HT in the hippocampus and its dependence on increased tryptophan availability: An *in vivo* electrochemical study. *Brain Research* **270**, 251–257.

Kawahara, H., Yoshida, M., Yokoo, H., Nishi, M., and Tanaka, M. (1993). Psychological stress increases serotonin release in the rat amygdala and prefrontal cortex assessed by in vivo microdialysis. *Neuroscience Letters* **162**, 81–84.

Kazakov, V.N., Kravtsov, P.Y., Krakhotkina, E.D., and Maisky, V.A. (1993) Sources of cortical, hypothalamic and spinal serotonergic projections: topical organization within the nucleus raphe dorsalis. *Neuroscience* **56**, 157–164.

Kennett, G.A., and Joseph, M.H. (1981). The functional importance of increased brain tryptophan in the serotonergic response to restraint stress. *Neuropharmacology* **20**, 39–43.

Knott, P.J., and Curzon, G. (1972). Free tryptophan in plasma and brain tryptophan metabolism. *Nature* **239**, 452–453.

Köhler, C., and Steinbusch, H. (1982). Identification of serotonin and non-serotonin-containing neurons of the mid-brain raphe projecting to the entorhinal area and the hippocampal formation. A combined immunohistochemical and fluorescent retrograde tracing study in the rat brain. *Neuroscience* **7**, 951–975.

Lowry, C.A., J.E. Rodda, S.L. Lightman, and C.D. Ingram (1999). Corticotropin-releasing factor (CRF) increases *in vitro* firing rates of serotonergic neurones in the rat dorsal raphe nucleus: Evidence for selective activation of a topographically organised mesolimbocortical serotonergic system. *Neuroscience Abstracts* **24**, 25, 75.

Lowry, C.A., Renner, K.J., and Moore, F.L. (1996). Catecholamines and indoleamines in the central nervous system of a urodele amphibian: A microdissection study with emphasis on the distribution of epinephrine. *Brain, Behavior and Evolution* **48**, 70–93.

Lucki, I. (1998). The spectrum of behaviors influenced by serotonin. *Biological Psychiatry* **44**, 151–162.

Mehler, W.R. (1980). Subcortical afferent connections of the amygdala in the monkey. *Journal of Comparative Neurology* **190**, 733–762.

Neckers, L., and Sze, P.Y. (1975). Regulation of 5-hydroxytryptamine metabolism in mouse brain by adrenal glucocorticoids. *Brain Research* **93**, 123–132.

Ottersen, O.P. (1981). Afferent connections to the amygdaloid complex of the rat with some observations in the cat. III. Afferents from the lower brain stem. *Journal of Comparative Neurology* **202**, 335–356.

Panksepp, J. (1990). The psychoneurology of fear: evolutionary perspectives and the role of animal models in understanding human anxiety. In *Handbook of Anxiety, Vol. 3: The Neurobiology of Anxiety* (G.D. Burrows, M. Roth, and R. Noyes, Jr., eds.). Elsevier Science Publishers, B.V., pp. 3–58.

Pardridge, W.M. (1998). Blood-brain barrier carrier-mediated transport and brain metabolism of amino acids. *Neurochemical Research* **23**, 635–644.

Pardridge, W.M., and Fierer, G. (1990). Transport of tryptophan into brain from the circulating, albumin-bound pool in rats and in rabbits. *Journal of Neurochemistry* **54**, 971–976.

Petrov, T., Krukoff, T.L., and Jhamandas, J.H. (1994). Chemically defined collateral projections from the pons to the central nucleus of the amygdala and hypothalamic paraventricular nucleus in the rat. *Cell and Tissue Research* **277**, 289–295.

Porrino, L.J., and Goldman-Rakic, P.S. (1982). Brainstem innervation of prefrontal and anterior cingulate cortex in the rhesus monkey revealed by retrograde transport of HRP. *Journal of Comparative Neurology* **205**, 63–76.

Rasmussen, K., Heym, J., and Jacobs, B.L. (1984). Activity of serotonin-containing neurons in the nucleus centralis superior of freely moving cats. *Experimental Neurology* **83**, 302–317.

Rueter, L.E., Fornal, C.A., and Jacobs, B.L. (1997). A critical review of 5-HT brain microdialysis and behavior. *Reviews in the Neurosciences* **8**, 117–137.

Ruggiero, D.A., Underwood, M.D., Rice, P.M., Mann, J.J., and Arango, V. (1999). Corticotropin-releasing hormone and serotonin interact in the human brainstem: behavioral implications. *Neuroscience* **91**, 1343–1354.

Taggart, P., and Carruthers, M. (1971). Endogenous hyperlipidaemia induced by emotional stress of racing driving. *Lancet* **1**, 363–366.

Tantaoui, H. (1986). Phylogenese et ontogenese des structures serotoninerguques de l'hypothalamus des vertebres. Diplome D'Etudes Approfondies de Neurosciences, Universite de Bordeaux, I. pp. 1–21.

Telegdy, G., and Vermes, I. (1975). Effect of adrenocortical hormones on activity of the serotoninergic system in limbic structures in rats. *Neuroendocrinology* **18**, 16–26.

Törk, I. (1985). Raphe nuclei and serotonin containing systems. In: G. Paxinos (Ed.), *The Rat Nervous System. Vol.2 Hindbrain and Spinal Cord*, pp. 43–78. Sydney: Academic Press.

Törk, I. (1990). Anatomy of the serotonergic system. *Annals of the New York Academy of Sciences* **600**, 9–35.

Towle, A.C., and P.Y. Sze (1983). Steroid binding to synaptic plasma membrane: Differential binding of glucocorticoids and gonadal steroids. *Journal of Steroid Biochemistry* **18**, 135–143.

Van Bockstaele, E.J., Biswas, A., and Pickel, V.M. (1993). Topography of serotonin neurons in the dorsal raphe nucleus that send axon collaterals to the rat prefrontal cortex and nucleus accumbens. *Brain Research* **624**, 188–198.

Van Loon, G.R., Shum, A., and Sole, M.J. (1981). Decreased brain serotonin turnover after short term (two-hour) adrenalectomy in rats: A comparison of four turnover methods. *Endocrinology* **108**, 1392–1402.

Vigh, B., Teichmann, I., and Aros, B. (1967). The "nucleus organi paraventricularis", as a neuronal part of the paraventricular ependymal organ of the hypothalamus: Comparative morphological study on various vertebrates. *Acta Biol. Sci. Hung.* **18**, 271–284.

Wyss, J.M., Swanson, L.W., and Cowan, W.M. (1979). A study of subcortical afferents to the rat hippocampal formation in the rat. *Neuroscience* **4**, 463–476.

Yoshioka, M., Matsumoto, M., Togashi, H., and Saito, H. (1995). Effects of conditioned fear stress on 5-HT release in the rat prefrontal cortex. *Pharmacology Biochemistry and Behavior* **51**, 515–519.

3. Neurocircuit Activation of the Hypothalamo-Pituitary-Adrenocortical Axis: Roles for Ascending Norepinephrine Systems

J.P. Herman, B.J. McCreary, K. Bettenhausen and D.R. Ziegler

Introduction

The hypothalamo-pituitary-adrenocortical axis is governed by a diverse set of neurocircuits that relay stressful information to the hypothalamic paraventricular nucleus (PVN) (Feldman *et al.*, 1995; Herman and Cullinan, 1997). The PVN causes release of corticotropin releasing hormone and associated secretagogs into the pituitary portal circulation, initiating the neuroendocrine cascade resulting in ACTH and hence, glucocorticoid release (Antoni, 1986; Whitnall, 1993).

Events controlling PVN activation likely involve brainstem NE pathways. Ascending NE/E systems are activated by interoceptive stimuli and induce cFOS expression in PVN neurons following peripheral injection of cytokines (Day and Akil, 1996; Larsen and Mikkelsen, 1995; Li *et al.*, 1996; Thrivikraman *et al.*, 1997). Lesions of the central noradrenergic bundle (VNAB), which destroy ascending axons from A1 and A2 medullary cell groups, attenuate HPA axis responses to numerous stressors (Gaillet *et al.*, 1993; Gaillet *et al.*, 1991), consistent with a role in PVN activation. Furthermore, CRH release into portal blood can be initiated by stimulation of the VNAB (Plotsky, 1987). Microdialysis experiments indicate release of NE in the region of the PVN following stress (Pacak *et al.*, 1995), and activation of the HPA axis by systemic stressors (hemorrhage or ether) can be blocked by i.c.v. injections of alpha 1 adrenergic receptor antagonists (Plotsky, 1987; Szafarczyk *et al.*, 1987).

The role of the locus coeruleus (LC) in HPA activation is considerably less clear. The LC is involved in mediating responses to stressful stimuli, perhaps through the influence of central CRH pathways (Valentino *et al.*, 1991). Acute restraint, swim, fear conditioning, footshock and hemorrhage reliably induce cFOS expression in the LC (Campeau and Watson, 1997; Cullinan *et al.*, 1995; Li and Sawchenko, 1998; Smith *et al.*, 1992; Thrivikraman *et al.*, 1997), and LC TH expression is increased with acute restraint, footshock or social stress (Smith *et al.*, 1991; Watanabe *et al.*, 1995). These findings are consistent with a role for the LC in processing of stressful information. The LC is also implicated in the etiology of major depression, which is accompanied by a dysfunctional HPA axis (Leonard, 1997).

Our group has used lesion analyses to assess the importance of candidate NE brain regions on HPA excitation. The present report represents a synthesis of our research efforts in this area. We have linked the results of previous published data (Ziegler *et al.*, 1999) with new experimental studies to provide an overview of the state of our thinking on NE regulation of HPA axis excitation and inhibition.

Materials and Methods

Subjects: Subjects in all studies were male Sprague-Dawley and F344 rats (Harlan), weighing between 250–300 g at the beginning of the stress experiments. All rats were maintained on a 12:12 h light:dark cycle in a constant temperature/humidity environment, and had access to food and water ad libitum. Animal procedures were approved by the University of Kentucky IACUC.

Surgery: Lesions of the ventral noradrenergic bundle (VNAB) were made by bilateral stereotaxic injection of 4 μg 6-hydroxydopamine in 0.5 μl saline containing 1 mg/ml ascorbic acid. Injections were centered 9.8 mm posterior to bregma, ± 1.4 mm lateral to midline, and 7.1 mm deep to the dura (after Maccari *et al.*, 1990). Sham lesions received vehicle only.

Acute Stress Protocol: Ether: Animals were allowed to recover from surgery for 1 week. Ether stress was induced by placing animals into a jar saturated with ether fumes for 2 minutes. Blood samples were collected by tail-nick at 30, 60 and 120 minutes following stress induction.

Restraint: Four days after ether stress, animals were placed in restraint cages for 30 min. Blood was sampled by tail-nick immediately upon placement in restrainers and immediately prior to release. Also, 60 min and 120 min samples were subsequently collected under light restraint.

Chronic Stress Protocol: Following acute stress exposure, groups of lesion and sham-lesion rats (n = 6/age) were subjected to a 14-day variable-stressor paradigm, which included the following: restraint (2 hour in plastic restraint cages (Plas Labs)), cold exposure (2 hour in a 4°C cold room), cold water swim (20 minutes in 16–18°C water), warm water swim (40 minutes in 26–30°C water), vibration (6 animals per cage placed on a shaker for 2 hour (1 cycle/sec)), crowding (6 rats per cage, overnight) and isolation (1 animal per cage, overnight). Handled (removed from their home cages and handled twice daily) groups were included as controls. Stressors were distributed randomly to avoid habituation, with two different stressors administered per day. Additional groups of non-lesioned Sprague-Dawley rats (n = 6/group) were run to assess TH regulation in LC and DVC following chronic stress.

After 14 days of variable intermittent stress exposure, all rats were sacrificed by rapid decapitation 16 h after the last stressor. Brains were rapidly removed, flash frozen in isopentane cooled to –40 or –50°C on dry ice and stored at –80°C until processing. Core blood samples were collected in heparinized tubes and spun at 1,500 g to separate plasma. Plasma was collected and stored at –20°C. Adrenal glands and thymi, and spleen were collected, cleaned, and weighed. Brains were sectioned at 15 μm in a Bright-Hacker cryostat, thaw-mounted onto Superfrost Plus slides, and stored at –20°C.

In Situ Hybridization: Hybridizations used probes complementary to CRH exon 2 mRNA (a 760-bp sequence encoding the coding region and proximal 3' untranslated region (UT) of pre-proCRH) or rat tyrosine hydroxylase (a 380 bp sequence through the coding region of rat TH mRNA). The probe was synthesized by *in vitro* transcription using T7 RNA polymerase (Boehringer Mannheim) and [33]P- or ^{35}S-UTP (Amersham). Hybridization protocols have been extensively detailed elsewhere (Seroogy and Herman, 1996).

Immunoautoradiography: Alternate series of tissue sections from rats processed for *in situ* hybridization were removed from the –20°C freezer, warmed to room temperature and ringed with a PAP pen (RBI). Immunoautoradiography was performed as described previously (Herman and Morrison, 1996), using primary antibodies against TH (1:2000) and DBH (1:1000) (Protos Labs) and ^{35}S-labeled anti-mouse or anti-rabbit IgG, diluted 1:500 in PBS. Sections were exposed to X-ray film for 5–7 days.

Hormone assays: Plasma samples were collected and stored at –20°C. Plasma CORT was assessed by radioimmunoassay, using a double-antibody kit from ICN (with ^{125}I-labeled CORT used as tracer). All plasma samples for each assay were processed at the same time.

Data Analysis. In situ hybridization data were analyzed using NIH Image 1.55 software for Macintosh. Grain counts were also performed using NIH image, as described previously (Bowers *et al.*, 1998). Acute stress time-course data were analyzed by repeated-measures ANOVA; chronic stress data were analyzed by two-way factorial ANOVA.

Results

Lesion Studies

Ventral Noradrenergic Bundle (VNAB) Lesions

Lesions of the VNAB were verified by 1) placement of the cannula tip and 2) marked reduction in DBH-immunoreactivity in the ventral midbrain and PVN regions by immunoautoradiographic analysis.

Acute Stress. Lesions of the VNAB did not elicit marked changes in CORT secretion following restraint stress (Figure 1A). In contrast, CORT responses to ether stress were substantially reduced following VNAB lesions, as noted by repeated measures ANOVA (effect of lesion: $F_{(1,33)} = 17.52$, $p < 0.05$; lesion by time interaction, $F_{(3,33)} = 3.39$, $p < 0.05$). Reduced CORT secretion was evident 30 and 60 minutes following initial exposure to ether.

Chronic Stress. Chronic stress data are summarized in Table 1. There was no effect of VNAB lesion on basal or post-stress ACTH or CORT secretion, and no effects on stress-induced weight loss, adrenal hypertrophy or thymic atrophy. Similarly, no VNAB-related changes in basal or post-stress CRH mRNA levels were recorded in the current study.

Locus Coeruleus Lesions

The effects of locus coeruleus lesions on the HPA axis have been published previously by our group (reproduced in Figure 1C and Table 1 (Ziegler *et al.*, 1999), with permission from Blackwell Science Ltd.). In contrast to the results of VNAB lesions, damage to the LC reduces CORT secretion elicited by acute exposure to restraint (Figure 1C). As was the case for VNAB lesions, LC damage did not affect CORT levels, weight loss, adrenal hypertrophy or CRH mRNA levels following chronic stress exposure (Table 1: data retabulated from Ziegler *et al.*, 1999, with permission). However, LC lesion attenuated ACTH secretion and stress-induced thymic atrophy.

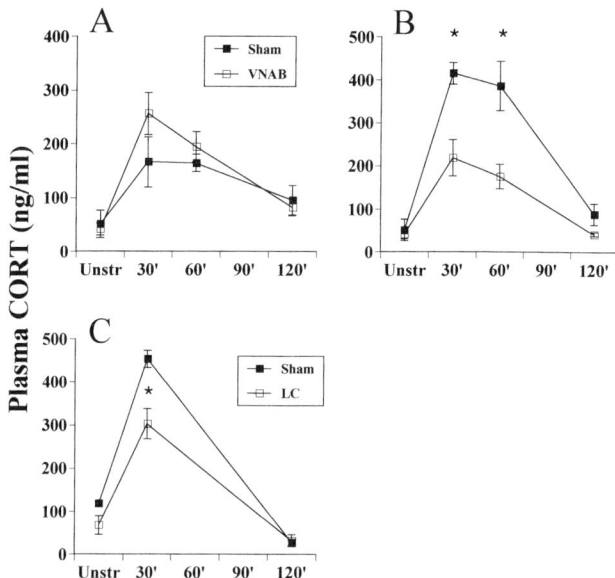

Fig. 1 Plasma corticosterone (CORT) responses to acute stress in ventral noradrenergic bundle (VNAB) or locus coeruleus (LC) lesion rats. A) In contrast, VNAB lesion animals do not hyposecrete CORT in response to restraint. B) Animals with lesions of the VNAB show markedly attenuated CORT responses to ether stress. C) Animals with LC lesions show reduced CORT release in response to restraint. Panel C is reprinted in modified form from (Ziegler *et al.*, 1999), with permission.

Table 1 Effects of VNAB or LC lesions on physiological responses to chronic intermittent stress.

Stress index	VNAB lesion	LC lesion*
Body Weight Decrease	n.c.	n.c.
Adrenal Hypertrophy	n.c.	n.c.
Thymic Atrophy	n.c.	↓
Plasma CORT	n.c.	n.c.
Plasma ACTH	n.c.	↓
CRH mRNA	n.c.	n.c.

* Locus Coeruleus data reprinted from Ziegler *et al.*, 1999.
↓ = significant decrease (p < 0.05) relative to sham-lesion rats
n.c. – no change

TH Regulation

The impact of chronic stress exposure on TH mRNA expression and TH immunoreactivity (TH-ir) were assessed by *in situ* hybridization (Figure 2) and immunoautoradiography (Figure 3), respectively. Grain count analysis revealed a significant effect of chronic stress exposure on TH mRNA expression in the LC (F (2,11) = 7.33, p < 0.05). Post-hoc analysis revealed that TH mRNA expression was lower in stressed animals than in unhandled controls (p < 0.0167, Bonferroni/Dunn test) (Figure 4A). In contrast, there was no significant effect of stress on TH mRNA in the dorsal vagal complex (F (2,11) = 2.80, p = 0.10), which contains the A2/C2 cell group. There was no effect of stress on TH-ir in either the LC or DVC (Figure 4B).

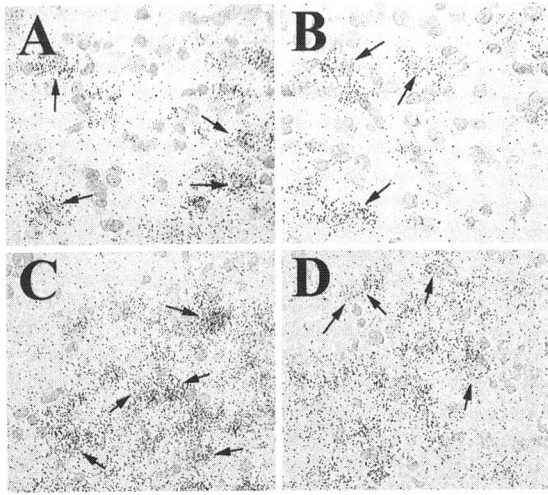

Fig. 2 Regulation of TH mRNA expression in the dorsal vagal complex (A,B) and locus coeruleus (C,D) following handling or chronic intermittent stress. There was no obvious difference in grain density/cell in neurons in the DVC between handled (A) and stressed (B) rats. In contrast, TH grain density was decreased in the LC of stressed rats (D) relative to unstressed controls (C).

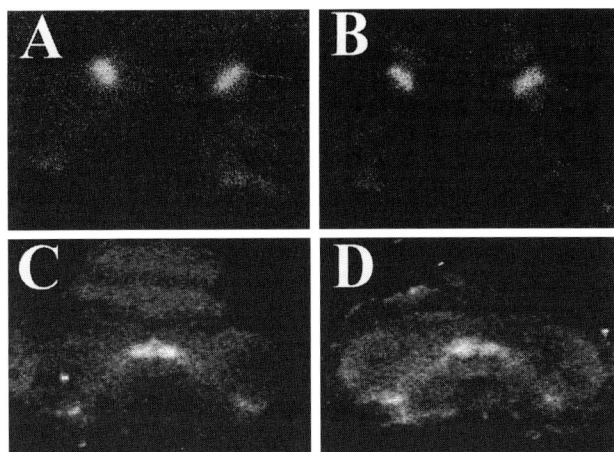

Fig. 3 TH immunoreactivity in the dorsal vagal complex (A,B) and locus coeruleus (C,D) following handling or chronic intermittent stress. There were no substantial changes in TH immunoreactivity in either the DVC or LC following chronic stress (compare images from handled (A,C) and stressed groups (B,D)).

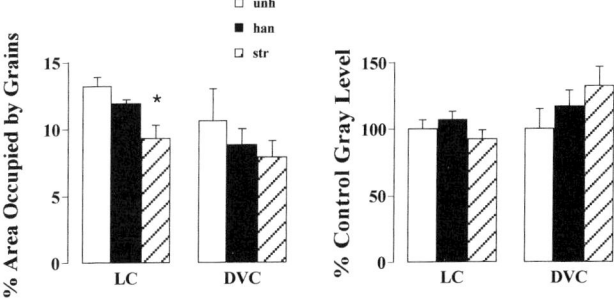

Fig. 4 Semi-quantitative analysis of TH mRNA and immunoreactivity in locus coeruleus (LC) and dorsal vagal complex (DVC). A) *In situ* hybridization analysis of TH mRNA, using grain density determinations. Stressed (str) animals showed significant reductions in grain density/cell relative to either unhandled (Unh) or handled (Han) controls, reflecting reductions in cellular mRNA. No changes were seen in the DVC. B) Immunoautoradiographic analysis of TH protein, using areal densitometry. No changes in TH immunoreactivity were seen in either region following stress.

Discussion

The present study indicates a role for medullary NE and perhaps E systems in integration of threats to systemic homeostasis. Damage to the VNAB, which contains ascending axons from NE and E containing neurons in the ventrolateral medulla/DVC, markedly decreases CORT responses to ether stress without affecting secretion following restraint. Thus, the integrity of this ascending system is required for relay of interoceptive cues related to respiratory ("interoceptive") distress, but is less relevant for initiation of responses requiring processing of exteroceptive cues (i.e., "processive" stressors), which are principally communicated by forebrain pathways (Herman and Cullinan, 1997).

Previous work supports this dichotomy of function within the VNAB system. Lesion analyses indicate that ascending information from the brainstem is required for PVN cFOS induction following IL-1 injection but not footshock (Li *et al.*, 1996). Similarly, previous studies employing a similar design indicate attenuated CORT responses in VNAB lesion groups exposed to ether viz. restraint (Gaillet *et al.*, 1991).

In contrast with the results of the VNAB lesion analysis, our previous data indicate that rats with lesions to the LC secrete less CORT following restraint stress exposure. These data suggest that, unlike medullary NE groups, the LC plays a role in regulation of HPA responses to processive stressors. This hypothesis is supported by the known anatomical distribution of ascending LC terminals, which are concentrated in limbic and neocortical structures. Indeed, the LC innervates limbic regions implicated in HPA integration, including the amygdala, bed nucleus of the stria terminalis, hippocampus and prefrontal cortex (Figure 5). Importantly, LC projections largely avoid the PVN proper (Cunningham and Sawchenko, 1988), making it likely that the influence of the LC on the HPA axis is indirect, working through modulation of descending input to the PVN.

Exposure to the chronic variable stress regimen did not significantly effect TH mRNA or protein expression in the DVC, the principle source of NE input to hypophysiotrophic neurons of the PVN (Cunningham and Sawchenko, 1988). These results contrast with data showing that chronic hypoxia increases TH expression in the medulla (Soulier *et al.*, 1992). The principal difference between the studies is in the nature of the

repeated stressor; our work uses a rotating paradigm involving a number of primarily processive stressors, whereas hypoxia represents repeated respiratory challenge. These studies are consistent with involvement of these ascending systems in responses to prolonged systemic but not processive stress exposure.

Lesions of the LC do not significantly affect CORT secretion, weight loss, or adrenal hypertrophy, key indicators of chronic stress in the rat. Similarly, elimination of LC neurons do not affect CRH mRNA expression in the PVN of unstressed or stressed rats. However, LC lesions decrease basal ACTH release and attenuate stress-induced thymic atrophy (Ziegler *et al.*, 1999), suggestive of a partial remediation of chronic stress-induced HPA hyperactivity. It is also notable that LC lesion rats does not attenuate CORT responses to restraint following repeated stress exposure, suggesting a role for the LC in the stress habituation process.

Regulatory studies indicate that TH mRNA expression is reduced in the LC following chronic stress, suggesting that active down-regulation occurs with repeated stress exposure. Taken together, our data suggest that the activational role of the LC in stress regulation may be principally manifest during initial exposures to stress. Repeated or prolonged exposure to stress may overcome or attenuate the stimulatory influence of the LC. This hypothesis is supported by data indicting marked habituation of LC responses over repeated exposure to social stress, restraint or vibration (Martinez *et al.*, 1998; Watanabe *et al.*, 1994). Indeed, down-regulation of TH mRNA in the face of normal TH protein levels suggests that repeated stress may reduce TH synthesis, transport or release, consistent with reduced activation of the NE projection neurons.

Mechanisms responsible for chronic stress modulation of LC activity are currently unclear. The LC expresses high levels of GR (Ahima and Harlan, 1990; Fuxe *et al.*, 1985; Herman, 1993), making it a potential target for stress-induced increases in circulating glucocorticoids. Notably, acute stress increases TH mRNA levels in the LC (Smith *et al.*, 1991), and glucocorticoids positively regulate TH gene expression in the adrenocortical cell lines (Fossom *et al.*, 1992). Responses of LC TH expression to repeated stress may be modality- or intensity-specific. For example, expression of TH mRNA is enhanced by chronic cold exposure or immobilization (Mamalaki *et al.*, 1993; Melia *et al.*, 1992); however, repeated immobilization, social stress or passive modulation of glucocorticoid levels are minimally effective in increasing TH levels in the LC (Smith *et al.*, 1991; Watanabe *et al.*, 1995). These data suggest that LC TH expression is susceptible to competing influences, perhaps balancing excitatory effects of afferent activity and/or glucocorticoids with indogenous down-regulatory processes (e.g., autoreceptor inhibition (c.f., (Grant and Redmond, 1981)). In the present case, chronic stress exposure is sufficient to reduce LC TH mRNA, perhaps as an endogenous mechanism to limit generalized stress activation of the CNS.

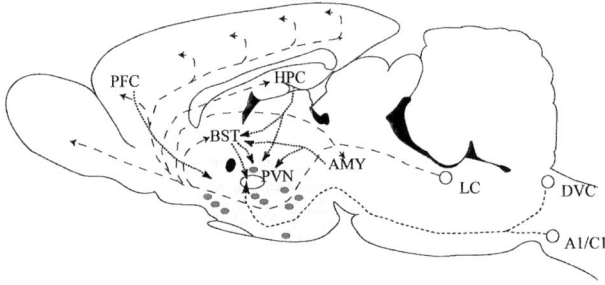

Fig. 5 Schematic diagram of potential NE interactions with the PVN. Neurons of the DVC and to a lesser extent A1/C1 regions project directly to hypophysiotrophic PVN neurons. This pathway likely integrates interoceptive information into appropriate HPA responses. In contrast, neurons of the LC project to neocortical and limbic regions, including the prefrontal cortex (PFC), hippocampus (HPC), amygdala (AMY) and bed nucleus of the stria terminalis (BST). The BST sends direct projections to the medial parvocellular PVN, whereas the HPA, AMY and PFC can interconnect with the PVN through the BST and/or any of a number of intrahypothalamic neuronal networks (the hypothalamus is depicted in light gray, with potential limbic-PVN interconnecting cell groups denoted by dark gray ovals). The LC likely affects the HPA axis circuitously through modulation of limbic outflow.

In summary, our data suggest very different roles for LC and medullary NE pathways in stress integration. The circuitry underlying these pathways is illustrated in Figure 5. The LC, which has extensive limbic connections, appears to mediate stress responses through its influence on suprahypothalamic structures. The LC is clearly involved in regulation of glucocorticoid secretion to novel stressors, perhaps through these limbic pathways. The multisynaptic nature of LC involvement in HPA activity suggests that this may be a by-product of LC activation, rather than a dedicated function. In contrast, the DVC/ventral medullary system is preferentially involved in integration of systemic stimuli, presumably through direct pathways to the PVN. This system appears well-suited for handling information from interoceptive brain pathways, thus directly mobilizing autonomic and neuroendocrine systems essential for survival.

Importantly, neither the LC nor the DVC are essential for elaboration of the chronic stress syndrome following chronic intermittent stress exposure. In the case of the LC, the attenuation of the excitatory influence of NE on forebrain stress pathways may be an adaptive change, serving to minimize mobilization of stress circuits and combat the negative effects of stress. This putative removal of excitatory NE drive to central stress pathways appears to be complemented by increases in GABA biosynthesis (Bowers *et al.*, 1998), and may be a component of global brain adaptation to stress exposure.

References

Ahima, R.S., and Harlan, R.E. (1990). Charting of type II glucocorticoid receptor-like immunoreactivity in the rat central nervous system. *Neuroscience* **39**, 579–604.

Antoni, F.A. (1986). Hypothalamic control of adrenocorticotropin secretion: Advances since the discovery of 41-residue corticotropin-releasing factor. *Endocrine Reviews* **7**, 351–378.

Bowers, G., Cullinan, W.E., and Herman, J.P. (1998). Region-specific regulation of glutamic acid decarboxylase (GAD) mRNA expression in central stress circuits. *Journal of Neuroscience* **18**, 5938–5947.

Campeau, S., and Watson, S.J. (1997). Neuroendocrine and behavioral responses and brain pattern of c-fos induction associated with audiogenic stress. *Journal of Neuroendocrinology* **9**, 577–588.

Cullinan, W.E., Herman, J.P., Battaglia, D.F., Akil, H., and Watson, S.J. (1995). Pattern and time course of immediate early gene expression in rat brain following acute stress. *Neuroscience* **64**, 477–505.

Cunningham, E.T., Jr and Sawchenko, P.E. (1988). Anatomical specificity of noradrenergic inputs to the paraventricular and supraoptic nuclei of the rat hypothalamus. *Journal of Comparative Neurology* **274**, 60–76.

Day, H.E., and Akil, H. (1996). Differential pattern of c-fos mRNA in rat brain following central and systemic administration of interleukin-1-beta: implications for mechanism of action. *Neuroendocrinology* **63**, 207–218.

Fossom, L.H. (1992). Regulation of tyrosine hydroxylase gene transcription rate and tyrosine hydroxylase mRNA stability by cyclic AMP and glucocorticoid. *Molecular Pharmacology* **42**, 898–908.

Fuxe, K., Wikstrom, A.C., Okret, S., Agnati, L.F., Harfstrand, A., Yu, Z.Y., Granholm, L Zol, M., Vale, W., and Gustafsson, J.A. (1985). Mapping of glucocorticoid receptor immunoreactive neurons in the rat tel- and diencephalon using a monoclonal antibody against rat liver glucocorticoid receptor. *Endocrinology* **117**, 1803–1812.

Gaillet, S., Alonso, G., Le Borgne, R., Barbanel, G., Malaval, F., Assenmacher, I., and Szafarczyk, A. (1993). Effects of discrete lesions in the ventral noradrenergic ascending bundle on the corticotropic stress response depend on the site of the lesion and on the plasma levels of adrenal steroids. *Neuroendocrinology* **58**, 408–419.

Gaillet, S., Lachuer, J., Malaval, F., Assenmacher, I., and Szafarczyk, A. (1991). The involvement of noradrenergic ascending pathways in the stress-induced activation of ACTH and corticosterone secretions is dependent on the nature of stressors. *Experimental Brain Research* **87**, 173–180.

Grant, S.J., and Redmond, D.E., Jr. (1981). The neuroanatomy and pharmacology of the nucleus locus coeruleus. *Progress in Clinical and Biological Research* **71**, 5–27.

Herman, J.P. (1993). Regulation of adrenocorticosteroid receptor mRNA expression in the central nervous system. *Cellular and Molecular Neurobiology* **13**, 349–372.

Herman, J.P., and Cullinan, W.E. (1997). Neurocircuitry of stress: Central control of the hypothalamo-pituitary-adrenocortical axis. *Trends in Neuroscience* **20**, 78–83.

Herman, J.P., and Morrison, D.G. (1996). Immunoautoradiographic and in situ hybridization analysis of corticotropin-releasing hormone biosynthesis in the hypothalamus paraventricular nucleus. *Journal of Chemical Neuroanatomy* **11**, 49–56.

Larsen, P.J., and Mikkelsen, J.D. (1995). Functional identification of central afferent projections conveying information of acute stress to the hypothalamic paraventricular nucleus. *Journal of Neuroscience* **15**, 2609–2627.

Leonard, B.E. (1997). The role of noradrenaline in depression: a review. *Journal of Psychopharmacology* **11**, S39–47.

Li, H.-Y., and Sawchenko, P.E. (1998). Hypothalamic effector neurons and extended circuitries activated in "neurogenic" stress: a comparison of footshock effects exerted acutely, chronically, and in animals with controlled glucocorticoid levels. *Journal of Comparative Neurology* **393**, 244–266.

Li, H.-Y., Ericsson, A., and Sawchenko, P.E. (1996). Distinct mechanisms underlie activation of hypothalamic neurosecretory neurons and their medullary catecholaminergic afferents in categorically different stress paradigms. *Proceedings of the National Academy of Sciences* **93**, 2359–2364.

Maccari, S., Le Moal, M., Angelucci, L., and Mormede, P. (1990). Influence of 6-OHDA lesion of central noradrenergic systems on corticosteroid receptors and neuroendocrine responses to stress. *Brain Research* **533**, 60–65.

Mamalaki, E. (1993). Repeated immobilization stress alters tyrosine hydroxylase, corticotropin-releasing hormone and corticosteroid receptor ribonucleic acid levels in rat brain. *Journal of Neuroendocrinology* **4**, 689–699.

Martinez, M., Phillips, P.J., and Herbert, J. (1998). Adaptation in patterns of c-fos expression in the brain associated with exposure to either single or repeated social stress in male rats. *European Journal of Neuroscience* **10**, 20–33.

Melia, K.R., Nestler, E.J., and Duman, R.S. (1992). Chronic imipramine treatment normalizes levels of tyrosine hydroxylase in the locus coeruleus of chronically stressed rats. *Psychopharmacology* **108**, 23–26.

Pacak, K., Palkovits, M., Kvetňanský, R., Yadid, G., Kopin, I.J., and Goldstein, D.S. (1995). Effects of various stressors on in vivo norepinephrine release in the hypothalamic paraventricular nucleus and on the pituitary-adrenocortical axis. *Annals of the New York Academy of Sciences* **771**, 115–130.

Plotsky, P.M. (1987). Facilitation of immunoreactive corticotropin-releasing factor secretion into the hypophysial-portal circulation after activation of catecholaminergic pathways or central norepinephrine injection. *Endocrinology* **121**, 924–934.

Seroogy, K., and Herman, J.P. (1996). In situ hybridization approaches to the study of the nervous system, In A.J. Turner and H.S. Bachelard (Eds.), *Neurochemistry A Practical Approach* (pp. 121–150): Oxford Universiy Press.

Smith, M.A., Banerjee, S., Gold, P.W., and Glowa, J. (1992). Induction of c-fos mRNA in rat brain by conditioned and unconditioned stressors. *Brain Research* **578**, 135–141.

Smith, M.A., Brady, L.S., Glowa, J., Gold, P.W., and Herkenham, M. (1991). Effects of stress and adrenalectomy on tyrosine hydroxylase mRNA levels in locus coeruleus by in situ hybridization. *Brain Research* **544**, 26–32.

Soulier, V., Cottet-Emard, J.M., Peguignot, J., Hanchin, F., Peyrin, L., and Peguignot, J.M. (1992). Differential effects of long-term hypoxia on norepinephrine turnover in brain stem cell groups. *Journal of Applied Physiology* **73**, 1810–1814.

Szafarczyk, A., Malaval, F., Laurent, A., Gibaud, R., and Assenmacher, I. (1987). Further evidence for a central stimulatory actions of catecholamines on adrenocorticotropin release in the rat. *Endocrinology* **121**, 883–392.

Thrivikraman, K.V., Su, Y., and Plotsky, P. (1997). Patterns of Fos-immunoreactivity in the CNS induced by repeated hemorrhage in conscious rats: Correlations with pituitary-adrenal axis activity. *Stress* **2**, 145–158.

Valentino, R.J., Page, M.E., and Curtis, A.L. (1991). Activation of noradrenergic locus coeruleus neurons by hemodynamic stress is due to local release of corticotropin-releasing factor. *Brain Research* **555**, 25–34.

Watanabe, Y., McKittrick, C.R., Blanchard, D.C., Blanchard, R.J., McEwen, B.S., and Sakai, R.R. (1995). Effects of chronic social stress on tyrosine hydroxylase mRNA and protein levels. *Molecular Brain Research* **32**, 176–180.

Watanabe, Y., Stone E., and McEwen, B.S. (1994). Induction and habituation of c-fos and zif/268 by acute and repeated stressors. *Neuroreport* **5**, 1321–1324.

Whitnall, M.H. (1993). Regulation of the hypothalamic corticotropin-releasing hormone neurosecretory system. *Progress in Neurobiology* **40**, 573–629.

Ziegler, D.R., Cass, W.A., and Herman, J.P. (1999). Excitatory influence of the locus coeruleus in hypothalamic-pituitary-adrenocortical axis responses to stress. *Journal of Neuroendocrinology* **11**, 361–369.

4. Pain and Stress: Convergence and Divergence of Neuronal Pathways

M. Palkovits

Introduction

Intense noxious stimuli elicit pain sensations and also act as stressors. Although pain and stress may have a common point of origin, they represent different functional mechanisms with different targets in the central nervous system. Signals evoked by somatosensory or viscerosensory stimuli are transferred to the central nervous system by primary sensory neurons and terminate in spinal cord or brainstem sensory neurons from where they diverge in different directions and activate neurons in different brain regions. Depending of the nature, the intensity, and the duration of noxious stimuli, they activate pain-related spinal reflexes and supraspinal discriminative-sensory and emotional-cognitive functions, as well as stress responses, such as activation of the hypothalamo-pituitary-adrenal (HPA) and sympathoadrenal axes.

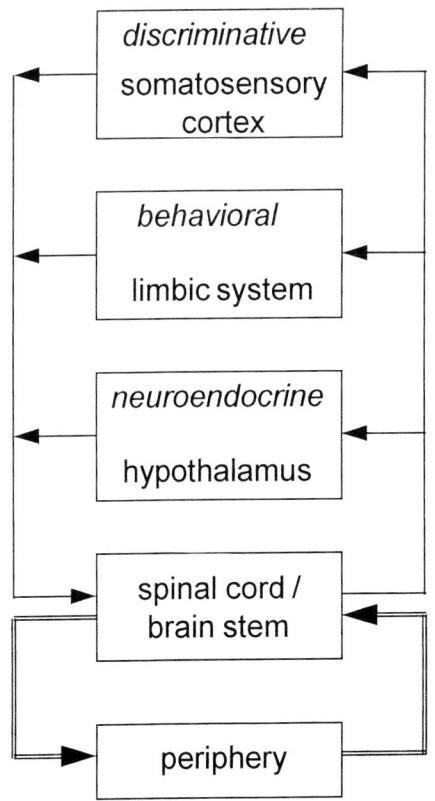

Fig. 1 Functional characteristics of pain-related pathways: spinal reflexes and supraspinal neuronal circuits.

From different brain regions, neuronal outputs responding to painful or stressful stimuli finally *converge* in common pathways to somatomotor, autonomic and endocrine hypothalamic neurons (Figure 1).

To investigate the fine topography and connections of pain- and stress-related neuronal pathways, two techniques have been applied. 1) Fos immunohistochemistry was used to localize the activated neuronal cells in response to a single, painful subcutaneous injection of 4% formalin. This stimulus induces immediate excitatory responses in relevant dorsal horn or sensory trigeminal neurons which are followed by a prolonged activated period (Dickenson and Sullivan, 1987; Abbadie *et al.*, 1997) that elevates plasma ACTH, corticosterone and catecholamine levels and induces marked corticotropin-releasing hormone (CRH) expression in the paraventricular nucleus (PVN) in the hypothalamus (Makara *et al.*, 1969; Pacak *et al.*, 1995; Palkovits *et al.*, 1995). 2) Neuronal projections to the spinal cord and higher brain areas have been visualized by a transneuronal tract-tracing technique. Pseudorabies virus (Bartha strain) is a useful marker for retrograde labeling of polysynaptic neuronal pathways. Injections of the virus subcutaneously in the paw produce immunohistochemically detectable infections in dorsal horn neurons of the spinal cord. After a short interval (while the virus is replicated in the cells), the virus infects secondary neurons, which are synaptically connected to infected dorsal horn neurons, and further transported retrogradely to higher brain areas.

Primary Sensory Neurons: Common Input of Pain and Stress Signals to the Nervous System

Somatosensory Pain Signals

Pain-related somatosensory signals are conveyed from the periphery to the dorsal horn of the spinal cord by $A\delta$ and C fibers and evoke a second wave of pain (Millan, 1999). They project to lamina I, II (outer), V and X neurons where painful stimuli, like a subcutaneous injection of formalin, produce *c-fos* expression, mainly ipsilateral to the site of the injection (Hunt *et al.*, 1987; Bullitt, 1990; Abbadie *et al.*, 1992; Senba *et al.*, 1993; Palkovits *et al.*, 1995; Palkovits *et al.*, 1999). Somatosensory signals from the head/neck region are carried by the trigeminal nerve, and their fibers terminate in a well-organized topographical pattern in the sensory trigeminal nucleus in the brainstem (Strassman and Vos, 1993; Strassman *et al.*, 1993). Secondary sensory fibers from this nucleus to the thalamus join the ascending sensory pathways of spinal cord origin.

Viscerosensory Pain Signals

Viscerosensory fibers are generally of a diffuse nature in contrast to the focal quality of somatosensory afferents (Millan, 1999). Most of the visceral pain-related signals are carried by vagal and glossopharyngeal nerves to the nucleus of the solitary tract (NTS) and by sympathetic chain and pelvic parasympathetic nerves to dorsal horn lamina I, V, VI and X neurons. After visceronociceptive stimuli, increased expression of *c-fos* was seen in these spinal and NTS neurons (Menétrey *et al.*, 1989; Lantéri-Minet *et al.*, 1993). From the spinal cord, viscerosensory inputs reach the NTS through the spinosolitary tract. Vagal and glossopharyngeal fibers terminate either on peptidergic NTS neurons which project directly or indirectly (mainly relayed by the parabrachial nuclei) to the hypothalamus and the limbic system (Riche *et al.*, 1990; Sawchenko *et al.*, 1990), or on catecholaminergic A2 neurons located in or around the NTS (Sumal *et al.*, 1983). From the NTS, topographically organized viecerosensory information is relayed to the ventral posterolateral thalamic nucleus and further up to the viscerosensory area of the sensory cortex.

Pain- and Stress-Conducting Neuronal Pathways

Painful stimuli from the spinal cord are transported by different pathways to higher brain areas. In the brainstem, 60 minutes after a unilateral formalin injection, strong bilateral *c-fos* labeling was seen in catecholaminergic neurons in the ventrolateral and dorsomedial medulla, in the locus coeruleus, the raphe and parabrachial nuclei and in the periaqueductal central gray (Bullitt, 1990; Senba *et al.*, 1993; Palkovits *et al.*, 1995; Baffi *et al.*, 1996; Palkovits *et al.*, 1997). Formalin activated *c-fos* expression in the midline and intralaminar thalamic nuclei and in the hypothalamus, particularly in the medial subdivision of the PVN (Figure 2).

Several pain-related neuronal pathways have been described. Unfortunately, their names show a great deal of variability and several of their synonyms are in use. Here, with certain simplifications, the basic topography and projections of seven neuronal pathways are briefly summarized. Five of them are considered as supraspinal pathways, while two of them represent spinal reflexes. Some of them are connected to discriminative-sensory or emotional-cognitive functions, while others represent pain-related stress pathways. The fine, discriminative localization of the pain is located in somatosensory cortical areas. Inputs to the cortex are carried by the spino- and trigeminothalamic tracts. Pain-induced behavioral responses are organized by neurons in limbic cortical areas. Neuronal inputs reach these areas (mainly in the piriform, entorhinal, and cingulate cortex and the hippocampus) through the spinoreticulothalamic tract,

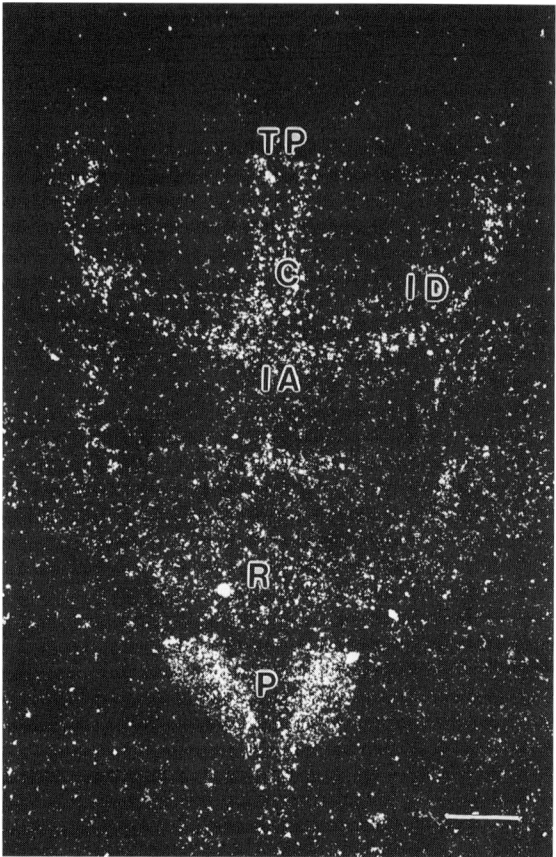

Fig. 2 Expression of c-fos mRNA in the diencephalon 60 minutes after a formalin injection (4%, 0.1 ml/100 g) into the hindlimb (a darkfield microphotograph). Intense labeling in the midline thalamic and hypothalamic paraventricular nuclei. Abbreviations: AD – anterodorsal thalamic nucleus, C – centromedian thalamic nucleus, IA – interanteromedian nucleus, ID – interanterodorsal nucleus, P – hypothalamic paraventricular nucleus, R – reuniens thalamic nucleus, TP – thalamic paraventricular nucleus. Bar scale: 400 μm.

which terminates in midline and intralaminar thalamic nuclei. Neuroendocrine and supraspinal autonomic responses to painful stress stimuli are organized by hypothalamic nuclei, which receive neuronal inputs through direct (spinohypothalamic) and indirect pathways (spinoreticular and/or spinosolitary relay to catecholaminergic and peptidergic neurons). The spinomesencephalic tract plays a role related to functioning of the spinal-midbrain pain modulatory loop.

Nociceptive stimuli elicit immediate spinal reflexes through signal transduction from dorsal horn sensory neurons to intermediolateral preganglionic neurons (sympathoadrenal reflex), or to motoneurons in the ventral horn (defense or withdrawal reflex).

1. Spinothalamic Tract (also called "ventral spinothalamic tract")

Spinothalamic neurons that respond to noxious stimuli exist in high concentration in the deep laminae of the dorsal horn (in lamina V and to a lesser extent in IV). These neurons encode the location, duration, intensity and quality of painful stimuli. Their ascending axons (in the ventrolateral funiculus of the spinal cord) finally terminate in the contralateral ventral posterolateral thalamic nucleus. From here, thalamocortical fibers reach the sensory cortex, i.e. the postcentral gyrus. This area represents the sensory-discriminative cortex.

2. Spinoreticulothalamic Tract (also called "dorsal spinothalamic", or "paleospinoreticular tract")

This multisynaptic pathway arises mainly from lamina I, and partly from lamina V and X neurons (Chaouch et al., 1983; Millan, 1999) which project to the intralaminar thalamic nuclear group including the centromedian, central lateral, paracentral, paraventricular and parafascicular nuclei (Peschanski and Besson, 1984) and are responsible for emotional-cognitive aspects of pain (Millan, 1999). Fibers of this tract have collaterals to brainstem reticular formation, including the gigantocellular reticular nucleus (Chaouch et al., 1983; Kevetter and Willis, 1983), the NTS (Otake et al., 1994), the parabrachial area (Hylden et al., 1989), and the periaqueductal central gray (Liu, 1986). The spinoreticulothalamic system is connected to limbic cortical structures: the intralaminar and midline thalamic nuclei project to the cingulate, piriform and entorhinal cortex. All of these thalamic and cortical areas show an immediate and intense c-fos expression in response to formalin injections (Bullitt, 1990; Senba et al., 1993; Palkovits et al., 1995). In contrast to the spinothalamic tract, which is almost completely contralateral, the spinoreticulothalamic system has both ipsi- and contralateral cortical projections. The limbic cortical areas likely play key roles in the affective rather than discriminative components of pain and are responsible for behavioral responses to noxious stimuli.

3. Spinoreticular and Spinosolitary Tracts

These monosynaptic pathways play a key role in the transfer of nociceptive stimuli from the spinal cord to neurons in the brainstem reticular formation, including catecholaminergic cells. Neurons in the dorsal horn project both to the reticular formation in the ventrolateral medulla (spinoreticular, or also called lateral spinoreticular tract) and to the dorsomedial medulla (spinosolitary tract) innervating mainly the peptidergic neurons in the NTS and A2 catecholaminergic neurons in this area (Menétrey et al., 1983; Shokunbi et al., 1985; Menétrey and Basbaum, 1987; Lima et al., 1991; Esteves et al., 1993; Tavares et al., 1993). Pain produced by formalin injected unilaterally into the hindlimb elicited c-fos activation on both sides of the medulla, in the A1 catecholaminergic cell group, the NTS and in the nucleus raphe pallidus (Senba et al., 1993; Palkovits et al., 1995; Baffi et al., 1996). Catecholaminergic neurons in the ventrolateral and the dorsomedial medulla are the major sources of adrenergic and noradrenergic nerve terminals in the hypothalamus (Palkovits et al., 1980; Sawchenko and Swanson, 1982; Cunningham and Sawchenko, 1988; Cunningham et al., 1990). In addition

to medullary catecholaminergic neurons, noradrenergic cells in the locus coeruleus also contribute to the innervation of the hypothalamus (Kobayashi et al., 1974; Sawchenko and Swanson, 1982; Cunningham and Sawchenko, 1988).

A high density of noradrenergic fibers and terminals is present in the dorsal and medial parvocellular subdivisions of the paraventricular nucleus (PVN), which contain the majority of CRH neurons of this nucleus (Swanson, 1987). Chemical lesioning of medullary catecholaminergic neurons or surgical transections of their ascending fibers reduce phenylethanolamine N-methyltransferase (PNMT) and tyrosine hydroxylase immunoreactivity and norepinephrine content in the PVN (Palkovits et al., 1980; Sawchenko, 1988; Mezey and Palkovits, 1991). Unilateral brainstem transection indicates that noradrenergic fibers travel ipsilaterally from the brainstem to the hypothalamus. After transection, the norepinephrine content in the extracellular fluid in the PVN is reduced markedly as measured by in vivo microdialysis (Pacak et al., 1993; Palkovits et al., 1999). CRH immunoreactivity (Mezey and Palkovits, 1991) and expression of CRH mRNA (Kiss et al., 1996; Pacak et al., 1996) are also reduced on the lesioned side of the PVN, but remained unchanged contralateral to the knife cut.

NTS neurons also project to the ventrolateral medulla innervating A1 noradrenergic and C1 adrenergic neurons (Ross et al., 1985; Hancock, 1988; Aicher et al., 1995; Chan et al., 1995) which, in turn, project to the forebrain, the locus coeruleus and the pontine parabrachial nuclei (Byrum et al., 1987; Woulfe et al., 1988; Kachidian et al., 1990). The parabrachial neurons provide an important relay for transfer of viscerosensory signals from the NTS and somatosensory signals from the spinal cord and the sensory trigeminal nucleus (Cechetto et al., 1985; Ma and Peschanski, 1988; Feil and Herbert, 1995) to the central nucleus of the amygdala. This pathway appears to contribute to autonomic regulation of the stress response to painful visceral stimuli.

4. Spinohypothalamic Tract

Various types of somato- and viscerosensory stimuli converge in the hypothalamus. Somatosensory signals from the spinal cord may be conducted directly to the hypothalamus by the spinohypothalamic tract (Burstein et al., 1987; Burstein et al., 1990; Cliffer et al., 1991), or by axon collaterals of fibers of the spinoreticulothalamic and the spinothalamic tract (Kevetter and Willis, 1983). The spinohypothalamic tract also conveys nociceptive information from the viscera (Giesler et al., 1994; Giesler, 1995; Burstein et al., 1996).

Axons arise in lamina I, IV and X neurons and from cells in the lateral spinal nucleus and parasympathetic cells in the sacral spinal cord (Burstein et al., 1990; Li et al., 1997). The spinohypothalamic tract ascends together with the spinothalamic tract through the contralateral spinal cord and brainstem to the lateral hypothalamus, where they terminate (Figure 3). Marked c-fos activation was seen in lateral hypothalamic neurons 60 minutes

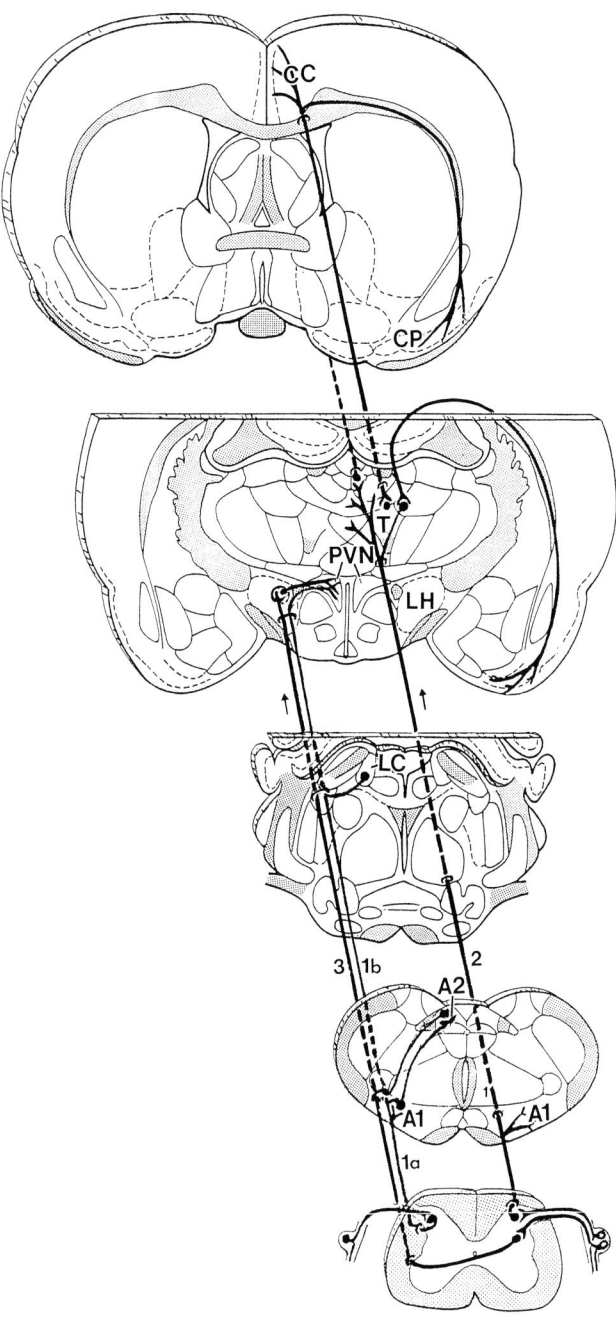

Fig. 3 Pain- and stress-related ascending pathways to the hypothalamic paraventricular nucleus (bilateral pathways are shown only on the left side) and to the midline thalamic nuclei (bilateral pathway is shown only on the right side). 1. Spinoreticular (a) – ascending noradrenergic pathway (b), 2. Spinoreticulothalamic pathway, 3. spinohypothalamic tract. Abbreviations: A1 – A1 noradrenergic cells in the caudal ventrolateral medulla, A2 – A2 noradrenergic cells in the dorsomedial medulla, CC – cingulate cortex, CP – piriform cortex, LC – locus coeruleus, LH – lateral hypothalamus, PVN – paraventicular nucleus, T – midline and intralaminar thalamic nuclei.

after formalin injections into the hind paw (Palkovits *et al.*, 1995). Cells from this area project to the ipsilateral PVN and terminate on CRH immunoreactive neurons

(M. Palkovits, K. Gallatz, A. Bratincsák, Z.J. Tóth, in preparation).

In the lateral hypothalamus, some axons of the spinohypothalamic tract have collaterals to the other side, providing bilateral innervation of the hypothalamus (Giesler *et al.*, 1994; Giesler, 1995; Burstein *et al.*, 1996; Li *et al.*, 1997). Furthermore, the spinohypothalamic tract also appears to provide fibers to ipsilateral limbic structures, as well as to the ventral pallidum (Giesler *et al.*, 1994). The ventral pallidum is thought to occupy a position as an interface between the limbic and motor systems. It has been hypothesized that the ventral pallidum mediates flow of motivationally relevant information to the motor system (Mogeson *et al.*, 1980).

Similar to spinal sensory neurons, nociceptive signals are also carried by the trigeminal nerve, and fibers also reach the lateral hypothalamus (Hamba *et al.*, 1990; Iwata *et al.*, 1992; Li *et al.*, 1997).

5. Spinomesencephalic Tract

The brainstem tonically inhibits nociceptive neurons in the spinal cord. This mechanism is based on ascending fibers carrying pain-related information from the spinal cord to the periaqueductal central gray (spinomesencephalic tract), and two-neuronal descending paths back to spinal nociceptive neurons (Figure 4). The cells of origin of the spinomesencephalic tract in rats are concentrated in lamina I, V and X (Menétrey *et al.*, 1982; Lima and Coimbra, 1989). Ascending painful stimuli activate endogeneous opioid peptide synthesizing neurons in the periaqueductal central gray. These neurons process that information and activate inhibitory neurons in the medial ventromedial medullary neurons and inhibit spinal nociceptive transmission *via* the spinal local enkephalinergic neurons (Fields and Basbaum, 1978; Fields *et al.*, 1991). The rostral ventromedial medulla includes the nucleus raphe magnus and adjacent reticular formation (gigantocellular, magnocellular and paragigantocellular reticular nuclei). Some of the neurons of these nuclei project directly to the dorsal horn. One day after viral infection of the dorsal horn of the spinal cord, virus-labeled cells are present throughout the rostral ventral medulla due to retrograde axonal transport of the virus (Figure 5).

In addition to medial ventromedial neurons, A5 noradrenergic and C1 adrenergic (Figure 6) cells in the rostral ventrolateral medulla also project to the spinal cord (Figure 4) and terminate in the intermediolateral cell column (Ross *et al.*, 1981; Tucker *et al.*, 1987). These A5 noradrenergic cells occupy a special position in the organization of responses to stress and painful stimuli: they receive neuronal input from the caudal ventrolateral medulla, from the viscerosensory NTS, and from the PVN (Hosoya *et al.*, 1990; Hosoya *et al.*, 1991). In turn, they innervate sympathetic preganglionic neurons in the intermediolateral cell column, thereby influencing sympathoadrenal outflow (Loewy *et al.*, 1979; Byrum and Guyenet, 1987), and their terminals in the dorsal horn facilitate the activity of nociceptive inhibitory interneurons.

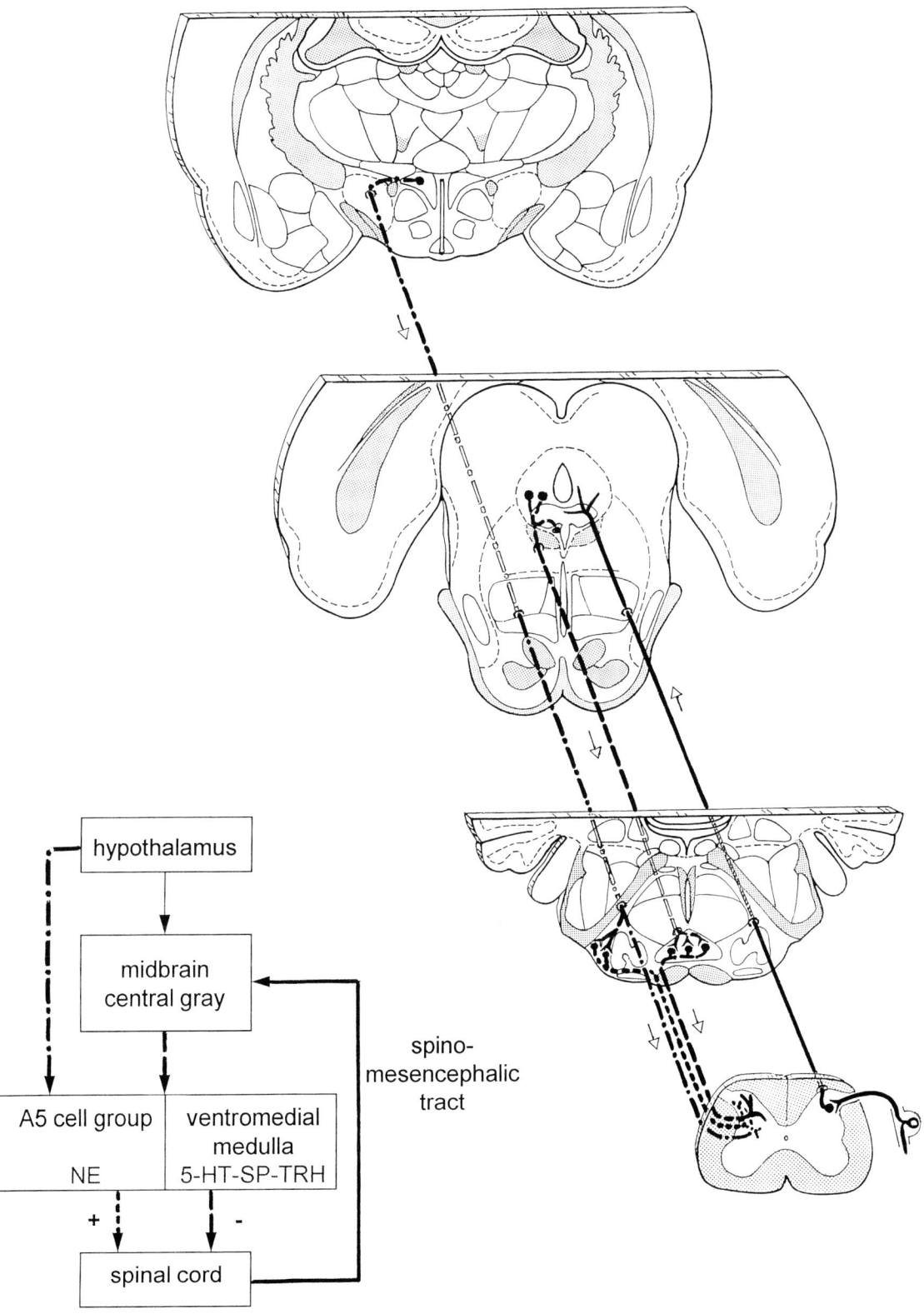

Fig. 4 Stress response and pain-inhibition (bilateral pathways are shown only on one side). 1. Spinomesencephalic tract to periaqueductal central gray neurons. 2. Central gray projections to rostral ventromedial medullary neurons. 3. Rostral ventomedial medullary projections to the dorsal horn. 4. Hypothalamic (paraventricular) projections to ventrolateral catecholaminergic (A5 and C1 cell groups) neurons and to the spinal cord. 5. A5 noradrenergic projections to the spinal cord. Abbreviations: A5–A5 catecholaminergic cell group, LH – lateral hypothalamus, PVN – paraventricular nucleus, SG – periaqueductal central gray, VM – medial ventromedial medulla.

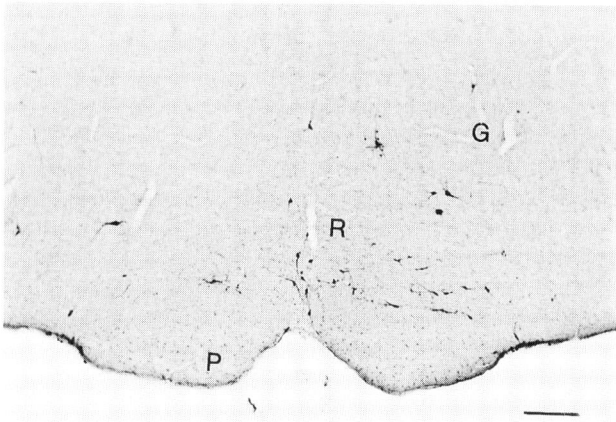

Fig. 5 Viral (pseudorabies, Bartha strain) labeling of rostral ventromedial medullary (VMM) neurons. Virus injection into the hindlimb, and retrograde infection through spinal neurons innervated by VMM neurons. Abbreviations: G – gigantocellular reticular nucleus. P – pyramidal tract, R – nucleus raphe magnus. Bar scale: 400 μm.

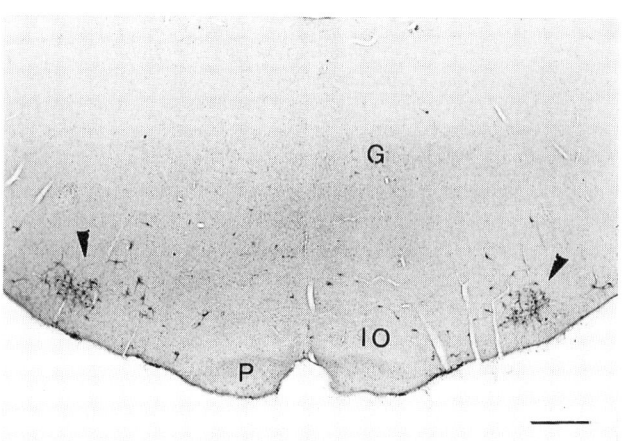

Fig. 6 Retrograde viral (pseudorabies, Bartha strain) labeling of spinal cord projecting ventrolateral (C1 adrenergic cell group) medullary neurons (arrowhead). Virus was injected into the hindlimb. Abbreviations: G – gigantocellular reticular nucleus, IO – inferior olive, P – pyramidal tract. Bar scale: 400 μm.

6. Spinal Sympathoadrenal Reflex

Pain may elicit immediate sympathetic activation. Sensory inputs from primary afferents converge upon sympathetic preganglionic neurons to regulate their activity. This connection appears to be monosynaptic, but interneurons also contribute to the spinal sympathoadrenal reflex.

7. Spinal Defense (withdrawal) Reflex

The withdrawal reflex (also called flexor reflex) is considered a protective function elicited by acute exposure to noxious stimuli. The afferent limb of this reflex arc consists of primary sensory afferents that carry noxious stimuli and terminate in superficial laminae neurons of the dorsal horn. Axon collaterals

of these neurons directly contact ipsilateral motoneurons providing a short, rapid pathway for eliciting limb withdrawal in response to noxious stimuli (Jasmin *et al.*, 1997). In addition to this monosynaptic reflex path, interneurons located in lamina V and VII link nociceptive afferents to motoneurons (Harrison *et al.*, 1984; Cabot *et al.*, 1994; Joshi *et al.*, 1995; Jasmin *et al.*, 1997). Interneurons can be subdivided into excitatory and inhibitory types. The withdrawal reflex appears to be a coordinated action of activated flexor and inhibited extensor responses (on the side of the painful stimulus and in an opposite arrangement on the contralateral side) conducted by inhibitory and excitatory spinal interneurons.

Common Pathways in Pain and Stress Responses: Humoral and Neuronal Outputs

Pain-evoked stress responses are organized by *converging neuronal inputs* to the final common output systems: the neurohumoral hypothalamo-pituitary and neuronal autonomic circuits (Figure 1). The presence of PVN neurons is considerable in both circuits. In addition to PVN projections to autonomic centers, some cortical (mainly from the insular cortex), limbic (mainly amygdaloid) and other hypothalamic (from the arcuate, perifornical, preoptic and lateral hypothalamic) neuronal projections terminate on parasympathetic (vagal) and spinal sympathetic preganglionic neurons and induce autonomic responses to painful stimuli (ref. see Swanson, 1987; Palkovits, 1999). Some descending supraspinal pathways may not project directly to sympathetic preganglionic neurons but may involve interneurons (Strack *et al.*, 1989). The intermediolateral sympathetic and vagal parasympathetic preganglionic neurons constitute the final sites of integration of central pain-evoked neuronal stress responses.

References

Abbadie, C., Lombard, M.-C., Morain, F., and Besson, J.M. (1992). Fos-like immunoreactivity in the rat superficial dorsal horn induced by formalin injection in the forepaw: effects of dorsal rhizotomies. *Brain Research* **578**, 17–25.

Abbadie, C., Taylor, B.K., Peterson, M.A., and Basbaum, A.I. (1997). Differential contribution of the two phases of the formalin test to the pattern of c-fos expression in the rat spinal cord: studies with remifentanil and lidocaine. *Pain* **69**, 101–110.

Aicher, S.A., Kurucz, O.S., Reis, D.J., and Milner, T.A. (1995). Nucleus tractus solitarius efferent terminals synapse on neurons in the caudal ventrolateral medulla that project to the rostral ventrolateral medulla. *Brain Research* **693**, 51–63.

Baffi, J., Pacak, K., Szabó, S., and Palkovits, M. (1996). Different stressors have different effects on c-fos activity of catecholaminergic neurons. In R. McCarty, G. Aguilera, E. Sabban, & R. Kvetňanský (Eds.), *Stress: Molecular Genetic and Neurological Advances* (pp. 143–155). New York: Gordon and Breach.

Bullitt, E. (1990). Expression of c-fos-like protein as a marker for neuronal activity following noxious stimulation in the rat. *Journal of Comparative Neurology* **296**, 517–530.

Burstein, R., Cliffer, K.D., and Giesler, G.J.J. (1987). Direct somatosensory projections from the spinal cord to the hypothalamus and telencephalon. *Journal of Neuroscience* **7**, 4159–4164.

Burstein, R., Cliffer, K.D., and Giesler, G.J.J. (1990). Cells of origin of the spinohypothalamic tract in the rat. *Journal of Comparative Neurology* **291**, 329–344.

Burstein, R., Falkowsky, O., Borsook, D., and Strassman, A. (1996). Distinct lateral and medial projections of the spinohypothalamic tract of the rat. *Journal of Comparative Neurology* **373**, 549–574.

Byrum, C.E., and Guyenet, P.G. (1987). Afferent and efferent connections of the A5 noradrenergic cell group in the rat. *Journal of Comparative Neurology* **261**, 529–542.

Cabot, J.B., Alessi, V., Carroll, J., and Ligorio, M. (1994). Spinal cord lamina V and lamina VII interneuronal projections to sympathetic preganglionic neurons. *Journal of Comparative Neurology* **347**, 515–530.

Cechetto, D.F., Standaert, D.G., and Saper, C.B. (1985). Spinal and trigeminal dorsal horn projections to the parabrachial nucleus in the rat. *Journal of Comparative Neurology* **240**, 153–160.

Chan, R.K.W., Peto, C.A., and Sawchenko, P.E. (1995). A1 catecholamine cell group: Fine structure and synaptic input from the nucleus of the solitary tract. *Journal of Comparative Neurology* **351**, 62–80.

Chaouch, A., Ménétrey, D., Binder, R., and Besson, J.M. (1983). Neurons at the origin of the medial spinoreticular tract in the rat: An anatomical study using horseradish peroxidase retrograde transport. *Journal of Comparative Neurology* **214**, 309–320.

Cliffer, K.D., Burstein, R., and Giesler, G.J.J. (1991). Distributions of spinothalamic, spinohypothalamic, and spinotelencephalic fibers revealed by anterograde transport of PHA-L in rats. *Journal of Neuroscience* **11**, 852–868.

Cunningham, E.T., and Sawchenko, P.E. (1988). Anatomical specificity of noradrenergic inputs to the paraventricular and supraoptic nuclei of the rat hypothalamus. *Journal of Comparative Neurology* **274**, 60–76.

Cunningham, E.T., Bohn, M.C., and Sawchenko, P.E. (1990). Organization of adrenergic inputs to the paraventricular and supraoptic nuclei of the hypothalamus in the rat. *Journal of Comparative Neurology* **292**, 651–667.

Dickenson, A.H., and Sullivan, A.F. (1987). Subcutaneous formalin-induced activity of dorsal horn neurones in the rat: Differential response to an intrathecal opiate administered pre or post formalin. *Pain* **30**, 349–360.

Esteves, F., Lima, D., and Coimbra, A. (1993). Structural types of spinal cord marginal (lamina-1) neurons projecting to the nucleus of the tractus solitarius in the rat. *Somatosensory and Motor Research* **10**, 203–216.

Feil, K., and Herbert, H. (1995). Topographical organization of spinal and trigeminal somatosensory pathways to the rat parabrachial and Kolliker-Fuse nuclei. *Journal of Comparative Neurology* **353**, 506–528.

Fields, H.I., and Basbaum, A.I. (1978). Brainstem control of spinal pain-transmission neurons. *Annual Review of Physiology* **40**, 217–248.

Fields, H.I., Heinricher, M.M., and Mason, P. (1991). Neurotransmitters in nociceptive modulatory circuits. *Annual Review of Neuroscience* **14**, 219–245.

Giesler, G.J. (1995). Evidence of direct nociceptive projections from the spinal cord to the hypothalamus and telencephalon. *Neuroscience*, 253–261.

Giesler, G.J., Katter, J.T., and Dado, R.J. (1994). Direct spinal pathways to the limbic system for nociceptive information. *Trends in Neuroscience* **17**, 244–250.

Hamba, M., Misumitsu, H., and Muro, M. (1990). Nociceptive projection from tooth-pulp to the lateral hypothalamus in rats. *Brain Research Bulletin* **25**, 355–364.

Hancock, M.B. (1988). Evidence for direct projections from the nucleus of the solitary tract onto medullary adrenaline cells. *Journal of Comparative Neurology* **276**, 460–467.

Harrison, P.J., Hultborn, H., Jankowska, E., Katz, R., Storai, B., and Zytnicki, D. (1984). Labelling of interneurones by retrograde transsynaptic transport of horseradish peroxidase from motoneurones in rats and cats. *Neuroscience Letters* **45**, 15–19.

Hosoya, Y., Sugiura, Y., Ito, R., and Kohno, K. (1990). Descending projections from the hypothalamic paraventricular nucleus to the A5 area, including the superior salivatory nucleus, in the rat. *Experimental Brain Research* **82**, 513–518.

Hosoya, Y., Sugiura, Y., Okado, N., Loewy, A.D., and Kohno, K. (1991). Descending input from the hypothalamic paraventricular nucleus to sympathetic preganglionic neurons in the rat. *Experimental Brain Research* **85**, 10–20.

Hunt, S.P., Pini, A., and Evan, G. (1987). Induction of c-fos-like protein in spinal cord neurons following sensory stimulation. *Nature* **328**, 632–634.

Hylden, J.L.K., Anton, F., and Nahin, R.L. (1989). Spinal lamina I projection neurons in the rat: Collateral innervation of parabrachial area and thalamus. *Neuroscience* **28**, 27–37.

Iwata, K., Kenshala, D.R.J., Dubner, R., and Nahin, R.L. (1992). Diencephalic projections from superficial and deep laminae of the medullary dorsal horn in the rat. *Journal of Comparative Neurology* **321**, 404–420.

Jasmin, L., Carstens, E., and Basbaum, A.I. (1997). Interneurons presynaptic to rat tail-flick motoneurons as mapped by transneuronal transport of pseudorabies virus: few have long ascending collaterals. *Neuroscience* **76**, 859–876.

Joshi, S., Levatte, M.A., Dekaban, G.A., and Weaver, L.C. (1995). Identification of spinal interneurons antecedent to adrenal sympathetic preganglionic neurons using trans-synaptic transport of herpes simplex virus type 1. *Neuroscience* **65**, 893–903.

Kachidian, P., Astier, B., Renaud, B., and Bosler, O. (1990). Adrenergic innervation of noradrenergic locus coeruleus neurons. A dual labeling immunocytochemical study in the rat. *Neuroscience Letters* **109**, 23–29.

Kevetter, G.A., and Willis, W.D. (1983). Collaterals of the spinothalamic cells in the rat. *Journal of Comparative Neurology* **215**, 453–464.

Kiss, A., Palkovits, M., and Aguilera, G. (1996). Neural regulation of corticotropin releasing hormone (CRH) and CRH receptor mRNA in the hypothalamic paraventricular nucleus in the rat. *Journal of Neuroendocrinology* **8**, 103–112.

Kobayashi, R.M., Palkovits, M., Kopin, I.J., and Jacobowitz, D.M. (1974). Biochemical mapping of noradrenergic nerves arising from the rat locus coeruleus. *Brain Research* **77**, 269–279.

Lantéri-Minet, M., Isnardon, P., De Pommery, J., and Ménétrey, D. (1993). Spinal and hindbrain structures involved in visceroception and visceronociception as revealed by the expression of Fos, Jun and Krox-24 proteins. *Neuroscience* **55**, 737–753.

Li, J.L., Kaneko, T., Shigemoto, R., and Mizuno, N. (1997). Distribution of trigeminohypothalamic and spinohypothalamic tract neurons displaying substance P receptor-like immunoreactivity in the rat. *Journal of Comparative Neurology* **378**, 508–521.

Lima, D., and Coimbra, A. (1989). Morphological types of spinomesencephalic neurons in the marginal zone (lamina I) of the rat spinal cord as shown after retrograde labelling with cholera toxin subunit B. *Journal of Comparative Neurology* **279**, 327–339.

Lima, D., Mendes-Ribeiro, J.A., and Coimbra, A. (1991). The spino-latero-reticular system of the rat: projections from the superficial dorsal horn and structural characterization of marginal neurons involved. *Neuroscience* **45**, 137–152.

Liu, R.P.C. (1986). Spinal neuronal collaterals to the intralaminar thalamic nuclei and periaqueductal gray. *Brain Research* **365**, 145–150.

Loewy, A.D., McKellar, S., and Saper, C.B. (1979). Direct projections from the A5 catecholamine cell group to the intermediolateral cell column. *Brain Research* **174**, 309–314.

Ma, W., and Peschanski, M. (1988). Spinal and trigeminal projections to the parabrachial nucleus in the rat: Electron-microscopic evidence of a spino-ponto-amygdalian somatosensory pathway. *Somatosensory Research* **5**, 247–257.

Makara, G.B., Stark, E., and Mihály, K. (1969). Corticotrophin release induced by injection of formalin in rats with hemisection of the spinal cord. *Acta Physiologica Academy of Sciences of Hungary* **35**, 331–333.

Ménétrey, D., and Basbaum, A.I. (1987). Spinal and trigeminal projections to the nucleus of the solitary tract. A possible substrate for somatovisceral and viscerovisceral reflex activation. *Journal of Comparative Neurology* **255**, 439–450.

Ménétrey, D., Chaouch, A., Binder, D., and Besson, J.M. (1982). The origin of spinomesencephalic tract in the rat: an anatomical study using the retrograde transport of horseradish peroxidase. *Journal of Comparative Neurology* **206**, 193–207.

Ménétrey, D., Gannon, A., Levine, J.D., and Basbaum, A.I. (1989). Expression of c-fos protein in interneurons and projection neurons of the rat spinal cord in response to noxious somatic, articular and visceral stimulation. *Journal of Comparative Neurology* **285**, 177–195.

Ménétrey, D., Roudier, F., and Besson, J.M. (1983). Spinal neurons reaching the lateral reticular nucleus as studied in the rat by retrograde transport of horseradish peroxidase. *Journal of Comparative Neurology* **220**, 439–452.

Mezey, E., and Palkovits, M. (1991). CRF-containing neurons in the hypothalamic paraventricular nucleus: Regulation, especially by catecholamines. *Frontiers in Neuroendocrinology* **12**, 23–37.

Millan, M.J. (1999). The induction of pain: An integrative review. *Progress in Neurobiology* **57**, 1–164.

Mogeson, G.J., Jones, D.L., and Sim, C.Y. (1980). From motivation to action: functional interface between the limbic system and the motor system. *Progress in Neurobiology* **14**, 69–97.

Otake, K., Reis, D.J., and Ruggiero, D.A. (1994). Afferents to the midline thalamus issue collaterals to the nucleus tractus solitarii: An anatomical basis for thalamic and visceral reflex integration. *Journal of Neuroscience* **14**, 5694–5707.

Pacak, K., Palkovits, M., Kopin, I.J., and Goldstein, D.S. (1995). Stress-induced norepinephrine release in the hypothalamic paraventricular nucleus and pituitary-adrenocortical and sympathoadrenal activity: *In vivo* microdialysis studies. *Frontiers in Neuroendocrinology* **16**, 89–150.

Pacak, K., Palkovits, M., Kvetňanský, R., Kopin, I.J., and Goldstein, D.S. (1993). Stress-induced norepinephrine release in the paraventricular nucleus of rats with brainstem hemisections: A microdialysis study. *Neuroendocrinology* **58**, 196–201.

Pacak, K., Palkovits, M., Makino, S., Kopin, I.J., and Goldstein, D.S. (1996). Brainstem hemisection decreases corticotropin-releasing hormone mRNA in the paraventricular nucleus but not in the central amygdaloid nucleus. *Journal of Neuroendocrinology* **8**, 543–551.

Palkovits, M. (1999). Interconnections between the neuroendocrine hypothalamus and the central autonomic system. *Frontiers in Neuroendocrinology* **20**, 270–298.

Palkovits, M., Baffi, J.S., and Dvori, S. (1995). Neuronal organization of stress response. Pain-induced c-fos expression in brain stem catecholaminergic cell groups. *Annals of the New York Academy of Science* **771**, 313–326.

Palkovits, M., Baffi, J.S., and Pacak, K. (1997). Stress-induced fos-like Immunoreactivity in the pons and the medulla oblongata of rats. *Stress* **1**, 155–168.

Palkovits, M., Baffi, J.S., and Pacak, K. (1999). The role of ascending neuronal pathways in stress-induced release of noradrenaline in the hypothalamic paraventricular nucleus of rats. *Journal of Neuroendocrinology* **11**, 529–539.

Palkovits, M., Záborszky, L., Feminger, A., Mezey, É., Fekete, M.I.K., Herman, J.P., Kanyicska, B., and Szabó, D. (1980). Noradrenergic innervation of the rat hypothalamus: Experimental biochemical and electron microscopic studies. *Brain Research* **191**, 161–171.

Peschanski, M., and Besson, J.-M. (1984). A spino-reticulo-thalamic pathway in the rat: an anatomical study with reference to pain transmission. *Neuroscience* **12**, 165–178.

Riche, D., De Pommery, J.D., and Menétrey, D. (1990). Neuropeptides and catecholamines in efferent projections of the nuclei of the solitary tract in the rat. *Neuroscience* **293**, 399–424.

Ross, C.A., Armstrong, D.M., Ruggiero, D.A., Pickel, V.M., Joh, T.H., and Reis, D.J. (1981). Adrenaline neurons in the rostral ventrolateral medulla innervate thoracic spinal cord: A combined immunocytochemical and retrograde transport demonstration. *Neuroscience Letters* **25**, 257–262.

Ross, C.A., Ruggiero, D.A., and Reis, D.J. (1985). Projections from the nucleus tractus solitarii to the rostral ventrolateral medulla. *Journal of Comparative Neurology* **242**, 511–534.

Sawchenko, P.E. (1988). Effects of catecholamine-depleting medullary knife cuts on corticotropin-releasing factor and vasopressin immunoreactivity in the hypothalamus of normal and steroid manipulated rats. *Neuroendocrinology* **48**, 459–470.

Sawchenko, P.E., Arias, C., and Bittencourt, J.C. (1990). Inhibin β, somatostatin, and enkephalin immunoreactivities coexist in caudal medullary neurons that project to the paraventricular nucleus of the hypothalamus. *Journal of Comparative Neurology* **291**, 269–280.

Sawchenko, P.E., and Swanson, L.W. (1982). The organization of noradrenergic pathways from the brainstem to the paraventricular and supraoptic nuclei in the rat. *Brain Research Review* **4**, 275–325.

Senba, E., Matsunaga, K., Tohyama, M., and Noguchi, K. (1993). Stress-induced c-fos expression in the rat brain: Activation mechanism of sympathetic pathway. *Brain Research Bulletin* **31**, 329–341.

Shokunbi, M.T., Hrycyshin, A.W., and Flumerfelt, B.A. (1985). Spinal projections to the lateral reticular nucleus in the rat: A retrograde labeling study using horseradish peroxidase. *Journal of Comparative Neurology* **239**, 216–226.

Strack, A.M., Sawyer, W.B., Hughes, J.H., Platt, K.B., and Loewy, A.D. (1989). A general pattern of CNS innervation of the sympathetic outflow demonstrated by transneuronal pseudorabies viral infections. *Brain Research* **491**, 156–162.

Strassman, A.M., and Vos, B.P. (1993). Somatotopic and laminar organization of fos-like immunoreactivity in the medullary and upper cervical dorsal horn induced by noxious facial stimulation in the rat. *Journal of Comparative Neurology* **331**, 495–516.

Strassman, A.M., Vos, B.P., Mineta, Y., Naderi, S., Borsook, D., and Burstein, R. (1993). Fos-like immunoreactivity in the superficial medullary dorsal horn induced by noxious and innocuous thermal stimulation of facial skin in the rat. *Journal of Neurophysiology* **70**, 1811–1821.

Sumal, K.K., Blessing, W.W., Joh, T.H., Reis, D.J., and Pickel, V.M. (1983). Synaptic interactions of vagal afferents and catecholaminergic neurons in the rat nucleus solitarius. *Brain Research* **277**, 31–40.

Swanson, L.W. (1987). The hypothalamus. In A. Björklund, T. Hökfelt, & L.W. Swanson (Eds.), *Handbook of Chemical Neuroanatomy* (Vol. 5: *Integrated Systems of the CNS, Part I, Hypothalamus, Hippocampus, Amygdala, Retina*, Chapt. I, pp. 1–124). Amsterdam-New York-Oxford: Elsevier.

Tavares, I., Lima, D., and Coimbra, A. (1993). Neurons in the superficial dorsal horn of the rat spinal cord projecting to the medullary ventrolateral reticular formation express c-fos after noxious stimulation of the skin. *Brain Research* **623**, 278–286.

Tucker, D.C., Saper, C.B., Ruggiero, D.A., and Reis, D.J. (1987). Organization of central adrenergic pathways: I. Relationships of ventrolateral medullary projections to the hypothalamus and spinal cord. *Journal of Comparative Neurology* **259**, 591–603.

Woulfe, J.M., Hrycushyn, A.W., and Flumerfelt, B.A. (1988). Collateral axonal projections from the A1 noradrenergic cell group to the paraventricular nucleus and bed nucleus of the stria terminalis in rat. *Experimental Neurology* **102**, 121–124.

5. Chronic Hypercortisolemia Inhibits Dopaminergic Activity in the Nucleus Accumbens

K. Pacak, O. Tjurmina, M. Palkovits, D.S. Goldstein, C.A. Koch, T. Hoff, P. Goldsmith and G.P. Chrousos

Introduction

Evidence from many preclinical studies indicates disturbances of mesolimbic dopamine (DA) function in depression and DA depletion in the nucleus accumbens and caudate nucleus occurs in animals with learned helplessness (Anisman et al., 1979). Treatment with DA agonists prevents development of this animal model of depression (Theohar et al., 1981). Acute immobilization-induced DA release in the mesolimbic system promotes behavioral activation and results in defensive responses toward the stressful stimulus (Cabib and Puglisi-Allegra, 1996; Imperato et al., 1989). In contrast, prolonged exposure to stress that reflects chronic hyperactivation of the hypothalamo-pituitary-adrenocortical (HPA) axis leads to inhibition of the mesolimbic DAergic system, associated with "coping failure" and cessation of defensive attempts, with subsequent development of depressive signs (for review, see Cabib and Puglisi-Allegra, 1996; Imperato et al., 1993). The nucleus accumbens, as part of the mesocorticolimbic dopaminergic system, has been suggested to play an important role in the pathogenesis of depression (Horger and Roth, 1996).

A substantial proportion of patients with major depression have abnormalities of the HPA axis, such as an increased frequency of ACTH-secretory episodes, elevated urinary free cortisol excretion and elevated cortisol and CRH levels in cerebrospinal fluid (Deuschle et al., 1997; Holsboer and Barden, 1996). Conversely, over 60% of patients with Cushing's syndrome suffer from major depression, which frequently abates after correction of the hypercortisolism (Dorn et al., 1995; Dorn et al., 1997; Starkman et al., 1981).

Recently, it has been demonstrated that chronic (7 days) elevation of plasma glucocorticoids decreases levels of DA and dihydroxyphenylacetic acid (DOPAC) in the shell of the nucleus accumbens and the medial prefrontal cortex (Lindley et al., 1999). In contrast, Rouge-Pont demonstrated that a ten minute period of tail-pinch-induced release of DA in the nucleus accumbens is facilitated by corticosterone (Rouge-Pont et al., 1998) and adrenalectomy was found to profoundly decrease basal DA release in the nucleus accumbens (Piazza et al., 1996a).

The present study examined the effect of chronic hypercortisolemia on DAergic activity in the nucleus accumbens using in vivo microdialysis. DA, DOPAC, and homovanillic acid (HVA) were simultaneously measured in extracellular fluid of the nucleus accumbens. By measuring dialysate concentrations of the catecholamine precursor, dihydroxyphenylalanine (DOPA), before and after local perfusion with NSD-1015 (m-hydroxybenzylhydrazine), an irreversible inhibitor of L-aromatic-amino acid decarboxylase, we also assessed whether chronic hypercortisolemia could affect tyrosine hydroxylation and therefore dopamine biosynthesis.

Materials and Methods

Animal Preparation

All procedures described in the present study were approved by the Animal Care and Use Committee of the National Institute of Neurological Disorders and Stroke.

Male Sprague-Dawley rats were obtained from Taconic Farms (Germantown, NY) and were group-housed in an animal room with ad libitum laboratory chow and tap water for at least 3 days prior to the acute experiments. The animal room was maintained at $22 \pm 1°C$ with a 12-h light-dark period (lights on from 0600–1800 h).

Minipump Insertion and Cortisol Treatment

One week before the acute experiment, rats were assigned randomly to one of two groups, for treatment with cortisol (CORT; n = 5, 253 ± 15 g final body weight BW) or saline vehicle (n = 6, 333 ± 16 g final BW). CORT or VEH was infused for 7 days via an osmotic minipump (model 2001, Alzet, Palo Alto, CA) inserted under the skin in the interscapular region during light pentobarbital anesthesia (30 mg/kg, ip). The CORT dose was 25 mg/kg/day (infusion rate 25 μl/day; 370 mg/ml hydrocortisone hemisuccinate sodium) as described previously by our group (Pacak et al., 1995b). This dose of cortisol is known to increase total glucocorticoid activity by about three-to four-fold (Szemeredi et al., 1988).

NSD Treatment

Each rat received 100 μM m-hydroxybenzylhydrazine dihydrochloride (NSD 1015) as previously described by our group (Pacak *et al.*, 1993). The decarboxylase inhibitor was administered via the implanted microdialysis probe after the third baseline microdialysate collection. The solution of NSD-1015 in artificial cerebrospinal fluid was prepared 1–2 hours before each experiment.

Microdialysis Probe Placement

Twenty-four hours prior to each acute experiment rats were anesthetized with pentobarbital (50 mg/kg, i.p.). Each rat was then placed in a stereotaxic apparatus (David Kopf Instruments, Tujunga, CA) with the incisor bar 3.1 mm below the interaural line. The skull was exposed by a midline incision, and a small hole was drilled over the right nucleus accumbens (stereotaxic coordinates with respect to bregma: posterior 1.2 mm, lateral 1.4 mm and vertical –7.9 mm according to the stereotaxic atlas of Paxinos and Watson (1986). The implanted microdialysis probe (1.0 mm in length, 0.24 mm diameter, molecular weight cut-off = 20,000 daltons, CMA/Microdialysis, Acton, MA) was anchored to the skull with acrylic dental cement. Following surgery, each animal was housed individually overnight in a cylindrical Plexiglas cage that contained laboratory chow, tap water, and bedding material.

Microdialysis Experiment

Next morning the 1 mm microdialysis probe was connected to a microinfusion pump (CMA 100, CMA/Microdialysis, Acton, MA) and was perfused continuously at a rate of 1.0 μl/min with artificial cerebrospinal fluid as described previously (Pacak *et al.*, 1995a, Tjurmina, 1999). After a 2-h equilibration period, three baseline 30-minute microdialysate samples were collected from each rat, followed by NSD-1015 administration via the implanted microdialysis probe and collection of eight consecutive 30-minute microdialysate samples. This time period was sufficient to reach steady-state levels of catechols after NSD-1015 perfusion.

Microdialysis Probe Location

The topographical location of the microdialysis probe in the nucleus accumbens was determined histologically; serial coronal sections of 12 μm thick were cut through the forebrain, rostral to the anterior commissure. Probes in the core region of the nucleus were accepted as correct location, others connected with the lateral ventricle or partly inside the caudate nucleus were excluded from the present study. Due to the size of the probe relative to the nucleus accumbens, the correctly placed probes collected extracellular fluid from both the core and the shell regions.

Assays

Microdialysate levels of DA, DOPAC, DOPA, and HVA were quantified using reverse phase liquid chromatography with electrochemical detection (Tjurmina, 1999).

Statistical Analysis

Results are presented as mean values ± SEM. Changes in microdialysate DA, DOPAC, DOPA, and HVA levels between CORT- and VEH-treated rats were analyzed by two-way analyses of variance (ANOVA) for repeated measures (the between-groups factor being treatment and the within-group factor being time). Where appropriate, post hoc Student-Newman-Keuls tests were utilized to compare CORT- and VEH-treated rats at specific time points. A P value less than 0.05 defined statistical significance.

Results

CORT-treated rats had significantly lower baseline mean microdialysate levels of DOPA (100.5 ± 25.0 versus 653.3 ± 226.0 pg/ml, P < 0.05) than did VEH-treated rats (Figure 1). Baseline mean microdialysate levels of DOPAC (69165 ± 13859 versus 116475 ± 18988 pg/ml), DA (803.3 ± 296 versus 1118 ± 497.0 pg/ml) and HVA (69136 ± 14829 versus 88939 ± 16541 pg/ml) tended to be lower in CORT-treated rats than did in VEH-treated rats.

Administration of NSD 1015 via the microdialysis probe increased microdialysate DOPA concentrations in all animals. Steady-state DOPA levels (defined by concentrations of DOPA that did not differ significantly in two consecutive samples) were attained within 90 min in CORT-treated and 120 min in VEH-treated rats. CORT-treated rats had significantly smaller increases in microdialysate DOPA levels than did VEH-treated rats (P < 0.05; Figure 1). Marked decreases

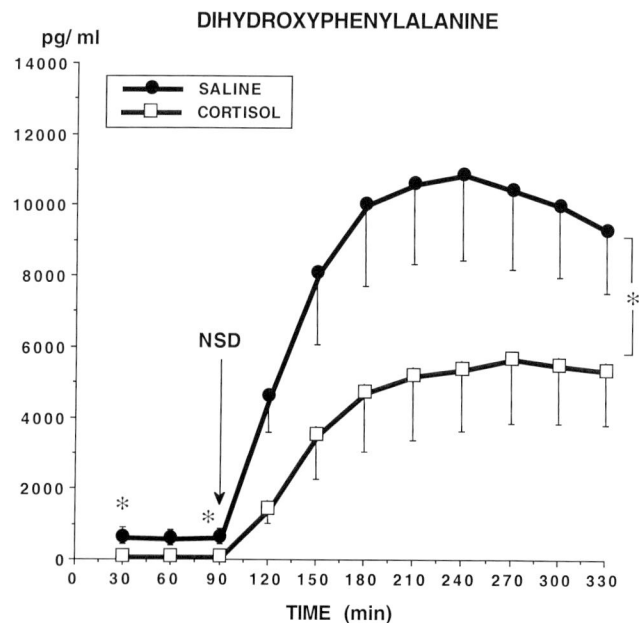

Fig. 1 Effects of local administration of NSD 1015 on microdialysate concentrations of dihydroxyphenylalanine in the nucleus accumbens in rats. Results are presented as means ± SEM. *, P < 0.05. (CORT-treated vs VEH-treated rats).

in dialysate DOPAC and HVA levels attended the increases of microdialysate DOPA concentrations after NSD 1015 administration in all animals (DOPAC data shown in Figure 2). DA levels increased significantly, with highest levels 210–240 min after administration of NSD 1015 (CORT-treated rats: 5206 pg/ml, VEH-treated rats: 4132 pg/ml, P < 0.01).

Discussion

Markedly decreased extracellular levels of DA, DOPA, DOPAC, and HVA in the nucleus accumbens and markedly decreased extracellular levels of DOPA after NSD 1015 in CORT-treated compared to VEH-treated rats suggest that chronic hypercortisolemia inhibits dopaminergic activity in the nucleus accumbens, and thus influences a major component of the reward system.

Dopaminergic neurons ascending from the A10 area (the ventral tegmental area) project to limbic and cortical areas via two pathways: mesolimbic and mesocortical (for review, see Kapur and Mann, 1992). The mesocorticolimbic DAergic system probably participates in the pathogenesis of affective disorders, schizophrenia, and substance abuse (Drevets et al., 1997; Willner, 1991) and in the reinforcing effects of cocaine, heroin, nicotine, and alcohol (Kapur and Mann, 1992; Piazza et al., 1996a; Piazza and Le Moal, 1997; Theohar et al., 1981).

After seven days, baseline microdialysate DOPA and DOPAC concentrations were significantly decreased and baseline microdialysate DA and HVA tended to be lower in CORT-treated than in VEH-treated rats. DOPA

is the immediate product of the rate-limiting enzymatic step in DA synthesis and changes in extracellular DOPA concentrations are known to closely reflect tyrosine hydroxylase activity (for review, see Pacak et al., 1993). Inhibition of L-aromatic amino acid decarboxylase increases DOPA levels; the rate of increase in DOPA levels is proportional to the rate of DOPA production by tyrosine hydroxylation (Pacak et al., 1993). Significantly lower baseline extracellular DOPA levels and the significantly smaller increases in microdialysate DOPA concentrations after decarboxylase inhibition in CORT-treated than in VEH-treated rats provide support for the view that, at least in the nucleus accumbens, tyrosine hydroxylation and therefore catecholamine biosynthesis is decreased after chronic cortisol treatment.

In exclusively dopaminergic regions such as the nucleus accumbens, extracellular DOPAC levels are mainly derived from newly synthesized intraneuronal DA (Zetterstrom et al., 1988). Glucocorticoids are also known to decrease monoamine oxidase activity, resulting in decreased conversion of DA to DOPAC (Caesar et al., 1970; Ho-van-Hap et al., 1967). Thus, the significantly decreased baseline microdialysate DOPAC concentrations in CORT-treated rats compared to controls further suggests that chronic hypercortisolemia decreases DA biosynthesis and its metabolism in this nucleus. Markedly decreasing extracellular fluid DOPAC levels after NSD 1015 administration in both experimental groups reflect NSD 1015-induced monoamine oxidase inhibition (Pacak et al., 1995b).

Although baseline extracellular fluid DA levels tended to be lower in CORT-treated than in VEH-treated rats, no statistical differences were found between these two groups. Several mechanisms may account for this observation. First, glucocorticoids decrease monoamine oxidase activity (Caesar et al., 1970; Ho-van-Hap et al., 1967) and second, glucocorticoids decrease DA reuptake (Gilad et al., 1987). These two mechanisms could account for nearly normal baseline extracellular fluid DA concentrations despite ongoing glucocorticoid-induced decreased DA biosynthesis.

Furthermore, NSD 1015 administration has been found to increase extracellular fluid DA concentrations in the nucleus accumbens in CORT-treated and VEH-treated rats. Several explanations can be offered for the pattern of high DA. NSD 1015 may inhibit neuronal reuptake of released DA, a buildup of cytoplasmic DA due to monoamine oxidase inhibition, or inhibit recycling of DA-containing vesicles fused with the axonal membrane.

Recently, several laboratories have focused on central dopaminergic systems as a link to cortisol-induced mood disturbances (for review, see Lindley et al., 1999). Depression is associated with hyperactivity of the HPA axis and deficits in DA function have been elucidated in depressed patients (Brown and Gershon, 1993; Roy et al., 1992; Sullivan and Gratton, 1997). Enhanced DA release in the nucleus accumbens parallels with reward-related activities, enhanced psychomotor activation

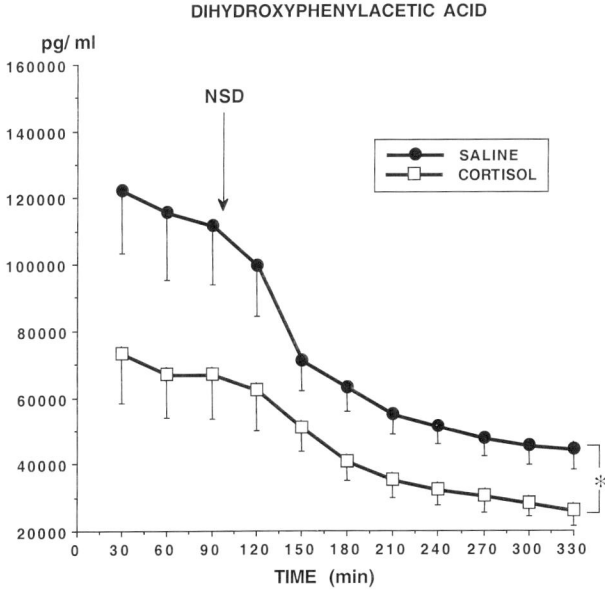

Fig. 2 Effects of local administration of NSD 1015 on microdialysate concentrations of dihydroxyphenylacetic acid in the nucleus accumbens in rats. Results are presented as means ± SEM. *, P < 0.05. (CORT-treated vs VEH-treated rats).

and decrease anxiety (Horger and Roth, 1996; Piazza and Le Moal, 1997). Increases in DA release following acute glucocorticoid treatment have been demonstrated in the nucleus accumbens (Piazza et al., 1996b) and this could be associated with euphoric states seen in the majority of people treated acutely with high doses of steroids. In contrast, chronic hypercortisolemia has been observed to increase, decrease, or not alter DA utilization and extracellular DA release in various brain regions (for review, see Lindley et al., 1999).

The present findings that chronic hypercortisolemia inhibits dopaminergic activity in the nucleus accumbens suggest a link between glucocorticoids and central dopaminergic systems in the pathogenesis of depression. Furthermore, the present results may explain the depression observed in Cushing's syndrome and normalization of the affect after correction of hypercortisolism.

References

Anisman, H., Remington, G., and Sklar, L.S. (1979). Effect of inescapable shock on subsequent escape performance: Catecholamine and cholinergic mediation of response initiation and maintenance. Psychopharmacology 61, 107–124.

Brown, A.S., and Gershon, S. (1993). Dopamine and depression. Journal of Neural Transmission 91, 75–109.

Cabib, S., and Puglisi-Allegra, S. (1996). Stress, depression and the mesolimbic dopamine system. Psychopharmacology 128, 331–342.

Caesar, P.M., Collins, G.G.S., and Sandler, M. (1970). Catecholamine metabolism and monoamine oxidase activity in adrenalectomized rats. Biochemistry and Pharmacology 19, 921–926.

Deuschle, M., Schweiger, U., Weber, B., Gotthardt, U., Korner, A., Schmider, J., Standhardt, H., Lammers, C.-H., and Heuser, I. (1997). Diurnal activity and pulsatility of the hypothalamus-pituitary-adrenal system in male depressed patients and healthy controls. Journal of Clinical Endocrinology and Metabolism 82, 234–238.

Dorn, L.D., Burgess, E.S., Dubbert, B., Simpson, S.-E., Friedman, T., Kling, M., Gold, P.W., and Chrousos, G.P. (1995). Psychopathology in patients with endogenous Cushing's syndrome: atypical or melancholic depression. Clinical Endocrinology 43, 433–442.

Dorn, L.D., Burgess, E.S., Friedman, T.C., Dubbert, B., Gold, P.W., and Chrousos, G.P. (1997). The longitudinal course of psychopathology in Cushing's syndrome after correction of hypercortisolism. Journal of Clinical Endocrinology and Metabolism 82, 912–919.

Drevet, W.C., Price, J.L., Simpson, J.R. Jr., Todd, R.D., Reich, T., Vannier, M., and Raichle, M.E. (1997). Subgenual prefrontal cortex abnormalities in mood disorders. Nature 386, 824–827.

Gilad, G.M., Rabey, J.M., and Gilad, V.H. (1987). Presynaptic effects of glucocorticoids on dopaminergic and cholinergic synaptosomes. Implications for rapid endocrine-neural interactions in stress. Life Science 40, 2401–2408.

Ho-van-Hap, A., Babineau, L.M., and Berlinguet, L. (1967). Hormonal action on monoamine oxidase activity in rats. Canadian Journal of Biochemistry 45, 355–362.

Holsboer, F., and Barden, N. (1996). Antidepressants and hypothalamic-pituitary-adrenocortical regulation. Endocrine Reviews 17, 187–205.

Horger, B.A., and Roth, R.H. (1996). The role of mesoprefrontal dopamine neurons in stress. Critical Reviews in Neurobiology 10, 395–418.

Imperato, A., Cabib, S., and Puglisi-Allegra, S. (1993). Repeated stressful experiences differently affect the time-dependent responses of the mesolimbic dopamine system to the stressor. Brain Research 601, 333–336.

Imperato, A., Puglisi-Allegra, S., Casolini, P., Zocchi, A., and Angelucci, L. (1989). Stress-induced enhancement of dopamine and acetylcholine release in limbic structures: role of corticosterone. European Journal of Pharmacology 165, 337–338.

Kapur, S., and Mann, J.J. (1992). Role of the dopaminergic system in depression. Biological Psychiatry 32, 1–17.

Lindley, S.E., Bengoechea, T.C., Schatzberg, A.F., and Wong, D.L. (1999). Glucocorticoid effects on mesotelencephalic dopamine neurotransmission. Neuropharmacology 21, 399–407.

Pacak, K., Palkovits, M., Kopin, I.J., and Goldstein, D.S. (1995a). Stress-induced norepinephrine release in the hypothalamic paraventricular nucleus and pituitary-adrenocortical and sympathoadrenal activity: In vivo microdialysis studies. Frontiers in Neuroendocrinology 16, 89–150.

Pacak, K., Palkovits, M., Kvetňanský, R., Matern, P., Hart, C., Kopin, I.J., and Goldstein, D.S. (1995b). Catecholaminergic inhibition by hypercortisolemia in the paraventricular nucleus of conscious rats. Endocrinology 136, 4814–4819.

Pacak, K., Yadid, G., Jakab, G., Lenders, J.W., Kopin, I.J, and Goldstein, D.S. (1993). In vivo hypothalamic release and synthesis of catecholamines in spontaneously hypertensive rats. Hypertension 22, 467–78.

Paxinos, G, and Watson, C. (1986). The Rat Brain in Stereotaxic Coordinates. Orlando: Academic Press.

Piazza, P.V., Barrot, M., Rouge-Pont, F., Marinelli, M., Maccari, S., Abrous, D.N., Simon, H., and Le Moal, M. (1996a). Suppression of glucocorticoid secretion and antipsychotic drugs have similar effects on the mesolimbic dopaminergic transmission. Proceedings of the National Academy of Sciences 93, 15445–15450.

Piazza, P.V., and Le Moal, M. (1997). Glucocorticoids as a biological substrate of reward: physiological and pathophysiological implications. Brain Research Reviews 25, 359–372.

Piazza, P.V., Rouge-Pont, F., Deroche, V., Maccari, S., Simon, S., and Le Moal, M. (1996b). Glucocorticoids have state dependent stimulant effects on the mesencephalic dopaminergic transmission. Proceedings of the National Academy of Sciences 93, 8716–8720.

Rouge-Pont, F., Deroche, V., Le Moal, M., and Piazza, P.V. (1998). Individual differences in stress-induced dopamine release in the nucleus accumbens are influenced by corticosterone. European Journal of Neuroscience 10, 3903–3907.

Roy, A., Karoum, F., and Pollack, S. (1992). Marked reduction in indexes of dopamine metabolism among patients with depression who attempted suicide. Archives in Genetic Psychiatry 49, 447–450.

Starkman, M.N., Schteingart, D.E., Schork, A.A. (1981). Depressed mood and other psychiatric manifestations of Cushing's syndrome: Relationship to hormone levels. Psychosomatic Medicine 43, 3–18.

Sullivan, R.M., and Gratton, A. (1997). Relationships between stress-induced increases in medial prefrontal cortical dopamine and plasma corticosterone levels in rats: Role of cerebral laterality. Neuroscience 83, 81–91.

Szemeredi, K., Bagdy, G., Stull, R., Calogero, A.E., Kopin, I.J., and Goldstein, D.S. (1988): Sympathoadrenomedullary inhibition by chronic glucocorticoid treatment in conscious rats. Endocrinology 123, 2585–2590.

Theohar, C., Fischer-Cornelssen, K., Akesson, H.O., Ansari, J., Gerlach, J., Harper, P., Ohman, R., Ose, E., and Stegink, A.J. (1981). Bromocriptine as antidepressant: double blind comparative study with imipramine in psychogenic and endogenous depression. Current Therapy Research 30, 830–841.

Tjurmina, O., Goldstein, D.S., Palkovits, M., Kopin, I.J (1999). α_2-adrenoceptor-mediated restraint of norepinephrine synthesis, release, and turnover during immobilization in rats. Brain Research 826, 243–252.

Willner, P. (1991): Animal models as simulations of depression. Trends in Pharmacological Science 12, 131–6.

Zetterstrom, T., Sharp, T., Collin, A.K., and Ungerstedt, U. (1988). In vivo measurement of extracellular dopamine and DOPAC in rat striatum after various dopamine-releasing drugs; implications for the origin of extracellular DOPAC. European Journal of Pharmacology 148, 327–334.

6. Distinctive Changes in Central Norepinephrinergic Transmission are Induced by Unconditioned and Conditioned Naturalistic Environmental Stimuli

S.C. Stanford and R. McQuade

Introduction

Stressful stimuli which involve somatosensory stimulation increase the activity of neurons projecting from the locus coeruleus (e.g., restraint: Abercrombie and Jacobs, 1987; footshock: Hirata and Aston-Jones, 1994) and the concentration of extracellular norepinephrine in their terminal field (Abercrombie et al., 1988; Pacak et al., 1995). Even an intraperitoneal injection of saline can have this effect (Dalley et al., 1996). However, little is known about the response to aversive environmental stressors that do not cause overt physical discomfort (e.g., exposure to a novel environment). Yet it is these forms of non-noxious stress which are commonly experienced by humans and which, by presenting a "threat" or "loss", are linked with anxiety or depression, respectively (see: Brown, 1993). When extrapolating the effects of stress in rats to predict their impact in humans it might be more valid to investigate the neurochemical response to naturalistic types of stress, therefore.

In pilot studies using microdialysis *in vivo*, we found that changes in the concentration of norepinephrine in dialysis samples taken from the rat cortex ("efflux") are extremely sensitive to subtle alterations of the animal's environment (Dalley and Stanford, 1995). Whether this activation reflects the rats' behavioral response to stress, or other feature(s) of the stimulus is unclear. In order to answer this question, the first aim of the present experiments was to investigate whether it is the *aversive* features of the environment which trigger the norepinephrine response, or whether other factors are relevant.

To answer this question, a modified light/dark shuttle-box was used to monitor the rats' behavioral response to different novel test environments. On the basis that rats tend to avoid an environment they perceive as aversive (Crawley and Goodwin, 1980) we used the rats' behavior to distinguish different test arenas in the shuttle-box. Microdialysis of freely-moving rats was then used to investigate how the rats' behavioral responses related to any changes in the concentration of extracellular norepinephrine ("efflux") caused by these different environmental conditions. Because there is evidence that the role of norepinephrine-releasing neurones in the cortex and hypothalamus have different roles in the response to stress (reviewed by Stanford, 1995), we also examined whether there were any differences in the neurochemical response in these two brain regions.

One unavoidable complication of this type of study is that the experiments require the implantation of a microdialysis probe and this procedure seems to disrupt the regulation of rats' body temperature and locomotor activity (Drijfhout et al., 1995). To see whether there might also be a disruption of rats' response to stress, we also investigated whether probe implantation had any effects on their behavior in the light/dark shuttle-box.

A final question was prompted by electrophysiological studies indicating that neurons in the rat locus coeruleus respond to formerly neutral stimuli which, through conditioning, assume the role of a cue predicting an aversive somatosensory stimulus (Rasmussen and Jacobs, 1986; Sara and Segal, 1991). Here, we investigated whether an increase in norepinephrine efflux could similarly be evoked by a previously neutral stimulus, the sound of a buzzer ("tone"), which has been repeatedly paired with subsequent transfer of the rats to an aversive environment (a brightly-lit, novel zone of the shuttle-box).

Methods

All procedures complied with the UK Animals (Scientific Procedures) Act, 1986. Male, outbred Sprague-Dawley rats weighing 270–300 g were obtained from the colony at University College London (UK). They were housed in groups of four under a 12-h light/dark cycle (lights on at 08.00 hours) and ambient temperature of 21–24°C. The rats had unlimited access to water and standard laboratory chow.

Surgical Procedures

Microdialysis probes were implanted into either the hypothalamus and/or frontal cortex of halothane-anesthetized rats. After securing the probe(s) with acrylic cement, the rats were allowed to recover from the anesthesia and then housed individually in a "home cage" with free access to food and water. Their behavior in the light/dark shuttle-box was evaluated on the following day.

Apparatus (Light/Dark Shuttle-Box)

The shuttle-box (length 90 cm; width 35 cm; depth 35 cm) comprised a narrow central zone (length 20 cm) and two larger end-arenas (length 35 cm). The walls and floor of the central, "neutral zone" were colored black. The end-arenas were adapted to produce one of the following test environments: a "dark" zone made of black, opaque walls and floor and illumated at 10 Lux (the same as the neutral zone); a "light" zone constructed of transparent walls and illuminated at 2500 Lux; and a light arena containing an unfamiliar rat ("L + UR"). This was set up in the same way as the "light" arena but also contained an unfamiliar rat which was placed behind a clear perspex partition in the corner of the test zone. The experimental rat could see and smell the resident, unfamiliar rat but could not make any physical contact. The floors of the test zones were marked so as to form a grid that was used to score locomotor activity. The partition between the neutral and test zones was made of black perspex and incorporated a guillotine-style door (height 12 cm, width 8 cm) which, when raised, enabled the experimental rat to traverse freely between the two compartments.

Behavioral Tests and Microdialysis

The first experiments were performed in both naïve (unoperated) and probe-implanted rats. Each experimental rat was first habituated to the neutral chamber of the shuttle-box for 100 min. After this, the rat was transferred (with minimal handling) to the novel, test arena. The guillotine door was opened immediately so as to enable the rat to traverse between the two compartments. The rats' behavior in the different test arenas was recorded on video for the following 2 h. These were scored for: latency to leave the test arena following forced entry; latency to return to the test arena following the first exit; the number of returns to the test arena; the total time spent in the test arena; and the number of lines crossed by the rat (as an index of locomotor activity).

In a parallel group of rats, microdialysis probes were implanted into the frontal cortex and hypothalamus and, on the following day, the probes were perfused with artificial CSF ("aCSF": 140 mM NaCl, 3 mM KCl, 1.2 mM $CaCl_2$, 1.0 mM $MgCl_2$, 1.2 mM Na_2HPO_4, 0.27 mM NaH_2PO_4, 7.2 mM glucose) at a flow rate of 2 μl/min. Each rat was placed in the neutral zone for 100 min after which they were transferred to, and confined within, one of the test arenas described above. Microdialysis samples were collected at 20-min intervals for the following 2 h and the norepinephrine content of the samples measured by HPLC with electrochemical detection.

In the conditioning trials, each naïve rat was confined in the neutral zone of the shuttle-box for 60 min. A tone (400 Hz; 70 dB at source), placed adjacent to the shuttle-box, was then sounded for 10 sec. Five seconds later, the rat was transferred to the test arena. The arena was set up for the "light" condition which, on the basis of results from the experiments already described, the rats perceived as aversive. After 30 min, the rats were returned to their home cage. This procedure was repeated twice-daily (at 09.30 h and 14.00 h) for three days. On the fourth day, a further conditioning trial was carried out at 09.30 h as usual. Immediately afterwards, the rats were anesthetized and microdialysis probes implanted in their frontal cortex and hypothalamus. After recovery from the anesthetic, rats were housed overnight in their home cages and test trials carried out on the following day.

On the fifth day, half the rats were used for microdialysis experiments and their probes were perfused with aCSF as described above. During the test trial, each of these rats was placed in the neutral zone of the shuttle-box and the tone sounded as usual. However, after this, half the rats normally destined for transfer to the novel arena remained in the neutral zone instead. The remainder of the rats were used for microdialysis and were transferred to the novel arena, as usual, to test for any change in their neurochemical response to this environment (i.e., sensitization or habituation) as described above. The behaviour of the rest of the probe-implanted rats on transfer to the test arena was evaluated as described above.

Finally, three further groups of probe-implanted rats were used to test for the behavioral and neurochemical effects of: 1) a single exposure to the tone alone (without transfer to the novel arena); 2) repeated exposure to tone alone (without transfer to the light arena); and 3) a single exposure to the light arena (without experiencing the tone).

Statistical Analysis

Split-plot ANOVA was carried out on raw data except for "total time spent in the test arena" which was log_{10} transformed. Because probe-implanted rats spent less time in the test arena than did naïve rats, rats' total locomotor activity was corrected for the total time spent in the test arena (i.e., activity per unit time). There were no differences in the behavior of rats with probes implanted in the frontal cortex or hypothalamus and so the behavioral data for all probe-implanted rats were pooled. The different experimental groups (with and without probe-implantation) were compared with "time" as the "within-subjects" and "surgery" and "arena" as "between subjects" factors. When a main effect of any of these factors, or an interaction between them, was evident, one-way ANOVA followed by the Tukey HSD test or the t-test with the Bonferroni correction was used to compare multiple experimental groups or individual pairs of data, respectively. Summary behavioral data were compared using 2-Way ANOVA with the Tukey HSD test or the t-test with the Bonferroni correction.

Similar statistical procedures were adopted to compare changes in norepinephrine efflux but, in this case, "brain region" was included as an additional "between subjects" factor. Net changes in this neurochemical measure were also analyzed in order to compensate for any differences in basal concentrations of norepinephrine across subjects. This involved subtracting the norepinephrine content of the last basal sample, taken from each rat while they were in the neutral zone, from all subsequent samples (see: Dalley *et al.*, 1996). Finally, basal efflux of norepinephrine was analyzed for main effects of "treatment" and "brain region" using independent and paired-sample *t*-tests, respectively.

Results

Single Exposure to a Novel Arena

Behavior

Comparison of the rats' behavioral response to the different test arenas indicated that the light arena was the most aversive. The most conspicuous change was that both naïve and probe-implanted rats spent less time in this arena than in the dark one ($P < 0.05$ in both cases) (Figure 1). The presence of an unfamiliar rat ameliorated the aversive impact of the light arena: both probe-implanted and naïve rats made more returns to ($P < 0.004$ and $P < 0.1$, respectively) and spent more time in the light arena ($P < 0.05$ in both cases; Figure 1) when this contained an unfamiliar rat. Also, for both groups of rats there was no longer any difference for either of these measures when compared with the response to the dark arena.

Implantation of a microdialysis probe modified behavior by exaggerating the behavioral response to the light arena: when compared with naïve rats, there were fewer returns to the test arena ($P < 0.001$), a longer

latency to return ($P < 0.05$) and a reduction in the total time spent within it ($P < 0.01$; Figure 1). These changes are not explained by any non-specific effects of probe-implantation on locomotor activity which, in the light arena, was greater in probe-implanted than naïve rats ($P < 0.01$).

Microdialysis

Transfer of rats to the light arena increased norepinephrine efflux in both the frontal cortex and hypothalamus whether or not the arena contained an unfamiliar rat ($P < 0.05$ in all cases) (Figures 2a and 2b). Transfer to the dark arena had no effect, however.

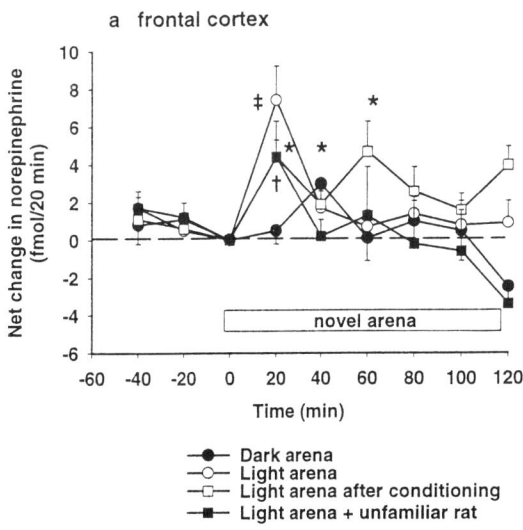

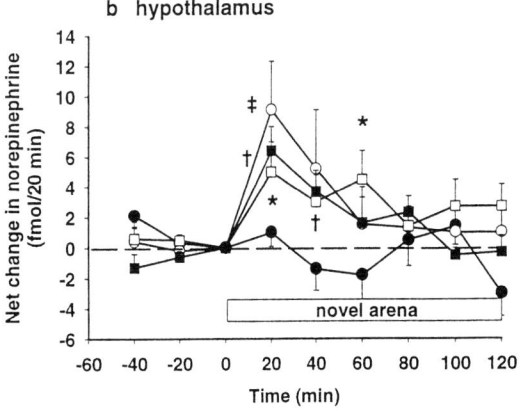

Fig. 2 The net change in the concentration of norepinephrine in microdialysis samples of the **a**) frontal cortex and **b**) hypothalamus on transfer of rats to a novel environment for a period of 2 h (marked). "O" (abscissa) indicates the time at which rats were moved from the neutral zone of the shuttle-box (i.e., the last basal sample) to the test arena. Points show mean ± SE. n = 6 or 7 for each group. ‡ $P < 0.05$ compared with final basal sample in naïve rats transferred to the light arena. * $P < 0.05$ compared with final basal sample in conditioned rats transferred to the light arena. † < 0.05 compared with final basal sample in naïve rats transferred to the light + unfamiliar rat arena.

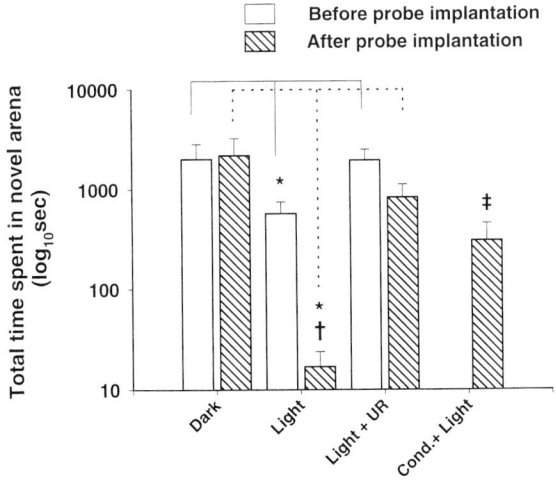

Fig. 1 Effect of environment on behaviour of rats pre- and post-probe implantation and of rats repeatedly exposed to a light arena (conditioned). Bars show mean total time spent in the novel arena ± SE (n = 7 or 8). * $P < 0.05$ for the difference across different arenas in either pre- or post-probe-implanted rats. † $P < 0.05$, for effect of probe implantation. ‡ $P < 0.05$, for differences between naïve and conditioned ("cond.") rats in light arena. "UR" indicates unfamiliar rat.

Conditioning

Behavior

Experience of the sequence of conditioning trials modified rats' behavior in the light arena. When compared with their behavior on their first exposure to the test arena, the rats' spent appreciably more time within it ($P < 0.05$) (Figure 1).

Microdialysis

When the animals remained in the neutral zone of the shuttle-box, neither a single nor repeated exposure to the tone evoked a norepinephrine response in either the frontal cortex (Figure 3a) or the hypothalamus (Figure 3b). However, after repeated pairings of the tone and subsequent transfer to the light arena, two changes were apparent: first, basal efflux of norepinephrine was reduced by approximately 50% in both brain regions (Table 1) and, secondly, the tone triggered a (+62%) increase in norepinephrine efflux in the frontal cortex (Figure 3a) but not the hypothalamus (Figure 3b).

For both the frontal cortex and the hypothalamus, the amplitude of the response to the light arena after the series of conditioning trials was not significantly different from that evoked by the first exposure. However, the duration of this response was prolonged in both brain regions (P < 0.05) (Figures 2a and 2b).

Discussion

These experiments confirm our earlier finding that exposure of rats to a novel environment activates the central noradrenergic system. Specific features of the novel environment seem to be crucial determinants of the neurochemical response. This is evident from the finding that a novel brightly-lit arena increased norepinephrine efflux in both the frontal cortex and the hypothalamus but a novel, dimly-lit arena evoked no response in either brain region. This distinction argues against novelty alone being the factor which determines norepinephrine efflux. However, the extent to which the rats find the arena aversive does not seem to define the response either. This is because, in both brain regions, the increase in norepinephrine efflux was the same whether or not the brightly-lit arena

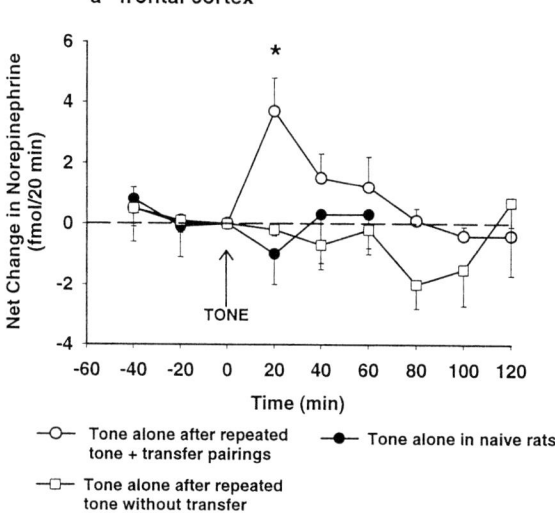

a frontal cortex

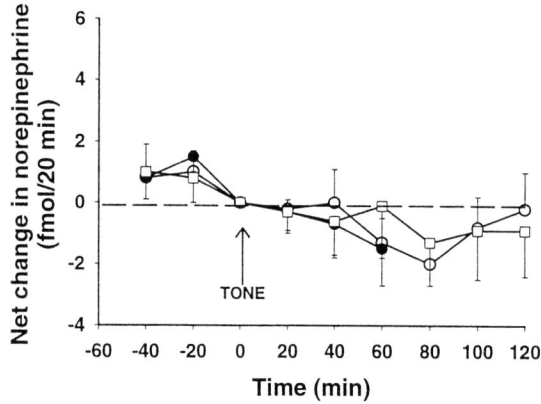

b hypothalamus

Fig. 3 The net change in the concentration of norepinephrine in microdialysis samples of the **a**) frontal cortex and **b**) hypothalamus in rats on exposure to a 10-sec conditioned cue (400 Hz tone; 70 dB at source) (marked by the arrow). "O" (abscissa) indicates the time at which rats were exposed to the tone (i.e., the last basal sample). Points show mean ± SE. (n = 7 & 8). * P < 0.05, compared with final basal sample.

contained an unfamiliar rat. Yet, the rats' behavior indicated that the presence of an unfamiliar rat appreciably diminished the aversive impact of a brightly-lit arena. Overall, the norepinephrine response seems to be determined by a combination of factors that cannot obviously be classified as either "aversive" or "novel".

The finding that there was no discernible difference in the norepinephrine response in the frontal cortex and hypothalamus is important for two reasons. First, the norepinephrine-releasing neurons projecting to these brain areas are largely derived from different sources: whereas those supplying the frontal cortex are thought to originate exclusively from the locus coeruleus, those in the hypothalamus originate predominately from the lateral tegmental nuclei (Holets, 1990). Secondly, extensive evidence suggests that these two groups of norepinephrinergic neurons make different contribu-

Table 1 Basal efflux of norepinephrine on the first and last of a series of conditioning trials in the light/dark shuttle-box.

	Before first exposure to shuttle-box (n = 11)	After repeated conditioning trials (n = 15)
Frontal cortex	11.0 ± 1.1	7.7 ± 0.6*
Hypothalamus	10.4 ± 1.1	8.7 ± 0.7*

Values show mean ± SE basal norepinephrine efflux (fmol/20 min). Sample sizes indicated in parenthesis. * P < 0.05 (c.f first exposure to the shuttle-box)

tions to the stress response (Glavin *et al.*, 1983; reviewed by Stanford, 1995).

The reason for the disparity between the present findings and previous reports is uncertain. However, it could be relevant that former studies have generally used laboratory stressors, such as footshock and immobilization, which are qualitatively quite different from the stimuli used here. In fact, a comparatively recent report suggests that the norepinephrine response in the hypothalamus is strongly dependent on the type of stressful stimulus imposed (Shibasaki *et al.*, 1995). This suggestion is entirely consistent with our contention that the neurochemical coding of the behavioral response to noxious and non-noxious stimuli are quite distinct (Salmon and Stanford, 1992; Stanford and Salmon, 1992). As a consequence of this, the neurochemical effects of stimuli that impose physical discomfort cannot be assumed to generalize to those of non-noxious, naturalistic stressors.

A further finding from the experiments described here was that the implantation of a microdialysis probe modified rats' behavioral response to some, but not all, test conditions. Moreover, it is important to note that not all features of the rats' behavioural profile were affected to the same extent. The most prominent effect of probe-implantation was the exaggeration of the behavioral response to the brightly-lit arena; the behavioral responses to other test conditions were unchanged. These findings underline the need for naïve (unoperated) control rats in experiments of this kind. Indeed, without these controls, the effects of different test environments (especially aversive ones) on behavior could easily be underestimated.

The next experiments investigated whether a formerly neutral environmental stimulus (a tone) could evoke a response in norepinephrine-releasing neurons if, after repeated pairings with exposure to an aversive environment, the tone assumed the role of a conditioned cue for an imminent aversive environmental stimulus. This line of study was prompted by reports that the firing-rate of norepinephrine-releasing neurons in the locus coeruleus is increased on presentation of a conditioned cue signaling footshock or an "air-puff" (Rasmussen and Jacobs, 1986; Sara and Segal, 1991). This raises the question of whether cues for non-noxious environmental stimuli can have the same effect.

Control experiments established that experience of either a single or repeated (neutral) tone evoked a norepinephrine response. However, after repeated exposure to the tone plus aversive arena contingency, the sound of the tone alone evoked a response, even though the animals remained in the neutral zone during this test trial. An important feature of this conditioned response is that it was evident in the frontal cortex, but not the hypothalamus. This suggests that norepinephrine-releasing neurons in the frontal cortex, but not those in the hypothalamus, are capable of adaptive changes that result in the development of a conditioned response to certain environmental stimuli.

It is clear that this distinction applies specifically to conditioned cues because, after repeated experience of this stimulus during the conditioning trials, basal efflux was reduced and the duration of the response to the light arena was prolonged in both the frontal cortex and the hypothalamus. These findings suggest that, in both these brain regions, at least two parameters of the norepinephrine response in rats show adaptive changes on repeated experience of an aversive environment. However, unlike the frontal cortex, norepinephrine efflux in the hypothalamus is not affected by conditioned stimuli. This suggests that these neurons are not activated by stimuli whose salience is altered as a results of rats' previous experiences.

The development of a conditioned response in the frontal cortex could be relevant to the problems experienced by anxious patients. In this condition, certain stimuli which are normally perceived as harmless can be interpreted as aversive and even trigger attacks of acute anxiety. A disorder of norepinephrinergic transmission has long been linked with anxiety (Redmond *et al.*, 1976), although the details of any causal links are far from clear (see, for example: Libet and Gleason, 1994). The present evidence, that a these neurones can develop a conditioned response to (normally neutral) environmental cues, supports the hypothesis that an inappropriate neurochemical response to external stimuli could contribute to the manifestations of anxiety disorders (Stanford, 1996; Mason *et al.*, 1998). However, the results of the present experiments also suggest that an increase in the norepinephrine response to an *unconditioned* aversive stimulus could contribute to the behavioral habituation to such stimuli.

Acknowledgments

This work was funded by the Wellcome Trust (UK).

References

Abercrombie, E.D., and Jacobs, B.L. (1987) Single-unit response of noradrenergic neurons in the locus coeruleus of freely-moving cats. Acutely presented stressful and nonstressful stimuli. *Journal of Neuroscience* **7**, 2837–2843.

Abercrombie, E.D., Keller, R.W., and Zigmond, M.J. (1988) Characterization of hippocampal norepinephrine release as measured by microdialysis perfusion: pharmacological and behavioral studies. *Neuroscience* **27**, 897–904.

Brown, G.W. (1993) The role of life events in the aetiology of depressive and anxiety disorders. In: S.C. Stanford and P. Salmon (Eds.), *Stress: from Synapse to Syndrome*. pp. 23–50. London: Academic Press.

Crawley, J., and Goodwin, F.K. (1980). Preliminary report of a simple animal behavior model for the anxiolytic effects of benzodiazepines. *Pharmacology Biochemistry & Behavior* **13**, 167–170.

Dalley, J.W., and Stanford, S.C. (1995). Incremental changes in extracellular noradrenaline availability in the frontal cortex induced by naturalistic environmental stimuli: a microdialysis study in the freely-moving rat. *Journal of Neurochemistry* **65**, 2644–2651.

Dalley, J.W., Mason, K., and Stanford, S.C. (1996). Increased levels of extracellular noradrenaline in the frontal cortex of rats exposed to naturalistic environmental stimuli: modulation by acute systemic administration of diazepam or buspirone. *Psychopharmacology* **127**, 47–54.

Drijfhout, W.J., Kemper, R.H.A., Meerlo, P., Koolhaas J.M., Grol, C.J., and Westerink, B.H.C. (1995). A telemetry study of the chronic effects of microdialysis probe implantation on the activity pattern and temperature rhythm of the rat. *Journal of Neuroscience Methods* **61**, 191–196.

Glavin, G.B., Tanaka, M., Tsuda, A., Kohno, Y., Hoaki, Y., and Nagasaki N. (1983) Regional rat brain noradrenaline turnover in response to restraint stress. *Pharmacology, Biochemistry & Behavior* **19**, 287–290.

Hirata, H., and Aston-Jones, G. (1994) A novel long-latency response of locus coeruleus neurons to noxious stimuli: mediation by peripheral C-fibers. *Journal of Neurophysiology* **71**, 1752–1761.

Holets, V.R. (1990). The anatomy and function of noradrenaline in the mammalian brain. In: D.J. Heal and C.A. Marsden (Eds), *The Psychopharmacology of Noradrenaline in the Central Nervous System*, pp. 1–40. Oxford, Oxford University Press.

Libet, B., and Gleason, C.A. (1994). The human locus coeruleus and anxiogenesis. *Brain Research* **634**, 178–180.

Mason, K., Heal, D.J., and Stanford, S.C. (1998). The anxiogenic agents, yohimbine and FG 7142, disrupt the noradrenergic response to novelty. *Pharmacology, Biochemistry & Behavior* **60**, 321–327.

Pacak. K., McCarty, R., Palkovits, M., Kopin, I.J., and Goldstein, D.S. (1995). Effects of immobilization on in vivo release of norepinephrine in the bed nucleus of the stria terminalis in conscious rats. *Brain Research* **688**, 242–246.

Rasmussen, K., and Jacobs, B.L. (1986). Single unit activity of locus coeruleus neurons in the freely moving cat. II. Conditioning and pharmacologic studies. *Brain Research* **371**, 335–344.

Redmond, D.E., Huang, Y.H., Snyder, D.R., and Maas, J.R. (1976) Behavioral effects of stimulation of the nucleus locus coeruleus in the stump-tailed monkey *Macaca arctoides*. *Brain Research* **114**, 502–510.

Salmon, P., and Stanford, S.C. (1992). Research strategies for decoding the neurochemical basis of resistance to stress. *Journal of Psychopharmacology* **6**, 1–7.

Sara, S.J., and Segal, M. (1991). Plasticity of sensory responses of locus coeruleus neurons in the behaving rat: implications for cognition. *Progress in Brain Research* **88**, 571–585.

Shibasaki, T., Tsumori, C., Hotta, M., Imaki, T., Yamada, K., and Demura, H. (1995). The response pattern of noradrenaline release to repeated stress in the hypothalamic paraventricular nucleus differs according to the form of stress in rats. *Brain Research* **670**, 169–172.

Stanford, S.C. (1995). Central noradrenergic neurones and stress. *Pharmacology & Therapeutics* **68**, 297–342.

Stanford, S.C. (1996) Stress: a major variable in the psychopharmacologic response. *Pharmacology, Biochemistry & Behavior* **54**, 211–217.

Stanford, S.C. and Salmon P. (1992) β-Adrenoceptors and resistance to stress: old problems and new possibilities. *Journal of Psychopharmacology* **6**, 15–19.

7. Interstitial Norepinephrine in the Hypothalamus of Rats Fed a Lysine-Deficient Diet

M. Smriga, M. Mori and K. Torii

Introduction

Acute amino acid deficiency is a nutritional stressful stimulus induced by feeding an essential amino acid-imbalanced diet to rats. Rapid anorectic response to such a diet is brain-mediated (Torii, 1998; Gietzen et al., 1998). The roles of anterior piriform cortex (Gietzen, 1993) and amygdala (Leung and Rogers, 1987) have been investigated in some detail. Importantly, several groups (Mori et al., 1991; Gietzen, 1993; Torii et al., 1996; Hawkins et al., 1995, 1998) showed that specific hypothalamic neuronal populations serve an indispensable role in the integration of signals about amino acid availability.

Our group has been using an L-lysine (Lys) deficient diet (Torii, 1998) as a model of amino acid deficiency. Lys has a neutral taste, it is not a precursor for the neurotransmitters and does not interact with the metabolism of central precursors. We already documented that Lys deficient rats selected Lys solution among other amino acid solutions (Mori et al., 1991) and that their blood and plasma Lys levels significantly declined during the deficiency (Mori et al., 1991; Torii, 1998). Moreover, we found enhanced neuronal activity in specific neurons of the lateral hypothalamus (LH), when Lys deficient rats were exposed to cues associated with Lys deficiency. Similarly, with the magnetic resonance imaging (MRI) technique, we detected activation of the LH during Lys deficiency. Using an operant paradigm, we identified the activin system in LH as a mediator of behavioral responsiveness to Lys deficiency (Hawkins et al., 1998). However, in spite of the above-summarized results, the neurochemical components of the hypothalamic response to Lys deficiency are still not fully elucidated.

Presently, we have used a combination of microdialysis and high performance liquid chromatography (HPLC) to measure the level of interstitial norepinephrine (NE) in LH and ventromedial hypothalamus (VMH) of Lys deficient rats. We have attempted to evaluate dynamic NE changes during a chronic (1-week) deficiency.

Methods

Male Wistar rats (body weight, 250–310 g) (Charles River, Japan) were individually housed in conventional hanging cages under a 12 h light/dark cycle. Distilled water and Lys sufficient (normal) powdered diet was available ad libitum. Normal diet contained 1.32% of Lys. Animals were handled daily and habituated to empty sound attenuated experimental boxes attached to the microdialysis apparatus. After 5 days of training, rats were implanted with guide cannulas for the microdialysis probes (Eicom, Japan). The cannulas were placed just above the left VMH, or the left LH, as described before (Hawkins et al., 1995). Animals were allowed 4–5 days for recovery.

Following recovery, a group of rats was provided with ad libitum Lys deficient diet (experimental Day 0–7). See Mori et al. (1991) for the detailed description of the powdered Lys deficient diet. Lys sufficient diet was reintroduced from experimental Day 8 onwards. Control groups were continuously fed Lys sufficient diet.

On experimental Day 1, a microdialysis probe (0.22 mm o.d., 50,000 MW cut-off, length, 1 mm) was inserted into the guide cannula and secured with a screw. After the probe insertion, a rat was transferred to a sound-proof box, connected to microdialysis tubes and left there for exactly 4 h. No food or drink was available in the box. Afterwards, dialysate NE concentrations were determined every 15 min for 1.5 h and averaged. The probes were perfused at a rate of 1.0 μl/min with a modified Ringer's solution (147 mM NaCl, 10 mM KCl, 1.1 mM CaCl$_2$ $\times$ 2H$_2$O, 1.1 mM MgCl$_2$ $\times$ 6H$_2$O, pH 6.0). The outflow was connected by a teflon tubing to an auto-injector system of HPLC (Eicom, Japan). After the end of the experiment, the probe was removed and rat was returned to its home cage. The same procedure was repeated after 48 h, at the same time of day.

NE was separated on an Eikompak CA-5 ODS column (5 mm, 4.6 $\times$ 150 mm) (Eicom, Japan), using a solution of phosphate buffer, octane-sulfonic acid, EDTA and 5% methanol as a mobil phase (flow rate 1.0 ml/min). The working electrode was set at $+$ 450 mV against a silver-silver chloride reference electrode.

After the end of the experiment (Day 10), the brains were removed, sliced and guide cannula placements were checked by examining slices under low power magnification. Results were evaluated by t-test.

Results

The principal results are summarized in Figure 1. The upper panel A shows that the food intake declined

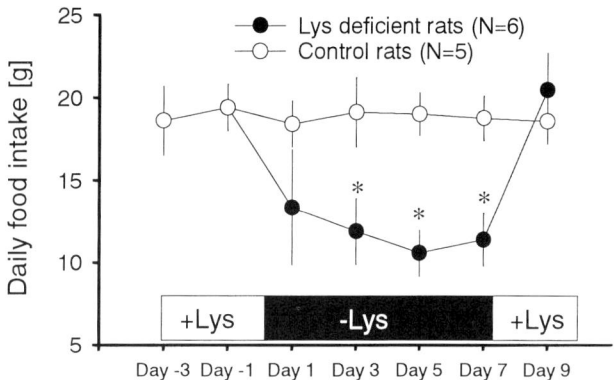

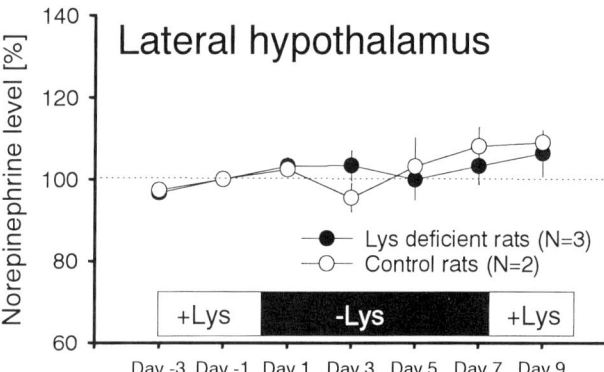

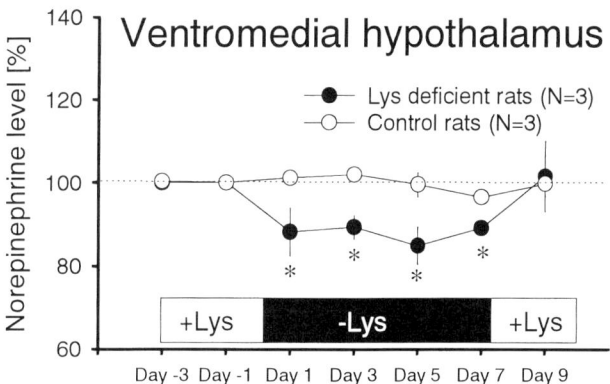

Fig. 1 Food intake and hypothalamic norepinephrine (NE) levels in male Wistar rats fed L-lysine (Lys) deficient diet for 1 week (Day 0–7).

Upper panel shows daily food intake of rats fed Lys deficient diet. Dietary Lys status is indirected at the bottom of each panel. (* $p < 0.05$ from control rats fed Lys sufficient diet; t-test)

Middle panel shows NE interstitial levels in ventromedial hypothalamus. Time-course of changes is compared to rats fed Lys sufficient diet (control rats). Results are expressed as percent differences from basal levels that were measured before the introduction of Lys deficient diet (Day -1).

Bottom panel shows NE interstitial levels in lateral hypothalamus. Time-course of changes is compared to rats fed Lys sufficient diet (control rats). Results are expressed as percent differences from basal levels that were measured before the introduction of Lys deficient diet (Day –1). (*p < 0.05 from control rats fed Lys sufficient diet; t-test)

within 24 h after introduction of a Lys deficient diet. The decline reached significant level on Day 3 and remained stable during the whole week. Immediately after the re-introduction of the Lys sufficient diet, rats increased their food intake to the previous, or even slightly higher, levels.

The middle panel demonstrates differential changes in LH and VMH interstitial NE. Lys deficiency did not result in significant LH NE changes, while it triggered a rapid decline of interstitial NE in the VMH (bottom panel). VMH NE was restored to its original values immediately after Lys sufficient diet was re-introduced.

Discussion

The present study extends our previous work on the hypothalamic modulation of an essential amino acid (Lys) deficiency. The main results point out that interstitial NE in the VMH significantly declines within the first 24 h after introduction of a Lys deficient diet. Our data on release are the opposite to the increased NE levels in the homogenized VMH tissue of threonine deficient rats found by Gietzen (1993). This increase in tissue NE content to extended synthesis or decreased release of NE. Although it is premature to generalize results from various amino acid deficiencies, our observations suggest the latter, because the microdialysis clearly showed that Lys deficiency resulted in decrease of VMH NE release. The lack of change in LH NE release, and the known involvement of the VMH in controlling food intake suggests that VMH NE is involved in the regulation of food intake during Lys deficiency.

We have previously shown that the LH has a role in the regulation of amino acid intake. Indeed, the MRI indicated that LH is activated during Lys deficiency (Torii et al., 1996) and the operant type paradigm showed that LH activin system is a regulator of behavioral responsiveness to the deficiency (Hawkins et al., 1998). Consequently, it appears that different neurotransmitters are involved in the regulation of responses to Lys deficiency, with monoamine setting in the VMH and setting in the LH.

VMH NE release remained inhibited during the whole period of Lys deficiency (1 week), but it recovered rapidly, when Lys intake was normalized. The inhibition of NE release corresponded with the significant suppression of food intake. During the first 24 h of the deficiency, the inhibition of NE release exceed the suppression of food intake. Thus, we assume that VMH NE participated in both the initiation and regulation of anorexia that characterized Lys deficiency. This assumption is consistent with the role of hypothalamic NE in general regulation of food intake (Bray, 1993). The mechanism of NE-triggered anorexia during Lys deficiency is not clear. It may include several peptidergic systems, such as cholecystokinin neuropeptide Y, which participate in control of appetite (Bray, 1993).

A problem with chronic microdialysis evaluation is a gradual formation of a glial barrier build up around the probe. This has been minimized by some groups by (e.g., Di Chiara *et al.*, 1996) reinsertion of the micro-dialysis probe every 24 hr. In the present study, we reinserted the probe every 48 h to minimize the stress of frequent re-insertion, while preventing formation of a glial barrier around the probe formed during a prolonged probe insertion.

In summary; a significant inhibition of VMH NE release was found during an essential amino acid (Lys) deficiency. This decrease lasted during the whole period of deficiency (1 week) and recovered immediately after the deficiency was normalized. No Lys-dependent changes in LH NE were recorded.

References

Bray, G.A. (1993). Food intake, sympathetic activity and adrenal steroids. *Brain Research Bulletin* **32**, 537–541.

Di Chiara, G., Tanda, G., and Carboni, E. (1996) Estimation of in vivo neurotransmitter release by brain microdialysis: the issue of validity. *Behavioral Pharmacology* **7**, 640–657.

Gietzen, D.W., Erecius, L.F., and Rogers, Q.R. (1998) Neurochemical changes after imbalanced diets suggest a brain circuit mediating anorectic responses to amino acid deficiency in rats. *Journal of Nutrition* **128**, 771–781.

Gietzen, D.W. (1993) Neural mechanisms in the responses to amino acid deficiency. *Journal of Nutrition* **123**, 610–625.

Hawkins, R.L., Inoue, M., Mori, M., and Torii, K. (1995) Effects of inhibin, follistatin, or activin infusion into the lateral hypothalamus on operant behavior of rats fed lysine deficient diet. *Brain Research* **704**, 1–9.

Hawkins, R.L., Murata, T., Inoue, M., Mori, M., and Torii, K. (1998) Activin antiserum infused into the lateral hypothalamic area affects operant type behavior of rats fed lysine-deficient diet. *Proceedings of the Society for Experimental Biology and Medicine* **219**, 149–153.

Leung, P.M.B. and Rogers, Q.R. (1987) The effects of amino acids and protein on dietary choice. In: Y. Kawamura and M. R. Kare (Eds.), *Umami, a basic taste*, pp. 565–610. New York: Marcel Dekker.

Mori, M., Kawada, T., Ono, T., and Torii, K. (1991) Taste preference, protein nutrition and L-amino acid homeostasis in male Sprague-Dawley rats. *Physiology and Behavior* **49**, 987–996.

Torii, K., Yokawa, T., Tabuchi, E., Hawkins, R.L., Mori, M., Kondoh, T. and Ono, T. (1996) Recognition of deficient nutrient intake in the brain of rats with the L-lysine deficiency monitored by functional magnetic resonance imaging, electrophysiologically and behaviorally. *Amino Acids* **10**, 73–81.

Torii, K. (1998) Central mechanisms of umami taste perception and effect of dietary protein on the preference for amino acids and sodium chloride in rats. *Food., Review International* **2**, 273–308.

8. A Possible Role of the Coeruleospinal Modulation System Under Painful Stress

Y. Yamakami, M. Usui, Y. Matsui, H. Osada, T. Tsutsumi and M. Tsuruoka

Introduction

The pontine locus coeruleus (LC) is the major source of norepinephrine (NE) in the central nervous system. LC neurons have been implicated in a variety of functions including regulation of vigilance and modulation of sensory processing (Aston-Jones et al., 1991; Foote et al., 1983). Anatomically, the LC innervates the spinal cord via descending pathways and almost every supraspinal region via ascending pathways (Dahlstrom et al., 1964). Descending NE-containing fibers from the LC have been demonstrated to be involved in analgesia.

Electrophysiological and neurochemical evidence indicates that stressful stimuli activate LC neurons (Stone, 1975). In this context, it is expected that painful stress activates the descending modulation system from the LC. The goal of our study is to clarify a possible role of the descending modulation system from the LC under conditions of painful stress.

Methods

Experiments were performed on male Sprague-Dawley (Harlan) rats weighing 250–320 g. One week before the experiment, the rats received bilateral electrolytic lesions of the LC. Stereotaxic coordinates were as follows; 9.5 mm caudal to the bregma; 1.1 mm lateral to the midline; 2.5 mm above the interaural axis. Lesions were induced by passing a cathodal current (1 mA, 20 s). In sham-operated animals, the electrode was lowered but no current was passed.

In a behavioral study, the paw withdrawal latency (PWL) to thermal stimuli was measured. In an electrophysiological study, extracellular recordings were made in rats anesthetized with sodium pentobarbital. In a neurochemical study using the microdialysis technique, a coaxial microdialysis probe (Carnegie Medicine, Sweden) was inserted into the dorsal horn to a vertical depth of 1.5 mm from the surface of the spinal cord at an angle of 45°. Artificial cerebrospinal fluid (ACSF) containing the monoamine oxidase inhibition, Pargyline, to prevent catecholamine metabolism was pumped through the microdialysis probe at a constant rate of 1 μl/min. Dialysates were collected every 20 min and NE concentration was measured by high performance liquid chromatography with electrochemical detection, using a mobile phase, pH 5.8 at a flow rate of 60 μl/min. Electrochemical detection was carried out at a glassy carbon electrode with the working potential set at 0.55 V. The NE content in the dialysate was quantified by comparison with a freshly prepared standard (1 pg/μl).

For induction of peripheral inflammation, carrageenan (2.0 mg in 0.15 ml saline) was injected subcutaneously into the plantar surface of the left hindpaw.

At the end of each experiment, the animals were deeply anesthetized with pentobarbital and perfused intracardially with a 10% formalin solution. Brains were removed, sectioned at a thickness of 50 μm, and stained with neutral red for histological verification of electrode placement.

A statistical analysis was carried out using the analysis of variance (ANOVA) with paired t-tests for the comparison of single groups before and after treatment. A difference was accepted as significant when $P < 0.05$.

Results

Following the injection of carrageenan, all rats developed edema and hyperalgesia in the injected left hindpaw (n = 15). Since hyperalgesia was maximal 4 h after injection, measures of PWLs, dorsal horn neuronal activities and NA content in the present study were recorded up to 4 h after the injection of carrageenan.

In the behavioral study, prior to injection of carrageenan, PWLs did not differ between the sham-operated (n = 20) and the LC-lesioned rats (n = 13). Four hours after carrageenan injection into the left hindpaw, PWLs in the inflamed paws decreased significantly in both groups. An overall analysis of variance on PWLs for sham-operated and LC-lesioned rats revealed a significant difference between groups (Figure 1A). In the contralateral uninjected (non-inflamed) paws, a decrease in PWL following injection of carrageenan was not observed in either of the two groups (Figure 1A). In rats receiving an injection of saline, no significant change in PWL was observed between before and 4 h after injection in either the sham-operated or LC-lesioned rats (Figure 1B).

Recordings were made from a total of 18 dorsal horn cells in the lumbar enlargement (14 cells in sham-operated rats and 4 cells in LC-lesioned rats). The background activity of dorsal horn cells in the LC-lesioned rats increased significantly 4 h after injection of carrageenan compared with the case of the sham-operated rats. (Figure 2). When thermal stimuli were

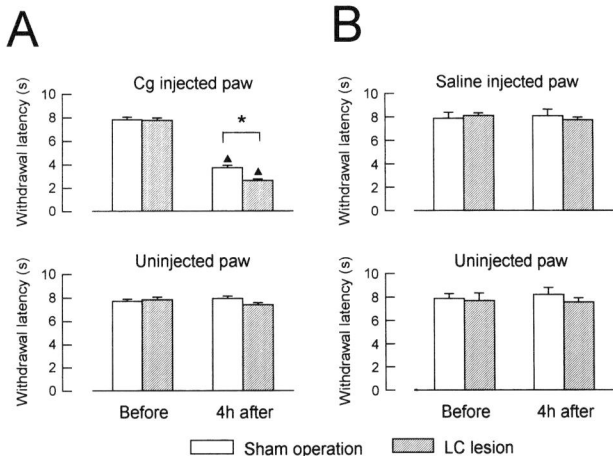

applied to the cutaneous receptive field after injection of carrageenan, there was a significant difference in the responses of the cells between the two groups, although the responses to heat increased in both groups when compared with the responses before carrageenan injection.

In the neurochemical study, changes in paw thickness following injection of carrageenan did not differ from that in the sham-operated rats (Figure 3A). Prior to carrageenan injection, the NE level (baseline level) did not differ between the LC-lesioned and the sham-operated groups. In the sham-operated group (n = 7), NE levels increased significantly compared to the baseline even 1 h after carrageenan injection (Figure 3B). In the LC-lesioned group (n = 7), following induction of inflammation, the NE concentration did not change significantly compared to the baseline level. Statistical analysis of NE levels for the LC-lesioned and the sham-operated rats revealed a significant difference between groups at 1, 2, 3 and 4 h. The NE levels did not change up to 4 h after carrageenan injection in the dorsal horn contralateral to the site of carrageenan injection (not shown).

Discussion

It is recognized that the LC is an important source of descending fibers that inhibit nociceptive transmission

Fig. 1 Changes in PWLs following unilateral injection of carrageenan or saline in LC-lesioned (n = 13) and sham-operated rats (n = 20). A: PWLs in rats with carrageenan injection (n = 9). B: PWLs in rats with saline injection (n = 4). Each column with a bar represents mean ± S.E.M. Cg, carrageenan. $P < 0.01$, significantly different from PWLs before injection. * $P < 0.01$, significantly different between two groups of rats. (Reprinted from Masayoshi Tsuruoka, William D. Willis Jr., (1996) Bilateral lesions in the area of the nucleus locus coeruleus affect the development of hyperalgesia during carrageenan-induced inflammation, Copyright (1996), *Brain Research*, 726, pp. 233–236, Figure 2, with permission from Elsevier Science).

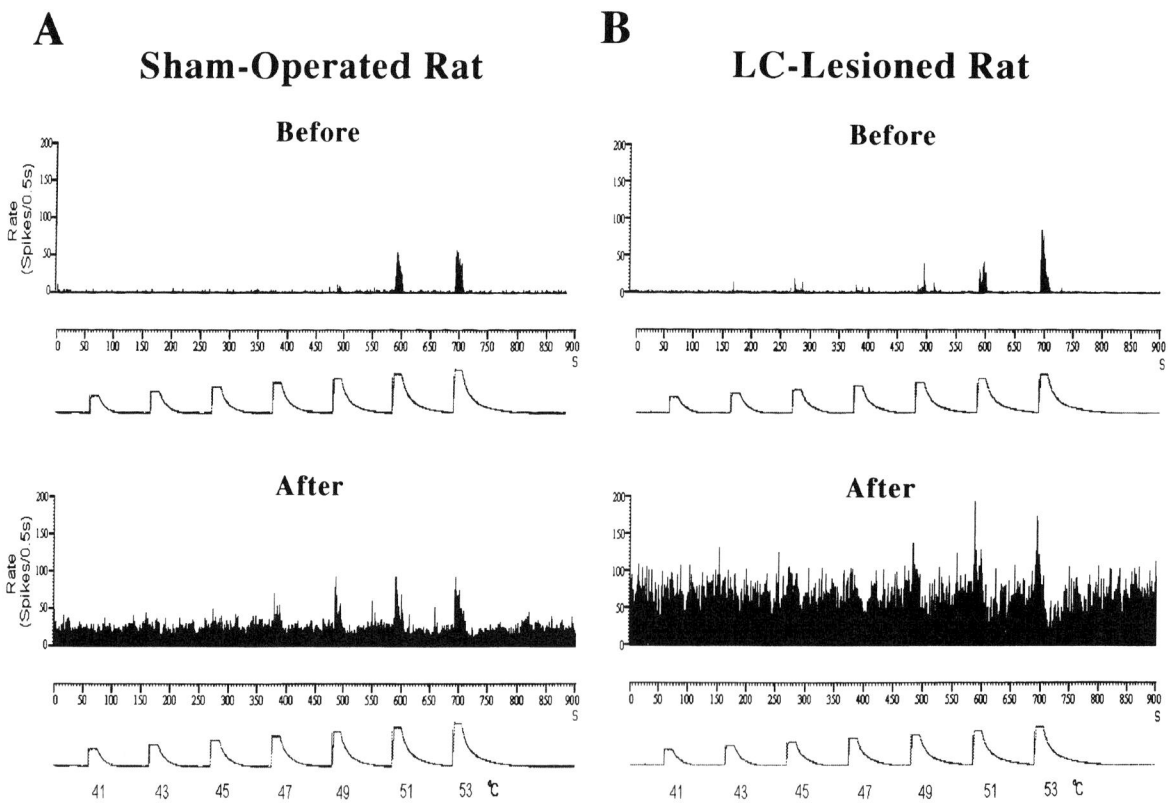

Fig. 2 Activity of dorsal horn neurons ipsilateral to the site of inflammation in the sham-operated and the LC-lesioned rats. The background and responses to thermal stimuli are represented as rate histograms.

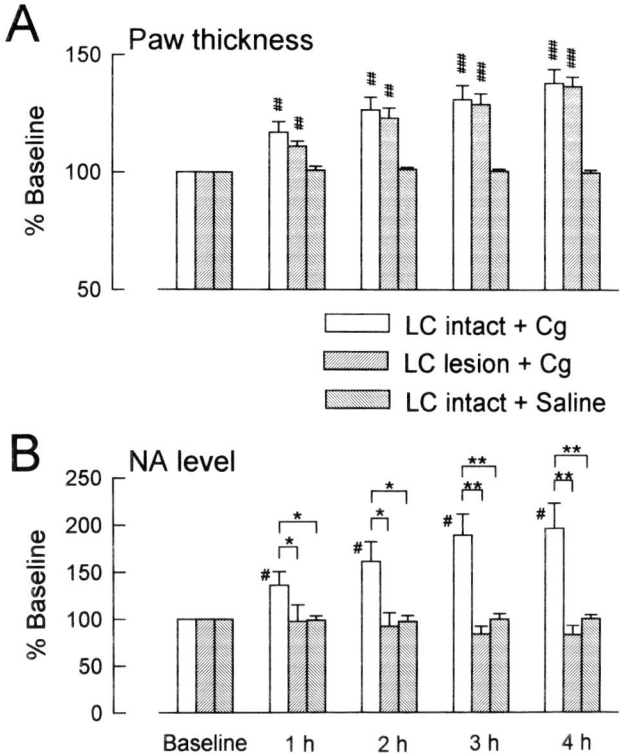

nociceptive activity of spinal cord dorsal horn neurons. This results in a decrease in the extent of the development of hyperalgesia associated with inflammation.

Another important finding in this work is that in the behavioral study, the descending modulation system from the LC is active only in the dorsal horn ipsilateral to the inflamed paw, and not in the non-inflamed paw. This is in apparent conflict with anatomical data showing bilateral retrograde labeling from the dorsal horn to the LC in the Harlan Sprague-Dawley rat used in the present study (Sluka *et al.*, 1993). However, in the neurochemical study, we found that descending NE-containing neurons from the LC which project to the dorsal horn contralateral to the site of inflammation were inactive during inflammation. This finding suggests that the failure of antinociception for the non-inflamed paw in the behavioral study is due to inactivation of descending NE-containing neurons from the LC which project to the dorsal horn contralateral to the site of inflammation.

Acknowledgements

This work was supported by a grant from The Science Research Promotion Fund.

References

Aston-Jones, G., Chiang, C., and Alexinsky, T. (1991). Discharge of noradrenergic locus coeruleus neurons in behaving rats and monkeys suggests a role in vigilance, *Progress in Brain Research* **88**, 760–769.

Dahlstrom, A., and Fuxe, K. (1964). Evidence for the exsistence of monoamine containing neurons in the central nervous system. I. Demonstration of monoamines in the cell bodies of brainstem neurons, *Acta Physiologica Scandinavica* **62** (Suppl 232), 1–155.

Foote, S.L., Bloom, F.E., and Aston-Jones, G. (1983). Nucleus locus coeruleus: new evidence of anatomical and physiological specificity, *Physiological Review* **63**, 844–913.

Jones, S.L., and Gebhart, G.F. (1986). Characterization of coeruleospinal inhibition of the nociceptive tail-flick reflex in the rat: mediation by spinal α2-adrenoceptors, *Brain Research* **364**, 315–330.

Sluka, K.A., and Westlund, K.N. (1992). Spinal projections of the locus coeruleus and the nucleus subcoeruleus in the Harlan and the Sasco Sprague-Dawley rat. *Brain Research* **579**, 67–73.

Stone, E.A. (1975). Stress and catecholamines. In: Friedhoff A.J. (Ed.), *Catecholamines and behavior*, vol. 2, pp. 31–72. New York: Plenum.

Tsuruoka, M., Hitoto, T., Hiruma, Y., and Matsui, Y. (1999). Neurochemical evidence for inflammation-induced activation of the coeruleospinal modulation system in the rat. *Brain Research* **821**, 236–240.

Tsuruoka, M., and Willis, W.D. (1996). Bilateral lesions in the area of the nucleus coeruleus affect the development of hyperalgesia during carrageenan-induced inflammation. *Brain Research* **726**, 233–236.

Fig. 3 Changes in paw thickness (A) and NE level (B) following injection of carrageenan or saline in LC-lesioned rats with carrageenan (n = 7), sham-operated rats with carrageenan (n = 7) and sham-operated rats with saline (n = 6). Each column with a bar represents mean ± S.E.M. Both paw thickness and NE level are represented as a percentage change from the baseline values (3.81 ± 0.84 pg/ sample) obtained immediately prior to the injection of carrageenan or saline. # $P < 0.05$, # # $P < 0.01$, # # # $P < 0.005$, significantly different from baseline. *$P < 0.05$, ** $P < 0.01$, significantly different between two groups of rats. (Reprinted from Masayoshi Tsuruoka *et al.* (1999) Neurochemical evidence for inflammation-induced activation of the coeruleospinal modulation system in the rat, Copyright (1999), *Brain Research*, **821**, pp. 236–240, Figure 2, with permission from Elsevier Science).

in the spinal and the medullary dorsal horn (Jones *et al.*, 1986). The results of the present study indicate that peripheral inflammation produces excitation of descending noradrenergic neurons from the LC. Activation of this modulation system induces tonic inhibition of

9. Responses of Catecholaminergic, β-Endorphinergic and Gonadoliberinergic Neuronal Systems in the Hypothalamus of Anestrous Ewes to Stressful Stimuli

D. Tomaszewska, F. Przekop and K. Mateusiak

Physiological regulation of LHRH secretion is associated with a complex interplay among excitatory and inhibitory neurotransmitter and neurohormone system activities within the hypothalamus (Kalra and Kalra, 1983; Yen, 1991; Kalra, 1993; Mohankumar *et al.*, 1994; Smith and Gallo, 1997). Changes in the activity of these systems under stress conditions alter LHRH/LH secretion (Przekop *et al.*, 1988; González-Quijano *et al.*, 1991; River and Rivest, 1991; Roozendaal *et al.*, 1995, 1997; Tomaszewska *et al.*, 1999) and disturb physiological ovulatory cycles (Przekop *et al.*, 1984; Rivier and Rivest, 1991). The precise central mechanism(s) that mediates LHRH secretion is still far from being understood. Current available results suggest that the immediate response of LHRH to stress depends primarily on central neuronal mechanisms, and that prolonged exposure to stressful stimuli additionally modulates this hormone secretion through neurotransmitters and hormones released during stress (Polkowska and Przekop, 1988; Rivier and Rivest, 1991; Polkowska and Przekop, 1992; Tomaszewska *et al.*, 1999). There is little doubt that the early effects of stress may, at least in some circumstances, be stimulatory with respect to LHRH/LH release (Aijka *et al.*, 1972; Euker *et al.*, 1975; Briski *et al.*, 1988), and that prolonged or chronic stressful stimulation is accompanied by suppression of LHRH release (Przekop *et al.*, 1988; Polkowska and Przekop, 1992; Tomaszewska *et al.*, 1999) and gonadotrophin secretion (González-Quijano *et al.*, 1991; Roozendaal *et al.*, 1995; 1997). Despite a number of studies that have been concerned with these aspects, there is as yet no coherent understanding of how the nature of the stressor modulates the neuroendocrine processes in the hypothalamus. The interplay between catecholaminergic and opioidergic systems in the mediobasal hypothalamus, including the nucleus infundibularis and median eminence (NI/ME) is considered to be a part of the neuronal mechanism(s) involved in the control of LHRH secretion (Domański *et al.*, 1991; Rosmussen, 1991; Conover *et al.*, 1993; Chomicka *et al.*, 1994; Goodman, 1996; Anselmo-Franci *et al.*, 1997; Tomaszewska *et al.*, 1999). Therefore, the objective of this study was to determine catecholaminergic, β-endorphinergic and gonadoliberinergic sys-tem activity in the nucleus infundibularis-median eminence of anestrous ewes during the time-course of footshock stimulation and, on the basis of extracellular concentrations of norepinephrine (NE), dopamine (DA) and β-endorphin, to pursue how the challenge of catecholaminergic and β-endorphinergic system activity in the stressed animals is linked with LHRH secretion. The state of stress was induced by applying electrical footshocks of an enduring and repetitive character, 5 hours daily during three consecutive days (Przekop *et al.*, 1985). We examined NE and DA release using an *in vivo* perfusion technique combined with electrochemical detection. LHRH and β-endorphin-like immunoreactivity (β-END-LI) were analysed in the perfusates by radioimmunoassay.

Footshock stimulation produced specific, time-dependent changes in the activities of all of the systems. During the first day it elicited a biphasic pattern in extracellular concentrations of NE and DA; the increase in their levels at the beginning of stimulation was followed by a return to baseline values. On the third day, a decrease of both amine concentrations in the perfusates was observed during the entire period of footshock application (Figure 1). The onset of stimulation induced short-lived peaks of extracellular β-END-LI on the first and third days (Figure 2); thereafter its elevated concentration was sustained during the entire period of stress on the first and third days. A brief facilitatory influence of stress on LHRH release was observed at the beginning of stimulation on the first and third days; thereafter on the first day LHRH returned to baseline, declining on the third day to below the baseline level (Figure 3).

Of particular importance in these findings is that onset of stimulation on the first day strongly activated all of these systems; thereafter they responded in a specific, time-dependent manner. Long-lasting stress facilitated β-END-LI release but suppressed NE and DA efflux in the NI/ME. The response of LHRH neurons in the NI/ME to stress consisted of a transient increase of this hormone output followed by marked suppression.

The time-dependent differences in the reactivity of these systems probably result from a number of

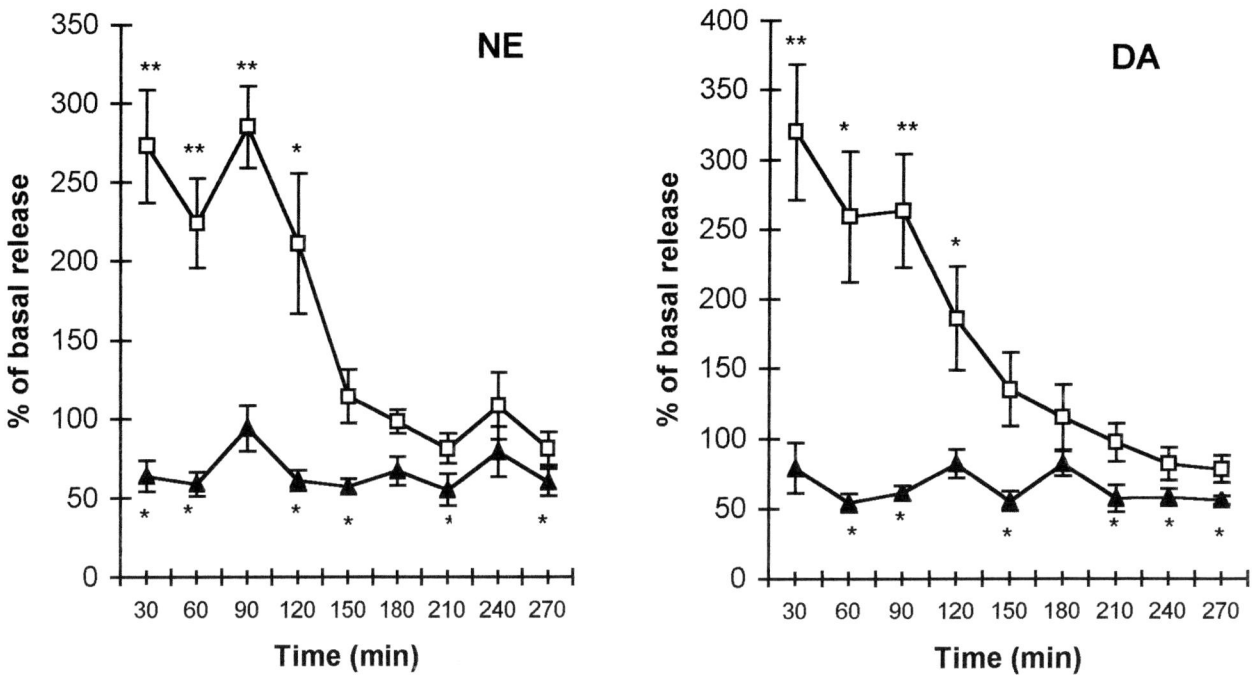

Fig. 1 Extracellular concentrations of NE and DA in the NI/ME of ewes during intermittent footshock stimulation on the first (opened square) and third days (solid triangle). The data are expressed as a percent of the control values in the respective pair fractions of non-stressed and stressed animals (*P ≤ 0.05; **P ≤ 0.01) (adapted from Tomaszewska and Przekop, in press).

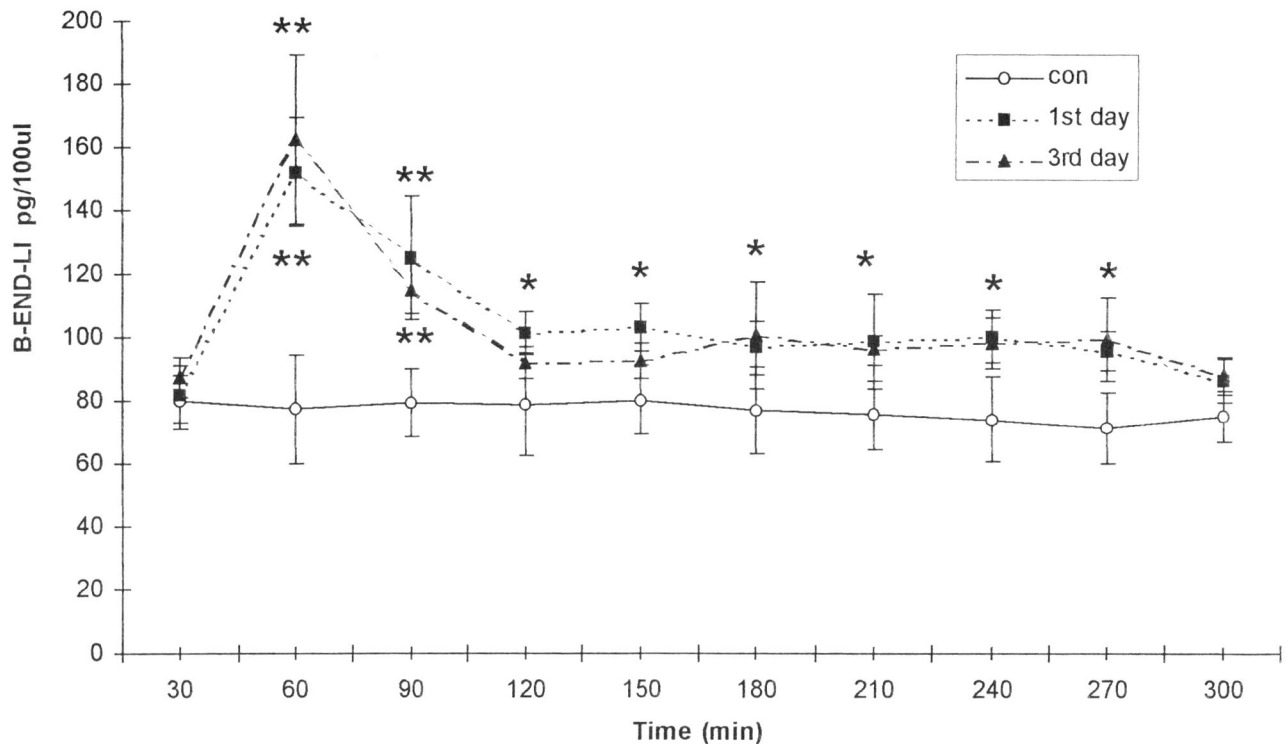

Fig. 2 β-END-LI levels in perfusates from NI/ME of anestrous ewes (n = 7) on the first (solid square) and third days (solid triangle) of stimulation. Data presented are mean ± SEM. Asterisks indicate values that differ significantly from control animals (open circle) according to Tukey's test (*P ≤ 0.05; **P ≤ 0.01) (Tomaszewska et al., 1999).

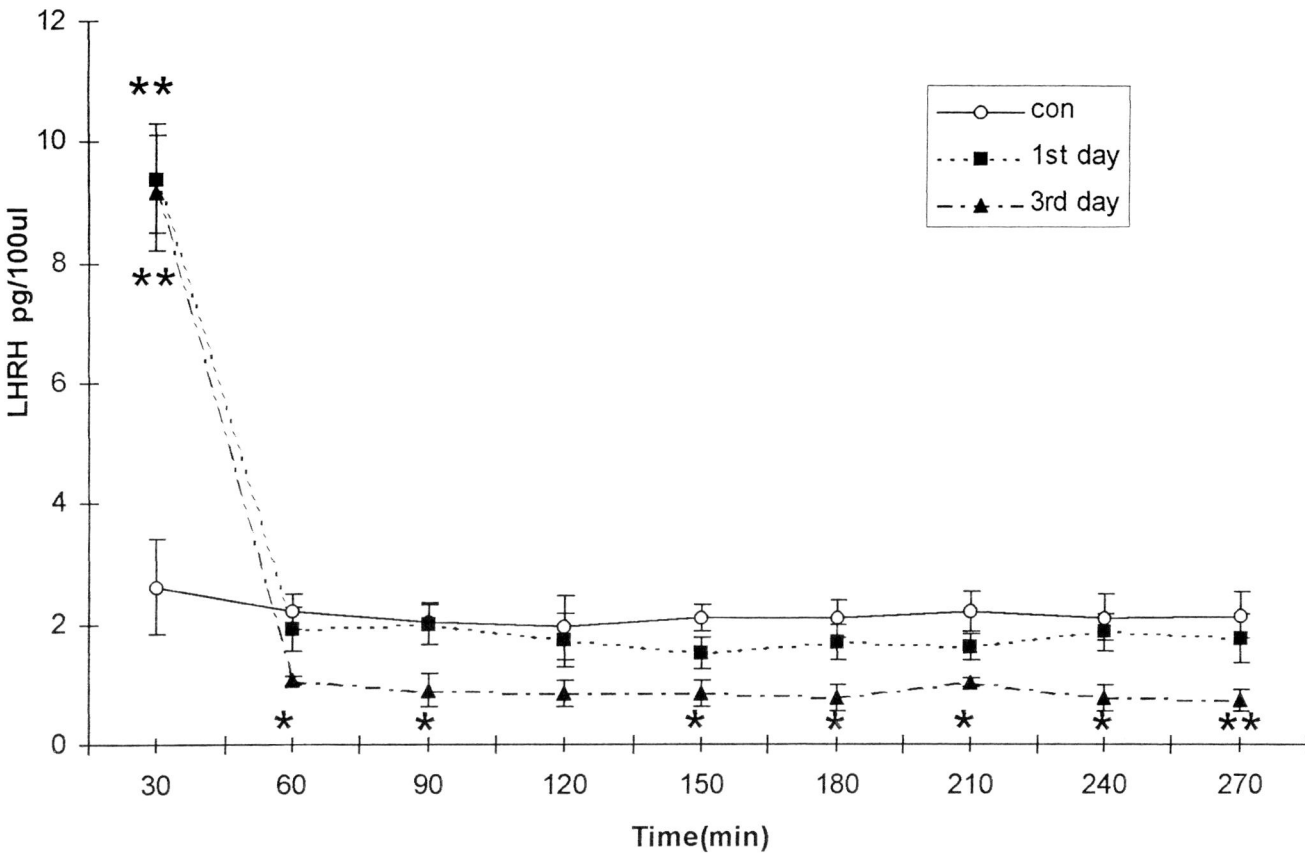

Fig. 3 LHRH levels in perfusates from the NI/ME of anestrous ewes (n = 7) on the first (solid square) and third day (solid triangle) of stimulation. Data presented are mean ± SEM. Asterisks indicate value that differ significantly from control animals (open circle) according to Tukey's test (*P ≤ 0.05; **P ≤ 0.01) (Tomaszewska *et al.*, 1999).

neurochemical processes including biosynthesis, release, metabolism, and changes in hypothalamic receptor sensitivities. For example the stressful stimuli elicit changes in release, reuptake, and metabolism of NE (Pacák *et al.*, 1993, 1995), degeneration of axon terminals of noradrenergic neurons (Kitayama *et al.*, 1996) leading to reduction of neurotransmission by lowering both synapsis and alpha and beta-adrenergic receptor density (Stone, 1983; Stanford, 1993). Also the dynamics of β-END-LI secretion by the hypothalamic nuclei have been observed in sheep under prolonged stress condition (Domański *et al.*, 1993). All of these events may modify the function of these systems, as has been previously documented for the role of the opioidergic system in the control of cortisol secretion in sheep (Przekop *et al.*, 1990; Domański *et al.*, 1993) and LHRH/LH in rats (González-Quijano *et al.*, 1991; Roozendaal *et al.*, 1997). Since under normal physiological conditions, in anestrous ewes both the catecholaminergic (Havern *et al.*, 1994; Goodman *et al.*, 1995; Goodman, 1996; Lehman *et al.*, 1996) and opioidergic pathways (Malven *et al.*, 1990; Whisnant *et al.*, 1991) have a predominantly inhibitory effect on LHRH/LH secretion, the pattern of LHRH secretion in stressed ewes in this study can not be explained adequately only on the basis of a neuropharmacological model of

catecholamine and β-endorphin actions on LHRH neurons in the hypothalamus of non-stressed animals. The inhibitory effect of the opioidergic and catecholaminergic systems on LHRH neurons under normal physiological conditions in anestrous ewes may by itself be inadequate proof for the action of these systems in stressed animals. The question is whether catecholamines and β-endorphin act on LHRH release in stressed ewes similarly as in non-stressed animals.

In light of the present results, it is reasonable to suggest that the roles of the catecholaminergic and β-endorphinergic systems in the NI/ME of anestrous ewes with respect to gonadotrophin secretion differ under normal and stress conditions.

Analysis of the relationship between CRF/ACTH-β-endorphin/cortisol and LHRH/LH release in anestrous ewes subjected to long-term intermittent stimulation demonstrated that the suppressive effect of stress on gonadotrophin secretion initiated at the beginning of stimulation persisted for a long time even after the hormones of the hypothalamic-pituitary-adrenal axis returned to baseline levels (Polkowska and Przekop, 1992). In estrous ewes, intermittent stressful stimuli during 3 consecutive days caused a long-term blockade of ovulation in some animals (Przekop *et al.*, 1984). This supports the hypothesis that the initial suppressive effect of stress on LHRH/LH release is sustained for a long period after the end of the stress procedure.

The present results indicate that catecholamine, β-endorphin and LHRH release change during the time-course of footshock stimulation. The onset of stimulation induces abrupt, brief increases of LHRH, β-endorphin and catecholamine release, whereas prolonged stimulation causes a dissociation in catecholaminergic, β-endorphinergic and gonadoliberinergic system activity; generally it activates β-endorphinergic, and suppresses catecholaminergic and gonadoliberinergic neural activity in the NI/ME. These responses suggest that prolonged intermittent footshock stress activates inhibitory components in the neuroendocrine control of LHRH release. Among others, the inhibition of catecholamine efflux and enhanced release of β-endorphin in the NI/ME in stressed animals may be considered as potential inhibitory factors responsible at least in part for LHRH suppression under prolonged stress conditions.

References

Aijka, D., Kalra, S.P., Fawcett, C.P., Krulich, L., and McCann, S.M. (1972). The effect of stress and nembutal on plasma levels of gonadotropins and prolactin in ovariectomized rats. *Endocrinology* **90**, 707–715.

Anselmo-Franci, J.A., Franci, L., Anturus-Rodriques J., and McCann, S.M. (1997). Locus coeruleus lesions decrease norepinephrine input into the medial preoptic area and medial basal hypothalamus and block the LH, FSH and prolactin preovulatory surge. *Brain Research* 289–296.

Brisky, K.P., and Sylvester, P.W. (1988). Effect of specific acute stressors on luteinizing hormone release in ovariectomized and ovariectomized estrogen-treated female rats. *Neuroendocrinology* **47**, 194–202.

Chomicka, L.K., Gajewska, A., and Przekop F. (1994). Release of luteinizing hormone-releasing hormone (LHRH) and dopamine (DA) by the nucleus infundibularis-median eminence (NI/ME) during seasonal anestrus in ewes. 3-rd International Congress of Neuroendocrinology, Budapest 1994, Karger, Neuroendocrinology. Abstracts p. 67.

Conover, C., Kuljis, R.O., Rabii, J., and Advis, J.-P. (1993). Beta-endorphin regulation of luteinizing hormone-releasing hormone release at the median eminence in ewes: immunocytochemical and physiological evidence. *Neuroendocrinology* **57**, 1182–1195.

Domański, E., Chomicka, L.K., Ostrowska, A., and Mateusiak K. (1991). Release of luteinizing hormone-releasing hormone, β-endorphin and noradrenaline by the nucleus infundibularis-median eminence during preovulatory period in the sheep. *Neuroendocrinology* **54**, 151–158.

Domański, E., Ostrowska, A., Rutkowski, J., and Luboińska, U. (1992). The dynamics of β-endorphin release by hypothalamic nuclei and pituitary gland in sheep under physiological and stress condition. *Experimental and Clinical Endocrinology* **99**, 39–44.

Domański, E., Romanowicz, K., and Kerdelhue, B. (1993) Enhancing effect of intracerebrally infused β-endorphin antiserum on the secretion of cortisol in footshocked sheep. *Neuroendocrinology* **57**, 147–153.

Euker, J.S., Meites, J., and Riegle, G.D. (1975). Effect of acute stress on serum LH and prolactin in intact, castrated and dexamethasone-treated male rats. *Endocrinology* **96**, 85–92.

Gonzĺez-Quijano, M.I., Ariznavarreta, C., Martin, A.I., Treguerres, J.A.F., and López-Calderón, A. (1991). Naltrexone does not reverse the inhibitory effect of chronic restraint on gonadotropin secretion in the intact male rat. *Neuroendocrinology* **54**, 447–453.

Goodman, R.L., Parfitt, D.B., Evans, N.P., Dahl, G.E., and Karsch, F.J. (1995). Endogenous opioid peptides control the amplitude and shape of gonadotropin-releasing hormone pulses in the ewe. *Endocrinology* **136**, 2412–2420.

Goodman, R.L., Robinson, J.E., Kendrick, K.M., and Dyes, R.G. (1995). Is the inhibitory action of estradiol on luteinizing hormone pulse frequency in anestrous ewes mediated by noradrenergic neurons in the preoptic area? *Neuroendocrinology* **61**, 284–292.

Goodman, R.L. (1996). Neural systems mediating the negative feedback actions of estradiol and progesterone in the ewe. *Acta Neurobiologiae Experimentalis* **56**, 727–741.

Havern, R.L., Whisnant, C.S., and Goodman, R.L. (1994). Dopaminergic structures in the ovine hypothalamus mediating estradiol negative feedback in anestrous ewes. *Endocrinology* **134**, 1905–1914.

Kalra, S.P., and Kalra, P.S. (1983). Neural regulation of luteinizing hormone secretion in the rat. *Endocrine Reviews* **4**, 311–351.

Kalra, S.P. (1993). Mandatory neuropeptide-steroid signaling for the preovulatory luteinizing hormone-releasing hormone discharge. *Endocrine Reviews* **14**, 507–538.

Kitayama, I., Yamashita, K., Nakamura, T., and Namura, J. (1996). Structural and functional changes of central noradrenergic neurons in chronic stress-induced animal model of depression. In: R. McCarty, G. Aquilera, E.L. Sabban, and R. Kvetňanský (Eds.), *Stress: Molecular Genetic and Neurobiological Advances*, Volume 1, pp. 135–141. New York: Gordon and Breach.

Lehman, M.N., Durham, D.M., Jansen, H.T., Adriom, B., and Goodman, R.L. (1996). Dopaminergic A14/A15 neurons are activated during estradiol negative feedback in anestrous, but not breeding season ewes. *Endocrinology* **137**, 4443–4450.

Mohankumar, P.S., Thyagarajom, S., and Quadri, S.K. (1994). Correlation of catecholamine release in the medial preoptic area with proestrus surges of luteinizing hormone and prolactin: effects of aging. *Endocrinology* **135**, 119–126.

Malven, P.V., Slanisiewski, E.P., and Hoglof, S.A. (1990). Ovine brain areas sensitive for naloxone-induced stimulation of luteinizing hormone release. *Neuroendocrinology* **52**, 373–381.

Pacák, K., Palkovits, M., Kvetňanský, R., Fukuhara, K., Armando, I., Kopin, I.J., and Goldstein, D.S. (1993). Effect of single or repeated immobilization on release of norepinephrine and its metabolites in the central nucleus of the amygdala in conscious rats. *Neuroendocrinology* **57**, 626–633.

Pacák, K., Palkovits, M., Kopin, I.J., and Goldstein, D.S. (1995). Stress-induced norepinephrine release in the hypothalamic paraventricular nucleus and pituitary-adrenal and sympathoadrenal activity: *in vivo* microdialysis studies. *Frontiers in Neuroendocrinology* **16**, 89–150.

Polkowska, J., and Przekop, F. (1988). Immunocytochemical changes in hypothalamic and pituitary hormones after acute and prolonged stressful stimuli in the anestrous ewes. *Acta Endocrinologica* **118**, 268–276.

Polkowska, J., and Przekop, F. (1992). Neuroendocrine responses to footshock stress in anestrous ewes. In: R. Kvetňanský, R. McCarty, and J. Axelrod (Eds.), *Stress: Neuroendocrine and Molecular Approaches*, Volume 1, pp. 467–472. New York: Gordon and Breach.

Przekop, F., Stupnicka, E., Wolińska-Witort, E., Mateusiak, K., Sadowski, B., and Domański, E. (1985). Changes in circadian rhythm and suppression of plasma cortisol level after prolonged stress in the sheep. *Acta Endocrinologica* **110**, 540–545.

Przekop, F., Wolińska, E., Mateusiak, K., Sadowski, B., and Domański, E. (1984). The effect of prolonged stress on the oestrous cycles and prolactin secretion in sheep. *Animal Reproduction Sciences* **7**, 333–342.

Przekop, F., Polkowska, J., and Mateusiak, K. (1988). The effect of prolonged stress on the hypothalamic luteinizing hormone-releasing hormone (LHRH) in the anestrous ewe. *Experimental and Clinical Endocrinology* **91**, 334–340.

Przekop, F., Mateusiak, K., Stupnicka, E., Romanowicz, K., and Domański, E. (1990). Suppressive effect of β-endorphin on the secretion of cortisol under stress conditions in sheep. *Experimental and Clinical Endocrinology* **95**, 210–216.

Rivier, C., and Rivest, S. (1991). Effect of stress on the activity of the hypothalamic-pituitary-gonadal axis: peripheral and central mechanisms. *Biology of Reproduction* **45**, 523–532.

Roozendaal, M.M., Swarts, H.J.M., Wiegant, V.M., and Mattheij, J.A.M. (1995). Effect of restraint stress on the preovulatory luteinizing hormone profile and ovulation in the rat. *European Journal of Endocrinology* **133**, 347–353.

Roozendaal, M.M., Swarts, H.J.M., Van Maanen, J.C., Wiegant, V.M., Mattheij, J.A.M. (1997). Inhibition of the LH surge in cyclic rats by stress is not mediated by opioids. *Life Sciences* **60**, 335–342.

Smith, M.J., and Gallo, R.V. (1997). Further studies on the suppression of luteinizing hormone release due to activation of medial preoptic-anterior hypothalamic area μ-opioid receptors. *Brain Research Bulletin* **42**, 1–7.

Stanford, S.C. (1993). Monoamines in response and adaptation stress. In: S.C. Stanford, and P. Salmon (Eds.), *Stress: from Synapse to Syndrome*, pp. 283–331. London: Academic Press.

Stone, E.A. (1983). Adaptation to stress and brain noradrenergic receptors. *Neuroscience and Biobehavioral Reviews* **7**, 503–509.

Tomaszewska, D., Mateusiak, K., and Przekop, F. (1999). Changes in extracellular LHRH and β-endorphin-like immunoreactivity in the nucleus infundibularis-median eminence of anestrous ewes under stress condition. *Journal Neural Transmission* **3/4**, 265–274.

Tomaszewska, D., and Przekop, F. (1999). Catecholaminergic activity in the medial preoptic area and nucleus infundibularis-median eminence of anestrous ewes in normal physiological state and under stress condition. *Journal of Neural Transmission*, **106**, 1031–1043.

Tortonese, D.J. (1999). Interaction between hypothalamic dopaminergic and opioidergic systems in the photoperiodic regulation of pulsatile luteinizing hormone secretion in sheep. *Endocrinology* **140**, 750–757.

Whisnant, C.S., Havern, R.R., and Goodman, R.L. (1991). Endogenous opioid suppression of luteinizing hormone pulse frequency and amplitude in the ewe: hypothalamic sites of actions. *Neuroendocrinology* **54**, 587–593.

Yen, S.S.C. (1991). Hypothalamic gonadotropin releasing hormone. Basic and clinical aspects. In: M. Motta (Eds), *Brain Endocrinology*, pp. 245–280. Raven Press Ltd, New York.

10. Effect of Various Stressors on Neuronal Activity in the Hypothalamic Paraventricular Nucleus of Long-term Repeatedly Immobilized Rats

M. Rusnák, R. Kvetňanský, J. Jeloková and M. Palkovits

Introduction

Corticotrophin-releasing hormone (CRH) produced in parvocellular neurons of the hypothalamic paraventricular nucleus (PVN) is known to play a crucial role in the activation of the hypothalamic-pituitary-adrenocortical (HPA) axis during stress (Vale *et al.*, 1981; Bruhn *et al.*, 1984; Boutilier *et al.*, 1995). Chronic, or repeated exposure of animals to some stressors may lead to desensitization or maintaining of the HPA axis response (Bhatnagar *et al.*, 1995; Aguilera, 1994; Hauger *et al.*, 1990; Hashimoto *et al.*, 1988; Dallman *et al.*, 1987; Rivier and Vale, 1987; De Goeij, 1992; Kant *et al.*, 1987). However, exposure of chronically stressed rats to acute novel stressors leads to normal or enhanced responses of this system (Ma and Aguilera, 1999; Bhatnagar and Dallman, 1998). Stress has been shown to induce expression of c-fos, considered a marker of neuronal activation (Bullit *et al.*, 1990), in the pituitary (Kononen *et al.*, 1992) as well as in the PVN (Smith *et al.*, 1992; Ceccatelli *et al.*, 1989; Senba and Ueyama, 1997; Bonaz and Rivest, 1998; Bhatnagar and Dallman, 1998), where also elevations of CRH gene expression by a number of stressors has been reported (Harbuz and Lightman, 1989, Bartanusz *et al.*, 1993; Mansi and Drolet, 1998). It has also been shown that in the parvocellular PVN c-fos induction preceded CRH gene expression (Imaki *et al.*, 1992). In repeatedly restrained rats (Bonaz and Rivest, 1998), a desensitization in c-fos activation in the parvocellular part of the PVN was observed. The adaptability of CRH mRNA gene expression in parvocellular PVN neurons of rats exposed to different types of chronic or repeated stress appears to be stressor specific. The desensitization in CRH mRNA expression has been shown in the responses of rats to milder stressors predominantly involving psychological components such as repeated restraint (Lightman and Harbuz, 1993; Ma and Aguilera, 1997) or chronic social stress (Albeck *et al.*, 1997). However, after repeated exposure to stronger stress stimuli such as intraperitoneal administration of hypertonic saline or immobilization, no desensitization in CRH mRNA levels has been reported (Kiss *et al.*, 1993; Bartanusz *et al.*, 1993; Makino *et al.*, 1995). Enkephalin (ENK), which is also expressed in parvocellular neurons of the PVN,

where it is partly colocalized with CRH (Lightman and Scott, 1987; Ceccateli *et al.*, 1989), may be involved in modulating the release of CRH as well as other important hypothalamic neurotransmitters in the median eminence during stress. Some stressors known to induce CRH gene expression also activate ENK gene expression in parvocellular neurons of the PVN (Harbuz and Lightman, 1989); however, there is a lack of evidence about ENK gene expression in the PVN of chronically stressed animals.

In the present study we investigated the effects of long-term repeated immobilization (IMMO) stress and various single novel stressors on activation of CRH and ENK mRNA gene expression in neurons of the hypothalamic PVN.

Materials and Methods

The protocol for this experiment was approved by the Animal Care Committee at the Institute of Experimental Endocrinology in Bratislava, Slovak Republic.

Stressors. Single stressors such as: immobilization (IMMO, taping the limbs of rats to immobilization boards for 2 h; Kvetňanský and Mikulaj, 1970), exposure to cold (4°C for 5 hours), insulin-induced hypoglycemia (insulin Actrapid, 5.0 I.U./kg ip) and cellular glucoprivation caused by 2-deoxy-D-glucose (2-DG, Sigma 500 mg/kg ip) were applied 24 hours after the last IMMO or to control (unstressed) rats (Sprague Dawley, 400–500 g, Charles River Viga, Germany). All stressors have been shown to strongly activate the HPA axis (Pacak *et al.*, 1995; Weidenfeld *et al.*, 1994, 1984). The effect of novel stressors on neuronal activity and neuropeptide gene expression was investigated in long-term repeatedly immobilized (2 h daily for 7 weeks) rats. Rats were sacrificed 5 h after the initiation of any single stressor, or 24 h after the last IMMO was completed.

Preparation of probes. The complete rat c-fos cDNA cloned in a pSP 65 plasmid was kindly provided by T. Curran. The 940 bp rat preproenkephalin A cDNA in pSP 65 plasmid and CRH c DNA in pGEM 3z plasmid were kindly provided by Dr. W.S. Young. To get a complementary probe to the c-fos cDNA sequence and the ENK cDNA sequence, for each, a pair of oligo-primers, containing the T7 RNA polymerase promoter site in antisense direction and the T3 RNA polymerase promoter site in sense direction, was designed and synthesized. A c-fos cDNA fragment from 114 bp to 521 bp and ENK cDNA fragment from 79 bp to 516 bp were amplified by polymerase chain reaction (PCR). The template for generating antisense cRNA probe for CRH was prepared by linearizing the CRH containing plasmid by *Eco*R1 (New England Biolabs, Beverly, MA).

In situ hybridization (ISH). For ISH, 12 μm thick frozen sections containing the hypothalamic PVN were cut on a cryostat (Reichert-Jung), and thaw-mounted onto silanized slides (Fischer Scientific, Pitsburgh, PA) on a hot plate at 37°C. for 15 min. The slides were then frozen and stored at –80°C until used. The frozen sections were air dried for 10 min, fixed in 4% formaldehyde in 1 × PBS for 10 min, washed in 1 × PBS (pH 7.4) twice for 5 min, acetylated for 10 min with 0.25% acetic anhydride in 0.1 mol/l triethanolamine-HCL (pH: 8.0), rinsed in 0.3 mol/L NaCl-mol/L sodium citrate solution, dehydrated in ethanol and air dried. Sections were hybridized overnight at 55°C with 1 × 106 cpm of UTP ^{35}S-labeled antisense probes, generated using T7 RNA polymerase for c-fos and ENK and SP6 RNA polymerase (Ambion, Austin, TX) for CRH, together with 80 μl of final hybridization solution containing: 50% formamide, 250 mM Tris-HCl (pH 7.4), 1 mM EDTA (pH 8), 300 mM NaCl, 10% Dextran Sulfate, 1 × Denhardt's, 25 mg/ml yeast tRNA, 100 μg/ml salmon sperm DNA, 250 μg/ml total yeast RNA, 100 mM dithiotreitol (DTT), 0.1% sodium thiosulfate and 0.1% SDS per 1 slide. After hybridization, the non-specifically hybridized probes were washed out as described (Bradley *et al.*, 1992). Sections were then dehydrated in ethanol, air dried and exposed to autoradiographic film (Kodak Biomax MR, Rochester, NY). After 3 days of exposure, slides with sections were coated with autoradiographic NTB3 emulsion (Kodak, Rochester, N.Y.) at 41°C After two weeks of incubation in the dark in dry boxes at 4°C, the reaction was developed using Kodak Dektol at 18°C and counter-stained with 0.5% Giemsa solution (Sigma, St. Louis, MO). The sections were air dried and mounted by Cytoseal 60 mounting medium (Stephens Scientific, NJ). The quantification of the reaction was performed under a microscope using the Scion Image program. The data were statistically evaluated by one-way analysis of variance (ANOVA). Fischer's protected LSD test was performed to determine differences between all investigated groups.

Results

c-fos Activation

Figure 1 shows that exposure of control rats to single stressors (except insulin-induced hypoglycemia) led to induction of c-fos mRNA in the parvocellular part (IMMO, cold) and the magnocellular part (2-DG) of the PVN. In long-term repeatedly IMMO rats, no c-fos

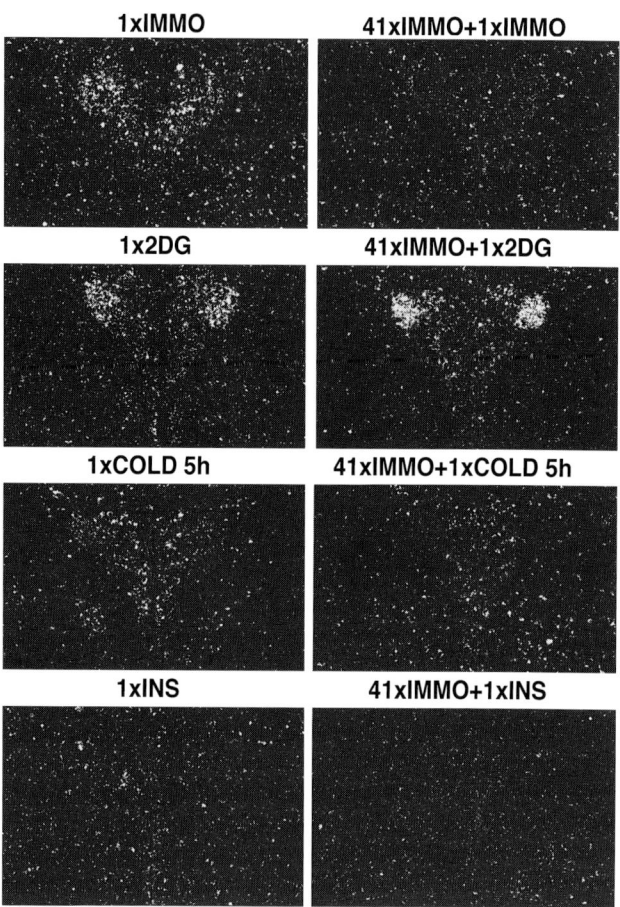

Fig. 1 Effect of a single immobilization (1 × IMMO) and various single novel stressors such as: cold for 5 h (1 × COLD 5 h), insulin-induced hypoglycemia (1 × INS) or a single 2-deoxy-D-glucose administration (1 × 2DG) on c-fos mRNA activation in the hypothalamic PVN of long-term repeatedly (41 × IMMO daily for 2 hours) immobilized rats. A representative dark-field photograph of *in situ* hybridization for c-fos mRNA in the PVN region. Magnification – 26×.

activation was observed whether rats were exposed to novel stressors or not, except for 41× IMMO animals treated with 2-DG, which showed c-fos activation in the same location as seen in controls exposed to the same stressor.

CRH and ENK Gene Expression

As shown in Figures 2 and 3, both single IMMO and single 2-DG increased CRH mRNA levels as well as ENK mRNA levels in parvocellular PVN neurons, while cold exposure and insulin administration failed to produce this effect. High elevation in mRNA levels of both neuropeptides was elicited also by long-term repeated IMMO (41 times). Twenty-four hours after the 41st IMMO a novel stressor (2-DG) as well as additional single IMMO induced further increases in already high ENK mRNA levels in the parvocellular part of the PVN, but were ineffective in causing a significant change in CRH mRNA levels.

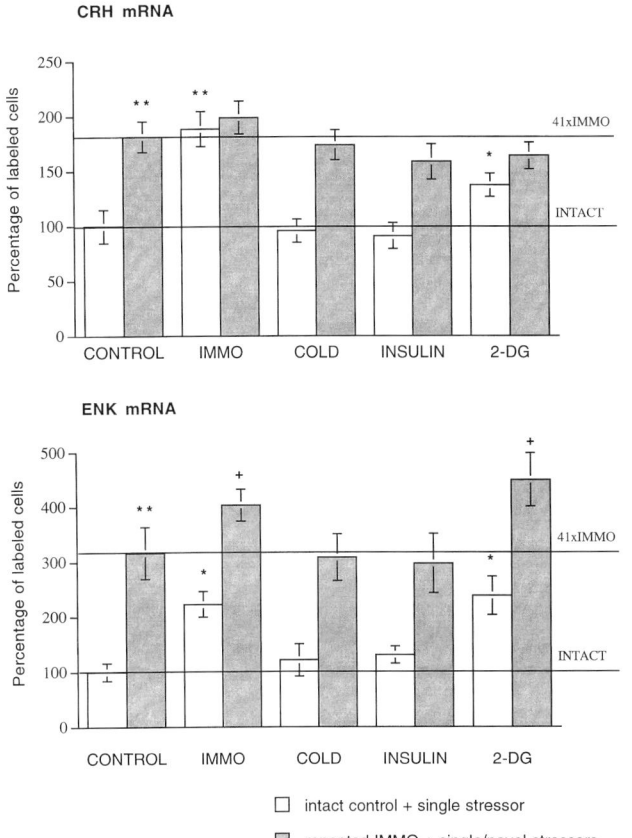

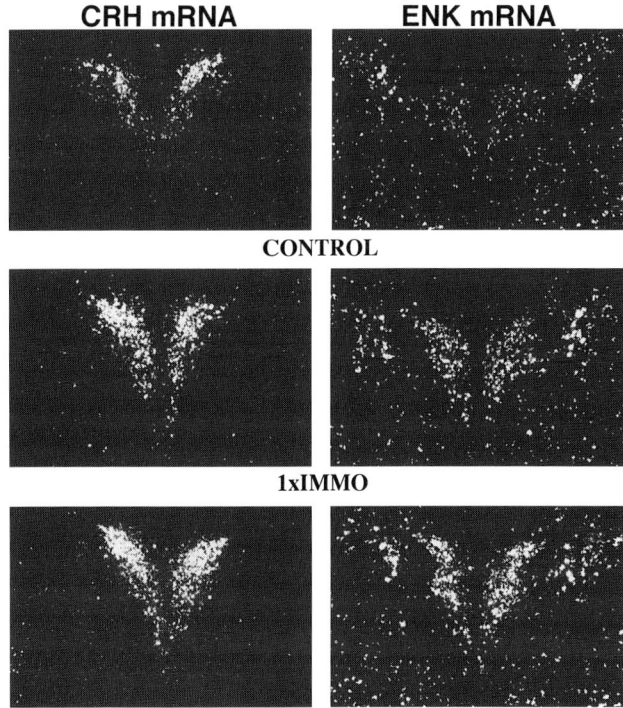

Fig. 3 Representative dark-field photographs of **in situ** hybridization for CRH mRNA and ENK mRNA in the PVN region of unstressed, 1 × immobilized (1 × IMMO) and 41 × immobilized rats exposed to another single immobilization (41 × IMMO + 1 × IMMO). Magnification – 25×.

Fig. 2 Effect of a single immobilization (IMMO) and various single novel stressors such as: cold for 5 h (COLD), insulin induced hypoglycemia (INSULIN) and 2-deoxy-D-glucose administration (2-DG), on corticotropin-releasing hormone (CRH)mRNA and enkephalin (ENK)mRNA levels in the hypothalamic PVN of control and long-term repeatedly immobilized rats (41 × IMMO). Data are expressed as percentage of number of labeled cells in intact controls (mean ± S.E.M.); compared to the number of labeled cells in unstressed controls; *p < 0.05, **p < 0.01; compared to long-term repeatedly immobilized rats decapitated 24 h after the 41st IMMO; #p < 0.05. Statistical differences were evaluated by ANOVA, followed by Fischer's Protected LSD Test.

Discussion

Our finding concerning the desensitization of c-fos expression in parvocellular neurons of the PVN of 41 times daily immobilized rats in response to a sustained stressor is in accordance with previously published data showing similar effects in rats exposed to other chronic stress models like social stress (Albeck et al., 1997) or repeated restraint (Bonaz and Rivest, 1998). Also immobilization for 2h daily for 6 days was shown to lead to desensitization of expression of c-fos and other immediate early genes like fos B, jun B, NGFI B and NGFI A in the PVN (Umemoto et al., 1997). Single stressors activated c-fos expression in different parts of the PVN (IMMO and cold in the parvocellular part and 2-DG in the magnocellular part). The diverse localization of c-fos expression by 2-DG vs. IMMO or cold

stress can be explained by different time course of c-fos activation in first parvo- and later in magnocellular neurons of the PVN by different stressors, as has been shown during exposure of rats to restraint, where the c-fos activation culminated in parvocellular PVN at the end of the 90 min stress exposure and was shifted to the magnocellular part of this nucleus after 6 hours (Bonaz and Rivest, 1998). Another possibility is that different stressors may predominantly activate different brain regions, including parvocellular versus magnocellular neurons of the PVN, as reviewed by Senba and Ueyama (1997). Our data show that long-term IMMO, despite desensitization to the same stressor, did not prevent the activation of c-fos in the PVN by a novel stressor (2-DG).

Even after 41 daily exposures of rats to 2 hours IMMO, CRH mRNA levels were still highly elevated as it was shown by shorter repeated exposure to the same stressor (Mamalaki et al., 1992; Bartanusz et al., 1993; Makino et al., 1995) or to hypertonic saline stress (Kiss et al., 1993; Ma and Aguilera, 1999)). In the same region, ENK mRNA levels were relatively low in control animals. Exposure of rats to a single IMMO as well as to a single dose of 2-DG significantly elevated ENK mRNA levels in the PVN. Intraperitoneal injection of hypertonic saline but not acute exposure to cold or swim stress has been shown to elevate ENK mRNA levels although each of these stressors induced CRH mRNA elevation in the PVN (Lightman and Scott, 1987;

Harbuz and Lightman, 1989; Harbuz et al., 1994). In our experiment, repeated IMMO produced a huge increase in expression of the ENK gene. It looks like the ENK gene is relatively silent under control conditions and is activated by some acute stressors, but it is highly expressed especially in long-term repeatedly IMMO animals. ENK mRNA levels in PVN parvocellular neurons of repeatedly IMMO rats were further increased following exposure to a novel stressor, e.g. after 2-DG administration. Changes in enkephalin mRNA in the PVN have been studied in acute or short-term stress models, but have not been examined in chronic or long-term stress models. The changes found in ENK mRNA levels suggest that ENK may play a significant role in repeatedly stressed rats.

Although all single stress models used have been shown to activate the HPA axis (Pacak et al., 1995; Weidenfeld et al., 1994), exposure of control or chronically IMMO rats to 5 hours of cold, or to insulin-induced hypoglycemia did not affect CRH- or ENK gene expression in the PVN. Our results concerning cold exposure are consistent with observations of Harbuz and Lightman (1989), who did not find induction of PVN CRH mRNA in rats exposed to cold, but did to other stressors (restraint, swimming, hypertonic saline). On the other hand, cold stress-induced CRH mRNA expression occurs in immature rats (Hatalski et al., 1998), which suggests age-related sensitivity to this stressor. Single exposure to IMMO or 2-DG highly elevated gene expression of both neuropeptides in paraventricular PVN neurons in naive rats, but only ENK mRNA levels were further elevated by these stressors in chronically IMMO animals. Thus ENK, versus CRH gene expression in the PVN of repeatedly IMMO rats appeared to be more sensitive to sustained stress as well as to a novel stressor. This may be caused by glucocorticoid feedback mechanisms regulating CRH gene expression. Repeated IMMO results in downregulation of anterior pituitary CRH receptors and reduction of corticotrope responsiveness to CRH (Hauger et al., 1988). It has been shown recently that vasopressin (synthesized in the parvocellular PVN), rather than CRH, may play a more important role in maintaining HPA axis activity during exposure to repeated stress (Ma and Lightman 1998, 1997). Our results demonstrate that enkephalin, synthesized in parvocellular neurons of the PVN (Ceccateli et al., 1989), may be another important neuropeptide predominantly involved in stress responses during chronic stress.

In Conclusion, a single two hour period of immobilization activates c-fos gene expression in the PVN, but long-term repeated exposure to the same stressor desensitize it. In contrast, long-term repeated IMMO does not prevent c-fos expression induced by a novel stressor, 2-deoxy-D-glucose administration.

Long-term repeated IMMO markedly increases both CRH and ENK mRNA levels in the parvocellular PVN. None of these novel stressors elicited further increases in CRH mRNA levels in PVN of rats. However, repeated IMMO does not desensitize the parvocellular PVN neurons in terms of ENK gene expression

response to a novel stressor, e.g. the administration of 2-DG, which further elevated ENK mRNA levels in the PVN. Our results suggest that enkephalin synthesized in parvocellular neurons of the PVN may play an important role in central responses to stress, especially in repeatedly stressed animals.

Acknowledgments

The authors thank Drs Eva Mezey and Zsuzsana Toth for providing the probes and for help with the in situ hybridization method.

This work was partly supported by Slovak Grant Agency for Science (2-610999) and FIRCA Grant (1R03TW00984-01).

References

Aguilera, G., (1994). Regulation of pituitary ACTH secretion during chronic stress. Frontiers in Neuroendocrinology 15, 321–350.

Albeck, D.S., McKittrick, Ch.R., Blanchard, C.D., Blanchard, R.J., Nikulina, J., McEwen, B.S., and Sakai. R.R. (1997). Chronic social stress alters levels of corticotropin-releasing factor and arginine vasopressin mRNA in rat brain. Journal of Neuroscience 17, 4895–4903.

Bartanusz, V., Jezova, D., Bertini, L.T., Tilders, F.H.J., Aubry, J.M. and Kiss, J.Z. (1993). Stress-induced increase in vasopressin and corticotropin-releasing factor expression in hypophysiotrophic paraventricular neurons. Endocrinology 132, 895–902.

Bhatnagar, S., and Dallman, M. (1998). Neuroanatomical basis for facilitation of hypothalamic-pituitary-adrenal responses to novel stressor after chronic stress. Neuroscience 84, 1025–1039.

Bhatnagar, S., Mitchell, J.B., Betito, K., Boksa, P., and Meaney, M.J. (1995). Effects of chronic intermittent cold stress on pituitary adrenocortical and sympathetic adrenomedullary functioning. Physiology and Behavior 57, 633–639.

Bonaz, B., and Rivest, S. (1998). Effect of a chronic stress on CRF neuronal activity and expression of its type 1 receptor in the rat brain. American Journal of Physiology 275, 1438–1449.

Boutilier, A.L., Monnier, D., Lorang, D., Lundblad, J.R., Roberts, J.L., and Loeffler, J.P. (1995). Corticotropin-releasing hormone stimulates proopiomenlanocortin transcription by c-fos-dependent and -independent pathways: characterization of an AP1 site in exon 1, Molecular Endocrinology 9, 745–755.

Bradley, D.J. Towle, C.H., and Young, W.S. (1992). Spatial and temporal expresssion of α- and β-thyroid hormone receptor mRNAs, including the β2-subtype, in the developing mammalian nervous system. Journal of Neuroscience 12, 2288–2302.

Bruhn, T.O., Sutton, R.E., Rivier, C.L., and Vale, W.W., (1984). Corticotropin-releasing factor regulates proopiomelanocortin messenger ribonucleic acid levels in vivo. Neuroendocrinology 39, 170–175.

Bullitt, E. (1990) Expression of c-fos like protein as a marker for neuronal activity following noxious stimulation in the rat. Journal of Comparative Neurology 296, 517–530.

Ceccatelli, S., Eriksson, M., and Hokfelt, T. (1989). Distribution and coexistence of corticotropin-releasing factor-, neurotensin-, enkephalin-, cholecystokinin-, galanin-, and vasoactive intestinal polypeptide/peptide histidine isoleucine-like peptides in the parvocellular part of the paraventricular nucleus. Neuroendocrinology 49, 309–323.

Ceccatelli, S., Villar, M.J., Goldstein, M., and Hokfelt, T. (1989). Expression of c-fos immunoreactivity in transmitter-characterized neurons after stress. Proceedings of the National Academy of Sciences 86, 9565–9573.

Dallman, M.F., Akana, S.F., Jacobson, L., Levin, N., Cascio, C.S., and Shinsako, J. (1987). Characterization of corticosterone feedback regulation of ACTH secretion. Annals of the New York Academy of Sciences 512, 402–414.

De Goeij, D.C.E., Binnekade, R., and Tilders, F.J.H. (1992). Chronic stress enhances vasopressin but not corticotropin-releasing factor secretion during hypoglycemia. American Journal of Physiology 263, 394–399.

Harbuz M.S., and Lightman S.L. (1989). Responses of hypothalamic and pituitary mRNA to physical and psychological stress in the rat. Journal of Endocrinology 122, 705–711.

Harbuz M.S., Jessop, D.S., Lightman S.L., and Chowdrey, H.S. (1994). The effects of hypertonic saline stress on corticotrophin-releasing factor, arginine vasopressin, and proenkephalin A mRNAs in the CFY, Sprague-Dawley and Wistar strains of rat. *Brain Research* **667**, 6–12.

Hashimoto, K., Suemaru, S., Takao, T., Sugawara, M., Makino, S., and Ota, Z. (1988). Corticotropin-releasing hormone and pituitary adrenocortical responses in chronically stressed rats. *Regulatory Peptides* **23**, 117–126.

Hatalski, C.G., Guirguis, C., and Baram, T.Z. (1998). Corticotropin releasing factor mRNA expression in the hypothalamic paraventricular nucleus and the central nucleus of the amygdala is modulated by repeated acute stress in the immature rat. *Journal of Neuroendocrinology* **10**, 663–669.

Hauger, R.L., Millan, M.A. Lorang, M., Harwood, J.P., and Aguilera, G. (1988). Corticotropin releasing factor receptors and pituitary-adrenal responses during immobilization stress. *Endocrinology* **123**, 396–405.

Hauger, R.L., Lorang, M., Irwin, M., Aguilera, G. and (1990). CRF receptor regulation and sensitization of ACTH response to acute ether stress during chronic intermittent immobilization stress. *Brain Research* **532**, 34–40.

Imaki, T., Shibasaki, T., Hotta, M., and Demura, H. (1992). Early induction of c-fos precedes increased expression of corticotropin-releasing factor messenger ribonucleic acid in the paravenrtricular nucleus after immobilization stress. *Endocrinology* **131**, 240–246.

Kant, G.J., Lue, J.R., Anderson, S.M., and Mougey, E.H. (1987). Effects of chronic stress on plasma corticosterone, ACTH and prolactin. *Physiology and Behavior* **40**, 775–779.

Kiss, A. and Aguilera, G. (1993). Regulation of the hypothalamic pituitary adrenal axis during chronic stress: responses to repeated intraperitoneal hypertonic saline injection. *Brain Research* **630**, 262–270.

Kononen, J., Honkaniemi, J., Alho, H., Koistinaho, J., Iadarola, M., and Pelto-Huiko, M. (1992). Fos-like immunoreactivity in the rat hypothalamic pituitary axis after immobilization stress. *Endocrinology* **130**, 3041–3047.

Kvetňanský, R., and Mikulaj, L. (1970). Adrenal and urinary catecholamines in rats during adaptation to repeated immobilization stress. *Endocrinology* **87**, 738–743.

Lightman, S.L., and Young, W.S. (1987). Changes in hypothalamic preproenkephalin A mRNA following stress and opiate withdrawal. *Nature* **328**, 643–645.

Lightman, S.L., and Harbuz, M.S. (1993). Expression of corticotropin-releasing factor mRNA in response to stress. *Ciba Foundation Symposia* **172**, 173–198.

Ma., X.M., and Aguilera, G. (1999). Transcriptional responses of the vasopressin and corticotropin-releasing hormone genes to acute and repeated intraperitoneal hypertonic saline injection in rats. *Molecular Brain Research* **68**, 129–140.

Ma., X.M., and Lightman, S.L. (1997). Emergence of an isolated arginine vasopressin (AVP) response to stress after repeated restraint: a study of both AVP and corticotropin-releasing hormone messenger ribonucleic acid (RNA) and heteronuclear RNA. *Endocrinology* **138**, 4351–4357.

Ma., X.M., and Lightman, S.L. (1998). The arginine vasopressin and corticotrophin-releasing hormone gene transcription responses to varied frequencies of repeated stress in rats. *Journal of Physiology* **510**, 605–614.

Makino, S., Smith, M.A., and Gold, P.W. (1995). Increased expression of corticotrophin-releasing hormone and vasopressin messenger ribonucleic acid (mRNA) in the hypothalamic paraventricular nucleus during repeated stress: association with reduction in glucocorticoid receptor mRNA levels. *Endocrinology* **136**, 3299–3309.

Mamalaki, E., Kvetňanský, R. Brady, L.S., Gold, P.W., and Harkenham, M. (1992). Repeated immobilization stress alters tyrosine hydroxylase, corticotrophin-releasing hormone and corticosteroid receptor messenger ribonucleic acid levels in rat brain. *Journal of Neuroendocrinology* **4**, 689–699.

Mansi J.A., Rivest, S., and Drolet, G. (1998). Effect of immobilization stress on transcriptional activity of inducible immediate-early genes, corticotropin-releasing factor, its type 1 receptor, and enkephalin in the hypothalamus of Borderline hypertensive rats. *Journal of Neurochemistry* **70**, 1556–1566.

Pacak, K., Palkovits, M., Kvetňanský, R., Yadid, G., Kopin, I.J., and Goldstein, D.S. (1995). Effects of various stressors on in vivo norepinephrine release in the hypothalamic paraventricular nucleus and on the pituitary-adrenocortical axis. *Annals of the New York Academy of Sciences* **771**, 115–130.

Rivier, C., and Vale, W. (1987). Diminished responsiveness of the hypothalamic-pituitary-adrenal axis of the rat during exposure to prolonged stress: A pituitary mediated mechanism. *Endocrinology* **121**, 1320–1328.

Senba, E., and Ueyama, T. (1997). Stress-induced expression of immediate early genes in the brain and peripheral organs of the rat. *Neuroscience Research* **29**, 183–207.

Smith, M.A., Banerjee, S., Gold, P.W., and Glowa, J. (1992). Induction of c-fos mRNA in rat brain by conditioned and unconditioned stressors. *Brain Research* **578**, 135–141.

Umemoto, S., Kawai, Y., Ueyama, T., and Senba, E. (1997). Chronic glucocorticoid administration as well as repeated stress affects the subsequent acute immobilization stress-induced expression of immediate early genes but not that of NGFI-A. *Neuroscience* **80**, 763–773.

Vale, W. Speiss, J., Rivier, C., and Rivier, J. (1981). Characterization of a 41-residue ovime hypothalamic peptide that stimulates secretion of corticotropin and β-endorphin. *Science* **213**, 1394–1397.

Weidenfeld, J., Corcos, A.P., Wohlman, A., and Feldman, S. (1994). Characterization of the 2 – Deoxyglucose effect on the adrenocortical axis. *Endocrinology* **134**, 1924–1931.

Weidenfeld, J., Siegel, R.A. Corcos, A.P., Heled, V., Conforti, N., and Chowers, I. (1984). ACTH and corticosterone secretion following 2-deoxyglucose administration in intact and in hypothalamic deafferentated male rats. *Brain Research* **305**, 109–113.

11. Gender Differences in Rats in Sympathoadrenal Activity at Rest and in Response to Stress

M. M. Weinstock, D. Schorer-Apelbaum, M. Razin, D. Men and R. McCarty

Introduction

Gender differences in the behavioral response of rats to stressful situations are well documented and include faster acquisition of active avoidance behavior in female (Beatty and Beatty, 1970; Fride et al., 1988) and a greater disruption of active and passive avoidance and taste aversion in male rats (Van Oyen et al., 1979; Steenbergen et al., 1989; Van Haaren and Van der Poll, 1984). Resting levels of plasma corticosterone (COR) are higher in female humans (Deuschle et al., 1996; Deuster et al., 1998) and rats (Kitay, 1961; Atkinson and Waddeil, 1997) than in males. Activation of the hypothalamic-pituitary-adrenal axis in response to restraint, novelty and footshock, as shown by elevations of plasma ACTH or COR, is greater in females than in males (Livezey et al., 1985; Weinstock et al., 1992). Footshock also produces a larger increase in catecholamine metabolites, but a smaller reduction in the amines in female than in male rat brains (Heinsbroek et al., 1990). This suggests that the greater utilization of amine neurotransmitters induced by stress in the female is compensated by a higher rate of synthesis. However, less is known about the influence of gender on the sympathoadrenal system at rest or after stress. In normotensive human subjects, no gender differences were found in plasma norepinephrine (NE) and epinephrine (EPI) at rest, but the increments in response to treadmill exercise at identical workloads were significantly greater in women (Lehmann et al., 1986). In rats, one study reported higher resting levels of NE and EPI in females than in males (Zukowaka-Grojec et al., 1991), while another found no sex differences in amine levels or COR. However, greater increases in EPI and COR, but not in NE, were obtained in females than in males after restraint stress (Livesey et al., 1985). Both studies were performed in males and females of similar weight but differing in age.

The levels of NE in plasma are the net effect of transmitter release and its removal by neuronal uptake and clearance (Eisenhoffer et al., 1989). The influence of neuronal uptake can be assessed by measurement of the change in plasma levels of dihydroxyphenylglycol (DHPG), which is derived from capture of the released NE into the neuron and from some leakage from storage vesicles. The degree of activation of tyrosine hydroxylase in nerve terminals and the adrenal medulla induced by the stressor may be reflected by increases in plasma dihydroxyphenylalanine (DOPA), the precursor of dopamine, NE and EPI (Goldstein et al., 1987). Dihydroxyphenyl acetic acid (DOPAC) in plasma may depend on the synthesis of dopamine in nerve terminals and may also give some indirect indication of tyrosine hydroxylase activation (Kvetňanský et al., 1992).

The aim of the present study was to compare blood pressure, plasma catechols and plasma COR in male and female rats, under basal conditions and following exposure to novelty and footshock. The rats were of the same age and were housed and reared under identical conditions from weaning in an attempt to reduce potential sources of variation.

Materials and Methods

Animals

Experiments were performed in 22 Sprague Dawley rats, aged 4.5–5 months, of which 6 pairs of males and females were littermates, while the remainder were born to different mothers, but were of the same age. Twenty-four hours before the experiment, rats were anesthetized with tribromoethanol (1 ml/kg i.p.) (Herskowitz et al., 1982) and implanted with indwelling tail artery catheters for blood pressure measurement and blood sampling as previously described (Chieuh and Kopin, 1978). The catheters were filled with heparinized saline (300 I.U./ml), exteriorized at the back of the neck and protected by a stainless steel extension spring wire that was secured by an adhesive tape collar which allowed complete freedom of movement within the cage. Following surgery, rats were housed individually with food and water ad libitum.

Experimental Protocol

On the day of the experiment, a baseline blood sample (1.5 ml) was taken from each animal between 0815–0930 hrs. The sample was centrifuged for 2 min at 12,000 rpm and the plasma removed and stored at –70°C until assayed. After this, and each subsequent blood sample, an equal volume of heparinized saline was added to the red cells which were slowly infused back into the catheter to maintain blood volume. The rats were transferred to the experimental room and allowed to rest for 15 min before their mean arterial blood pressure (MAP) and heart rate (HR) were measured. They were then placed individually into the shock chamber and a further blood sample taken after 3 min. Three footshocks 1.5 mA, 4 sec duration were delivered at 25 sec intervals and another blood sample was taken immediately after the last shock. Two further samples were collected 5 and 15 min later. Each animal was then returned to its home cage and another blood sample collected 45 min after the footshock. Ninety min after footshock a final blood sample of 0.5 ml was taken for measurement of COR only.

Assays

Catechols, NE, EPI, DOPA, DOPAC and DHPG were extracted from 0.2–0.5 ml plasma by batch alumina extraction and measured by high performance liquid chromatography with electrochemical detection (Holmes *et al.*, 1994). COR was measured by radio-immunoassay using a rabbit antiserum raised against corticosterone-21-thyroglobulin (Biomakor, Kiryat Weizmann, Rehovot, Israel) and [1,2,6,7]-^{3}H-N-corti-costerone, (0.016 Ci/mmol) (New England Nuclear, Boston, MA, USA). The sensitivity of the COR assay was 5 ng/ml.

Statistical Analyses

The results are expressed as means ± SEM. Resting values of MAP, HR, and plasma levels of COR, catecholamines and metabolites were analyzed by ANOVA and Levine's test for equality of variances. Since it was found that the variances of the concentrations of COR and catecholamines were higher in females than in males, gender differences between mean values were analyzed by the non parametric Mann-Whitney test. The criterion of significance was set at p < 0.05.

Results

Male rats were about 150 g heavier than females at the same age, and had significantly higher MAP at rest. Heart rates tended to be lower in males than in females, but this did not reach statistical significance (Table 1). Basal levels of COR, NE and DOPAC were significantly higher (Figures 1a and 2) and the ratio of DHPG/NE lower, in females than in males (Figure 3), but concentrations of EPI, DHPG and DOPA did not differ in the two sexes (Figure 1a).

Table 1 Resting values for bodyweight, MAP, and HR in male and female rats.

Parameter	Males (11)	Females (11)
Body weight (g)	404 ± 9	252 ± 3*
MAP (mmHg)	112 ± 4	94 ± 3*
HR (beats/min)	419 ± 20	444 ± 12

* Significantly different from males, p < 0.05.

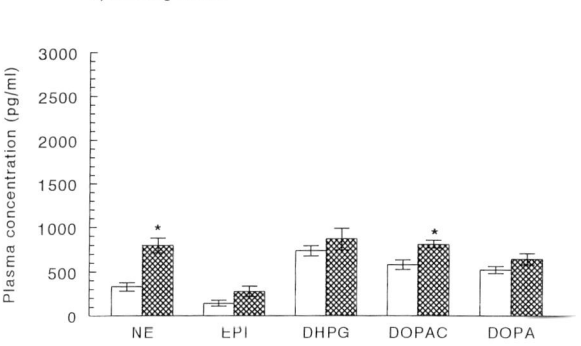

a) Resting levels

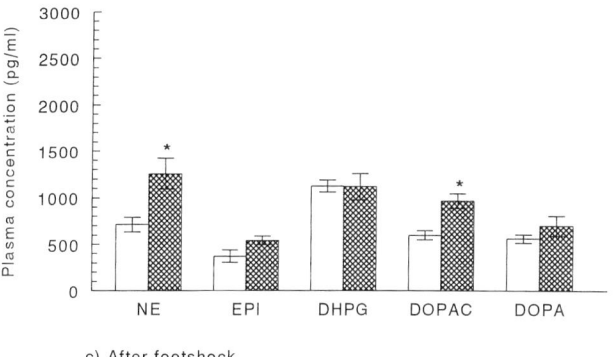

b) After transfer to shockbox

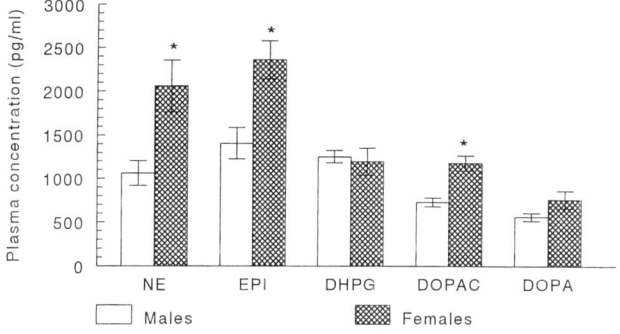

c) After footshock

☐ Males   ▨ Females

Fig. 1 Plasma levels of catecholamines and their metabolites in male and female rats at rest and after exposure to novelty and footshock stress.
* Significantly different from males, P < 0.05.

Exposure to the novel chamber in which the rats subsequently received footshock resulted in significant elevations within 3 min in COR, NE, EPI and DHPG but not in DOPA or DOPAC, in rats of each gender (Figures 1b and 2). Levels of COR, NE and DOPAC were still higher in females than in males after this novelty stress. The increase in DHPG was significantly greater in males than in females, while that of COR was greater in females (Table 2).

Amine levels and their metabolites were still elevated above baseline when the footshock was

Table 2 Change in COR, catecholamines and their metabolites in response to novelty and footshock.

Parameter	Transfer to cage		1–2 or 5 min after footshock	
	Males	Females	Males	Females
COR (ng/ml)[b]	$113 \pm 30^{\S}$	$277 \pm 50^{\S *}$	65.8 ± 34.6	64.4 ± 35.9
NE (pg/ml)[a]	$380 \pm 66^{\S}$	$572 \pm 151^{\S}$	$349 \pm 90^{\P}$	$687 \pm 209^{\P}$
EPI (pg/ml)[a]	$228 \pm 57^{\S}$	$289 \pm 63^{\S}$	$1031 \pm 163^{\P}$	$1791 \pm 238^{\P *}$
DHPG (pg/ml)[a]	$387 \pm 57^{\S}$	$189 \pm 62^{\S *}$	$125 \pm 28^{\P}$	136 ± 86
DOPA (pg/ml)[a]	38.9 ± 28.9	$87.1 \pm 27.2^{\S}$	54.2 ± 69.2	$62.3 \pm 24.8^{\P}$
DOPAC (pg/ml)[b]	18.8 ± 18.4	158 ± 78	$166 \pm 16^{\P}$	$316 \pm 57^{\P *}$
DHPG/NE[a]	$1.12 \pm 0.09^{\S}$	$0.52 \pm 0.18^{\S *}$	$0.34 \pm 0.08^{\P}$	0.20 ± 0.10

[a] 1–2 min after footshock; [b] 5 min after footshock; § significant change from resting levels, * significantly different from males, $p < 0.05$; ¶ significant change from value obtained after transfer, $p < 0.05$.

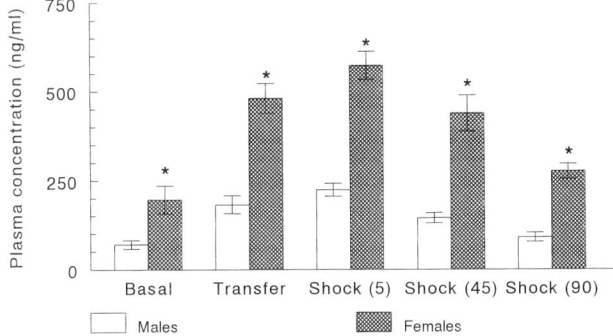

Fig. 2 Plasma levels of corticosterone in male and female rats at rest and after exposure to novelty and footshock stress.
( ) Time in min after footshock.
* Significantly different from males, $P < 0.05$.

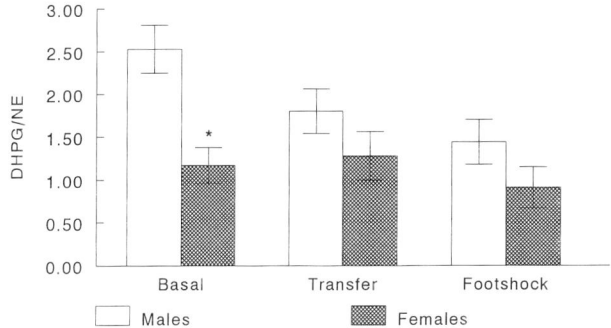

Fig. 3 DHPG/NE ratios in male and female rats before and after exposure to novelty and footshock stress.
* Significantly different from males, $P < 0.05$.

administered. Within 2 min of its completion, NE increased further by a similar amount in both sexes, while the increment in plasma EPI was at least 5-fold greater than after exposure to the novel environment and EPI concentrations became significantly higher in females than in males (Figure 1c and Table 2). The increase in DHPG was of similar magnitude in females and smaller in males. The increase in DOPAC levels was larger in both sexes than after cage transfer (Figure 1c) and significantly higher in females than in males (Table 2). Amine levels were significantly lower 5 min after stress than at their peak but took more than 15 min to return to pre-stress values. The increase in COR was slower than that of the catecholamines and reached peak levels in both sexes at 5 min post shock, and declined to basal levels in males but not in females by 90 min (Figure 2). The DHPG/NE ratio, which was significantly higher in male than in female rats at rest, decreased sequentially in males in response to novelty and footshock stress, since the increase in NE levels was greater than that of DHPG. The DHPG/NE ratio was not significantly influenced by the two stressful procedures (Figure 3).

Discussion

The present study found that male rats had significantly higher resting blood pressures and a tendency to lower heart rates than females of the same age, in spite of the fact that their plasma NE concentrations were significantly lower. This suggests that resistance vessels in males may be more sensitive than those of females to its vasoconstrictor effect, and accords with the observation of a greater contractile response to NE in isolated tail arteries from male rats after correcting for gender differences in arterial mass (Li *et al.*, 1997). The greater sensitivity to NE in males is dependent on the presence of testosterone, since it can be eliminated by gonadectomy, while estradiol can decrease contractility of vascular smooth muscle by inhibiting Ca^{++} influx (McNeill *et al.*, 1996; Kitazawa *et al.*, 1997). The finding of higher resting levels of NE in females is in agreement with that of Zukowska-Grojek *et al.*, (1991), and is therefore not due to differences in body weight, since these were similar in the two sexes in their study. In accordance with previous observations, female rats also had significantly higher resting levels of COR at rest and after stress.

Although plasma NE concentrations in females were almost double those in males at rest, DHPG did not differ significantly. This suggests that the leakage of NE from vesicles and its deamination by monoamine oxidase is lower in females, or that males may have a more efficient neuronal uptake system, or both. Monoamine oxidase activity may be higher in males under resting conditions, as it is significantly increased by administration of testosterone after gonadectomy (Luine *et al.*, 1975), but direct information is lacking about possible gender differences in the NE transporter.

Handling of rats and transfer to a novel environment in which they were to receive footshock produced a greater increase in COR in females, but similar increments in NE and EPI to those in males. However, the increment in DHPG was significantly greater in males, supporting the existence of a more efficient re-uptake mechanism. Footshock caused a much greater increase in EPI than NE from the level they had reached after cage transfer, testifying to greater adrenal activation by this stressor. Females were more sensitive to the effects of footshock stress as shown by the larger increment in their plasma EPI than that in males, and supports the observation in humans after stressful exercise (Lehmann *et al.*, 1986).

Measurement of the concentration of the catecholamines and their metabolites in plasma enabled us to detect possible gender differences in the synthesis and re-uptake of NE. The ratio of DHPG/NE was significantly greater in males at rest, and declined after each form of stress, because of the proportionately larger increase in NE than in DHPG. This suggests that the rate of reuptake of NE decreased in response to stress in males in order to provide adequate amounts of NE for receptor activation. In females, the DHPG/NE ratio did not change in response to stress. On the other hand, they had a significantly greater increase than males in DOPAC, an indirect indicator of tyrosine hydroxylase activity (Kvetňanský *et al.*, 1992). This suggests that the tissue levels and release of NE are maintained by *de novo* synthesis, and support the data of Heinsbroek *et al.*, (1990), who found that brain catecholamine levels remained unchanged after stress in females but declined in males.

References

Atkinson, H.C., and Waddell, B.J. (1997). Circadian variation in basal plasma corticosterone and adrenocorticotropin in the rat: sexual dimorphism and changes across the estrous cycle. *Endocrinology*. **138**, 3842–3848.

Beatty, W.W., and Beatty, P.A. (1970). Hormonal determinants of sex differences in avoidance behavior and reactivity to electric shock in the rat. *Journal of Comparative and Physiological Psychology* **73**, 446–455

Chieuh, C.C., and Kopin, I.J. (1978). Hyperresponsivity of spontaneously hypertensive rats to indirect measurement of blood pressure. *American Journal of Physiology* **234**, H690–H695.

Deuschle, M., Schweiger, U., Standhardt, H., Weber, B., and Heuser, I. (1996). Corticosteroid-binding globulin is not decreased in depressed patients. *Psychoneuroendocrinology* **21**, 645–649.

Deuster, P.A., Petrides, J.S., Singh, A., Lucci, E.B., Chrousos, G.P., and Gold, P.W. (1998). High intensity exercise promotes escape of

adrenocorticotropin and cortisol from suppression by dexamethasone: sexually dimorphic responses. *Journal of Clinical Endocrinology and Metabolism* **83**, 3332–3338.

Elsenhofer, G., Goldstein, D.S., and Kopin, I.J. (1989) Plasma dihydroxyphenglycol for estimation of noradrenaline neuronal reuptake in the sympathetic nervous system *in vivo*. *Clinical Science* **76**, 171–182.

Fride, E., Dan, Y., Feldon, J., Halevy, G., and Weinstock, M. (1986). Effects of prenatal stress on vulnerability to stress during development and at adulthood. *Physiology and Behavior* **37**, 681–687.

Goldstein, D.S., Udelsman, R., Eisenhofer, G., Stull, R., Keiser, H.R., and Kopin, I.J. (1987). Neuronal sources of plasma dihydroxyphenylalanine. *Journal of Clinical Endocrinology and Metabolism* **64**, 856–861.

Heinsbroek, R.P.W., van Haaren, F., Feenstra, M.G.P. van Galen, H., Boer, G., and van de Poll, N.E. (1990). Sex differences in the effects of inescapable footshock on central catecholaminergic and serotonergic activity *Pharmacology, Biochemistry and Behavior* **37**, 539–550.

Herskowitz, M., Eliash, S., and Cohen, S. (1982) The muscarinic cholinergic receptors in the posterior hypothalamus of hypertensive and normotensive rats. *European Journal of Pharmacology* **86**, 229–236.

Holmes, C., Eisenhofer, G., and Goldstein, D.S. (1994). Improved assay for plasma dihydroxyphenylacetic acid and other catechos using high performance liquid chromatography with electrochemical detection. *Journal of Chromatography* **653**, 131–138.

Kitay, J.I. (1961). Sex differences in adrenal cortical secretion in the rat. *Endocrinology* **68**, 818–824.

Kitazawa, T., Hamada, E., Kitazawa, K., and Gaznabi, A.K. (1997). Non-genomic mechanism of 17 beta-oestradiol-induced inhibition of contraction in mammalian vascular smooth muscle. *Journal of Physiology* **499** (pt 2), 497–511.

Kvetňanský, R., Armando, I., Weise, V.K., Holmes, C., Fukuhara, K., Deka-Starosta, A., Kopin, I.J., and Goldstein, D.S. (1992). Plasma DOPA responses during stress: dependance on sympathoneural activity and tyrosine hydroxylation. *Journal of Pharmacology and Experimental Therapeutics* **261**, 899–908.

Lehmann, M., Berg, A., and Keul, J. (1986). Sex-related differences in free plasma catecholamines in individuals of similar performance ability during graded ergometric exercise. *European Journal of Applied Physiology* **55**, 54–58.

Li, Z., Krause, D.N., Doolen, S., and Duckles, S.P. (1997). Ovariectomy eliminates sex differences in rat tail artery response to adrenergic nerve stimulation. *American Journal of Physiology* **272**, H1819–H1825.

Livezey, G.T., Miller, J.M., and Vogel, W.H. (1985). Plasma norepinephrine, epinephrine and corticosterone stress responses to restraint in individual male and female rats, and their correlations. *Neuroscience Letters* **62**, 51–58.

Luine, V.N., Khylchevskaya, R.I., and McEwen, B.S. (1975). Effect of gonadal steroids on activities of monoamine oxidase and choline acetylase in rat brain. *Brain Research* **86**, 293–306.

McNeill, A.M., Duckles, S.P., and Krause, D.N. (1996). Relaxant effects of 17 beta-estradiol in the rat tail artery are greater in females than males. *European Journal of Pharmacology* **308**, 305–309.

Steenbergen, H.L., Heinsbroak, R.P.W., van Haaren, F., and van der Poll, N.E. (1989). Sex-dependent effects of inescapable shock administration on behavior and subsequent escape performance in rats. *Physiology and Behavior* **45**, 781–787.

Van Haaren, F., and Van der Poll, N.E. (1984). The number of pre-shock trials affects sex differences in passive avoidance behavior. *Physiology and Behavior* **33**, 269–272.

Van Oyen, H.G., Van der Poll, N.E. and de Bruin, J.P.C. (1979). Sex, age and shock-intensity as factors in passive avoidance. *Physiology and Behavior* **23**, 915–918.

Weinstock, M., Matlina, E., Maor, G.I., Rosen, H., and McEwen, B.S. (1992). Prenatal stress selectively alters the reactivity of the hypothalamic-pituitary adrenal system in the female rat. *Brain Research* **595**, 195–200.

Zukowska-Grojec, Z., Shen, G.H., Capraro, P.A., and Vaz, C.A. (1991). Cardiovascular, neuropeptide Y, and adrenergic responses in stress are sexually differentiated. *Physiology and Behavior* **49**, 771–777.

12. Sex Differences in Diurnal Variations in Corticosterone Responses to Emotional Stressors

N.B. Igosheva and T.G. Anishchenko

Introduction

It is well established that significant sex-related differences exist in stress-induced activation of hypothalamo-pituitary-adrenal (HPA) system with females demonstrating higher HPA system reactivity to stress (Lesniewska et al., 1990; Jezova et al., 1996). In studies performed on males it has been found that corticosterone (CS) and adrenocorticotropin (ACTH) responses to stress exhibited marked diurnal variations (Abe et al., 1979). At the same time, data on the diurnal pattern of CS stress responses in females are scarce and controversial (Dunn et al., 1972; Guo et al., 1994).

Therefore, it was of interest to compare CS stress response patterns at different periods of the light-dark cycle in female and male rats.

Materials and Methods

Female and male albino rats weighing 180–250 g were used. Animals were housed 4 per cage under standard light (lights on 0700–1900 h) and temperature (23°C) conditions. Food and water were freely available. To avoid the hormonal surge elicited by uncontrolled environmental factors, the animal room was not entered during an 18–24 hour period before taking basal samples. The investigation conforms with Guide for the Care and Use of Laboratory Animals published by US National Institute of Health (NIH publication N 85–23, revised 1985).

Female and male rats were subjected to mild emotional stressors. The model used has been developed in our laboratory and is called "victim-spectator". According to the stress model, two rats were immobilized in the supine position and placed in front of an open cage. Two other rats of the same sex left in the cage and experienced emotional stress caused by the sight and vocalizations of their immobilized counterparts. To compare the CS response to stressors applied at different stages of the 24 hour light-dark cycle, experiments were carried out in the morning (09.00–10.00) and afternoon (15.00–16.00). Rats were decapitated and blood and adrenal samples were taken 10 min following the onset of stress and 20, 40 and 60 min after the termination of stress. Vaginal smears were taken from each female after decapitation to assess the estrous cycle. CS was determined by means of a modification of the fluorometric method (De Moor et al., 1960).

Statistical analysis of the data was performed by unpaired and paired Student t-tests. Significance was defined at the 0.05 level. All data are represented as the mean $\pm$ SEM.

Results

Morning and Afternoon CS Responses to the Stressful Situation Caused by Sight and Vocalizations of Immobilized Counterparts in Female and Male Rats

Basal plasma and adrenal CS levels did not differ between females and males. In the morning basal CS levels in females were 7.4 ± 1.4 μg/g and 5.5 ± 1.2 μg/ 100 ml in adrenal and plasma, respectively, whereas corresponding values for males were 8.6 ± 1.1 μg/g and 6.6 ± 0.9 μg/100 ml. Afternoon plasma and adrenal CS levels were twice as high as the morning values in both female and male rats. Thus, in unstressed females, plasma and adrenal CS concentrations were 14.5 ± 2 μg/100 ml and 12.8 ± 1.4 μg/g, respectively, whereas unstressed males exhibited plasma and adrenal CS levels of 16.9 ± 1.8 μg/100 ml and 12.9 ± 1.6 μg/g.

As shown in Figure 1A, in the morning, sight and vocalizations of immobilized counterparts caused in female "spectators" a significant average elevation in plasma and adrenal CS levels by 264% (P < 0.05) and 154% (P < 0.05) respectively. Adrenal levels of CS fell thereafter and reached baseline at 20 min after the cessation of the stressful situation while the plasma values approached basal levels at 40 min after the stress. In the afternoon stress-induced increments in plasma CS levels in female "spectators" were comparable to those observed in the morning (Figure 1B). Changes in adrenal CS concentrations were not statistically significant Figure 2A shows that morning male "spectators" were mostly unaffected by sight and vocalizations of immobilized counterparts. In the afternoon male "spectators" failed to show a rise in plasma and adrenal CS levels during the stressful situation. However, they exhibited a marked elevation in the plasma CS concentration by 60% (P < 0.05) at 20 min after the stress without a concomitant elevation in adrenal concentration of this steroid. In male

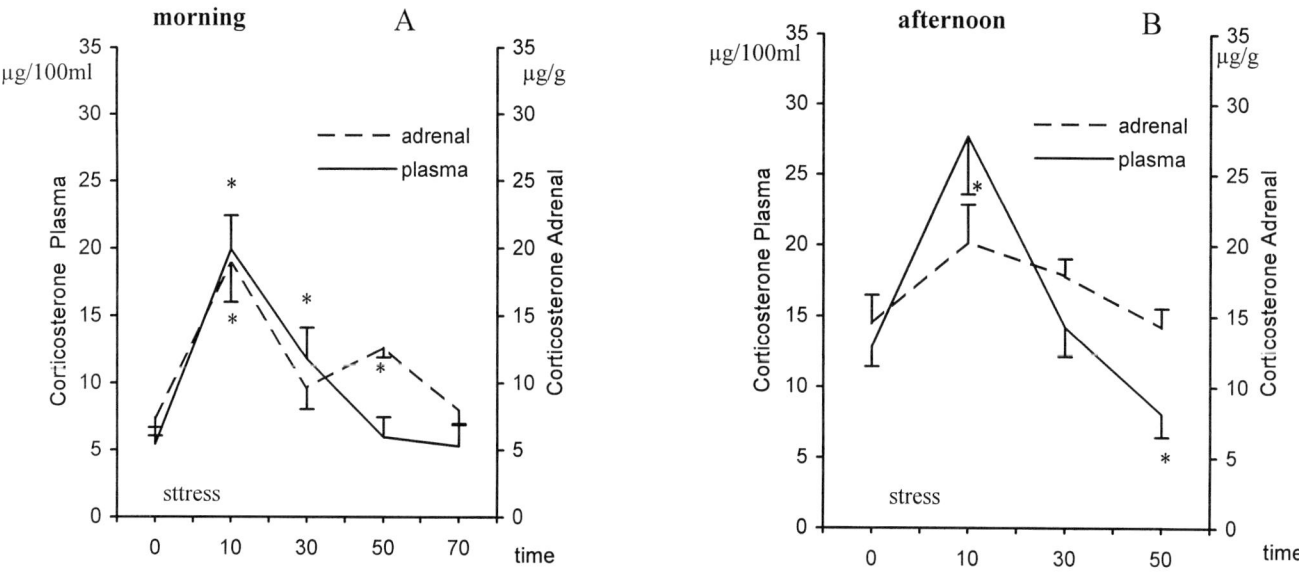

Fig. 1 Plasma and adrenal CS responses in female rats ("spectators") caused by the sight and vocalizations of immobilized counterparts in the morning (A) and afternoon (B).
Each point represents mean values for 12 rats; vertical lines indicate ± SEM. * – P < 0.05 vs. corresponding control.

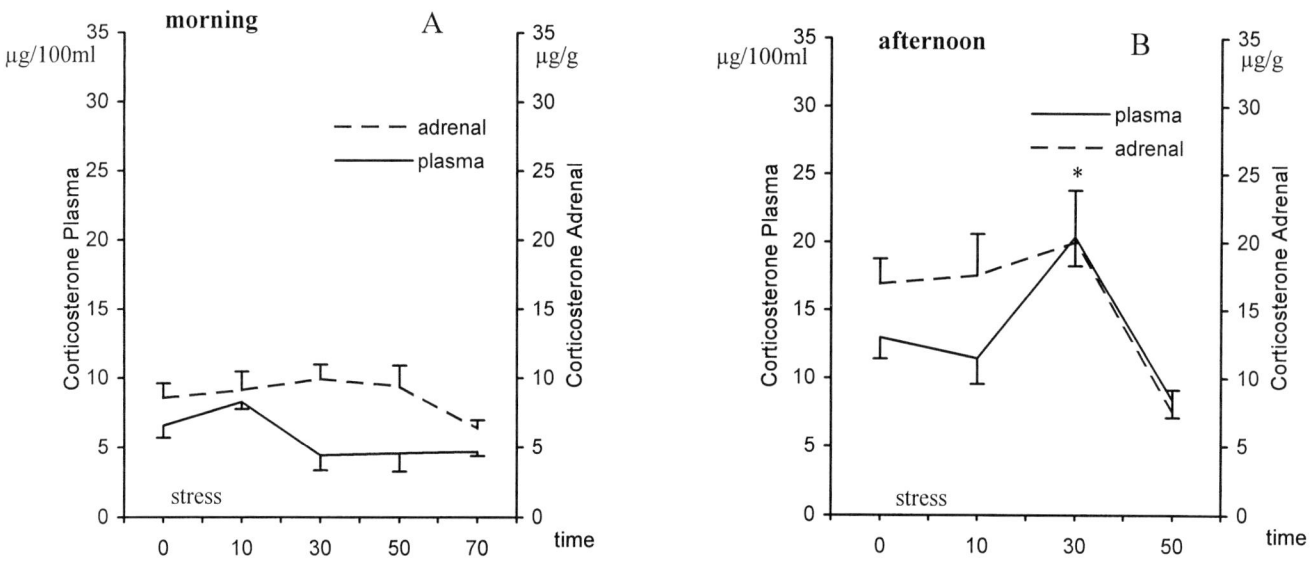

Fig. 2 Plasma and adrenal CS responses in male rats ("spectators") caused by the sight and vocalizations of immobilized counterparts in the morning (A) and afternoon (B).
Each point represents mean values for 12 rats; vertical lines indicate ± SEM. * – P < 0.05 vs. corresponding control.

"spectators" the delayed CS responses to the weak emotional stress was followed by a marked decline in CS levels (P < 0.05) at 40 minutes of the poststress period (Figure 2B).

Morning and Afternoon CS Responses to Immobilization Stress in Female and Male Rats

As illustrated in Figure 3AB immobilized females demonstrated significant increments in CS levels and the magnitude of this increment was almost the same in the morning and afternoon. Morning exposure to

immobilization caused in females a pronounced elevation in both plasma and adrenal CS levels by 294% (P < 0.05) and 202% (P < 0.05), respectively. CS values remained elevated within 40 min of the poststress period and returned to basal levels at 60 min after stress (Figure 3A). Immobilization stress applied in the afternoon increased plasma and adrenal CS concentrations by 254% (P < 0.05) and 134% (P < 0.05), respectively. Restoration processes were rapid and CS levels returned to basal values at 40 min after the stress (Figure 3B).

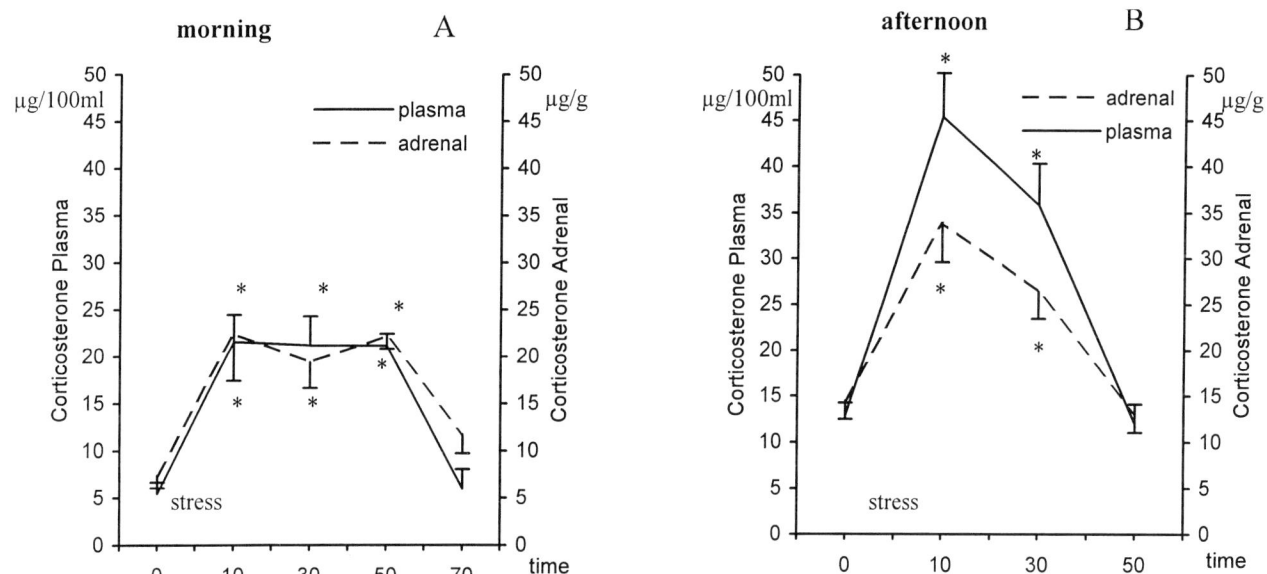

Fig. 3 Plasma and adrenal CS responses to immobilization stress (10 min) in female rats in the morning (A) and afternoon (B). Each point represents mean values for 12 rats; vertical lines indicate ± SEM. * – P < 0.05 vs. corresponding control.

In immobilized males CS responsiveness was lower than in females and unlike females differed significantly at the two times of day (Figure 4AB). In the morning immobilization stress increased plasma and adrenal levels by 111% (P < 0.05) and 142% (P < 0.05), respectively. At 40 min after the termination of immobilization CS levels remained elevated and reached basal levels at 60 min after stress (Figure 4A). Immobilization stress applied in the afternoon increased plasma and adrenal CS levels by 51%

(P < 0.05) and 63% (P < 0.05), respectively. However, the highest plasma CS rise by 117% (P < 0.05) occured in males at 20 min after stress. Post-stress activation of adrenal function was followed by a drastic decrease in both plasma and adrenal CS content.

Discussion

In our experiments, basal CS levels measured in the morning and afternoon did not differ significantly

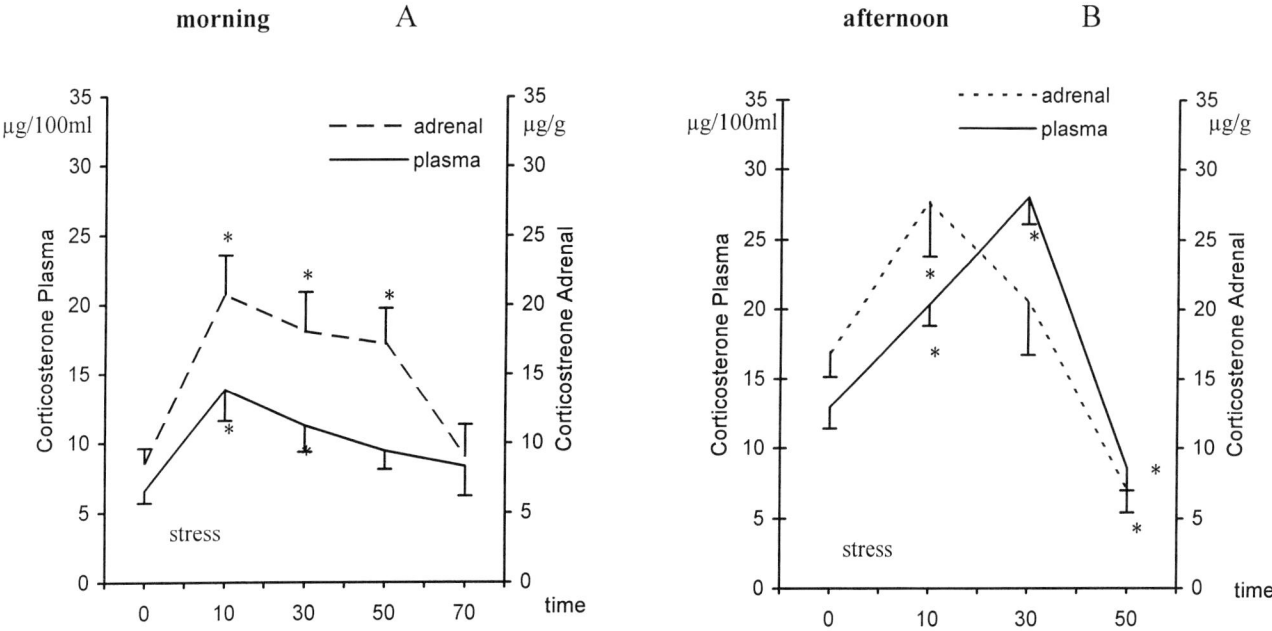

Fig. 4 Plasma and adrenal CS responses to immobilization stress (10 min) in male rats in the morning (A) and afternoon (B). Each point represents mean values for 12 rats; vertical lines indicate ± SEM. * – P < 0.05 vs. corresponding control.

between females and males. These findings are in agreement with those reported previously (Dunn *et al.*, 1971; Cizza *et al.*, 1996) but contradict the results of other works (Jezova *et al.*, 1996; Galea *et al.*, 1997). These discrepancies may be explained by underestimation of heightened responsiveness of females to weak environmental stressors and experimental manipulations (Kugler *et al.*, 1988).

Despite similar basal CS levels in males and females the amplitude of the CS stress response was almost twice as high in females than in males in both stressful situations. It seems that sex-related differences in CS stress-reactivity are likely to originate from widespread sex-specific distinctions in neuroendocrine systems involved in activation of the HPA system by stress (Wardlaw *et al.*, 1986; Siddiqui *et al.*, 1988; Jezova *et al.*, 1996).

Together with heightened CS responsiveness to stress, females exhibited rapid restoration of basal CS levels after cessation of the stress. In the afternoon restoration processes went faster since the higher absolute CS levels in plasma would provide more effective feedback inhibition of ACTH and CRF secretion at that period of the day compared with morning. Finally, in females the CS stress response pattern did not differ significantly in the morning and afternoon. These results are consistent with those reported in an earlier study (Zimmermann *et al.*, 1967). In males the amplitude of CS stress responses was lower compared with females and altered significantly by diurnal variations in basal CS levels. Sex differences in the diurnal pattern of the CS stress response may be explained at least in part by stability of the diurnal pattern of stress-induced secretion of pituitary hormones (Carter and Lightman, 1986) in females but not in males. It is worth noting that in male rats restoration of CS basal levels after stress was delayed. Both post-stress activation and post-stress suppression of steroidogenesis observed in males might have unfavorable consequences for male's health. This suggestion is based on findings that prolonged disturbances of steroid homeostasis are implicated in etiology of various diseases (Sapolsky *et al.*, 1985; Miller *et al.*, 1993).

To summarize, the patterns of CS stress responses differed significantly in female and male rats with females demonstrating more stable and adequate CS responses to weak emotional stressors.

Acknowledgements

This work was partly supported by a INTAS grant (N 96-0305) and the Royal Society of London.

References

Abe, K., Kroning, J., Greer, M., and Crichlow, V. (1979). Effect of destruction of the suprachiasmatic nuclei on the circadian rhythms in plasma corticosterone, body temperature, feeding and plasma thyrotropin. *Neuroendocrinology* **29**, 119–131.

Carter, D. and Lightman, S. (1986) Diurnal pattern of stress-evoked neurohypophyseal hormones: sexual dimorphism in rats. *Neuroscience Letters* **71**, 252–255.

Cizza, G., Brady, L., Esclapes, M., Blackman, M., Gold, P., and Chrousos, G. (1996) Age and gender influence basal and stress-modulated hypothalamic-pituitary-thyroidal function in Fischer 344/N rats. *Neuroendocrinology* **64**, 440–448.

De Moor, P., Steeno, O., Raskin, M., and Hendrix A. (1960) Fluorometric determination of free plasma 11-hydroxycorticosteroids in man. *Acta Endocrinology* **33**, 297–307.

Dunn, J., Scheving, L., and Millet P. (1972) Circadian variation in stress-evoked increases in plasma corticosterone. *American Journal of Physiology* **223**, 402–406.

Galea, L., McEwen, B., Tanapat, P, Deak, T., Spencer, R., and Dhabhar, F. (1997) Sex differences in dendritic atrophy of CA3 pyramidal neurons in response to chronic restraint. *Neuroscience* **81**, 689–697.

Jezova, D., Jurankova, E., Mosnarova, A., Kriska, M., and Skultetyova, I. (1996) Neuroendocrine response during stress with relation to gender differences. *Acta Neurobiologica Experimentalis (Warszava)* **56**, 779–785.

Kugler, J., Lange, K., and Kalveram, K. (1988) Influence of bleeding order on plasma corticosterone concentration in the mouse. *Experimental and Clinical Endocrinology* **91**, 241–243.

Miller, A., Spencer, R., McEwen, B., and Stein, M. (1993). Depression, adrenal steroids and the immune system. *Annals of Medicine* **25**, 481–487.

Lesniewska, B., Miskowiak, B., Nowak, M., and Malendowicz, L. (1990) Sex differences in adrenocortical structure and function. XXVII The effect of ether stress on ACTH and corticosterone in intact, gonadoectomized, and testosterone- or estrodiol – replaced rats. *Research in Experimental Medicine* (Berlin) **190**, 95–103.

Sapolsky, R., and Donnelly, T. (1985) Vulnerability of stress-induced tumor growth increases with age in rats: role of glucocorticoids. *Endocrinology* **117**, 662–666.

Wardlaw, S. (1986) Regulation of β-endorphin, corticotropin-like intermediate lobe peptide and α-melanotropin stimulating hormone in the hypothalamus by testosterone. *Endocrinology* **119**, 19–24.

13. Hemodynamic Response Variability to Behavioral Stress Correlates with Predisposition to Disease

M.M. Knuepfer, Q. Gan, J.R. Muller, G.M. Matuschak and A.J. Lechner

While it is commonly believed that stress is detrimental to one's health (Cacioppo *et al.*, 1998; Cohen *et al.*, 1991; Eliot, 1988; Friedman and Irwin, 1993; Pike *et al.*, 1997; Turner *et al.*, 1992), direct evidence for stress causing or exacerbating specific diseases and the mechanisms by which this occurs are not understood. For example, it has been suggested that cardiovascular disease results from behavioral stress in susceptible individuals (Galosy and Gaebelein, 1977; Natelson, 1983). Clinical studies revealed that a subset of individuals had a decrease in cardiac output in response to stress whereas the majority of subjects studied responded to stress with an increase in cardiac output (Brod, 1962; Herd, 1991; Turner *et al.*, 1991). Despite this, arterial pressure responses were similar, suggesting that individuals with a decrease in cardiac output have greater elevations in systemic vascular resistance and, therefore, were referred to as vascular responders. Some individuals, particularly vascular responders, are more prone to develop cardiac disease (Eliot, 1992; Lown, 1979) and hypertension (Kasprowicz *et al.*, 1990; Turner *et al.*, 1991). These data suggest that there is considerable individual variation in functional and pathological cardiovascular responses to stress in humans, yet few investigators address variability in their animal models.

Intrinsic susceptibility to cocaine-induced cardiotoxicity also varies widely in humans. There is a poor dose-response relationship between cocaine use and the incidences of myocardial ischemia and related ECG abnormalities, coronary vasoconstriction, cardiomyopathies, and morbidity (Lange *et al.*, 1989; Minor *et al.*, 1991; Smart and Anglin, 1988). Since the symptoms of cocaine cardiotoxicity vary greatly and are fatal in only a fraction of those using cocaine (Mittleman *et al.*, 1999), accurate estimates of individual risk are difficult to determine.

Stress also enhances the susceptibility of some individuals to opportunistic infections (Cacioppo *et al.*, 1998; Cohen *et al.*, 1991; Pike *et al.*, 1997). The autonomic nervous system regulates the progression of host defense responses to septicemia (Felten *et al.*, 1998; Friedman and Irwin, 1993). In addition, innate host defense against infection varies substantially in the population (Cacioppo *et al.*, 1998) yet animal studies are rarely designed to examine the causes of variability.

Animal Model of Hemodynamic Response Variability

Hemodynamic Responses to Cocaine

We have described a model of varying hemodynamic sensitivity to stress and cocaine. Using rats instrumented with pulsed Doppler flow probes on the ascending aorta (Branch and Knuepfer, 1994a), we noted changes in cardiac output (CO) to 5 mg/kg cocaine that were highly reproducible within animals but variable in the outbred population (Branch and Knuepfer, 1994a,b; Knuepfer and Mueller, 1999). Across this continuum of responses in CO, we distinguished two principal groups: *vascular responders* in which CO decreased (at least 8%) and *mixed responders* which showed little decrease or even an increase in CO. Despite equivalent pressor responses, vascular responders had larger increases in systemic vascular resistance (SVR) after cocaine compared to mixed responders. The arbitrary separation of groups above and below a change of –8% coincides with a natural nadir in the distribution of experimental results. CO response variability was also noted at 1 mg/kg cocaine i.v. (injected over 45s) or with rapid administration of 0.5 mg/kg of cocaine. The responses did not appear to result from a direct effect of cocaine on the heart (Branch and Knuepfer, 1994b). Our data suggest that the initial pressor response depends on a burst of sympathetic activity that occurs only in vascular responders (Branch and Knuepfer, 1994a).

Nonspecificity of Hemodynamic Response Pattern to Cocaine

Several lines of evidence indicate that the hemodynamic responses to cocaine are mediated by the arousal and behavioral stress evoked by cocaine. First, cardiovascular responses to cocaine are reduced by anesthesia (Knuepfer and Branch, 1992). Second, the unique hemodynamic response pattern in individual rats was mimicked by amphetamine or ethanol administration and could be predicted by cocaine-induced responses (Branch and Knuepfer, 1994a; Mueller *et al.*, 1997). Third, CO responses to brief (1–2s) air jet stress or to a 15s tone preceding a 1s footshock elicits CO responses in rats that parallel the CO response pattern noted after cocaine administration (Knuepfer *et al.*, 1993b). As noted with cocaine, rats exhibit vascular or mixed responses to conditioned or unconditioned behavioral stress, despite showing uniform patterns of elevations in arterial pressure to air jet stress or tone preceding footshock in both subsets of animals (Haines *et al.*, 1996; Knuepfer *et al.*, 1993b). These observations substantiate that hemodynamic response patterns to multiple psychoactive agents and behavioral stressors are remarkably consistent within individuals while varying widely in the population.

Recently, we sought a more convincing model of acute stress that does not produce habituation. We exposed rats to cold water (1 cm deep for 1 min, 3–11 trials) by rapidly pouring ice water into a cage. The initial pressor response was associated with an increase in cardiac output (CO) in some rats (n = 7) and a decrease in other rats (n − 8, Figure 1). Since the change in arterial pressure (MAP) was not different between these rats, the increase in systemic vascular resistance (SVR) was considerably greater in vascular responders compared to mixed responders (Figure 1). These data confirm the existence of variable initial hemodynamic response patterns to exposure to cold stress.

Subsequently, we characterized these rats using cocaine (5 mg/kg, i.v., 4–6 trials, n = 11) as vascular or mixed responders to cocaine as previously described (Knuepfer and Mueller, 1999). When we compared the initial cardiac output responses of individual rats to cocaine and to cold water stress by linear regression, we noted a close correlation (R = 0.685, p = 0.02). Likewise, we reported a correlation between cardiac output responses to cocaine and to a startle response (air jet) in a previous report (Knuepfer *et al.*, 1993b). Therefore, it appears as though these stimuli evoke a similar pattern of cardiovascular responses to diverse stimuli that varies between individuals.

Initial Effects of Cold Stress

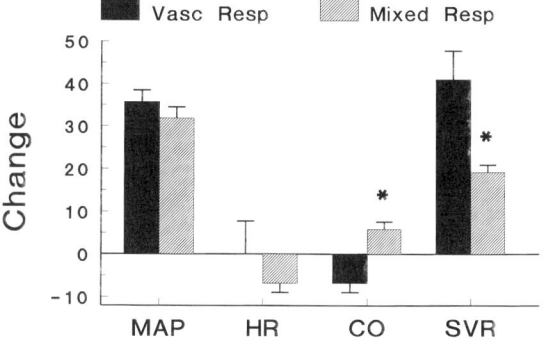

Fig. 1 The initial responses (during the peak pressor response) to acute cold stress on arterial pressure (MAP in mmHg), heart rate (HR in beats/min), cardiac output (CO expressed as percent) and systemic vascular resistance (SVR expressed as percent). Ice cold water was added rapidly to the cage to a depth of 1 cm. One minute later, the water was drained. Rats were divided into groups according to their initial CO response such that vascular responders had an initial decrease in CO whereas mixed responders had an increase. Asterisks denote significant differences (p < 0.05) at the time of the peak response as determined by Student's test. Data are mean +/− standard error.

Significance of Response Variability to Predisposition to Disease

Relationship Between Hemodynamic Response Pattern and Cardiomyopathies

We sought to determine if this intrinsic pattern of hemodynamic responses to stress correlated with the incidence of cardiomyopathies. Rats (8 mixed responders and 11 vascular responders) instrumented for ascending aortic blood flow received cocaine (5 mg/kg, i.v.) twice daily for 2 weeks. Postmortem analysis of the myocardia revealed ultrastructural alterations including focal hypercontraction of myofibrils, subsarcolemmal and intramyocytic dilations of the sarcoplasmic reticulum (frequently perinuclear) and myofibrillar derangement. We noted greater cardiac ultrastructural changes in vascular responders compared to mixed responders (Knuepfer *et al.*, 1993a). Essentially identical ultrastructural changes were noted after repeated treatments with phenylephrine (8 μg/kg, i.v.), isoproterenol (10 μg/kg, i.v.) or restraint stress (1–7 hr), emphasizing a potentially similar pathophysiology between catecholamine-, stress- and cocaine-induced lesions.

Relationship Between Hemodynamic Response Pattern and Hypertension

We reported a sustained elevation in vascular responders but not mixed responders after repeated cocaine treatment (Table 1, Branch and Knuepfer, 1994a). Since conditioned or unconditioned stress evokes similar acute hemodynamic responses in rats (Haines *et al.*, 1996), we examined the effects of repeated stress on arterial pressure. We characterized individual response patterns in 23 rats to acute air jet stress (10–20 min apart with 6 trials) and to tone followed by brief footshock (10 min apart with 12 trials) while recording MAP and CO. Subsequently, rats recovered 4–6 days before exposing them daily to one of three stressors twice a week for 4 weeks (6 days of stress per week). The stressors were 1 hour restraint stress in a plastic restraining tube, 1 hour cold water stress (1 cm deep ice cold water) and 3 hr of repeated tone/footshock at five min intervals (36 total/day). Arterial pressure was chronically elevated in vascular responders for at least 3 weeks after stress presentation whereas it returned to normal within 1–3 days in mixed responders (Table 1).

Relationship Between Hemodynamic Response Pattern and Endotoxemia

We performed experiments to examine the effects of acute cocaine administration on host inflammatory responses to Gram-negative bacterial endotoxin. As a component of the cell wall of *E. coli* and other Gram-negative bacteria, lipopolysaccharide (LPS) induces an acute inflammatory response that may cause circulatory collapse and myocardial failure, virtually identical to that elicited by exposure to live *E. coli*. We cannulated Sprague-Dawley rats and allowed them to recover overnight. Purified LPS from *E. coli* (serotype O55:B5, Sigma Chemicals, Inc.) was administered over

Table 1 Effects of repeated stress or cocaine on resting mean arterial pressure.

	Duration	Mixed Responder (mmHg)		Vascular Responder (mmHg)	
		Control	After	Control	After
Cocaine					
Twice daily (5 mg/kg) [Branch and Knuepfer, 1994a]	5 days	116 ± 3 (14)	118 ± 2	117 ± 2 (14)	125 ± 2*
Stress					
Once daily (1–3 hrs.) [Haines *et al.*, 1996]	4 weeks	111 ± 2 (10)	118 ± 4	107 ± 2 (15)	123 ± 2**

* Significantly different from mixed responder value; ** significantly different from control value.
N values are in parentheses whereas the reference number for the first and third set of data is in brackets.
NOTE: Twice daily cocaine data (after) were obtained one day after last dose whereas four week stress data (after) were obtained 2 days after last stress exposure, respectively.

15 minutes in varying doses (10, 20 or 40 mg/kg, n = 2–5 for each dose) while recording arterial pressure and heart rate continuously for the next 24 hours. LPS elicited a dose-dependent hypotension but all rats survived the 24 hour protocol.

In a subsequent experiment, we examined the effects of using cocaine as a stressor before exposure to LPS to determine whether the combination of agents synergistically enhanced the cardiovascular responses and toxicity and whether vascular responders were more sensitive to toxicity. We administered cocaine (5 mg/kg, iv, twice daily with 4–6 trials) to characterize individual rats as vascular or mixed responders to cocaine. After complete recovery (1–4 months later), the same rats were cannulated for LPS challenge. The following morning, cocaine (5 mg/kg, i.v.) was administered while recording hemodynamic responses. Five minutes later, lipopolysaccharide (LPS from *E. coli*, serotype 055:B5, 20 mg/kg, i.v.) was infused over 15 minutes. Depressor responses to LPS at 60 min were similar in all rats but the increase in SVR was enhanced in non-survivors (Figure 2). In addition, nonsurvivors had greater pupillary mydriasis compared to survivors (257 ± 46 vs. 123 ± 24% change, respectively). Therefore, cocaine enhances the morbidity to endotoxemia in some rats apparently due to greater sympathoexcitatory responses (Knuepfer *et al.* 1999).

This fundamental difference in lethality to endotoxemia was predicted in the response patterns to cocaine that were noted 1–4 months earlier. All six cocaine-pretreated rats that succumbed to LPS within 24 hours had been previously classified as vascular responders with >12% decrease in CO to cocaine (5 mg/kg, iv). There was a highly significant difference in the average change in CO to cocaine alone in survivors vs. non-survivors (−6.2 ± 2.6% vs. −15 ± 2.1%, p < 0.005). Survivors of cocaine plus LPS demonstrated greater bradycardia, smaller decreases in CO and smaller increases in SVR to cocaine alone than did the nonsurvivors. Survivors to cocaine plus LPS also had significantly less mydriasis compared to nonsurvivors

(p < 0.0038), both to cocaine alone and to cocaine plus LPS (Figure 2). These data suggest that differential sympathetic responsiveness modulates hemodynamic alterations *and* acute inflammatory responses to endotoxemia that are exacerbated by cocaine.

We assayed serum samples from rats treated with 20 mg/kg LPS alone (described above) and to survivors and non-survivors of cocaine (5 mg/kg) plus LPS (20 mg/kg). Using rat-specific ELISA's (Biosource International, Inc.), immunoreactive IL-1β, TNF-α, IL-6, and IL-10 levels were significantly elevated at 1.5 or 3.5 h among nonsurvivors after cocaine plus LPS administration compared to survivors or to LPS alone. These results confirm that differences in circulating cytokine levels correlated with lethality induced by cocaine plus LPS, exaggerating these LPS-induced

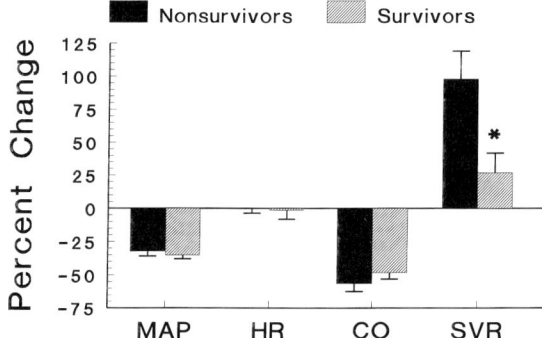

Effects of Cocaine and LPS

Fig. 2 Hemodynamic responses to cocaine (5 mg/kg, i.v., infused over 45s) followed (5 min later) by LPS (20 mg/kg, from *E. coli*, serotype 055:B5, infused over 15 min). Data were obtained 60 minutes after LPS administration during the hypotensive response. Six of twelve rats survived for 24 hours (Survivors) whereas the others died (Nonsurvivors). Groups were compared with a Student's t-test. Nonsurvivors had significantly greater systemic vasoconstriction and pupillary dilation (data not shown) compared to nonsurvivors.

increases primarily in vascular responders. It is not clear whether cocaine pretreatment directly enhanced LPS-initiated cytokine expression, or acted indirectly by producing ischemia in cytokine- generating tissues, an event that can itself potently modulate such expression (Lechner *et al.*, 1993).

Summary and Conclusions

These data demonstrate that a variety of pharmacological and behavioral stimuli evoke unique hemodynamic response patterns within individuals. The response patterns are similar to responses noted in humans. We suggest that these studies in this model may elucidate the autonomic mechanisms by which individuals vary in their sensitivity to stress-induced disease.

References

Branch, C.A. and Knuepfer, M.M. (1994a). Causes of differential cardiovascular sensitivity to cocaine I: studies in conscious rats. *Journal of Pharmacology and Experimental Therapeutics* **269**, 674–683.

Branch, C.A. and Knuepfer, M.M. (1994b). Causes of differential cardiovascular sensitivity to cocaine II: sympathetic metabolic and cardiac effects. *Journal of Pharmacology and Experimental Therapeutics* **271**, 1103–1113.

Brod, J. (1963). Haemodynamic basis of acute pressor reactions and hypertension, *British Heart Journal* **25**, 227–45.

Cacioppo, J.T., Berntson, G.G., Malarkey, W.B., Kiecolt-Glaser, J.K., Sheridan, J.F., Poehlmann, K.M., Burleson, M.H., Ernst, J.M., Hawkley, L.C., and Glaser, R. (1998). Autonomic, neuroendocrine, & immune responses to psychological stress: the reactivity hypothesis. *Annals of the New York Academy of Science* **840**, 664–673.

Cohen, S., Tyrell, D.A.J., and Smith, A.P. (1991). Psychological stress and susceptibility to the common cold. *New England Journal of Medicine* **325**, 606–612.

Eliot, R.S. (1988). "Stress and the Heart: Mechanisms, Measurements, Management," Futura, NY.

Felten, S.Y., Madden, K.S., Bellinger, D.L., Kruszewska, B., Moynihan, J.A., and Felten, D.L. (1998). The role of the sympathetic nervous system in the modulation of immune responses. *Advances in Pharmacology* **42**, 583–587.

Friedman, E.M. and Irwin, M.R. (1997). Modulation of immune cell function by the autonomic nervous system. *Pharmacology and Therapeutics* **74**, 27–38.

Galosy, R.A. and Gaebelein, C.J. (1977). Cardiovascular adaptation to environmental stress: its role in the development of hypertension, responsible mechanisms, and hypotheses. *Biobehavioral Reviews* **1**, 165–175.

Haines, W.R., Muller, J.R., Gan, Q., Lemker, J., and Knuepfer, M.M. (1996). Differential hemodynamic response to behavioral stress predicts the development of stress-induced hypertension in rats. *FASEB Journal* **10**, A629.

Herd, J.A. (1991). Cardiovascular response to stress. *Physiological Review* **71**, 305–330.

Kasprowicz, A.L., Manuck, S.B., Malkoff, S.B., and Krantz, D.S. (1990). Individual differences in behaviorally evoked cardiovascular response: temporal stability and hemodynamic patterning. *Psychophysiology* **27**, 605–619.

Knuepfer, M.M. and Branch, C.A. (1992). Cardiovascular responses to cocaine are initially mediated by the central nervous system in rats. *Journal of Pharmacology and Experimental Therapeutics* **263**, 734–741.

Knuepfer, M.M., Branch, C.A., Gan, Q., and Fischer, V.W. (1993a). Relationship between cocaine-induced cardiac ultrastructural alterations and cardiac output responses in rats. *Experimental Molecular Pathology* **59**, 155–168.

Knuepfer, M.M., Branch, C.A., Mueller, P.J., and Gan, Q. (1993b). Stress or cocaine induce a decrease in cardiac output in a subset of rats. *American Journal of Physiology* **265**, H779–H782.

Knuepfer, M.M., Gan, Q., Chen, Z., Matuschak, G.M., and Lechner, A.J. (1999). Variable sympathetic responses to cocaine correlate with susceptibility to endotoxemia in rats. *FASEB J* **13**, A1018.

Knuepfer, M.M. and Mueller, P.J. (1999). Review of evidence for a novel model of cocaine-induced cardiovascular toxicity. *Pharmacology Biochemistry and Behavior*, **63**, 489–500.

Lange, R.A., Cigarroa, R.G., Yancy, C.W., Willard, J.E., Popma, J.J., Sills, M.N., McBride, W., Kim, A.S., and Hillis, L.D. (1989). Cocaine-induced coronary artery vasoconstriction. *New England Journal of Medicine* **321**, 1557–1562.

Lechner, A.J., Ryerse, J.S. and Matuschak, G.M. (1993). Acute lung injury during bacterial or fungal sepsis. *Microsurgical Research Technician* **26**, 444–456.

Lown B. (1979). Sudden cardiac death: the major challenge confronting contemporary cardiology. *American Journal of Cardiology* **43**, 313–328.

Minor, R.L., Scott, B.D., Brown, D.D., and Winniford, M.D. (1991). Cocaine-induced myocardial infarction in patients with normal coronary arteries. *Annals of Internal Medicine* **115**, 797–806.

Mittleman, M.A., Mintzer, D., Maclure, M., Tofler, G.H., Shrewood, J.B., and Muller, J.E. (1999). Triggering of myocardial infarction by cocaine. *Circulation* **99**, 2737–2741.

Mueller, P.J., Gan, Q., and Knuepfer, M.M. (1997). Ethanol alters hemodynamic responses to cocaine in conscious rats. *Drug and Alcohol Dependence* **48**, 17–24.

Natelson, B.H. (1983). Stress, predisposition and the onset of serious disease: implications about psychosomatic etiology. *Neuroscience and Biobehavioral Review* **7**, 511–527.

Pike, J.L., Smith, T.L., Hauger, R.L., Nicassio, P.M., Patterson, T.L., McClintick, J., Costlow, C., and Irwin, M.R. (1997). Chronic life stress alters sympathetic, neuroendocrine, and immune responsivity to an acute psychological stressor in humans. *Psychosomatic Medicine* **59**, 447–457.

Smart, R.G. and Anglin, L. (1987). Do we know the lethal dose of cocaine? *Journal of Forensic Science* **32**, 303–312.

Turner, J.R., Sherwood, A., and Light, K.C. (1992). *Individual Differences in Cardiovascular Response to Stress*. Plenum Press, NY.

14. Immobilization Stress Affects Calcium Homeostasis in Rat Kidney by Modulation of Calcium Transport Systems

L'. Žáčiková, K. Ondriaš, R. Kvetňanský and O. Križanová

Introduction

Intracellular calcium is an important second messenger that regulates a wide variety of metabolic processes in the cell. Regulation of cell function by treurel or bomones is usually associated with rapid elevations in intracellular calcium followed by a subsequent decrease to basal levels. Several Ca^{2+}-transport systems, localized either on plasma membrane or in intracellular stores, are able to selectively increase or decrease intracellular calcium concentrations (Figure 1). Calcium can enter the cell mainly through voltage-dependent channels (VOC) and/or receptor-operated channels (ROC). Among VOC, the L-type of VOC, also called the dihydropyridine receptor (DHPR), is of special interest since it is involved in excitation-contraction coupling. Release of calcium ions from intracellular stores is the key step in many signal transduction processes and thus plays a role

in the etiology of certain pathological states (Berridge, 1993). Two major mechanisms through ryanodine receptor (RYR) and inositol 1,4,5-trisphosphate receptor (IP_3R) mediate marked release of calcium from intracellular stores into the cytosol. On the other hand, other mechanisms localized on the plasma membrane (Ca^{2+}-ATPase, Na^+/Ca^{2+} exchanger) can effectively release calcium from the cell. Activity of all these Ca^{2+}-transport systems can be modulated by various exogenous and endogenous factors, e. g. insulin, phosphorylation, stress, etc. (Moore, 1983; Thurzová et al., 1995; Yang and Tsien, 1993; Hudecova et al., 1996), and also by a number of more or less selective ligands, e.g. dihydropyridines, ryanodine, etc. (for reviews see Križanová, 1996; Marks, 1997; Zahradníková and Lacinová, 1998).

One of the important factors which can alter calcium homeostasis by modulation of calcium transport systems is stress. Stress is defined as the nonspecific response of the body to any demand (Selye, 1950). Numerous studies have demonstrated differential neuroendocrine responses during exposure to different stressors (Pacák et al., 1998). Immobilization was shown to be one of the most potent stressors, since it activates both components of the sympathoadrenal system (Kvetňanský and Mikulaj, 1970).

This study tests the proposal that immobilization stress could affect calcium homeostasis by modulating some of the calcium transport systems. Since IP_3-mediated calcium signaling controls important cell functions such as contraction, secretion, gene expression and synaptic plasticity (Berridge, 1993), we focused on the effect of immobilization stress on type 1 IP_3 receptors in the rat kidney.

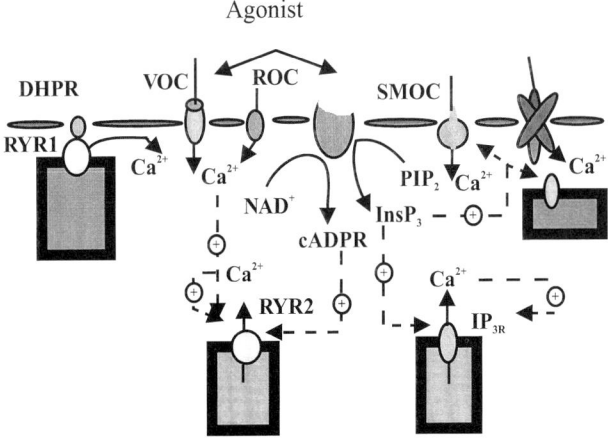

Fig. 1 Calcium transport systems in the cell. Calcium can enter a cell through the plasma membrane and can also be released from intracellular stores. The important calcium transport systems, which transport calcium through the plasma membrane, are voltage operated channels (VOC) or receptor operated channels (ROC). Intracellular channels, such as 1,4,5-inositol trisphosphate receptor (IP_3R) or ryanodine receptor (RYR), communicate with plasma membrane transporters either directly (e.g. RYR communicates with dihydropyridine receptor; DHPR), or indirectly (through calcium which enters the cell, or through the 1,4,5-inositol trisphosphate produced; InsP₃).

Methods

Animals

The protocol used was approved by the Animal Care Committee of the Slovak Academy of Sciences, Bratislava, Slovak Republic. Normotensive Wistar-Kyoto (WKY) three-month-old male rats were used for all experiments. Before initiation of the experimental procedures, the animals were housed 3–4 per cage for at least 7 days in a room at $23 \pm 2°C$ and a light-dark cycle of 12-h with food and water available *ad libitum*.

Immobilization Stress

As described by Kvetňanský and Mikulaj (1970). Rats were subjected to a single immobilization, either for 30 min or two hours, or repeated daily immobilization for two hours daily for 7 days, except of a control group. Animals were killed either immediately after the last immobilization, or three or twenty four hours later. Renal cortex and medulla were dissected, washed in ice cold physiological saline and frozen in liquid nitrogen.

Western Blot Analysis

Tissues were homogenized and membrane fractions collected. Aliquots of each protein extract (20 μg of total protein) was separated by electrophoresis on a 6.5% sodium dodecyl sulphate-polyacrylamide gel (Laemmli, 1970) and then transferred to a nitrocellulose membrane (Hybond C, Amersham UK), using semi-dry blotting. The blot was blocked in 5% nonfat dry milk diluted in TBST, and incubated with a polyclonal antibody against IP$_3$ type 1 receptor (gift of Dr. L.S. Haug, University of Oslo, Norway). The protein bands (240 kDa) were detected using a second anti-rabbit antibody labeled by peroxidase (Calbiochem) and a sensitive chemiluminiscent detection system-ECL (Amersham, UK).

Calcium Transport into the Membrane Vesicles from Rat Kidneys

Membranes from renal cortex and medulla were prepared by ultracentrifugation as described previously in Krizanova *et al.* (1990). The fractions were resuspended in buffer (50 mM Tris-HCl, pH 7.4). Fifty μl aliquots were subjected to two cycles of freezing and thawing in liquid nitrogen and then sonicated 2 × 30 s on ice to obtain unilamellar particles. Calcium transport into the proteoliposomes, as well as calcium IP$_3$ induced release, were measured as described in Duesbery and Masui, (1996). Briefly, proteoliposomes were diluted with 375 μl buffer (20 mM NaN$_3$, 20 mM MgCl$_2$, 200 mM KCl, 40 mM HEPES and 0.042 mM EGTA pH 7.8) and 316 μl H$_2$O and 1.5 μl ^{45}Ca^{2+} (20 mCi/mg Ca^{2+}) were added. Uptake was initiated by addition of 6.9 μl 0.1 M Na$_2$-ATP (Merck) at time 0. Intravesicular ^{45}Ca^{2+} content was determined just before the addition of ATP and at specified times (5, 10, 15 min) by filtration of 100 μl reaction medium through filters (Millipore Whatman GF/B) and washing with 5 ml 10 mM EGTA (pH 7.8). At time 15 min, 2 μl IP$_3$ were added to the samples and incubated for an additional 5 min. Radioactivity was measured after addition of Bray's scintillation cocktail in a Beta counter (Rackbeta, LKB).

Statistical Analysis

Each value represents an average for 6 animals. Results are presented as means ± S.E.M. Statistical differences among groups were determined by one way analysis of variance (ANOVA). Statistical significance was defined as p < 0.05. For multiple comparisons, an adjusted t-test with P values

corrected by the Bonferroni method was used (Instat, GraphPad Software, USA).

Results

Immobilization stress for 30 min had no significant effect on the level of type 1 IP$_3$R (from 6.44 ± 0.45 arbitrary units to 5.80 ± 0.41 arbitrary units; Figure 2). However, after 2 hours of immobilization, levels of type 1 IP$_3$ receptor decreased markedly to a value of 0.25 ± 0.02 arbitrary units in the renal medulla (Figure 2). Three hours after immobilization, the amount of this protein was still significantly down regulated (1.71 ± 0.12 arbitrary units) compared to controls, and returned to control levels after twenty-four hours rest (to 6.47 ± 0.45 arbitrary units). After repeated immobilization for seven times, down-regulation to the same extent as during a single immobilization was observed (Figure 3). Microsomes from each fraction were tested for ATP dependent IP$_3$ release (Figure 4). For 15 minutes microsomes were loaded with ATP dependent calcium uptake. Afterwards, IP$_3$ was added to each fraction and a decrease

A   time (min)

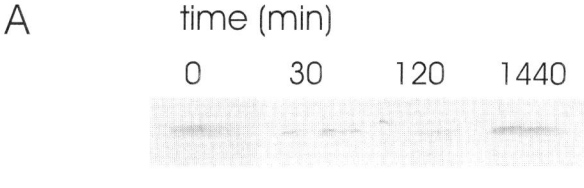

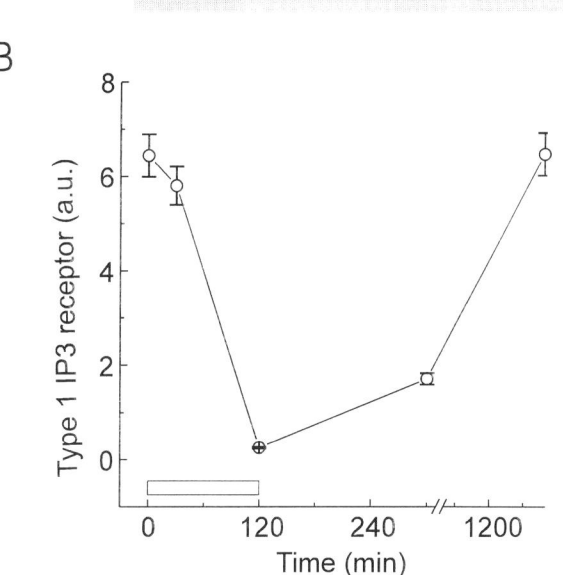

Fig. 2 The effect of immobilization stress on type 1 IP$_3$-receptor in renal medulla of normotensive WKY rats. Gel (A) represents a typical result observed with binding of polyclonal antibody against type 1 IP$_3$R and detection by ECL kit. Graph (B) represents an average from 6 animals. Each result is displayed as mean ± S.E.M. In renal medulla of normotensive rats, immobilization decreased the amount type 1 IP$_3$-receptor after 120 min of immobilization. After 3 hours of rest, a significant decrease in the amount of this receptor was still observed. After 24 hours of rest, protein amount of the type 1 IP$_3$-receptor returned to the control levels.

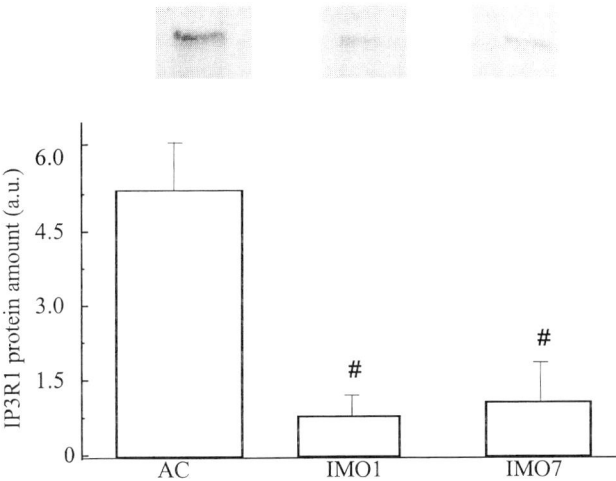

Fig. 3 The effect of repeated immobilization for 7 times on the amount of IP$_3$R1 protein in renal medulla. Typical result of Western blot and subsequent hybridization with polyclonal antibody against this receptor is in the upper part of the Figure. Graph represents an average from 6 animals and results are displayed as mean ± S.E.M. IP$_3$R protein amount decreased significantly after repeated stress (IMO7) (# –p < 0.001) compared to absolute controls (AC) and corresponds to the amount of this protein after the single immobilization (IMO1).

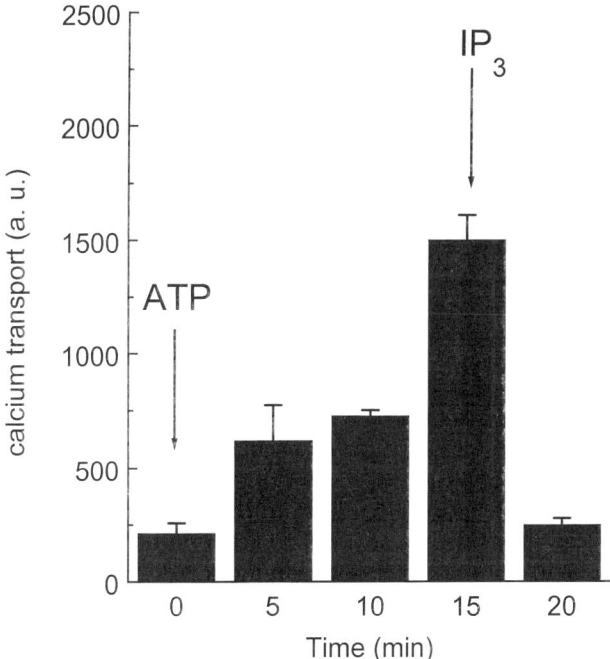

Fig. 4 Calcium ATP-dependent uptake with subsequent IP$_3$ release in microsomes of renal medulla of WKY rats. Calcium ATP-dependent uptake with subsequent IP$_3$ release was measured in microsomes from the renal medulla of WKY rats. Arrows represent the time of addition of ATP and/or IP3. Results are displayed as mean ± S.E.M. and each column represents an average of at least 3 animals. IP$_3$ induced calcium release from microsomes of renal medulla indicates that calcium was accumulated into IP$_3$-sensitive stores.

of $^{45}Ca^{2+}$ in microsomes was measured to determine the calcium release through IP$_3$R.

Discussion

This study shows that the type 1 IP$_3$-receptor is reversibly down-regulated by immobilization stress. Down-regulation occurs to the same extent after repeated immobilization for seven times. However, a functional consequence as well as the mechanism of the IP$_3$-receptor down-regulation remain to be elucidated.

Discrete intracellular calcium controls discrete pools of calcium and thus may regulate different functions in response to intracellular calcium mobilizing agonists. Because of the different sensitivity of the various subtypes of IP$_3$Rs to IP$_3$, the rate of calcium release was enhanced by ATP in IP$_3$R1 expressing cells, while a less significant effect of ATP on IP$_3$R3 and no effect on IP$_3$R2 was observed (Miyakawa et al., 1999), it is likely that each subtype has distinct functions in the cells. The mechanisms of type 1 IP$_3$R down-regulation may involve phosphorylation. Phosphorylation occurs in response to hormones that activate cAMP formation. On the other hand type 2 and 3 receptors, which lack the relevant consensus sequences (Joseph and Ryan, 1993; Yamamoto-Hino et al., 1994) are unlikely to be phosphorylated, and this may explain the down-regulation of the subtypes.

Down-regulation might also be a result of changes in mRNA levels, or a profound acceleration of IP$_3$ receptor degradation. It has been suggested that the down-regulation of IP$_3$-receptors is caused by IP$_3$ binding (Zhu et al., 1999). Down-regulation of type 1 IP$_3$R was observed also in glomerular and vascular cells in diabetic kidney (Sharma et al., 1999). It has been proposed that this down-regulation may contribute to renal vasoregulation and renal hypertrophy of diabetes. The physiological relevance of down-regulation of type 1 IP$_3$ receptor in rat kidney during adaptation to stress and the mechanism of this change remain to be elucidated.

References

Berridge, M. (1993). Inositol trisphosphate and calcium signalling. *Nature* **361**, 315–325.

Duesbery, N.S., and Masui, Y. (1996). The role of microtubules and inositol triphosphate induced Ca^{2+} release in the tyrosine phosphorylation of mitogen-activated protein kinase in extracts of Xenopus laevis oocytes. *Zygote* **4**, 21–30.

Hudecova, S., Kvetňanský, R., and Krizanova, O. (1996). Rapid stress-induced expression of the L-type voltage-dependent calcium channels in rat kidney. In: R. McCarty, G. Aguilera, E.L. Sabban, and R. Kvetňanský (Eds.), *Stress: Molecular Genetic and Neurobiological Advances*, pp. 571–562. New York: Gordon and Breach.

Joseph, S.K., and Ryan, S.V. (1993). Phosphorylation of the inositol trisphosphate receptor in isolated rat hepatocytes. *Journal of Biological Chemistry* **268**, 23059–23065.

Krizanova, O., Novotova, M., and Zachar, J. (1990). Characterization of DHP binding protein in crayfish striated muscle. *FEBS Letters* **267**, 311–315.

Krizanova, O. (1996). Structural implications in the function of L-type voltage-dependent calcium channels. *General Physiology and Biophysics* **15**, 79–87.

Kvetňanský, R., and Mikulaj, L. (1970). Adrenal and urinary catecholamines in rat during adaptation to repeated immobilization stress. *Endocrinology* **87**, 738–743.

Laemmli, U.K. (1970). Cleavage of structural proteins during the assembly of the head of bacteriophage T4. *Nature* **227**, 680–685.

Lowry, O.H., Rosenbrough, N.J., Farr, A.I., and Randal, R.J. (1951). Protein measurement with the Folin phenol reagent. *Journal of Biological Chemistry* **193**, 256–272.

Marks, A.R. (1997). Intracellular calcium-release channels: regulators of cell life and death. *American Journal of Physiology*, **272**, H597-H605.

Miyakawa, T., Maeda, A., Yamazawa, T., Hirose, K., Kurosaki, T., and Iino, M. (1999). Encoding of Ca^{2+} signals by differential expression of IP_3 receptor subtypes. *EMBO Journal* **18**, 1303–1308.

Moore, R.D. (1983). Effect of insulin upon ion transport. *Biochemica et Biophysica Acta* **737**, 1–49.

Pacák, K., Palkovits, M., Yadid, G., Kvetňanský, R., Kopin, I.J., and Goldstein, D.S. (1998). Heterogenous neurochemical responses to different stressors: a test of Selye's of doctrine of nonspecifity. *American Journal of Physiology* **275**, R1247–R1255.

Sharma, K., Wang, L., Zhu, Y., DeGuzman, A., Cao, G.Y., Lynn, R.B., and Joseph, S.K. (1999). Renal type I inositol 1,4,5-trisphosphate receptors is reduced in streptozocin-induced diabetic rats and mice. *American Journal of Physiology* **276**, 54–61.

Selye, H. (1950). *The Physiology and Pathology of Exposure to Stress. A Treatise Based on the Concepts of the General-Adaptation Syndrome and the Diseases of Adaptation* Montreal: Acta, Inc.

Thurzová, M., Kvetňanský, R., and Križanová, O. (1995). Modulation of the L-type Ca-channels by insulin treatment in rat aorta. *General Physiology and Biophysics* **14**, 217–224.

Wojcikiewicz, R.J., Furuichi, T., Nakada S., Mikoshiba, K., and Nahorski, S.R. (1994). Muscarinic receptor activation down-regulates the type I inositol 1,4,5-trisphosphate receptor by accelerating its degradation. *Journal of Biological Chemistry* **269**, 7963–7969.

Wojcikiewicz, R.J. (1995). Type I, II and III inositol 1,4,5-trisphosphate receptors are unequally susceptible to down-regulation and are expressed in markedly different proportions in different cell types. *Journal of Biological Chemistry* **270**, 11678–11683.

Yamamoto-Hino, M., Sugiyama, T., Hikichi, K., Mattei, M.G., Hasegawa, K., Sekine, S., Sakurada, K., Miyawaki, A., Furuichi, T., Hasegawa, M., and Mikoshiba, K. (1994). Cloning and characterization of human type 2 and type 3 inositol 1,4,5-trisphosphate receptors. *Receptors and Channels* **2**, 9–22.

Yang, J., and Tsien, R.W. (1993). Enhancement of N- and L-type calcium channel currents by protein kinase C in frog sympathetic neurons. *Neuron* **10**, 127–136.

Zahradníková, A., and Lacinová, L. (1998). Molecular determinants of the interaction of calcium channels with calcium channel drugs. *Experimental and Clinical Cardiology* **3**, 121–127.

Zhu, Ch.Ch., Furuichi, T., Mikoshiba, K., and Wojcikiewicz, R.J.H. (1999). Inositol 1,4,5-trisphosphate receptor down-regulation is activated directly by inositol 1,4,5-trisphosphate binding. Studies with binding – defective mutant receptors. *Journal of Biological Chemistry* **274**, 3476–3484.

15. Osteocalcin Response to Immobilization: Possible Involvement of Parathyroid Hormone

P.E. Patterson-Buckendahl, S. Droniak, M. Rusnák and R. Kvetňanský

Introduction

Osteocalcin, a calcium binding peptide of bone origin, is closely associated with bone mineral, and, in plasma, reflects osteoblastic activity. We have previously shown (Patterson-Buckendahl et al., 1995) that acute immobilization of rats by forced restraint for two hours (IMMO), a well-characterized model of fight/flight response (Kvetňanský and Mikulaj, 1970), resulted in a rapid and consistent increase in plasma osteocalcin (pOC). Return of pOC to basal levels required the presence of both corticosterone (CS) and norepinephrine (NE), or other sympathetic neural influences. In some instances we observed a slight hypocalcemia in these rats.

A number of investigators have reported hypocalcemia in response to various stressors, including immobilization (IMMO). Morimoto et al. (1986) investigated a possible link to calcitonin (CT), one of the major calciotropic hormones. However, there was no causal relationship between CT and hypocalcemia during either acute or repeated IMMO. Subsequently, Aou et al. (1995) reported that stimulation of various areas of the hypothalamus could induce hypocalcemia. Other work by the same researchers showed that parathyroid hormone (PTH) was involved in the hypocalcemic response to urethane injection (Matsui et al., 1995). Stern et al. (1993) reported increased PTH and decreased calcitonin in response to injection of turpentine oil. The response of PTH was eliminated by superior cervical gangliectomy while hypocalcemia and calcitonin response was unaltered. The present experiments were designed to test PTH response to hypocalcemia in IMMO rats, and whether surgical removal of the parathyroid might abolish the pOC response.

Methods

Sprague-Dawley male rats obtained from either Charles River Laboratories, Wiga, Germany, or Iffa Credo Laboratories, Lyon, France, were housed in single cages for one week prior to experiments. IMMO followed well-established protocols of Kvetňanský and Mikulaj (1970). Animals were restrained in the prone position by taping the feet to raised supports with bandage tape. For all IMMO experiments, tail arteries of male rats were cannulated under pentobarbital anesthesia 24 hours prior to IMMO to allow blood sampling without handling. Approximately 0.5 ml of blood was collected into heparinized microfuge tubes and placed on ice prior to separation of plasma. Blood was replaced with equal volumes of isotonic saline containing 50 IU/ml heparin. All samples were stored frozen at $-70°C$ prior to assay. Experiment 1 (n = 8) provided samples at 0, 5, 20, 60, and 120 min. of IMMO for PTH and pCa. Experiment 2 compared parathyroidectomized (PTX) and sham PTX rats. Surgeries were performed at Iffa Credo Laboratories 1 week prior to shipment to the Institute of Experimental Endocrinology, Bratislava. Animals were supplemented with calcium lactate in their drinking water. IMMO was imposed two weeks post surgery at which time PTX rats (n = 11) weighed 284 ± 4 g and sham rats (n = 9) weighed 326 ± 6 g (p << 0.01). Blood was sampled in the same manner as for Exp. 1 and assayed for pOC, PTH, catecholamines, and corticosterone. PTH was assayed by IRMA (Immutopics), OC by RIA (Patterson-Allen, 1982), and Ca by atomic absorption (Perkin-Elmer). Corticosterone was determined by radioimmunoassay (ICN Diagnostics) and catecholamines by HPLC with electrochemical detection (Eisenhofer et al., 1986).

Multiple analysis of variance with repeated measures (SAS) was used for statistical comparisons between PTX and sham levels of hormones and osteocalcin. Student's t-test was used to compare time points.

Results

In the first experiment immobilization of intact rats caused significant elevation in plasma PTH levels within 5 min. of IMMO (p < 0.001, Figure 1). This was accompanied by a significant decrease in plasma total calcium at 5 min. IMMO (P < 0.01). PTH returned toward normal after 20 min., but remained slightly elevated throughout the remainder of IMMO.

In a second experiment comparing PTX rats with sham PTX controls, plasma calcium was significantly lower in PTX rats than in controls prior to IMMO (p < 0.05 to 0.001). Both PTX and control rats exhibited calcium levels that were significantly decreased throughout the IMMO period compared with pre-IMMO values (Figure 2A). PTH in PTX rats was less than 4 pg/ml, approximately the limit of sensitivity for the assay. In control rats, PTH response confirmed findings in experiment one, with significant elevation after 5 min. IMMO (p < 0.01), a return to basal levels after 20 min. and a subsequent slight, but significant rise after 120 min (p < 0.01, Figure 2B). As shown in Figure 3A, the expected osteocalcin increase after 5 min. IMMO was not eliminated by PTX (p < 0.05 vs T = 0).

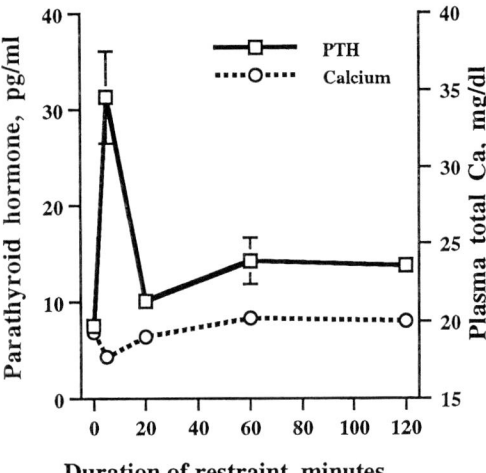

Fig. 1 Parathyroid hormone and plasma total calcium concentrations in immobilized rats. Values are mean ± SEM. Where bars are not seen, they are within the symbol for the data point.

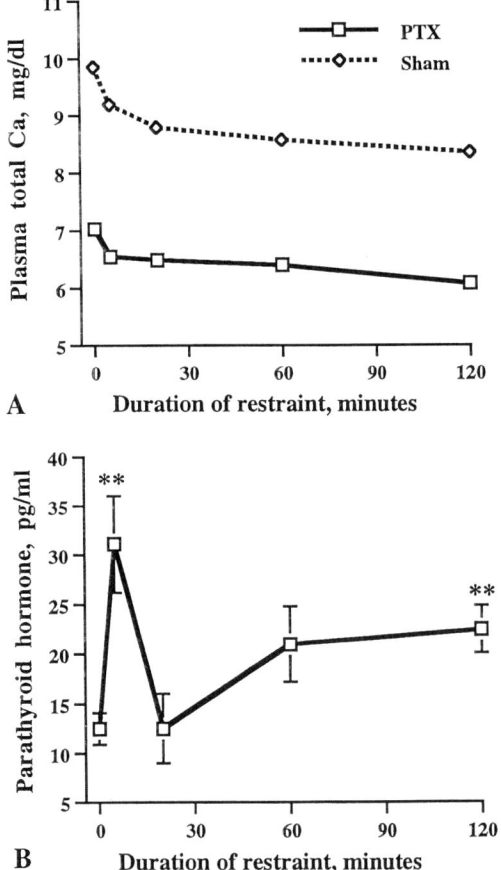

Fig. 2 (A) Response of plasma total calcium concentration in parathyroidectomized and sham operated rats subjected to two hours immobilization. (B) Parathyroid hormone in sham operated rats subjected to two hours immobilization. PTH in PTX rats was below limits of detection of the assay. Values are mean ± SEM. Where error bars are not seen, they are within the symbol for the data point.

However, in PTX rats pOC was significantly lower than in sham controls after 60 min. IMMO (p < 0.01).

Stress hormone levels in PTX and sham rats are shown in figure 3B-D. Corticosterone levels were elevated during IMMO in both groups (p < 0.001), but did not differ between groups (Figure 3B). Plasma catecholamines were also significantly elevated at all times during IMMO (p < 0.001). Epinephrine levels in PTX rats were significantly higher than sham operated rats after 60 min. IMMO (p < 0.05, Figure 3C). However, norepinephrine response did not differ between groups (Figure 3D).

Discussion

These experiments confirmed previous reports indicating the occurrence of hypocalcemia during IMMO as well as in response to a wide variety of stressors. PTH was acutely elevated 2.5 fold over basal levels in an appropriate response to this hypocalcemia. In the absence of PTH, hypocalcemia was present prior to IMMO, but displayed the same general pattern of decrease during IMMO. Previous researchers established that elevated calcitonin is not the cause of the hypocalcemia (Morimoto *et al.*, 1986).

Stress hormone levels in both PTX and sham operated rats followed predictable patterns typical of this model, with the exception of a significant difference in epinephrine between the two groups after 60 min. IMMO. Aou *et al.* (1995) have reported hypothalamic regulation of calcium and PTH levels in stressed rats. Stern *et al.* (1993) reported an interaction of superior cervical ganglia with thyroid and parathyroid tissues.

OC mRNA is stabilized by PTH (Noda *et al.*, 1988) and hyperparathyroidism is characterized by greatly elevated pOC (Hauschka *et al.*, 1989). This is consistent with the low basal OC levels in PTX rats. However, the parallel changes in OC in intact and PTX rats during immobilization suggest additional mechanisms induced by stress. Other researchers suggested that elevated pOC during hypocalcemia induced by citrate infusion originates from the bone surface rather than from increased cellular synthesis or secretion (Gundberg *et al.*, 1991). A similiar explanation could apply to the acute elevation of pOC during IMMO. It is probable that release of OC from bone, along with well known PTH effects on bone resorption, allows the release of exchangeable bone calcium contributing to restore plasma calcium levels. What remains unknown is the stimulus for the hypocalcemia of stress or the destination of the plasma calcium. Possible mechanisms are increases in urinary calcium excretion, or a shift in calcium stores in various tissues in response to restraint or other stressors.

Acknowledgements

This research was supported in part by NIH/NIAAA Grant AAI2705 (PPB) and by Slovak Grant Agency for Science Grant 2-610999 (R.K.).

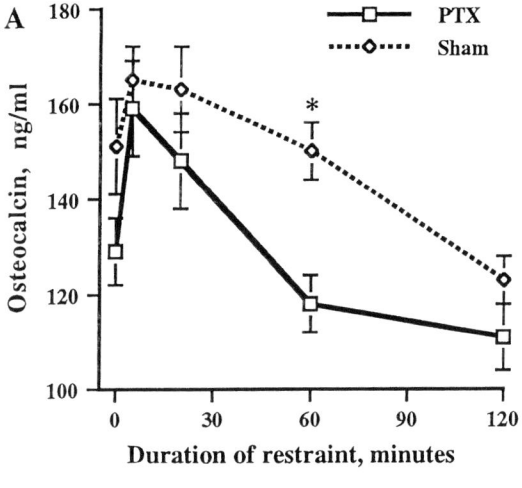

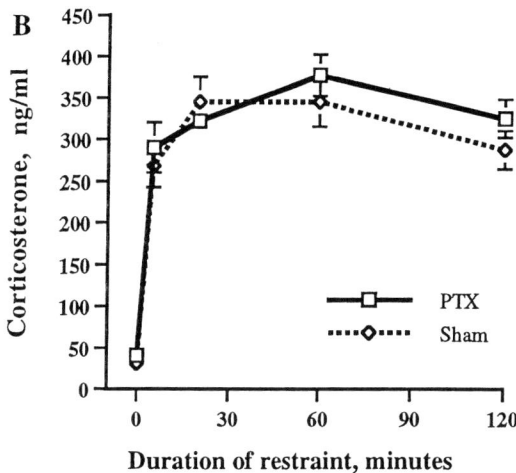

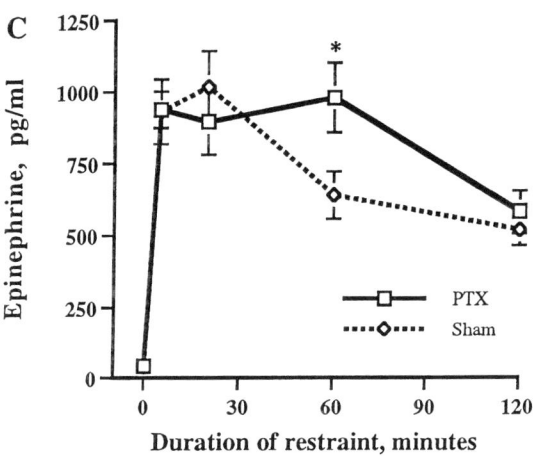

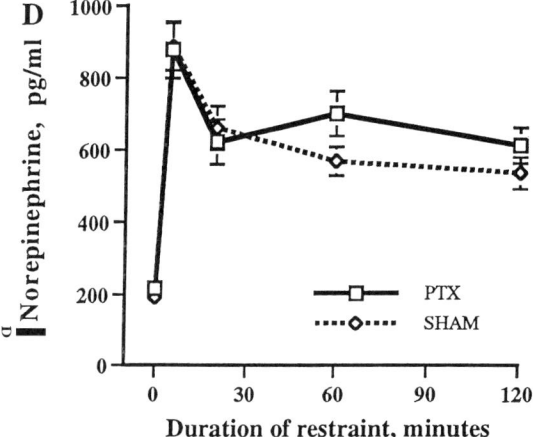

Fig. 3 Plasma osteocalcin (A), corticosterone (B), epinephrine (C), and norepinephrine (D) concentrations in parathyroidectomized and sham-operated rats subjected to two hours immobilization. Values are mean ± SEM. All four parameters were significantly elevated over pre-IMMO levels after 5 min. restraint. Osteocalcin in PTX rats was slightly below baseline after 60 min, and differed significantly from sham rats at that time. Epinephrine in PTX remained elevated longer in PTX rats than in sham, differing significantly at 60 min. There were no differences between PTX and sham operated rats in corticosterone or norepinephrine concentrations at any time point.

References

Aou, S., Shiramine, K., Ma, J., Matsui, H., and Hori, T. (1995) Hypothalamus regulates calcium metabolism in rats. *Neurobiology* **3**, 339–350.

Eisenhofer, G., Goldstein, D.S., Stull, R., Heiser, H.R., Sunderland, T., Murphy, D.L., and Kopin, I.J. (1986) Simultaneous liquid-chromatographic determination of 3,4-Dihydroxyphenylglycol, catecholamines, and 3,4-dihydroxyphenylalanine in plasma, and their responses to inhibition of monoamine oxidase. *Clinical Chemistry* **32**, 2030–2033.

Gundberg, C.M., Grant, F.D., Conlin, P.R., Chen, C.J., Brown, E.M., Johnson, P.J., and LeBoff, M.S. (1991). Acute changes in serum osteocalcin during induced hypocalcemia in humans. *Journal of Clinical Endocrinology and Metabolism* **72**, 438–443.

Hauschka, P.V., Lian, J.B., Cole, D.E., and Gundberg, C.M. (1989). Osteocalcin and Matrix gla protein: Vitamin K-dependent proteins in bone. *Physiological Reviews* **69**, 990–1047.

Kvetňanský, R., and Mikulaj, L. (1970) Adrenal and urinary catecholamines in rats during adaptation to repeated immobilization stress. *Endocrinology* **87**, 738–743.

Matsui, H., Aou, S., Ma, J., and Hori, T. (1995) Central actions of parathyroid hormone on blood calcium and hypothalamic neuronal activity in the rat. *American Journal of Physiology* **268**, R21–R27.

Morimoto, S., Fausto, A., Birge, S.J., and Avioli, L.V. (1986) Effect of short- and long-term stress on plasma calcium and calcitonin in the rat. *Hormone and Metabolism Research* **18**, 818–20.

Noda, M., Yoon, K., Rodan, G.A. (1988) Cyclic AMP-mediated stabilization of osteocalcin mRNA in rat osteoblast-like cells treated with parathyroid hormone. *Journal of Biological Chemistry* **263**, 18574–18577.

Patterson-Allen, P., Brautigam, C.E., Grindeland, R.E., Asling, C.W., and Callahan, P.X. (1982) A specific radioimmunoassay for osteocalcin with advantageous species crossreactivity. *Analytical Biochemistry* **120**, 1–7.

Patterson-Buckendahl, P.E., Kvetňanský, R., Fukuhara, K., Cizza, G., and Cann, C. (1995) Regulation of plasma osteocalcin by corticosterone and norepinephrine during restraint stress. *Bone* **17**, 467–472.

Stern, J.E., Ladizesky, M.G., Keller Sarmiento, M.I., and Cardinali, D.P. (1993) Involvement of the cervical sympathetic nervous system in the changes of calcium homeostasis during turpentine oil-induced stress in rats. *Neuroendocrinology* **57**, 381–387.

16. Effects of Novel Stressors on Plasma Catecholamine Levels in Rats Exposed to Long-term Cold

S. Dronjak, M. Ondriska, D. Svetlovska, D. Jezova and R. Kvetňanský

Introduction

Stress is defined as a state of threatened homeostasis which is re-established by a complex repertoire of physiological and behavioral adaptive responses of the organism (Chrousos, 1998). Activation of the sympathoadrenal system (SAS) leads to the outflow of catecholamines (CA) from the adrenal medulla and the sympathetic nerve endings. Plasma levels of norepinephrine (NE) and epinephrine (EPI) are increased during sympathetic activation but some studies have demonstrated that the two branches of the SAS (sympathoneural and adrenomedullary) can be activated independently by various stressors of different quality or intensity (Kvetňanský et al., 1998). Significant release of CA from the adrenomedullary system is elicited by insulin-induced hypoglycemia (Goldstein et al., 1996; Vietor et al., 1996; Kvetňanský et al., 1998), cellular glucoprivation caused by 2-deoxy-D-glucose (2DG) administration (Scheuring et al., 1996), and by immobilization (Kvetňanský et al., 1996). The administration of insulin results in extracellular glucoprivation via intracellular glucose uptake and utilization, whereas 2DG induces cellular glucoprivation and elevates plasma glucose levels (Breier et al., 1993). Cold exposure was shown to produce an activation of the sympathoneural system as evidenced by increased plasma NE levels and NE metabolites (Fukuhara et al., 1996). Even if the animals were exposed to temperatures of 4°C or –3°C, the adrenomedullary system was not noticeably affected, whereas the sympathoneural system was markedly activated by both temperatures.

In previous studies, we have shown that repeated exposure to some stressors (immobilization, acclimatization to mountain conditions) leads to reduced plasma CA levels (Kvetňanský et al., 1984). Repeatedly immobilized rats, however, responded to acute experimental trauma with substantially higher plasma EPI and NE levels than intact rats that had never been stressed before (Kvetňanský et al., 1984). Rats acclimated to high mountain conditions (including also cold) showed significantly higher levels of plasma CA in response to immobilization (IMO) than control rats (Balaz et al., 1980, Kvetňanský et al., 1984).

Similar changes have been found in activation of the HPA axis (Bhatnagar and Dallman, 1998) and prolactin secretion (Kvetňanský et al., 1984) in chronically stressed rats exposed to novel stressors. Rats exposed to long-term cold showed an increase in adrenal medullary CA levels (Kvetňanský et al., 1971a).

The aim of this work was to investigate the activity of the sympathoadrenal system, by measurement of plasma epinephrine and norepinephrine levels, in rats exposed to chronic cold (28 days at 4°C), and in such rats exposed to various novel stressors: insulin-induced hypoglycemia, 2DG-induced cellular glucoprivation, and immobilization stress.

Materials and Methods

The experiments were carried out using male Sprague-Dawley rats weighing 400–450 g (Charles River, Wiga, Germany). The rats were divided into two groups. The first group was exposed to long-term cold (LTC) and the second group was kept at room temperature (control) and every operation was done at 24°C. Rats were housed two per cage in a cold chamber at 4°C for 28 days on a 12 hr light–12 hr dark cycle. Food and water were available ad libitum.

Animals were cannulated both in the tail artery (collection of blood samples) and in the peritoneum (administration of drugs) to prevent any influence of unwanted stressors. Blood samples were taken through a polyethylene catheter which had been inserted under pentobarbital anasthesia 24 h before the experiment started. After the baseline blood collection, animals were administered insulin, 2DG, or were immobilized, or exposed to cold stress. Blood was collected 15, 30, 60 and 120 minutes after drug administration or after the immobilization started. The animals exposed for the first time to cold for 2 h were at room temperature and after the baseline blood collection were carefully transferred within their home cages into the cold chamber and blood was collected at 30, 60 and 120 min of the cold exposure. Long-term cold exposed rats were in the cold chamber even during application of novel stressors.

Insulin (Actrapid, Novo Nordisk, Denmark, 5IU/kg b.w.) and 2-deoxy-D-glucose (Sigma, St. Louis, USA, 500 mg/kg b.w.) dissolved in saline were injected via an i.p. indwelling catheter without handling the animals. Saline alone was also injected via the i.p. catheter. All novel stressors were administered to long-term cold exposed rats and the blood collection was

83

performed while the animals were in the cold chamber. Immobilization stress was provided as described by Kveťanský and Mikulaj (1970). Plasma catecholamines were assayed by our modification of the radioenzymatic method (Peuler and Johnson, 1977).

Data were statistically evaluated by two-way ANOVA results are presented as means ± SEM.

Results

Plasma Levels of Catecholamines in Rats Exposed to Acute and Long-term Cold

Plasma levels of NE and EPI in rats exposed for the first time to cold for 2h (control) and in long-term cold exposed rats (LTC) are shown in Figure 1. Basal

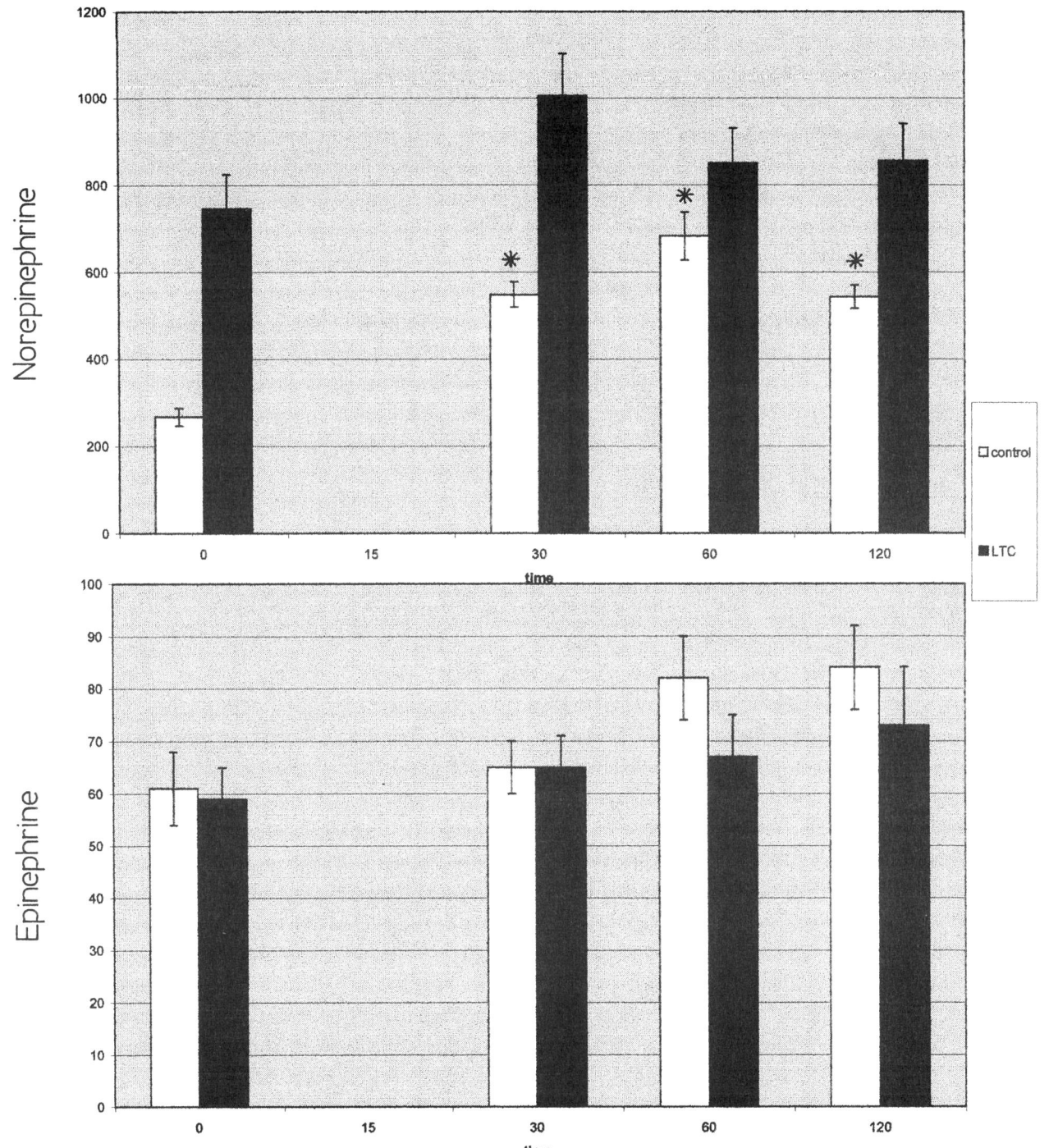

Fig. 1 The effect of cold exposure on plasma levels of norepinephrine and epinephrine (pg/ml) in naive control rats and in rats exposed to long-term cold (LTC). The values represent means + SEM of 7–10 animals. Statistical significance *p < 0.01 compared to basal levels.

plasma NE levels were significantly elevated in LTC exposed rats while plasma EPI levels were not significantly different from the control group. The first cold exposure of rats induced a significant gradual increase in plasma NE levels. Increased NE levels were seen in all intervals in LTC rats (Figure 1). No significantly different changes were found in plasma EPI levels in acutely and chronically cold exposed animals.

Effect of Immobilization on Plasma Catecholamine Levels in Rats Exposed to Acute and Long-term Cold

IMO stress was associated with a significant elevation of plasma NE levels in both groups of rats but the concentration of NE was increased substantially more in LTC than in naive control rats (Figure 2). Plasma EPI levels were highly significantly elevated in both groups; however, the reduced EPI levels in control rats at 60 and 120 min of IMO, were not seen in LTC exposed rats (Figure 2).

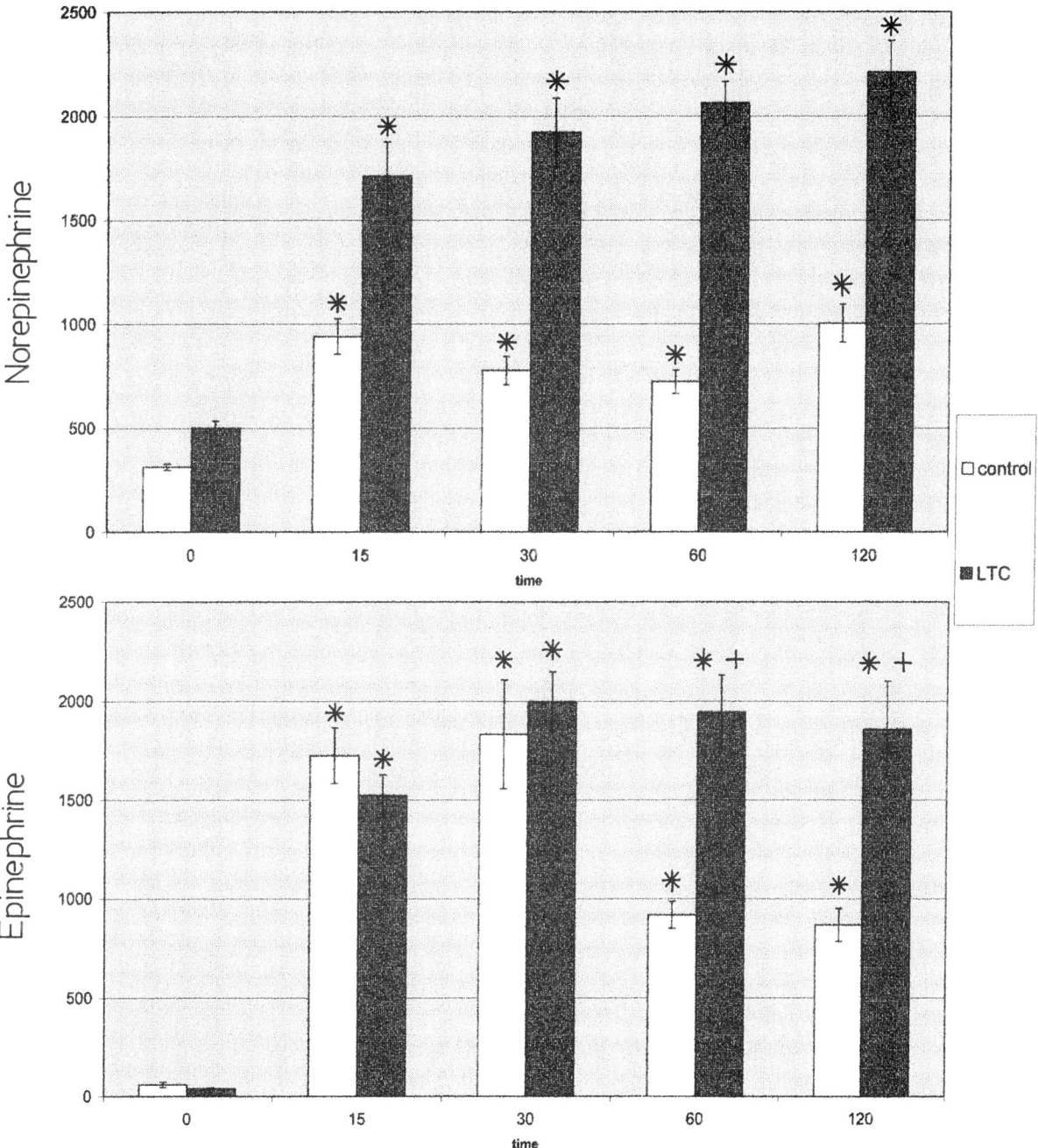

Fig. 2 The effect of immobilization stress on plasma levels of norepinephrine and epinephrine (pg/ml) in naive control rats and in rats exposed to long-term cold (LTC). The values represent means + SEM of 7–10 animals. Statistical significance *p < 0.01 compared to basal levels; + p < 0.01 compared to control group.

Effect of 2DG on Plasma Catecholamines in Rats Exposed to Acute and Long-term Cold

The results presented in Figure 3 show that exposure to 28 day cold stress produces a significant increase in basal plasma NE levels but not EPI levels. After administration of 2DG, plasma NE levels gradually significantly increased in both groups. This increase, however, was proportional to the significantly higher NE baseline levels found in LTC rats (Figure 3). The administration of 2DG produced highly significant increases in plasma EPI levels in both groups of rats. In the LTC group, however, plasma EPI levels were more pronounced at 60 and 120 min after 2DG administration (Figure 3). The administration of 2DG also produced an increase in plasma glucose levels (data not shown).

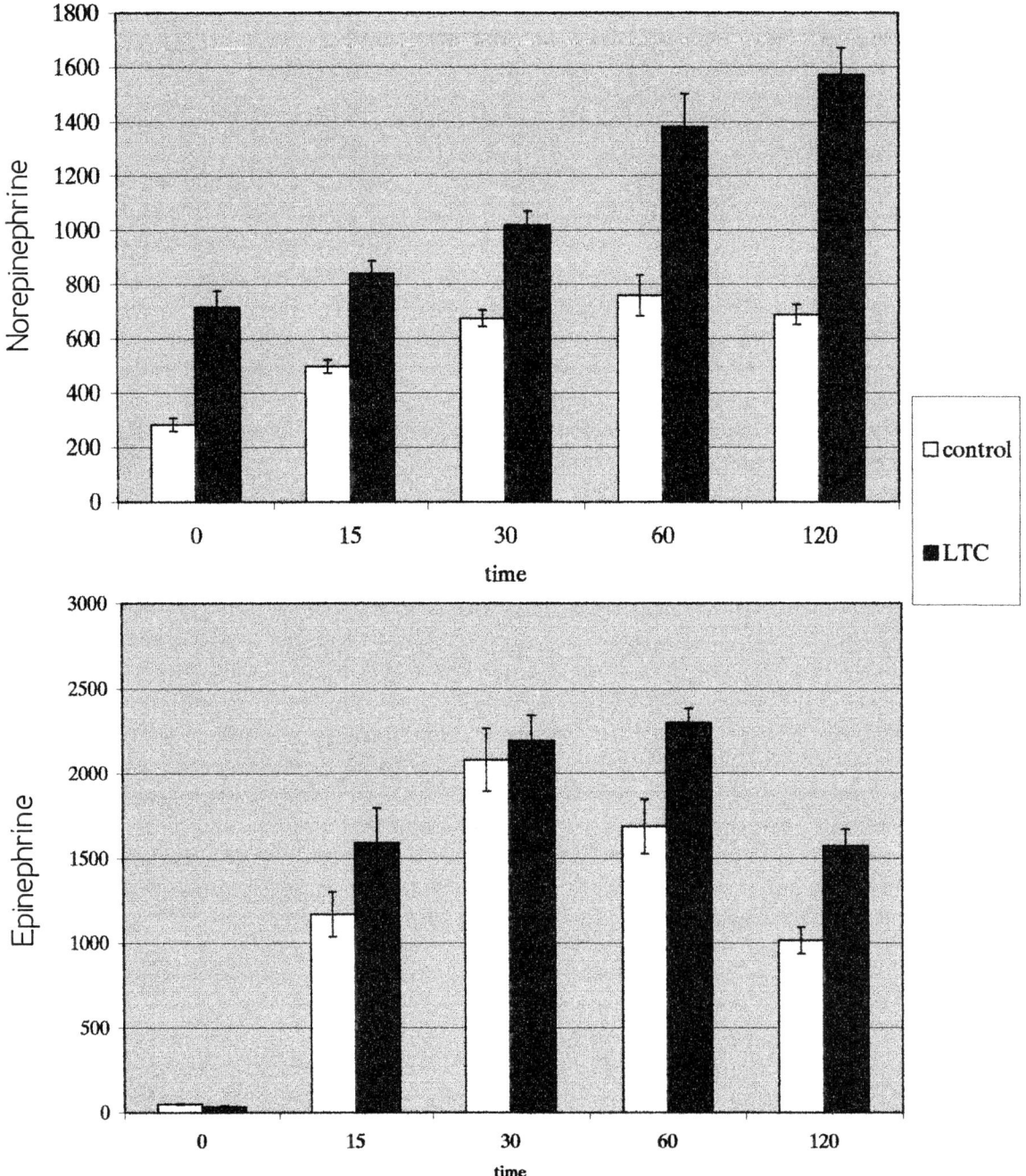

Fig. 3 The effect of 2-deoxy-D-glucose administration (500 mg/kg) on plasma levels of norepinephrine and epinephrine (pg/ml) in naive control rats and in rats exposed to long-term cold (LTC). The values represent means + SEM of 7–10 animals. Plasma EPI and NE levels were significantly elevated at all intervals after 2DG administration when compared to basal levels.

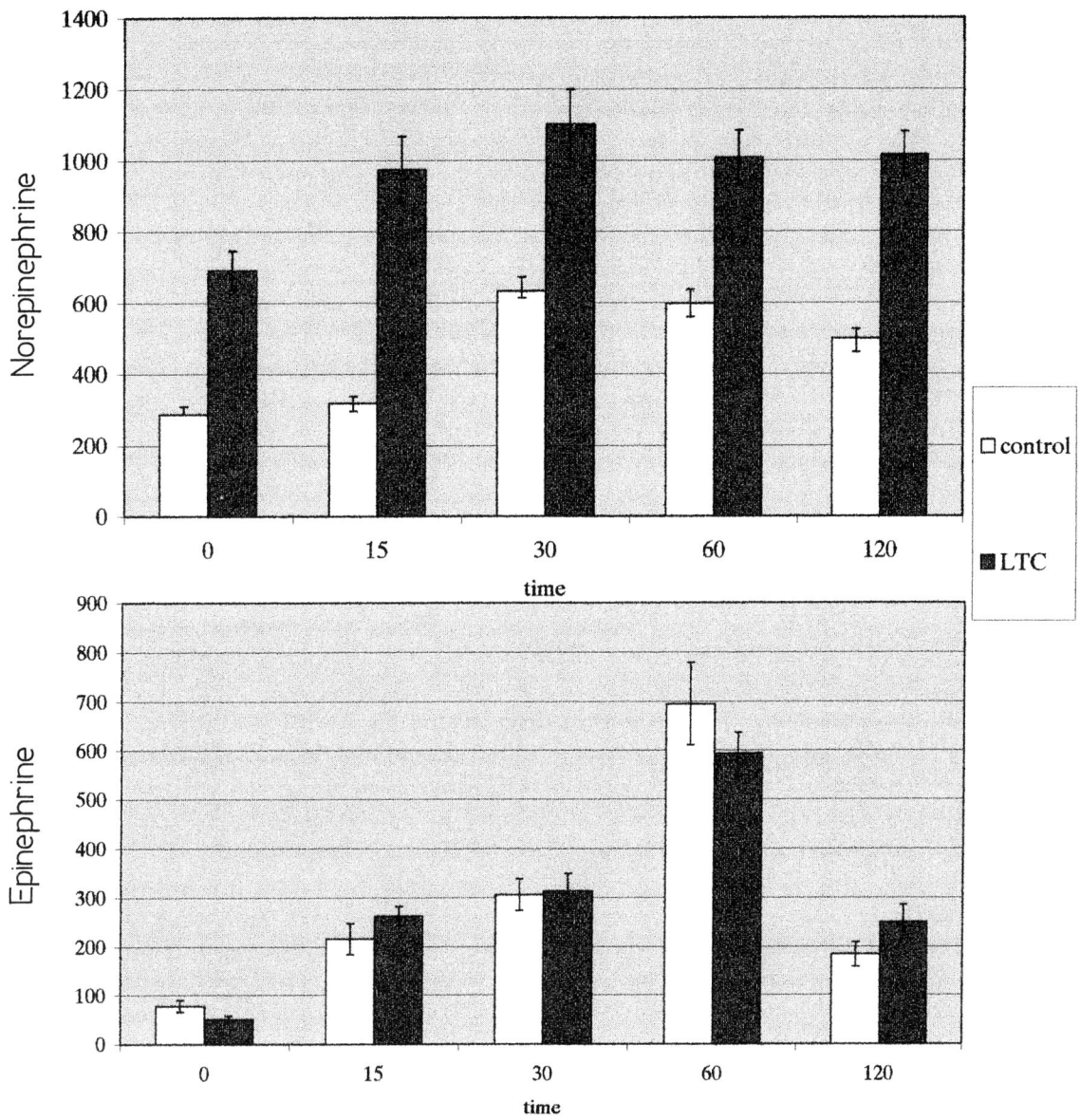

Fig. 4 The effect of insulin administration (5IU/kg) on plasma levels of norepinephrine and epinephrine (pg/ml) in naive control rats and in rats exposed to long-term cold (LTC). The values represent means + SEM of 7–10 animals.

Effect of Insulin on Plasma Catecholamines in Rats Exposed to Acute and Long-term Cold

The effect of insulin-induced hypoglycemia on plasma levels of NE and EPI is shown in Figure 4. Insulin-induced hypoglycemia produced a very similar activation of the adrenomedullary system in both LTC and control rats. The highest levels of EPI were found 60 min after insulin administration and reached about a 10-fold increase in both groups of rats. In contrast, plasma NE levels increased only about two-fold in both groups. Thus, plasma NE levels were only a little affected by insulin administration and the increases were proportional to the highly increased NE baseline in LTC exposed rats (Figure 4). The administration of

insulin produced a significant, more than 50% reduction in plasma glucose levels (data not shown).

Discussion

In this study we compared the effect of novel stressors, i.e immobilization, cold, insulin-induced hypoglycemia, and 2DG-induced glucoprivation, on plasma catecholamine levels in rats exposed to long-term cold (28 days). Insulin administration caused about a 50% decrease of plasma glucose levels whereas 2DG administration caused an increase. During long-term cold exposure, basal plasma NE levels increased several fold, whereas basal plasma EPI levels were not changed. This indicates that long-term cold stress is a specific stimulus for the sympathoneural system, without significant effect on activation of the adrenomedullary system. Exposure to cold has been previously shown to increase sympathoneural activity without significant activation of the adrenomedullary system and EPI release (Fukuhara et al., 1996). These authors compared sympathoadrenal responses to intermittent cold with exposure to continuous cold stress. Neither intermittent nor continuous cold exposure altered plasma EPI levels. However, NE and dihydroxyphenylglycol (DHPG) levels increased markedly during exposure to this stressor. Previous studies have shown that cold exposure decreases not only plasma NE levels but also levels of plasma DHPG, DOPA, and DOPAC (Kvetňanský et al., 1998; Pacak et al., 1998).

The increase in basal plasma NE levels in LTC rats is a result of the adaptation process to cold. The function of elevated NE levels found in LTC rats is most probably in the thermogenic and calorigenic effects of this neurotransmitter. Liu and coworkers (1994) showed that NE infusion increased the concentration of circulating free fatty acids which are important energy substrates. The two branches of the SAS are not only functionally but also metabolically dissociated. Changes in plasma EPI levels are always followed by parallel alterations in blood glucose. Changes in plasma NE levels are generally accompanied by similar changes in plasma free fatty acid levels (Scheurink et al., 1990).

Plasma CA responses to glucoprivation were similar to those reported previously (Kvetňanský et al., 1998). Both 2DG and the hypoglycemic dose of insulin increased adrenomedullary secretion as demonstrated by the marked increases in plasma EPI concentrations. The present results show that even basal levels of plasma EPI in LTC and control rats are not very different. The values of plasma EPI in LTC rats injected with insulin are elevated practically to the same extent as control rats, indicating that the sensitivity of the adrenal medullary cells to insulin is not changed in cold-adapted rats. The administration of 2DG, which seems to be a stronger stimulus for the release of adrenomedullary EPI than insulin, resulted in elevated plasma EPI responses at 60 and 120 min after 2DG administration. This might be an effect of almost complete depletion of adrenal medullary CA stores produced by 2DG administration (Kvetňanský et al., 1971b).

Plasma NE levels were also increased during glucoprivation but to a much smaller extent than EPI levels. It has been shown that during stress only about 15–20% of circulating NE was released from the adrenal medulla in rats (Kvetňanský et al., 1979). Thus, even if 2DG does not specifically activate the sympathoneural system its effect on the adrenal medulla might be responsible for the elevated plasma NE levels. Previous results have demonstrated that high circulating levels of EPI increase plasma NE concentrations via β_2-adrenoceptor mediated stimulation of neural NE outflow (Scheurink et al., 1989, 1990).

Nuclei of the ventromedial hypothalamus (VMH) play a key role in the detection of counterregulatory responses to hyperglycemia. Local perfusion of 2DG into the VMH caused a prompt two-fold increase in plasma glucose in association with an elevation of plasma EPI (30-fold) and plasma NE (3.5-fold) (Borg et al., 1995).

Immobilization stress increased levels of plasma CA and activated both the sympathoneural and adrenomedullary systems (Kvetňanský et al., 1998; Pacak et al., 1998). In LTC rats, IMO exaggerated the values of plasma NE and partly also plasma EPI especially at longer IMO intervals (60–120 min). These results are similar to data obtained by Brimijoin et al. (1994) who found that IMO increased levels of CA up to 35-fold. In LTC rats plasma levels of EPI, after reaching peak values around 30 min after the start of IMO, remained at elevated levels during the whole period of IMO, while after IMO of the control group of rats kept at room temperature the EPI levels were significantly reduced. The explanation for these findings might be in greater CA stores, a larger or longer period of CA secretion, or reduced CA degradation or reuptake in cold-adapted rats. The exaggerated levels of plasma NE and also EPI in long-term cold exposed rats might also be a consequence of the readiness of such animals to respond to altered quality or quantity of a novel stressor. Plasma CA levels in long-term cold exposed rats certainly reflect a new homeostasis due to adaptive mechanisms in some central regulatory areas. Novel stressors might interfere with these mechanisms and thus produce the exaggerated responses.

Based on these results, it may be concluded that rats exposed to long-term cold are able to respond to novel stressors, especially to immobilization and 2DG administration, by higher activation of the sympathoneural and adrenomedullary systems compared to rats kept at room temperature. The present findings emphasize the specificity of catecholaminergic systems' response patterns to various stressors.

Acknowledgments

The present work was supported by Fogarty International Research Collaboration Award, grant No TW 00984, and by Slovak Science Grant Agency (VEGA), Grant No 2–610999.

References

Balaz, V., Balazova, E., Blazicek, P., and Kvetňanský, R. (1980). The effect of one-year acclimatization of rats to mountain conditions on plasma catecholamines and dopamine-β-hydroxylase activity. In *Catecholamines and Stress: Recent Advances*, E. Usdin, R. Kvetňanský & I.J. Kopin, Eds, 259–264. Elsevier, New York.

Bhatnagar, S., and Dallman, M. (1998). Neuroanatomical basis for facilitation of hypothalamic-pituitary-adrenal responses to a novel stressor after chronic stress. *Neuroscience* **84**, 1025–1039.

Borg, W.P., Sherwin, R.S., During, M.J., Borg, M.A., and Shulman, G.I. (1995). Local ventromedial hypothalamus glucopenia triggers counter-regulatory hormone release. *Diabetes* **44**, 180–184.

Brimijoin, J., Hammond, P., Khraibi, A.A., and Rochester, M.N. (1994). Catecholamine release and excretion in rats with immunologically induced preganglionic sympathectomy. *Journal of Neurochemistry* **62**, 2195–2204.

Chrousos, G.P. (1998). Stress as a medical and scientific idea and its implications. In *Catecholamines: Bridging Basic Science with Clinical Medicine*, Advances in Pharmacology **42**, pp. 552–556. Academic Press, San Diego.

Fukuhara, K., Kvetňanský, R., Weise, V.K., Ohara, H., Yoneda, R., Goldstein, D.S., and Kopin, I.J. (1996). Effects of continuous and intermittent cold (SART) stress on sympathoadrenal system activity in rats. *Journal of Neuroendocrinology* **8**, 65–72.

Goldstein, D.S., Pacak, K., and Kopin, I.J. (1996). Nonspecificity versus primitive specificity of stress responses. In *Stress: Molecular Genetic and Neurobiological Advances*, R. McCarty, G. Aguilera, E. Sabban & R. Kvetňanský, Eds. pp. 3–22. Gordon and Breach, New York.

Kvetňanský, R., Gewirtz, G.P., Weise, V.K., and Kopin, I.J. (1971a). Catecholamine-synthesizing enzymes in the rat adrenal gland during exposure to cold. *American Journal of Physiology* **220**, 928–931.

Kvetňanský, R., and Mikulaj, L. (1970). Adrenal and urinary catecholamines in rat during adaptation to repeated immobilization stress. *Endocrinology* **87**, 738–743.

Kvetňanský, R., Nankova, B., Hiremagalur, B., Viskupic, E., Vietor, I., Rusnak, M., McMahon, A., Kopin, I.J., and Sabban, E.L. (1996). Induction of adrenal tyrosine hydroxylase mRNA by single immobilization stress occurs even after splanchnic transection and in presence of cholinergic antagonists. *Journal of Neurochemistry* **66**, 138–146.

Kvetňanský, R., Nemeth, S., Vigas, M., Oprsalova, Z., and Jurcovicova, J. (1984). Plasma catecholamines in rats during adaptation to intermittent exposure to different stressors. In *Stress: The Role of Catecholamines and Other Neurotransmitters*, Vol 1. E. Usdin, R. Kvetňanský & J. Axelrod, Eds, 537–562. Gordon and Breach Sci. Publishers, New York.

Kvetňanský, R., Pacak, K., Sabban, E.L., Kopin, I.J., and Goldstein, D.S. (1998). Stressor-specificity of peripheral catecholaminergic activation. In *Catecholamines: Bridging Basic Science with Clinical Medicine*, Advances in Pharmacology **42**, pp. 556–560. Academic Press, San Diego.

Kvetňanský, R., Silbergeld, S., Weise, V.K., and Kopin, I.J. (1971b). Effect of restraint on rat adrenomedullary response to 2-deoxy-D-glucose. *Psychopharmacologia* (Berlin) **20**, 22–31.

Kvetňanský, R., Weise, V.K., Thoa, N.B., and Kopin, I.J. (1979). Effects of chronic guanethidine treatment and adrenal medullectomy on plasma levels of catecholamines and corticosterone in forcibly immobilized rats. *Journal of Pharmacology and Experimental Therapeutics*, **209**, 287–291.

Liu, X., Perusse, F., and Bukowiecki, L.J. (1994). Chronic norepinephrine infusion stimulates glucose uptake in white and brown adipose tissue. *American Journal of Physiology* **226**, 914–920.

Pacak, K., Palkovits, M., Yadid, G., Kvetňanský, R., Kopin, I.J., and Goldstein, D.S. (1998): Heterogenous neurochemical responses to different stressors: a test of Selye's doctrine of nonspecificity. *American Journal of Physiology* **275**, R1247–R1255.

Peuler, J.D., and Johnson, G.A. (1977). Simultaneous single isotope radioenzymatic assay of plasma norepinephrine, epinephrine and dopamine. *Life Science* **21**, 625–636.

Scheurink, A.J.W., Steffen, A.B., Bouritius, H., Dreteler, G.H., Brutnik, R., Remie, R., and Zaagsma, J. (1989). Adrenal and sympathetic catecholamines in exercising rats. *American Journal of Physiology* **250**, 1003–1006.

Scheurink, A.J.W., and Steffens, A.B. (1990). Central and peripheral control of sympathoadrenal activity and energy metabolism in rats. *Physiology and Behavior* **48**, 909–920.

Scheurink, A.J.W., DeBoer, S.F., Van Dijk, G., and Steffens, A.B. (1996). Central and peripheral mechanisms involved in the regulation of sympathoadrenal outflow. In *Stress: Molecular Genetic and Neurobiological Advances*, R. McCarty, G. Aguilera, E. Sabban & R. Kuetnansky, Eds, pp. 227–241. Gordon and Breach, New York.

Vietor, I., Rusnak, M., Viskupic, E., Blazicek, P., Sabban, E.L., and Kvetňanský, R. (1996). Glucoprivation by insulin leads to trans-synaptic increase in rat adrenal tyrosine hydroxylase mRNA levels. *European Journal of Pharmacology* **313**, 119–127.

17. Transcriptional and Post-Transcriptional Regulation of Corticotropin Releasing Hormone and Vasopressin Expression by Stress and Glucocorticoids

G. Aguilera, S.L. Lightman and X-M. Ma

Introduction

Acute stress activates the hypothalamic-pituitary-adrenal (HPA) axis, resulting in a rapid and self-limited increase in both circulating ACTH and glucocorticoids (Dallman et al., 1992; Aguilera, 1994). Depending on the stress paradigm, a repeated stressor of the same type (homotypic) may or may not result in adaptation of the ACTH response to the stimulus. On the other hand there is almost invariably a facilitation of the response to a novel or heterotypic stress (Aguilera, 1994). The main regulators of ACTH secretion in the pituitary corticotroph are corticotropin releasing hormone (CRH) and vasopressin (VP). These neuropeptides are synthesized by parvocellular neurons in the hypothalamic paraventricular nucleus (PVN), and secreted into the pituitary portal circulation through nerve terminals in the external zone of the median eminence (Vale et al., 1983; Antoni, 1993). Although VP is a weak ACTH secretagogue on its own, it acts synergistically with CRH, and plays an important role in sustaining pituitary responsiveness during chronic stress (Gilles et al., 1982; Antoni, 1993; Aguilera, 1994). There are two populations of CRH neurons in the PVN, one in which only CRH can be detected, and another in which both CRH and VP co-exist (Whitnall, 1993). Studies based on the levels of immunoreactive peptide and mRNA for CRH and VP have shown that while the expression of both peptides is activated by stress, a preferential stimulation of VP occurs in conditions associated with sustained activation of the HPA axis, such as chronic stress and adrenalectomy (Antoni, 1993; Whitnall, 1993; Aguilera, 1994). This suggests that differential regulation of CRH and VP can modulate corticotroph responsiveness. Important questions such as the relative roles of CRH and VP in mediating responses to acute and chronic stress, and the molecular mechanisms regulating the expression of these peptides have only recently been addressed with the development of intronic in situ hybridization procedures (Herman, 1990; Kovacs and Sawchenko, 1996; Ma et al., 1997a,b,c). Since introns are rapidly spliced out, intronic hybridization detects newly transcribed or heteronuclear (hn) RNA, which reflects changes in gene transcription. Because of the very low levels detected under basal conditions, measurements of hnRNA are a much more sensitive index of neuronal response than assessment of steady state mRNA or immunoreactive peptide levels. This article will describe recent studies on transcriptional and posttranscriptional mechanisms by which stress and glucocorticoids regulate CRH and VP mRNA levels.

CRH and VP Responses to Acute Stress

Acute stress stimulates release of CRH and VP from the median eminence into the pituitary portal circulation and increases the expression of both peptides in parvocellular neurons of the PVN (Berkenbosh et al., 1989; Plotsky, 1991). Measurement of the changes in primary transcript using intronic probes for CRH and VP reveal not only different kinetics of transcriptional activation for both peptides, but that the pattern of activation varies depending on the stress paradigm (Ma et al., 1997a; Kovacs and Sawchenko, 1996; Ma and Aguilera, 1999). For example, ip injection of hypertonic saline results in rapid activation of the HPA axis with a 5-fold increase in plasma corticosterone within 15 min and peak levels at about 2 h. This is associated with a rapid increase in CRH hnRNA which falls back to normal levels by 1 h, followed by a more gradual increase in CRH mRNA peaking at 2 h. VP transcripts respond in a different time frame with no change in hnRNA at 15 min, a peak response at 2 h with levels remaining elevated up to 4 h (Ma and Aguilera, 1999). While a similar pattern of increase in CRH hnRNA preceding the increase in VP has been described in response to ether exposure (Kovacs and Sawchenko, 1996), CRH and VP hnRNA increase simultaneously during restraint stress, with significant rises at 1 and 2 h, returning to basal levels by 4 h (Ma et al., 1997a). The different time course of CRH and VP transcript responses imply different regulatory mechanisms for the transcription of these genes as has been suggested in a number of reports (Herman et al., 1990; DeGoeij et al., 1992; Kovacs et al., 1998). Differential sensitivity to corticosterone feedback (Kovacs and Sawchenko, 1996; Ma et al., 1997b), second messengers and transcription factors (Kovacs et al., 1998) may all be involved.

In all stress paradigms studied, the increases in CRH and VP primary transcripts were relatively rapid, while the increases in mRNA failed to reach significant levels until one or two hours after the initiation of the stress or, at which time the levels of hnRNA had started to decline or had returned to basal levels (Imaki et al., 1995; Kovacs and Sawchenko, 1996; Ma et al., 1997b; Ma and Aguilera, 1999). It should be noted that there is not only an increase in the prevalence of VP transcripts in the parvocellular PVN, but the number of parvocellular cells containing VP hnRNA and mRNA transcripts increases markedly following acute stress (Ma et al., 1997b; Ma and Aguilera, 1999). This shows that hypothalamic neurons can rapidly change their properties and indicates that most parvocellular cells have the capacity to cosecrete CRH and VP under appropriate conditions.

CRH and VP Responses to Repeated Stress

Depending on the stress paradigm, repeated stress may or may not result in habituation of the HPA axis responses to the homotypic stimulus. While adaptation of plasma ACTH and corticosterone responses is observed during cold exposure, restraint stress and water immersion (Vernikos et al., 1982; Hauger et al., 1990; Hauger and Aguilera, 1993), responses are preserved during repeated painful stimuli such as ip hypertonic saline injection or foot shock, or metabolic disturbance by repeated insulin-induced hypoglycemia (DeGoeij et al., 1992; Kiss and Aguilera, 1993). In contrast to the variable response to the homotypic stressor, there is good evidence that during chronic stress there is preservation or even sensitization of the hormonal response to a heterotypic stressor in spite of desensitization to the repeated stimulus. It has been shown that parvocellular VP mRNA levels rise significantly after repeated immobilization or restraint (Bartanusz et al., 1993; Ma et al., 1997c), foot shock (Sawchenko et al., 1993), or ip hypertonic saline injection (Ma and Aguilera, 1999), while CRH mRNA levels are elevated only in stress paradigms which show preserved plasma hormone responses to the repeated stimulus, such as repeated ip hypertonic saline injection or foot shock (Imaki et al., 1991; Kiss and Aguilera, 1993; Ma and Aguilera, 1999). Similarly, daily repeated immobilization does not affect irCRH stores but causes a progressive increase in VP stores, and in the number of CRH nerve endings containing VP in the external zone of the median eminence (DeGoeij et al., 1991). In addition, in vivo and in vitro studies have shown that the rate of release of immunoreactive VP from median eminence terminals increases in response to repeated or chronic stress (DeGoeij et al., 1992; Aguilera et al., 1993). Even a transient activation of hypothalamic CRH neurons by a single stressor can cause long lasting increases in VP co-expression irrespective of the nature of the stressor, which in most cases is not accompanied by changes in CRH (Schmidt et al., 1997).

Analysis of the changes in CRH hnRNA during repeated stress indicates that adaptation of hypothalamic CRH responses occurs at least in part at the transcriptional level. Thus, the decrease in plasma ACTH and corticosterone and hypothalamic CRH mRNA responses to repeated restraint are associated with adaptation of the CRH hnRNA response, which are proportional to the frequency of the stimulus (Ma et al., 1998, Figure 1). After 2 weeks of daily restraint for one hour there is no change in basal corticosterone, CRH hnRNA or CRH mRNA, and there is a complete lack of CRH hnRNA response to an additional restraint episode (Ma et al., 1998). On the other hand, increased CRH mRNA levels in the PVN and preserved plasma hormone responses to repeated ip hypertonic saline injection, are accompanied by preserved CRH hnRNA responses (Ma and Aguilera, 1999). In contrast to the varying CRH responses, VP hnRNA and mRNA responses to a repeated homotypic stressor are usually preserved or increased, even when there is desensitization of CRH responses (Ma et al., 1997c; Ma et al., 1999).

Parvocellular neurons with CRH transcription responses showing adaptation to a repeated homotypic stressor are capable of responding to a novel stressor. When rats adapted to 14 days of daily restraint for one hour underwent the heterotypic stressor of ip hypertonic saline injection, CRH hnRNA and mRNA responses were preserved, and the peak response of both was greater than that seen in the control naive rats

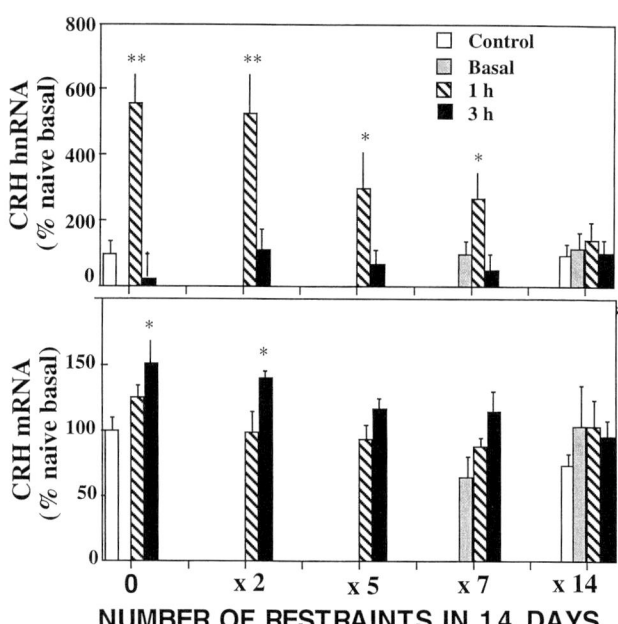

Fig. 1 Effect of the frequency of repeated restraint stress on CRH hnRNA (upper panel) and CRH mRNA (lower panel) responses to an additional episode of restraint. Rats subjected to one hour of restraint for the number of days indicated were sacrificed 1 and 3 h after an additional restraint session. Bars represent the mean and SEM of 6 rats per experimental group. **, p < 0.01 vs controls; * p < 0.05 vs controls.

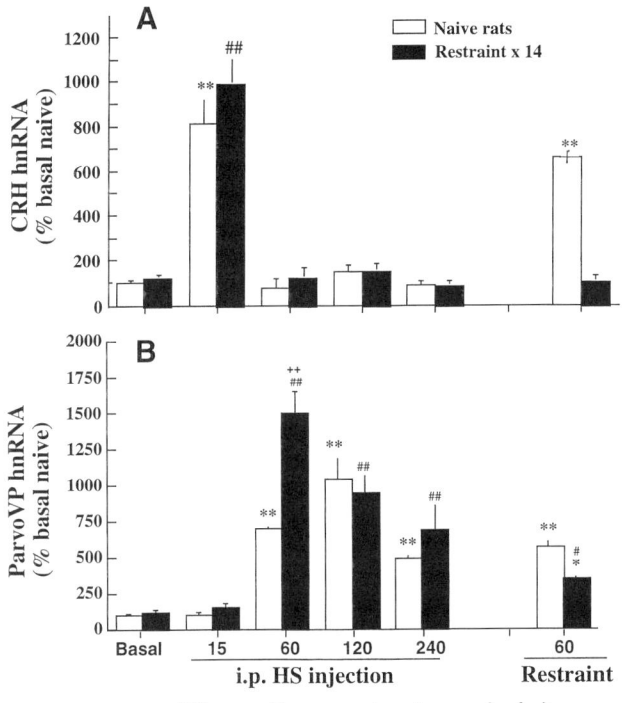

Fig. 2 Levels of CRH hnRNA (A) and VP hnRNA (B) in the parvocellular PVN in response to an acute homotypic restraint or heterotypic ip hypertonic saline injection (ipHS) in naïve rats or rats subjected to repeated restraint for 1 h per day for 14 days (restraint × 14). Bars represent the mean and SEM of the values obtained by *in situ* hybridization in 5 or 6 rats per experimental group, expressed as the percent of change from basal in naïve rats. *, p < 0.05 vs basal naïve; ** p < 0.01 vs basal naïve; #, p < 0.05 vs basal restraint × 14; # #, p < 0.01 vs basal restraint × 14.

(Figure 2-A). In response to ip hypertonic saline, VP hnRNA also reaches peak levels earlier in the repeatedly restrained animals than in the controls. VP mRNA, despite starting at much higher prevalence in the repeatedly restrained animals, increases in a similar proportion as in control rats, achieving much higher absolute levels (Figure 2-B). Both control and repeatedly restrained rats showed a similar increase in the number of parvocellular PVN cells containing VP hnRNA transcripts, but the relative increase in the number of cells expressing VP mRNA was greater for the control rats since the repeatedly restrained animals started from a higher basal level.

It is clear from these experiments that animals which have adapted to repeated restraint stress maintain a normal or increased CRH hnRNA and mRNA response to a heterotypic stressor. The differential responses of the parvocellular neuron to repeated homotypic stress and novel heterotypic stimuli suggest the activation of distinct stimulatory and inhibitory pathways by the different stressors. It is possible that adaptation to the repeated homotypic stressor is due to desensitization of afferent pathways to the PVN at the synaptic level or at the parvocellular neuron itself, and that the novel

stressor uses different pathways and neurotransmitters. However, despite the habituation, there is little evidence of desensitization, since VP responses to the repeated stressor are preserved (Ma *et al.*, 1997a; Ma *et al.*, 1998; Ma *et al.*, 1999), and microdialysis experiments have shown that norepinephrine turnover in the PVN is increased rather than decreased during repeated immobilization (Pakac *et al.*, 1992). In contrast to restraint, the stress of ip hypertonic saline does not result in habituation of the HPA axis response after repeated stimuli (Aguilera, 1994; Ma and Aguilera, 1999), suggesting that different pathways are involved. Facilitation of the HPA axis response, however, has been described for a variety of stress combinations, irrespective of the ability of the stimuli to cause adaptation during repeated exposure (Dallman *et al.*, 1992), suggesting that the mechanism of differential responses is more complex than simply different pathways and neurotransmitters. For example, processing and integration of stimuli in the limbic system could activate or suppress inhibitory pathways to the PVN, enhancing or inhibiting parvocellular neuron responses depending on previous experience and type of stimulus (Li *et al.*, 1996; Cullinan *et al.*, 1994). The exact mechanisms of the differential responses to homotypic and heterotypic stressors remain to be elucidated but the data suggest that the neurotransmitter repertoire (and pathways) activated by different forms of stressors is distinct enough to be distinguished by the stress responsive cells in the parvocellular PVN.

Effect of Glucocorticoids

Basal levels of CRH and VP expression are under negative feedback inhibition by glucocorticoids. Immunoreactive (ir) CRH and CRH mRNA levels in the PVN decrease after glucocorticoid administration or implantation of dexamethasone in the PVN (Kovacs *et al.*, 1986; Lightman and Young, 1989; Sawchenko, 1987), but increase markedly following adrenalectomy, the latter effect being prevented by glucocorticoid replacement (Jingami *et al.*, 1985; Young *et al.*, 1986; Beyer *et al.*, 1988; Swanson and Simons, 1989). The increases in CRH mRNA after adrenalectomy are slow, requiring 48 h to become detectable and reaching stable levels of 2- to 3-fold above controls after 72 hr (Lightman and Young, 1989; Brown and Sawchenko, 1997; Ma *et al.*, 1997b). Consistent with the delayed mRNA responses, activation of transcription is also slow, with CRH hnRNA starting to increase only 12 h after surgery in spite of the fact that glucocorticoids had been cleared from the circulation by 3 h. In another report, increases in CRH transcription occurred within 30 min of chemical adrenalectomy by injection of the steroid 11-beta-hydroxylase inhibitor, metyrapone (Herman *et al.*, 1992). The reason for this discrepancy is not clear, but it is possible that metyrapone injection is sufficiently stressful to cause an increase in CRH transcription under conditions in which circulating glucocorticoids start to decrease due to blockade of the steroidogenic pathway. A slow action of glucocorticoids on transcription would be consistent with the lag time observed between glucocorticoid administration and inhibition of CRH mRNA (Lightman and Young, 1989; Swanson and Simmonds, 1989; Ma *et al.*, 1997b).

After long-term adrenalectomy (6 days) the levels of CRH hnRNA are very low, similar to basal levels in intact rats (Imaki *et al.*, 1995, Ma *et al.*, 1997b). Recent analysis of the time course of changes in CRH expression during adrenalectomy show that the increases in CRH hnRNA reach maximal levels by 48 h before starting to decline (Ma and Aguilera, 1999b; Table 1). This suggests that the increase in CRH transcription is not the only mechanism responsible for the sustained increases in CRH mRNA and that changes in CRH mRNA stability may contribute to the regulation of CRH mRNA levels during adrenalectomy. Supporting this possibility, *in situ* hybridization studies in which hypothalamic sections were incubated *in vitro* before fixation, show a longer half-life of CRH mRNA in 6-day adrenalectomized rats than in sham-operated controls. In contrast, corticosterone administration markedly reduces CRH mRNA stability in both adrenalectomized and sham-operated rats, and there is a marked decrease after corticosterone injection (Table 2).

In contrast to the 12 h lag period preceding the rise in CRH hnRNA, VP hnRNA is already elevated 3 h after adrenalectomy. This would be consistent with the notion that VP transcription is more sensitive to glucocorticoid feedback than CRH transcription. However, the pattern of change in VP transcription after adrenalectomy is biphasic, decreasing markedly by 18 h, before increasing again progressively between 24 and 72 h. It is possible that the early rise in VP hnRNA is the result of an exaggerated response to surgical stress in the absence of glucocorticoids rather than to

Table 1 Time course of the changes in VP hnRNA and CRH hnRNA in parvocellular neurons of the PVN following adrenalectomy. Rats were adrenalectomized (ADX) or sham operated and killed at indicated times. Values are the mean and SEM of the data obtained in 6 rats per group. No significant changes in CRH or VP hnRNA levels were observed at any time after sham operation.

Time after ADX (h)	CRH hnRNA (% sham ADX)	VP hnRNA (% sham ADX)
3	111.6 ± 16.8	535.5 ± 67.5**
6	103.8 ± 11.5	1151.4 ± 158.3**
12	176.0 ± 32.4**	674.9 ± 188.3**
18	242.8 ± 5.7**	219.9 ± 4.8* #
24	275.9 ± 41.2**	507.7 ± 73.3**
48	329.8 ± 8.9**	1024.5 ± 117.1**
72	262.2 ± 43.4**	1270.7 ± 207.7**

*, $p < 0.05$ vs sham ADX; **, $p < 0.01$ vs sham ADX; #, $p < 0.05$ lower than 12 h or 48 h.

Table 2 Effect of 6 days adrenalectomy or *in vivo* administration of glucocorticoids (900 μg/day) on CRH mRNA stability in the PVN. Sets of hypothalamic sections from individual rats were incubated at 37°C for 15 to 120 min before fixation in 4% formaldehyde and measurement of CRH mRNA levels by *in situ* hybridization. CRH mRNA half-life was calculated from the values in 4 rats per experimental group. *, $p < 0.05$ vs sham controls; #, $p < 0.01$ vs sham operated rats and adrenalectomized controls.

Experimental condition	CRH mRNA half life (min)
Sham operation	12.5 ± 1.2
Sham operation + corticosterone	5.0 ± 0.3*
6 day ADX	28.5 ± 2.4 #
6 day ADX + corticosterone	6.5 ± 0.8*

*, $p < 0.01$ vs no corticosterone; #, 0.01 vs sham operation.

glucocorticoid withdrawal per se. Although the effect of stress could mask the onset of the VP response to glucocorticoid withdrawal, the low VP hnRNA levels at 18 hr suggest that, similar to CRH, the release of VP transcription from glucocorticoid inhibition is not immediate after removal of glucocorticoids (Table 1).

Glucocorticoids influence not only basal levels of CRH and VP expression but also modulate the responsiveness of parvocellular neurons to stress. Recent experiments in adrenalectomized rats show that a minor stressor such as a vehicle injection causes marked increases in CRH hnRNA and VP hnRNA in the absence of glucocorticoids (Ma and Aguilera, 1999b). The same injection caused only minor increases in plasma corticosterone and had no detectable effect on CRH or VP hnRNA levels in sham-operated rats, indicating that the parvocellular neuron becomes much more sensitive to stress in the absence of glucocorticoids. This is in agreement with studies showing enhanced CRH and VP mRNA responses to stress in adrenalectomized compared with intact rats (Makino *et al.*, 1995), and suggests that under normal conditions, low prevailing levels of glucocorticoids restrain the responsiveness of the parvocellular neuron to stress. This action of low levels of glucocorticoids may act as a protective mechanism to prevent inappropriate activation of the HPA axis in response to minor stimuli that occur frequently in physiological conditions.

While glucocorticoid withdrawal sensitizes responses of both CRH and VP to stress, exogenous glucocorticoids have differential effects on these genes, markedly inhibiting VP but not CRH transcription. In 48-h adrenalectomized rats, a time at which CRH hnRNA levels are elevated, a corticosterone injection does not inhibit CRH transcription, either basal or stimulated by vehicle injection, in spite of rapid and massive increases in plasma corticosterone levels (Figure 3-A; Ma and Aguilera, 1999b). This lack of CRH hnRNA inhibition by corticosterone suggests that feedback inhibition of CRH mRNA levels by glucocor-

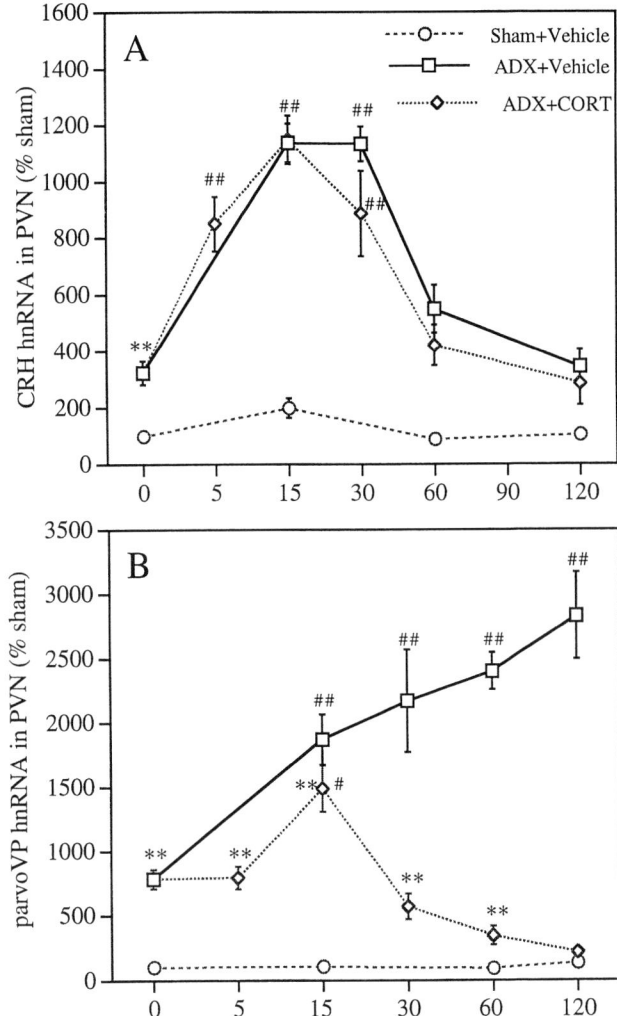

Fig. 3 Time course of the changes in CRH hnRNA (A) and VP hnRNA (B) after injection of corticosterone (2.8 mg/100 g BW, i.p.), or vehicle in 48-h adrenalectomized rats (ADX) or sham-operated rats. Data points are the mean and SE of the values obtained by *in situ* hybridization in 6 rats per experimental group. ** p < 0.01 vs sham basal 0 min; #, p < 0.05 vs ADX basal 0 min; ##, p < 0.01 vs ADX basal 0 min.

ticoids may depend more on regulation of mRNA stability rather than on transcriptional inhibition. In addition, the similar kinetics of CRH hnRNA irrespective of the levels of circulating corticosterone, suggests that glucocorticoids are not responsible for the usually self-limiting CRH hnRNA observed during stress (Kovacs and Sawchenko, 1996; Ma *et al.*, 1997b; Ma and Aguilera, 1999).

In striking contrast to the lack of effect of corticosterone on CRH expression, there is a rapid and marked inhibition of VP hnRNA responses to the stress of vehicle injection (Figure 3-B). Corticosterone decreases the elevated basal VP hnRNA levels observed after 48-h or 6 day adrenalectomy (Ma *et al.*, 1997b; Ma and Aguilera, 1999b), and also reduces the effect of vehicle injection stress. This supports the hypothesis that VP

mRNA levels in the parvocellular PVN are largely controlled at the level of transcription.

It has been postulated that differential sensitivity of CRH and VP to glucocorticoid feedback explains the differential regulation in expression of both peptides during stress. This hypothesis is based on the demonstration that the declining phase of CRH hnRNA responses to ether exposure are preceded by peak corticosterone response, while the VP hnRNA responses coincide with the decrease in plasma corticosterone (Kovacs and Sawchenko, 1996). The demonstration in 48-h adrenalectomized rats that CRH hnRNA responses to the stress of vehicle-injection decline to basal levels by 1 h in the absence of the stress-induced glucocorticoid surge, indicates that this self-limiting pattern of response is independent of glucocorticoid feedback. The fact that at the time of the plasma corticosterone peak VP hnRNA reaches levels similar to vehicle-injected rats, goes against the possibility that the stress-induced glucocorticoid surge causes the delay in VP transcription response described during ether exposure and ip hypertonic saline injection (Kovacs and Sawchenko, 1996; Ma and Aguilera, 1999). In stress models, such as restraint and ip hypertonic saline injection, the onset of VP transcriptional responses also occurs at the time of maximal increases in plasma corticosterone (Ma *et al.*, 1997a; Ma and Aguilera, 1999), and VP hnRNA can be sustained for up to 4 h, despite equally sustained elevations in plasma corticosterone (Ma and Aguilera 1999). While it is clear from these experiments and previous reports (Ma *et al.*, 1997b) that VP transcription is highly sensitive to glucocorticoid inhibition, stress can overcome this inhibition to different degrees, probably depending on the glucocorticoid levels and intensity of the stressor.

The mechanism responsible for the modulatory effect of stress on glucocorticoid inhibition may involve interaction of glucocorticoids with neurotransmitters and neuropeptides activated by stress (Herman and Cullinan, 1997; Kiss, 1998). At the molecular level, second messenger-induced expression of intermediate early genes or transcription factors could modify glucocorticoid receptor activity by heterodimerization with the receptor, or by binding to regulatory elements, changing the conformation of the promoter (Diamond *et al.*, 1990; Yan-Yen *et al.*, 1990; Autelitano *et al.*, 1996). In the parvicellular cell, increases in VP hnRNA induced by ether exposure are closely preceded by the induction of c-fos mRNA and fos protein (Kovac and Sawchenko, 1996), which has been shown to interact with the glucocorticoid receptor, modifying its activity at the DNA responsive element (Pearce *et al.*, 1998).

Summary and Conclusions

Corticotropin releasing hormone (CRH) and vasopressin (VP), the main hypothalamic regulators of pituitary ACTH secretion, co-exist in parvocellular neurons of the PVN but their levels of expression are differentially regulated during manipulations of the HPA axis. Acute somatosensory stressors cause rapid and pronounced increases in CRH and VP transcription in parvocellular PVN neurons. On the other hand, after repeated homotypic stress, VP mRNA and hnRNA levels are elevated, but CRH responses vary in parallel with the ability of the stressor to cause desensitization of ACTH responses to the repeated stimulus. Despite adaptation to a repeated stimulus, CRH responsiveness to a heterotypic stimulus is preserved or even increased. Adrenalectomy increases CRH hnRNA only transiently but also increases CRH mRNA stability, which may account for the long-term increases in CRH mRNA. In contrast, VP hnRNA remained elevated during long-term adrenalectomy, suggesting that transcriptional activation is largely responsible for maintaining the high VP mRNA levels. VP mRNA and hnRNA levels are always elevated after a repeated homotypic stressor, whereas CRH responses vary in parallel with the ability of the stressor to cause desensitization of ACTH responses to the repeated stimulus. Despite adaptation to a repeated stimulus, CRH responsiveness to a heterotypic stimulus is preserved or even increased. Adrenalectomized but not control rats show CRH and VP hnRNA responses to the mild stress of vehicle injection. While CRH hnRNA responses are refractory to glucocorticoid feedback, VP hnRNA responses are rapidly inhibited. In conclusion, parvocellular neurons show plasticity to differentially respond to repeated homotypic and novel stimuli. While the prevailing levels of glucocorticoids restrain the responsiveness of parvocellular neurons, increases in glucocorticoids alone cannot explain the differential regulation of CRH and VP during stress.

References

Aguilera, G., Kiss, A., Hauger, R.L., and Tizabi, Y. (1993). Regulation of the hypothalamic pituitary adrenal axis during stress: role of neuropeptides and neurotransmitters. In Kvetňanský, R., McCarty, R., and Axelrod, J., eds. *Stress: Neuroendocrine and molecular approaches.* Gordon and Breach, New York, 365–381.

Aguilera, G. (1994). Regulation of pituitary ACTH secretion during chronic stress. *Frontiers in Neuroendocrinology* **15**, 321–350.

Aguilera, G. (1998). Corticotropin releasing hormone, receptor regulation and the stress response. *Trends in Endocrinology and Metabolism* **9**, 329–336.

Antoni, F.A. (1993). Vasopressinergic control of adrenocorticotropin secretion comes of age. *Frontiers in Neuroendocrinology* **14**, 76–122.

Autelitano, D.J., and Cohen, D.R. (1996). CRF stimulates expression of multiple fos and jun related genes in the AtT-20 corticotroph cell. *Molecular and Cellular Endocrinology* **119**, 25–35.

Bartanusz, V., Aubry, J.-M., Jevoza, D., Baffi, J., and Kiss, J.Z. (1993). Upregulation of vasopressin mRNA in paraventricular hypophysiotrophic neurons after acute immobilization stress. *Neuroendocrinology* **58**, 625–629.

Berkenbosh, F., deGoeij, D., and Tilders, F.J.H. (1989). Hypoglycemia enhances turnover of corticotropin releasing factor and vasopressin in the zona externa of the rat median eminence. *Endocrinology* **125**, 28–34.

Beyer, H.S., Matta, S.G., and Sharp, B.M. (1988). Regulation of the messenger ribonucleic acid for corticotropin releasing factor in the paraventricular nucleus and other brain regions. *Endocrinology* **123**, 2117–2123.

Brown, E.R., and Sawchenko, P.E. (1997). Hypophyseotropic CRF neurons display a sustained immediate-early gene response to chronic stress but not to adrenalectomy. *Journal of Neuroendocrinology* **9**, 307–316.

Cullinan, W.E., Herman, J.P., Battaglia, D.F., Akil, H., and Watson, S.J. (1994). Pattern and time course of immediate early gene expression in rat brain following acute stress. *Neuroscience* **64**, 477–505.

Dallman, M.F., Akana, S.F., Scribner, K.A., Bradbury, M.J., Walker, C.D., Strack, A.M., and Casio, C.S. (1992). Stress, feedback and facilitation in the hypothalamus pituitary adrenal axis. *Journal of Neuroendocrinology* **4**, 517–526.

De Goeij, D.C.E., Binnekade, R., and Tilders, F.J.H. (1992). Chronic stress enhances vasopressin but not corticotropin-releasing factor secretion during hypoglycemia. *American Journal of Physiology* **263**, E394–E399.

De Goeij, D.C.E., Kvetňanský, R., Whitnall, M.H., Jezova, D., Berkenbosch, F., and Tilders, F.J.H. (1991). Repeated stress-induced activation of corticotropin-releasing factor neurons enhances vasopressin stores and colocalization with corticotropin releasing factor neurons in the median eminence of rats. *Neuroendocrinology* **53**, 150–159.

Diamond, M.I., Miner, N.J., Yoshinaga, S.K., and Yamamoto, K.R. (1990). Transcription factor interactions: selector of positive and negative regulation from a single DNA element. *Science* **249**, 1266–1272.

Gilles, G., Linton, E.A., and Lowry, P.F. (1982). Corticotropin releasing activity of the new CRF is potentiated several times by vasopressin. *Nature* **299**, 355–357.

Harbuz, M.S., and Lightman, S.L. (1989). Responses of hypothalamic and pituitary mRNA to physical and physiological stress in the rat. *Journal of Endocrinology* **122**, 705–711.

Hauger, R. L., Lorang, M., Irwin, M., and Aguilera, G. (1990). CRF receptor regulation and sensitization of ACTH responses to acute stress during chronic intermittent immobilization stress. *Brain Research* **532**, 34–40.

Hauger, R.L., and Aguilera, G. (1993). Regulation of pituitary corticotropin releasing hormone (CRH) receptors by CRH: Interaction with vasopressin. *Endocrinology* **133**, 1708–1714.

Herman, J.P., and Cullinan, W.E. (1997). Neurocircuitry of stress: central control of the hypothalamo-pituitary-adrenocortical axis. *Trends in Neuroscieuce* **20**, 78–84.

Herman, J.P., Schafer, M.K.H., Thompson, R.C., and Watson, S.L. (1992). Rapid regulation of corticotropin releasing hormone gene transcription in vivo. *Molecular Endocrinology* **6**, 1061–1069.

Herman, J.P., Wiegand, S.J., and Watson, S.J. (1990). Regulation of basal corticotropin-releasing hormone and arginine vasopressin messenger ribonucleic acid expression in the paraventricular nucleus: effects of selective hypothalamic deafferentations. *Endocrinology* **127**, 2408–2417.

Imaki, T., Wang, X.-Q., Shibasaki, T., Yamada, K., Chikada, N., Naruse, M., and Demura, H. (1995). Stress induced activation of neuronal activity and corticotropin releasing factor gene expression in the paraventricular nucleus is modulated by glucocorticoids in the rat. *Journal of Clinical Investigation* **96**, 231–238.

Imaki, T., Nahan, J.-l., Rivier, C., Sawchenko, P.E., and Vale, W.W. (1991). Differential regulation of corticotropin releasing factor mRNA in rat brain regions by glucocorticoids and stress. *Journal of Neuroendocrinology* **11**, 585–599.

Jingami, H., Matsukura, S., Numa, S., and Imura, H. (1985). Effects of adrenalectomy and dexamethasone administration on the levels of prepro-corticotropin releasing factor messenger ribonucleic acid (mRNA) in the hypothalamus and adrenocorticotropin/B-lipotropin processor mRNA in the pituitary in rats. *Endocrinology* **117**, 1314–1320.

Kiss, A., and Aguilera, G. (1993). Regulation of the hypothalamic pituitary adrenal axis during chronic stress: responses to repeated intraperitoneal hypertonic saline injection. *Brain Research* **630**, 262–270.

Kiss, J.Z. (1998). Dynamism of chemoarchitecture in hypothalamic paraventricular nucleus. *Brain Research Bulletin* **20**, 699–708.

Kovács, K.J., Arias, C., and Sawchenko, P.E. (1998). Protein synthesis blockade differentially affects the stress-induced transcriptional activation of neuropeptide genes in parvocellular neurosecretory neurons. *Molecular Brain Research* **54**, 85–91.

Kovacs, K., Kiss, J.Z., and Makara, G.B. (1986). Glucocorticoid implants around the hypothalamic paraventricular nucleus prevent the increase of corticotropin releasing factor and arginine vasopressin immunostaining induced by adrenalectomy. *Neuroendocrinology* **44**, 229–234.

Kovacs, K.J., and Sawchenko, P.E. (1996). Sequence of stress induced alterations in indices of synaptic and transcriptional activation in parvocellular secretory neurons. *Journal of Neuroscience* **16**, 262–273.

Li, H.-Y., Erickson, A., and Sawchenko, P.E. (1996). Distinct mechanisms underlie activation of hypothalamic neurosecretory neurons and their medullary catecholaminergic afferents in categorically different stress paradigms. *Proceedings of the National Academy of Science USA* **93**, 2359–2364.

Lightman, S.L., and Young, W.S. (1989). Influence of steroids on the hypothalamic corticotropin releasing factor and proenkephalin mRNA responses to stress. *Proceedings of the National Academy of Sciences USA* **86**, 4306–4310.

Ma, X.-M., and Aguilera, G. (1999a). Transcriptional responses of the vasopressin and corticotropin releasing hormone genes to acute and repeated intraperitoneal hypertonic saline injection. *Molecular Brain Research* **68**, 129–140.

Ma, X.-M., and Aguilera, G. (1999b). Differential regulation of corticotropin releasing hormone and vasopressin transcription by glucocorticoids. *Endocrinology* **140**, 5642–5650.

Ma, X.-M., Levy, A., and Lightman, S.L. (1997a). Rapid changes in heteronuclear RNA for corticotropin releasing hormone and arginine vasopressin in response to acute stress. *Journal of Endocrinology* **152**, 81–89.

Ma, X.-M., Levy, A., and Lightman, S.L. (1997b). Rapid changes of heteronuclear RNA for arginine vasopressin but not for corticotropin releasing hormone in response to acute corticosterone administration. *Journal of Neuroendocrinology* **9**, 723–728.

Ma, X.-M., Levy, A., and Lightman, S.L. (1997c). Emergence of an isolated VP response to stress following repeated restraint: a study of both VP and CRH mRNA and hnRNA. *Endocrinology* **138**, 4351–4357.

Ma, X.-M., and Lightman, S.L. (1998). The arginine vasopressin and corticotrophin-releasing hormone gene transcription responses to varied frequencies of repeated stress in rats. *Journal of Physiology (London)* **510**, 605–614.

Ma, X.-M., Lightman, S.L., and Aguilera, G. (1999). Vasopressin and corticotropin releasing hormone gene responses to novel stress in rats adapted to repeated restraint. *Endocrinology* **140**, 3623–3632.

Makino, S., Smith, M.A., and Gold, P.W. (1995). Increased expression of corticosterone-releasing hormone and vasopressin messenger ribonucleic acid (mRNA) in the hypothalamic paraventricular nucleus during repeated stress: association with reduction in glucocorticoid receptor mRNA levels. *Endocrinology* **136**, 3299–3309.

Pacak, K., Armando, I., Fukuhara, K., Kvetňanský, R., Palkovits, M., and Goldstein, D.S. (1992). Noradrenergic activation of the paraventricular nucleus during acute and chronic immobilization stress in rats: An in vivo microdialysis study. *Brain Research* **589**, 91–96.

Pearce, D., Matsui, W., Miner, J.N., and Yamamoto, K.R. (1998). Glucocorticoid receptor transcriptional activity determined by spacing of receptor and non-receptor DNA sites. *Journal of Biological Chemistry* **273**, 30081–30085.

Plotsky, P.M. (1991). Pathways to the secretion of adrenocorticotropin: a view from the portal. *Journal of Neuroendocrinology* **3**, 1–9.

Sawchenko, P.E. (1987). Evidence for a local site of action for glucocorticoids in inhibiting CRF and vasopressin expression in the paraventricular nucleus. *Brain Research* **403**, 213–224.

Sawchenko, P.E., Arias, C.A., and Mortrund, M.T. (1993). Local tetradotoxin blocks chronic stress effects on corticotropin releasing factor and vasopressin messenger ribonucleic acid in hypophysiotropic neurons. *Journal of Neuroendocrinology* **5**, 341–348.

Schmidt, E.D., Janszen, J.W., Binnekade, R., and Tilders, F.J.H. (1997). Transient suppression of resting corticosterone levels induces sustained increase of VP stores in hypothalamic CRH-neurons of rats. *Journal of Neuroendocrinology* **9**, 69–77.

Swanson, L.W., and Simmonds, D.M. (1989). Differential steroid hormone and neural influences on peptide mRNA levels in CRH cells of the paraventricular nucleus: A hybridization histochemistry study in the rat. *Journal of Compative Neurology* **285**, 413–435.

Vale, W.W., Rivier, C., Brown, M.R., Spiess, J., Koob, G., Swanson, L., Bilezikjian, L., Bloom, F., and Rivier, J. (1983). Chemical and biological characterization of corticotropin releasing factor. *Recent Progress in Hormone Research* **39**, 245–270.

Vernikos, J., Dallman, M.F., Bonner, C., Katzen, A., and Shinsako, J. (1982). Pituitary-adrenal function in rats chronically exposed to cold. *Endocrinology* **110**, 413–420.

Whitnall, M.H. (1993). Regulation of the hypothalamic corticotropin releasing hormone (CRH) and other neuropeptide genes. *Progress in Neurobiology* **40**, 573–629.

Yang-Yen, H.F., Chambard, J.-C., Sun, J.-L., Smeal, T., Schmidt, T.J., Drouin, J., and Karin, M. (1990). Transcriptional interference between c-jun and the glucocorticoid receptor: mutual inhibition of DNA binding due to direct protein-protein interaction. *Cell* **62**, 1205–1215.

Young, W.S., Mezey, E., and Siegel, R.E. (1986). Quantitative in situ hybridization histochemistry reveals increased levels of corticotropin releasing factor mRNA after adrenalectomy in rats. *Neuroscience Letters* **70**, 198–203.

18. Transcriptional Mechanisms of Stress-Triggered Activation of Neurotransmitter Gene Expression

E.L. Sabban, B.B. Nankova, L.I. Serova and R. Kvetňanský

Introduction

Exposure of animals to stress rapidly activates the hypothalamic corticotropin-releasing hormone (CRH) and the locus coeruleus norepinephrine (NE)/autonomic systems along with their peripheral effectors, the pituitary-adrenal axis and the limbs of the autonomic system (Goldstein, 1995). These systems enable the organism to overcome immediate threats to its biological homeostasis (Sapolsky, 1996; McEwen, 1998). However, in cases of extreme or repeated stress (the definition of which is influenced by genetics as well as prior experience), the wear and tear that results from use of the allostatic system can be physiologically damaging (McEwen, 1998).

Stress-elicited changes in the expression of the genes encoding catecholamine biosynthetic enzymes in the adrenal medulla have been intensively studied (Kvetňanský and Sabban, 1993; Kvetňanský and Sabban, 1998). It has been shown that immobilization stress differentially affects tyrosine hydroxylase (TH) and dopamine β-hydroxylase (DBH) mRNA levels (McMahon et al., 1992; Nankova et al., 1993). While a single immobilization event is sufficient to trigger maximal accumulation of TH mRNA, it does not result in significant alterations in steady-state DBH mRNA levels (McMahon et al., 1992; Nankova et al., 1994; Sabban et al., 1995; Sabban et al., 1996). The changes in steady-state TH mRNA levels elicited by a single episode of stress are transient and return to near basal levels when measured one day later. Repeated stress does not further elevate TH mRNA levels, but maintains them high even a week after the last episode of stress. These results confirm that the previously reported increases in TH enzymatic activity triggered by repeated immobilization stress (Kvetňanský et al., 1970) are largely a consequence of increased mRNA levels. Other types of stress, such as chronic cold stress, insulin-induced hypoglycemia, glucoprivation by 2-deoxy-glucose, and foot shock, were also shown to elevate adrenomedullary TH mRNA levels and subsequently if stress was sufficiently prolonged, enzymatic activity increased (Stachowiak et al., 1985; Baruchin et al., 1990; Kvetňanský and Sabban, 1993).

In contrast to a single immobilization, multiple immobilizations led to DBH mRNA levels about 400%–500% higher than in adrenal medulla of the control rats (McMahon et al., 1992; Nankova et al., 1993). Similarly, sustained stimulation of the neural axis by chronic reserpine treatment induces DBH mRNA levels (Wong and Wang, 1994). In both cases, the accumulation of DBH mRNA is followed by elevation of immunoreactive DBH protein and enzyme activity (Kvetňanský et al., 1971; Kvetňanský and Sabban, 1993; Wong and Wang, 1994).

It has become increasingly apparent that transcription is a major mechanism of regulation of adrenomedullary TH gene expression by stress (Nankova and Sabban, 1999). Administration of actinomycin D prevented the stress-elicited rise in adrenal TH mRNA and activity levels, suggesting involvement of transcriptional mechanisms (Kvetňanský and Sabban, 1993; Nankova et al., 1994). These findings provided clear evidence for transcriptional activation during both single and repeated stress. This group using transgenic animals in which the TH promoter directed chloramphenicol acetyl transferase (CAT) reporter gene Osterhout et al. (1997) found that cold or immobilization stress elevated CAT reporter activity. The effect of a single immobilization was transient, returning to control levels one day later. The rise in CAT activity with two daily immobilizations was still evident one day after the cessation of the stress and paralleled the changes that we observed in TH mRNA levels (Nankova et al., 1994). Recently, nuclear run-on assays of transcription directly implicated transcriptional activation of both TH and DBH gene expression in adrenal medulla and locus coeruleus (Nankova et al., 1999b; Serova et al., 1999).

A crucial question is how a transient transcriptional signal is converted to a long lasting alteration in neurotransmission. Here we test the hypothesis that different cascades of transcriptional activation mediated by distinct sets of transcription factors with different kinetics or induction are employed, depending on the duration and reiteration of the stress signal.

Methods

Animals

Male, murine pathogen-free, Sprague-Dawley rats weighing 280 to 320 g were obtained from Taconic Farms (Germantown, NY). The animal care and the immobilization stress were carried out as previously described (Nankova et al., 1994). The adrenal medulla (AM) and the punched region of the locus coeruleus

(LC) were taken from individual rats for RNA isolation or pooled from 8 animals per group to isolate nuclei for the run-on assay of transcription.

Nuclear Run-on Assay

The rates of TH and DBH gene transcription in AM and LC of control rats and in rats exposed to immobilization stress for different times were measured by the method of McKnight and Palmiter (1979), as described in detail (Nankova *et al.*, 1999b; Serova *et al.*, 1999). In each reaction, equal amounts of nuclei based on the DNA quantity were used. The *in vitro*-labeled nascent transcripts were hybridized to plasmid cDNAs encoding rat TH, DBH and cyclophilin (CPH); as well as vector DNA (pBluescript) that were previously immobilized on nitrocellulose membranes. After proper washing, the levels of hybridization were analyzed by autoradiography.

Northern Blot Analysis

Total RNA was isolated by using RNAzol (Tel-Test, Inc.) and Northern blot analysis was performed as described previously (Nankova *et al.*, 1993; Nankova *et al.*, 1994). The autoradiographs were then scanned and the abundance of each mRNA was expressed relative to concomitantly measured 18S rRNA levels in the same samples. The results are presented as fold induction compared to the levels obtained in the corresponding control group on the same Northern blot.

Western Blot Analysis

Total protein extracts from pooled AM (4 to 8 animals from each experimental group) were prepared as

follows. The frozen tissues were homogenized on ice in 20 mM HEPES pH 7.5, 350 mM NaCl, 25% glycerol, 0.25% NP-40, 1 mM Na_2VO_3, 0.25 mM PMSF, 5 mM $MgCl_2$, 1 $\mu g/ml$ – aprotinin, pepstatin, leupeptin, 1 mM EGTA, 1mM DTT and clarified by centrifugation. The proteins were separated on 10% SDS-PAGE and electroblotted onto a nitrocellulose membrane. After blocking, the membranes were incubated with appropriate amounts of primary and secondary antibodies and analyzed by chemiluminescence.

Statistical Analysis

Data from at least three independent experiments are summarized. Differences between the experimental groups were evaluated by performing Analysis Of Variance (ANOVA) followed by the Fisher's Least Significant Difference Test for experiments with more than two groups. A level of $P \leq .05$ was accepted as statistically significant.

Results

Transient Activation of Transcription after a Single Stress in Adrenal Medulla

Brief Stress

Animals were exposed to a single immobilization stress for 5 min. The relative rates of transcription for TH and DBH in adrenal medulla were determined immediately after the stress or at the end of the 2 hour period. Even this short period of immobilization stress resulted in increased relative rate of transcription of both genes (Figure 1A). The effect

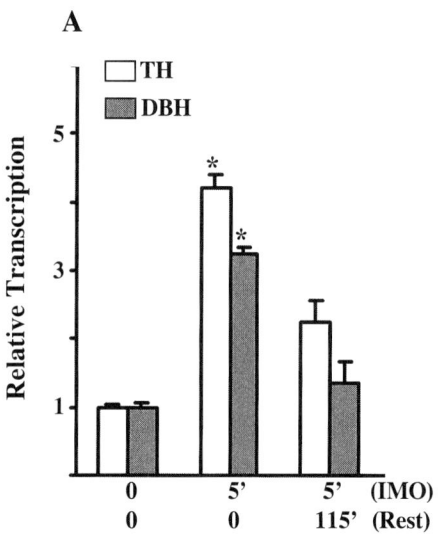

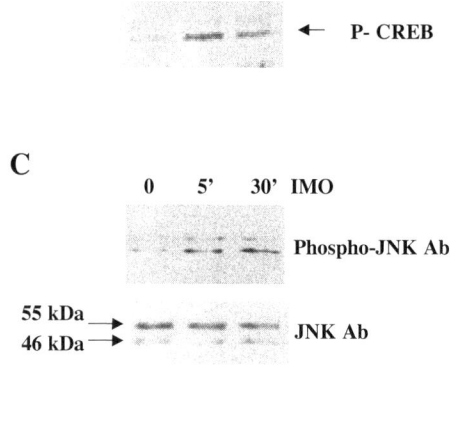

Fig. 1 Rapid transcriptional activation in response to brief immobilization. The relative rates of TH and DBH transcription were measured in AM of controls and rats subjected to 5′ immobilization stress (IMO) and sacrificed immediately or after 115′ rest. The hybridization signal due to vector sequences was subtracted and the mean levels in the controls were taken as 1.0. The results (mean ± SEM) are presented as the fold-increase over control. * p < 0.05 versus control. B. Time course of CREB phosphorylation induced by immobilization stress. Animals were exposed to a single immobilization for 5 or 30 min and sacrificed immediately. Immunoblots of total AM-homogenates with an antibody specific for CREB phosphorylated at Ser 133 (UBI) are shown. C. Time-course of JNK activation by stress. The changes in JNK activity in AM of rats in response to immobilization stress were measured by Western blot analysis using anti-JNK1 (PharMingen) and anti ACTIVE JNK (Promega) antibodies.

was transient and the relative rate of transcription was found to be similar to control levels when examined at the end of the 2 hours interval.

This short period of stress is most probably not sufficient to enable *de novo* synthesis of transcription factors. The rapid induction of TH and DBH transcription is likely mediated by post-translational activation of pre-existing factors. To test this hypothesis, we analyzed the impact of short exposures to IMO stress on the extent of phosphorylation of CREB (cAMP Regulatory Element Binding Factor), a convergence point of cAMP, Ca^{2+}, growth factors and stress-triggered signaling pathways (De *et al.*, 1999). Using an antibody specific for the Ser^{133} phosphorylated form of CREB (P-CREB), maximal immunoreactivity was detected after 5′ of immobilization (Figure 1B). With longer exposure to stress the amount of immunoreactive P-CREB decreased gradually and at the 2 hour time point, CREB phosphorylation was similar to levels seen in control animals (not shown). For all time points tested, immobilization stress did not alter the amount of CREB immunoreactive protein in the adrenal medulla.

Given the potential role of MAP kinase cascades in the regulation of AP1 activity (Karin, 1995) the effect of immobilization stress on c-Jun-N-terminal kinase (JNK) activity in rat AM was also studied. Western blot analysis using JNK antibody, recognizing all isoforms of the enzyme, detected similar immunoreactivity in adrenomedullary homogenates from control and stressed animals. In contrast, when the same membranes were incubated with antibody for dually phosphorylated active form of JNK, the amount of phosphorylated immunoreactive JNK proteins was significantly increased after 5′ or 30′ of immobilization stress (Figure 1C). Since JNK phosphorylates and up-regulates c-Jun, ATF2 and Elk1 (Force *et al.*, 1996), these results suggest that both CREB and Jun are among the likely candidates for mediating the transcriptional responses of TH to short episodes of stress.

The DBH promoter was shown not to bind CREB in adrenomedullary nuclear extracts, but to interact with c-Jun (Sabban and Nankova, 1998). Therefore phosphorylated c-Jun/JNK pathway is likely to mediate the transcriptional activation of DBH by a single immobilization stress.

Intermediate Duration of Stress

Next, we examined if longer exposures to a single immobilization stress will cause a more persistent rise in the transcription in adrenal medulla. Immobilization stress of intermediate duration (30 min) or 2 hours triggered similar increases in the relative rates of transcription of both genes (Figure 2). However, only the rise in the relative rate of transcription of TH gene was evident 90 min after cessation of the stress episode (Figure 2A). When examined one day later, TH transcription rate was no longer elevated beyond the control levels (Figure 2B). In contrast to TH, the rise in DBH transcription was still transient with intermediate duration of a single IMO. After 30 min of IMO, DBH transcription had declined towards control levels when examined 90 min later (Figure 2A).

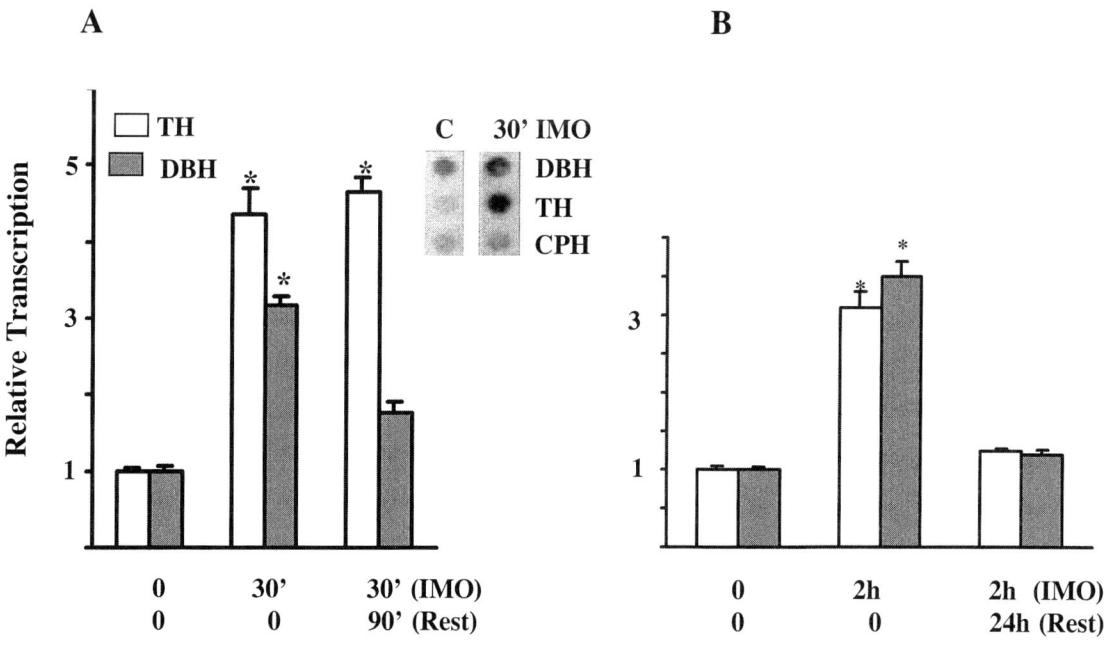

Fig. 2 Differential effect of intermediate duration of immobilization stress on TH and DBH transcription. The relative rates of transcription were measured in control animals and in rats immobilized for 30 min (A) or 2 hours (B) and sacrificed immediately or at the indicated time after the beginning of the stress. * p < 0.05 versus control. The insert shows autoradiograph of nascent transcripts in control (C) and nuclei isolated after 30 min stress (30′ IMO).

Sustained Activation of TH and DBH Transcription by Repeated Stress in the Adrenal Medulla

Previous results revealed that there are sustained increases in TH and DBH mRNA levels with repeated immobilization stress that remain high for several days after the last episode of stress. These changes might reflect activated transcription, enhanced stability of the corresponding transcripts or both. Nuclear run-on assays were used to evaluate changes in the relative rates of transcription of TH and DBH genes in rat adrenal medulla following repeated episodes of stress. Exposure to seven daily repeated immobilizations resulted in almost 4-fold activation of transcription (Figure 3) compared to the corresponding handled controls. In contrast, the relative rate of transcription of CPH gene, used as control, was not affected under these conditions (Figure 3, left insert), consistent with the lack of alteration in adrenal CPH mRNA levels with immobilization stress (Nankova and Sabban, 1999). The observed changes in the transcription and steady-state mRNA levels of TH and DBH were sustained when examined one day after the last episode of stress. Therefore the transcriptional mechanisms are also involved in the sustained up-regulation of neurotransmitter gene expression in response to repeated stress.

Transcriptional Changes in Response to Immobilization Stress in Locus Coeruleus

Recently we have shown that immobilization stress concomitantly alters the expression of the TH and DBH genes in the locus coeruleus, the major noradrenergic nucleus in the brain (Serova et al., 1999). To evaluate whether these changes are transcriptionally mediated, we developed suitable conditions to isolate nuclei from LC tissue (Serova et al., 1999). The incorporation of ^{32}P-labeled UTP into nascent transcripts of all genes of interest was higher in the LC from the immobilized rats than from the control rats (Figure 4). Even a single 30 min immobilization stress triggered increased transcription of the TH and DBH genes, while no significant effect was found on the transcription of constitutively expressed CPH gene. Transcription of the TH and DBH genes remained 2- to 3-fold above control levels in rats after 2 hours of a single stress or after exposure to stress for 2 hours, on four consecutive days (Figure 4). Taken together, our results suggest that the concomitant induction of TH and DBH genes in LC by stress involve transcriptionally-mediated mechanisms.

Discussion

The present study directly demonstrates changes in the transcription rates of the genes encoding the catecholamine biosynthetic enzymes TH and DBH in vivo, in response to immobilization stress in the rat adrenal medulla and locus coeruleus.

Notably, different phases of transcriptional activation were observed in adrenal medulla, depending on the duration and reiteration of the stress signal. With single immobilization stress depending on the duration of the stress signal, consecutive phases of transient activation were observed. 1. Brief stress (5 min) elicited about 3-fold activation of transcription for both TH and DBH genes without an accumulation of the corresponding mRNAs. 2. During intermediate stress (30 min to 2 hours immobilization) the duration of transcriptional activation differs for TH and DBH and only TH mRNA levels increased. The rise in transcription is not

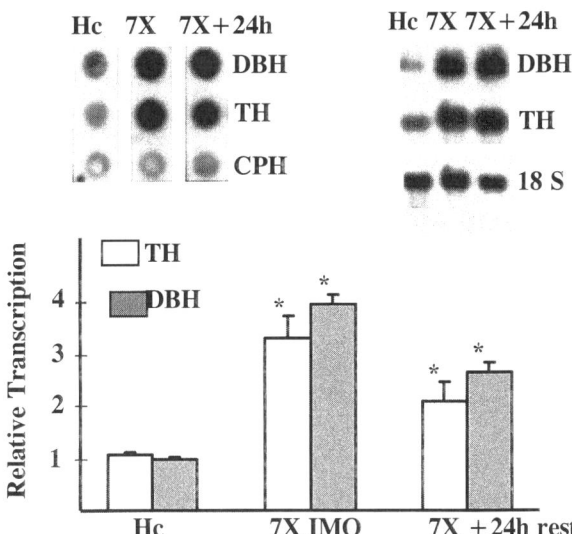

Fig. 3 Sustained activation of transcription by repeated stress. Animals were exposed to 7 daily repeated immobilizations for 2 hours each and sacrificed immediately (7x) or 24 hours after the last stress episode (7x + 24h). The controls were handled (Hc). The insert shows autoradiograms of run on assay (left) and typical Northern blot from the same experiment (right). * p < 0.05 versus handled control.

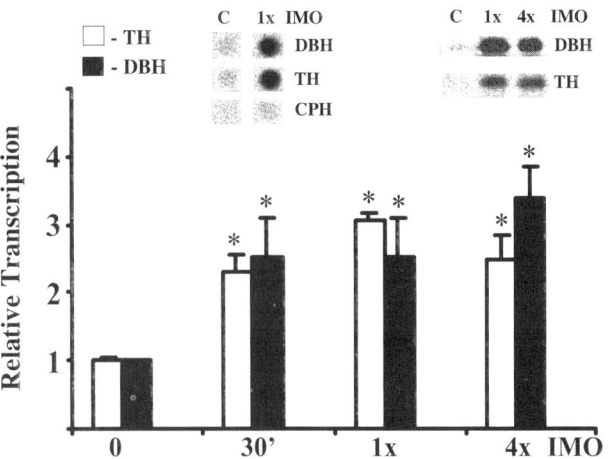

Fig. 4 Transcriptional induction of TH and DBH genes in LC by immobilization stress. Rats were exposed to a single 30 min or 2 hours (1x) immobilization or to 4 daily immobilizations (4×). Insert represents autoradiographs of run on assay (left) and a Northern blot (right). * p < 0.05 versus control.

sustained by the following day. In contrast, repeated stress triggers persistent activation for both genes, evident even 24 hours after the last episode of stress.

We hypothesize that different sets of transcription factors, with different kinetics of induction/activation are involved in the various phases of transcriptional regulation of TH and DBH by stress. The initial burst in adrenal TH and DBH transcription in response to short immobilization stress may be controlled by phosphorylation-dependent (e.g., MAP kinases; AP-1; PKA; CREB/ATF) signaling pathways/factors. It coincides temporally with the maximal levels of CREB phosphorylation at Ser 133 and JNK activation, as revealed by the findings reported in this study. The effects of stress of intermediate and longer duration may involve *de novo* synthesis of transcription factors, e.g. immediate early genes. Exposure to repeated stress triggers long-term adaptation programs, most probably mediated by long-lasting changes in the expression of stress-responsive transcription factors.

The differential effect of stress on TH and DBH transcription may reflect differences in the promoter context. The 5′ flanking region of the TH gene contains functional AP1 and perfect consensus CRE sites and recently discovered Egr1 motif (Sabban, 1996; Fung *et al.*, 1992; Nagamoto-Combs *et al.*, 1997; Yang *et al.*, 1998; Papanikolaou and Sabban, 1999). They may participate in the *in vivo* regulation of TH gene expression. In this regard, increased binding of c-Jun/c-fos to the TH AP1-site is observed in AM extracts from animals, exposed to cold or to immobilization stress (Miner *et al.*, 1992; Nankova *et al.*, 1994). Nevertheless, c-fos is not indispensable for the induction of TH. We found that repeated immobilization of c-fos knockout mice did not block the induction of adrenal TH mRNA (Serova *et al.*, 1998). In addition, our recent study suggests for the first time that the Sp-1 and overlapping Egr motif may function in the regulation of TH transcription by stress (Papanikolaou and Sabban, 1999). In this regard, increased Egr1 expression is observed in AM of rats following neural stimulation (Morita *et al.*, 1996; Wong *et al.*, 1998). Therefore stress-induced Egr1 is a candidate, alone or together with c-fos, in mediating the transcriptional activation of TH by intermediate durations of a single immobilization stress.

Importantly the functioning of several second messenger pathways converges on a perfect consensus cAMP response element (CRE, TGACGTCA) within the TH promoter. CREB, ATF1 and Jun family members have been identified in specific DNA-protein complexes with the TH-CRE in AM nuclear extracts (Sabban, 1997), but are not altered by immobilization stress. The increased AM-P-CREB observed in the present study may ensure the sustained rise of TH gene transcription and the accumulation of TH mRNA after stress. The importance of CREB for the transcriptional regulation of TH has been recently shown. In cell lines expressing antisense CREB mRNA, the response of the TH gene promoter to cAMP is dramatically inhibited (Nagamoto-Combs *et al.*, 1997). Alternatively,

phosphorylation of CREB may serve as a signal to promote the transcription of other CRE-regulated genes (e.g., c-fos and/or fos-family and Jun-family members), which could produce long lasting effects on the expression of stress-responsive genes.

In contrast to TH gene expression, even though DBH transcription rate is stimulated rapidly after a single episode of stress, the effect remains transient and persists for less than 2 hours. DBH promoter contains the sequence TGCGTCA (identical to the regulatory motif in Enk-CRE2, in the proenkephalin promoter and DYN-CRE3 in the prodynorphin promoter), that is similar to both AP-1 and CRE sequences (McMahon and Sabban, 1992; Shaskus *et al.*, 1992). It has been shown to mediate both basal and second messenger-inducible transcription of the gene (Shaskus *et al.*, 1992; Zellmer *et al.*, 1995; Ishiguro *et al.*, 1993). The effect of a single immobilization stress on DBH transcription might be mediated by activation of JNK pathway/c-Jun phosphorylation (Nankova *et al.*, 1999a). DBH mRNA levels do not rise significantly after a single episode of stress. Possibly, the amount of newly transcribed DBH mRNA in response to immobilization is relatively small. Hence, longer or repeated exposure to stress is required to alter the large pre-existing adrenal pool of DBH transcripts. We have shown previously that repeated immobilizations result in increased binding of c-Jun and a fos-related factor, but not c-fos, to the DB1 cis-regulatory element of the promoter (Nankova *et al.*, 1993). Our results suggest that long-lasting changes in the expression of AP1-like proteins may mediate the effects of repeated immobilization stress on DBH transcription.

In the brain noradrenergic system the increased synthesis of NE in response to stress may result from a coordinated up-regulation of the expression of TH and DBH genes. Even a single immobilization stress elevated TH and DBH expression levels, as was previously reported (Rusnak *et al.*, 1996; Serova *et al.*, 1999). Using run-on assay of transcription, a single 30 min immobilzation was found to lead to substantial rises in the transcription rates of the two genes examined. Increased transcription in this short period of time suggests that post-translational modification of pre-existing factors may play a role. Such modifications may include phosphorylation of CREB in LC by cAMP-dependent protein kinase (protein kinase A), or calcium-calmodulin-dependent protein kinase (CaM kinase). Notably, the coordinate regulation of the cAMP system with the NE firing rate and the expression of TH in the LC was reported (Melia *et al.*, 1992; Nestler and Aghajanian, 1997). Not only TH, but also DBH gene expression can be activated by the cAMP-mediated pathway in cultured cells (McMahon and Sabban, 1992; Shaskus *et al.*, 1992). In this regard, increased phosphorylation of CREB was observed by conditions that activate TH in the LC (Nestler and Aghajanian, 1997). Moreover, the infusion of CREB antisense oligonucleotides directly into the LC completely blocked the morphine-induced up-regulation of TH, indicating the importance of transcriptional activation mediated by CREB (Lane-Ladd *et al.*, 1997).

Taken together our results indicate that activation of transcription may play an important role in the stress-elicited induction of the genes encoding catecholamine biosynthetic enzymes in rat AM and LC. Different signaling pathways and/or transcription factors are activated depending on the duration and the frequency of the stress signal.

Acknowledgments

We gratefully acknowledge support by grants NS28869 and NS32166 from National Institutes of Health, grant 0251 from Smokeless Tobacco Research Council, Fogarty International Collaboration Award TW00984 and grant 2-610999 from Slovak Grant-Agency for Science.

References

Baruchin, A., Weisberg, E.P., Miner, L.L., Ennis, D., Nisenbaum, L.K., Naylor, E., Stricker, E.M., Zigmond, M.J., and Kaplan, B.B. (1990). Effects of cold exposure on rat adrenal tyrosine hydroxylase: An analysis of RNA, protein, enzyme activity, and cofactor levels. *Journal of Neurochemistry* **54**, 1769–1775.

De, C.D., Fimia, G.M., and Sassone-Corsi, P. (1999). Signaling routes to CREM and CREB: plasticity in transcriptional activation. *Trends in Biochemical Sciences* **24**, 281–285.

Force, T., Pombo, C.M., Avruch, J.A., Bonbentre, J.V., and Kyriakis, J.M. (1996). Stress-activated protein kinases in cardiovascular disease. *Circulation Research* **78**, 947–953.

Fung, B.P., Yoon, S.O., and Chikaraishi, D.M. (1992). Sequences that direct rat tyrosine hydroxylase gene expression. *Journal of Neurochemistry* **58**, 2044–2052.

Goldstein, D.S. (1995). *Stress, Catecholamines, and Cardiovascular Disease.* New York, New York: Oxford University.

Ishiguro, H., Kim, K.T., Joh, T.H., and Kim, K.S. (1993). Neuron-specific expression of the human dopamine (β-hydroxylase gene requires both the cAMP-response element and a silencer region. *Journal of Biological Chemistry* **268**, 17987–17994.

Karin, M. (1995). The regulation of AP-1 activity by mitogen-activated protein kinases. *Journal of Biological Chemistry* **270**, 16483–16486.

Kvetňanský, R., and Sabban, E.L. (1993). Stress-induced changes in tyrosine hydroxylase and other catecholamine biosynthetic enzymes. In: M. Naoi, and S.H. Parvez (Eds.) *Tyrosine Hydroxylase*, pp. 253–281. Utrecht, The Netherlands: VSP

Kvetňanský, R., and Sabban, E.L. (1998). Stress and molecular biology of neurotransmitter-related enzymes. *Annals of the New York Academy of Sciences* **851**, 342–356.

Kvetňanský, R., Weise, V.K., and Kopin, I.J. (1970). Elevation of adrenal tyrosine hydroxylase and phenylethanolamine-N-methyltransferase by repeated immobilizations of rats. *Endocrinology* **87**, 744–749.

Kvetňanský, R., Gewirtz, G.P., Weise, V.K., and Kopin, I.J. (1971). Enhanced synthesis of adrenal dopamine β-hydroxylase induced by repeated immobilization of rats. *Molecular Pharmacology* **7**, 81–86.

Lane-Ladd, S.B., Pineda, J., Boundy, V.A., Pfeuffer, T., Krupinski, J., Aghajanian, G.K., and Nestler, E.J. (1997). CREB (cAMP response element-binding protein) in the locus coeruleus: biochemical, physiological, and behavioral evidence for a role in opiate dependence. *Journal of Neuroscience* **17**, 7890–7901.

McEwen, B.S. (1998). Protective and damaging effects of stress mediators. *New England Journal of Medicine* **338**, 171–179.

McKnight, G.S., and Palmiter, R. D. (1979). Transcriptional regulation of the ovalbumin and conalbumin genes by steroid hormones in chick oviduct. *Journal of Biological Chemistry* **254**, 9050–9058.

McMahon, A., and Sabban, E. (1992). Regulation of expression of dopamine β-hydroxylase in PC12 cells by glucocorticoids and cyclic AMP analogs. *Journal of Neurochemistry* **59**, 2040–2047.

McMahon, A., Kvetňanský, R., Fukuhara, K., Weise, V.K., Kopin, I., and Sabban, E.L. (1992). Regulation of tyrosine hydroxylase and dopamine

β-hydroxylase mRNA levels in rat adrenals by a single and repeated immobilization stress. *Journal of Neurochemistry* **58**, 2124–2130.

Melia, K.R., Rasmussen, K., Terwilliger, R.Z., Haycock, J.W., Nestler, E.J., and Duman, R.S. (1992). Coordinate regulation of the cyclic AMP system with firing rate and expression of tyrosine hydroxylase in rat locus coeruleus: Effects of chronic stress and drug treatments. *Journal of Neurochemistry* **58**, 494–502.

Miner, L.L., Pandalai, S.P., Weisberg, E.P., Sell, S.L., Kovacs, D.M., and Kaplan, B.B. (1992). Cold induced alterations in the binding of adrenomedullary nuclear proteins to the promoter region of the tyrosine hydroxylase gene. *Journal of Neuroscience Research* **33**, 10–18.

Morita, K., Bell, R.A., Siddall, B.J., and Wong, D.L. (1996). Neural stimulation of Egr-1 messenger RNA expression in rat adrenal gland: possible relation to phenylethanolamine N-methyltransferase gene regulation. *Journal of Pharmacology and Experimental Therapeutics* **279**, 379–385.

Nagamoto-Combs, K., Piech, K.M., Best, J.A., Sun, B., and Tank, A.W. (1997). Tyrosine hydroxylase gene promoter activity is regulated by both cyclic AMP-responsive element and AP1 sites following calcium influx. Evidence for cyclic AMP-responsive element binding protein-independent regulation. *Journal of Biological Chemistry* **272**, 6051–6058.

Nankova, B., Devlin, D., Kvetňanský, R., Kopin, I.J., and Sabban, E.L. (1993). Repeated immobilization stress increases the binding of c-fos-like proteins to a rat dopamine β-hydroxylase promoter enhancer sequence. *Journal of Neurochemistry* **61**, 776–779.

Nankova, B., Kvetňanský, R., McMahon, A., Viskupic, E., Hiremagalur, B., Frankle, G., Fukuhara, K., Kopin, I.J., and Sabban, E.L. (1994). Induction of tyrosine hydroxylase gene expression by a nonneuronal nonpituitary-mediated mechanism in immobilization stress. *Proceedings of the National Academy of Sciences USA* **91**, 5937–5941.

Nankova, B.B., Fuchs, S.Y., Serova, L.I., Ronai, Z., Wild, D., and Sabban, E.L. (1999a). Selective *in vivo* stimulation of stress-activated protein kinase in different rat tissues by immobilization stress. *Stress* **2**, 289–298.

Nankova, B.B., and Sabban, E.L. (1999). Multiple signaling pathways exist in the stress-triggered regulation of gene expression for catecholamine biosynthetic enzymes and several neuropeptides in the rat adrenal medulla. *Acta Physiologica Scandanavica* **167**, 1–9.

Nankova, B.B., Tank, A.W., and Sabban, E.L. (1999b). Transient or sustained transcriptional activation of the genes encoding rat adrenomedullary catecholamine biosynthetic enzymes by different durations of immobilization stress. *Neuroscience*, **94**, 803–808.

Nestler, E.J., and Aghajanian, G.K. (1997). Molecular and cellular basis of addiction. *Science* **278**, 58–63.

Osterhout, C.A., Chikaraishi, D.M., and Tank, A.W. (1997). Induction of tyrosine hydroxylase protein and a transgene containing tyrosine hydroxylase 5′ flanking sequences by stress in mouse adrenal gland. *Journal of Neurochemistry* **68**, 1071–1077.

Papanikolaou, N.A., and Sabban, E.L. (1999). Sp1/Egr1 motif: a new candidate in the regulation of rat tyrosine hydroxylase gene transcription by immobilization stress. *Journal of Neurochemistry* **73**, 433–436.

Rusnak, M., Zorad, S., Smith, M.A., Hiremagalur, B., Nankova, B., Sabban, E.L., and Kvetňanský, R. (1996). Immobilization stress induces different changes in tyrosine hydroxylase vs. proenkephalin gene expression in the locus coeruleus of rats. In: R. McCarty, G. Aguillera, E.L. Sabban, and R. Kvetňanský (Eds.), *Stress: Molecular, Genetic and Neurobiological Advances*, pp. 711–722. New York: Gordon and Breach.

Sabban, E.L. (1996). Synthesis of dopamine and its regulation. In: T.W. Stone (Ed.), *CNS Neurotransmitters and Neuromodulators Dopamine*, pp. 1–20. Boca Raton, Florida: CRC.

Sabban, E.L., Hiremagalur, B., Nankova, B., and Kvetňanský, R. (1995). Molecular biology of stress-elicited induction of catecholamine biosynthetic enzymes. *Annals of the New York Academy of Sciences* **771**, 327–338.

Sabban, E.L., Nankova, B., Hiremagalur, B., Orrin, S., Rusnak, M., Viskupic, E., and Kvetňanský, R. (1996). Molecular mechanisms in immobilization stress elicited rise in expression of genes for adrenal catecholamine biosynthetic enzymes, neuropeptide Y and proenkephalin. In: R. McCarty, G. Aguillera, E.L. Sabban, and R. Kvetňanský (Eds.), *Stress: Molecular, Genetic and Neurobiological Advances*, pp. 611–628. New York: Gordon and Breach.

Sabban, E.L. (1997). Control of tyrosine hydroxylase gene expression in chromaffin and PC12 cells. *Seminars in Cell and Developmental Biology* **8**, 101–111.

Sabban, E.L., and Nankova, B.B. (1998). Multiple pathways in regulation of dopamine β-hydroxylase. *Advances in Pharmacology* **42**, 53–56.

Sapolsky, R.M. (1996). Stress, glucocorticoids, and damage to the nervous system: The current state of confusion. *Stress* **1**, 1–19.

Senba, E., Umemoto, S., Kawa, Y., and Noguchi, K. (1994). Differential expression of fos family and jun family mRNAs in the rat hypothalamo-pituitary-adrenal axis after immobilization stress. *Molecular Brain Research* **24**, 283–294.

Serova, L.I., Saez, E., Spiegelman, B.M., and Sabban, E.L. (1998). c-Fos deficiency inhibits induction of mRNA for some, but not all, neurotransmitter biosynthetic enzymes by immobilization stress. *Journal of Neurochemistry* **70**, 1935–1940.

Serova, L.I., Nankova, B.B., Feng, Z., Hong, J.S., Hutt, M., and Sabban, E.L. (1999). Heightened transcription for enzymes involved in norepinephrine biosynthesis in the rat locus coeruleus by immobilization stress. *Biological Psychiatry* **45**, 853–862.

Shaskus, J., Greco, D., Asnani, L.P., and Lewis, E.J. (1992). A bifunctional genetic regulatory element of the rat dopamine β-hydroxylase gene influences cell type specificity and second messenger-mediated transcription. *Journal of Biological Chemistry* **267**, 18821–18830.

Stachowiak, M.K., Sebbane, R., Stricker, E.M., Zigmond, M.J., and Kaplan, B.B. (1985). Effect of chronic cold exposure on tyrosine hydroxylase mRNA in rat adrenal gland. *Brain Research* **3**, 193–196.

Wong, D.L., and Wang, W. (1994). Neuronal control of dopamine β-hydroxylase in vivo: acute and chronic effects. *Molecular Brain Research* **25**, 57–66.

Wong, D.L., Ebert, S.N. and Morita, K. (1998) Neural control of phenylethanolamine-N-methyltransferase via cholinergic activation of Egr-I. *Advances in Pharmacology* **42**, 77–81.

Yang, C., Kim, H.-S., Seo, H., and Kim, K.-S. (1998) Identification and characterization of potential cis-regulatory elements governing transcriptional activation of the rat tyrosine hydroxylase gene. *Journal of Neurochemistry* **71**, 1358–1368.

Zellmer, E., Zhang, Z., Greco, D., Rhodes, J., Cassel, S., and Lewis, E.J. (1995). A homeodomain protein selectively expressed in noradrenergic tissue regulates transcription of neurotransmitter biosynthetic genes. *Journal of Neuroscience* **15**, 8109–8120.

19. Regulation of Tyrosine Hydroxylase Gene Transcripton Rate by Stress: Use of Transgenic Mice

C. Osterhout, B. Sun and A.W. Tank

Introduction

Based on groundbreaking work over 30 years ago by Thoenen, Axelrod, Kvetňanský, Kopin and coworkers (Kvetňanský et al., 1970; Thoenen, 1970), it is now well-established that different forms of stress elicit induction of tyrosine hydroxylase (TH) expression in the adrenal medulla, sympathetic ganglia and locus coeruleus (see Kvetňanský and Sabban, 1993; Kumer and Vrana, 1996 for reviews of the work supporting this hypothesis). In the adrenal medulla this response is generally dependent on presynaptic input and can be initiated by agonist occupation of multiple adrenal chromaffin cell receptors that are linked to multiple intracellular signaling pathways. Some of these receptors and signaling pathways are presented in Figure 1.

As reviewed by Kumer and Vrana (Kumer and Vrana, 1996), TH expression is regulated by numerous types of intracellular mechanisms. Many of these mechanisms involve regulation of TH gene transcrip-

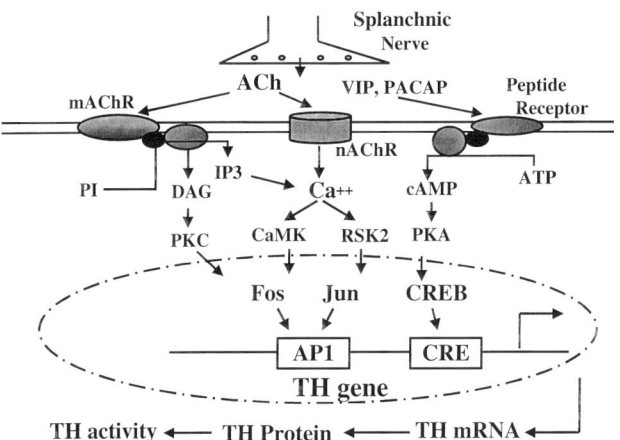

Fig. 1 Schematic diagram of signaling pathways involved in regulation of TH gene expression in adrenal medulla. Abbreviations are: ACh, acetylcholine; VIP, vasoactive intestinal peptide; PACAP, pituitary adenylyl cyclase activating peptide; mAChR, muscarinic acetylcholine receptor; nAChR, nicotinic acetylcholine receptor; PI, phosphatidyl-inositol; DAG, diacylglycerol; IP3, inositol trisphosphate; PKC, protein kinase C; PKA, cAMP-dependent protein kinase; CaMK, calcium/calmodulin-dependent protein kinase; RSK2, pp90 ribosomal S6 kinase.

tion rate, but post-transcriptional mechanisms that regulate TH mRNA stablity and/or TH protein degradation rate have also been identified. The intracellular mechanisms that participate in the induction of TH elicited by different types of stress have not been well-elucidated. The initial question posed when addressing this issue is whether the response to a particular type of stress in a particular type of catecholaminergic cell is mediated by transcriptional or post-transcriptional mechanisms. This question is not always trivial to answer, because the conventional and most reliable method to measure changes in gene transcription rate is the nuclear run-on assay. This assay is direct and precise, but it is technically challenging and requires sufficient amounts of sample tissue to isolate enough nuclei to obtain a significant signal. This assay has been used successfully in small tissue samples like rat adrenal medulla to demonstrate that nicotine induces TH via an initial stimulation of TH gene transcription rate (Fossom et al., 1991). More recent studies have used nuclear run-on assays to demonstrate that immobilization stress activates the TH gene in rat adrenal medulla and locus coeruleus (Nankova et al., 1999; Serova et al., 1999). Nevertheless, measuring changes in gene transcription rate in small brain nuclei using nuclear run-on assays remains a difficult, if not impractical endeavor.

Sherman and Moody (Sherman and Moody, 1995) have successfully used an alternative procedure, measurement of TH mRNA nuclear precursors, to estimate changes in TH gene transcription rate in midbrain neurons. This approach is technically less demanding, but requires the presence of detectable steady-state levels of these precursor RNAs and assumes that changes in the levels of these precursors is dependent only on changes in transcription rate, not on changes in precursor processing.

Recently, transgenic mice have been produced which express transgenes encoding TH gene 5' flanking sequences inserted upstream of a reporter gene that is easily detected in mammalian cells. This approach has been pioneered by Chikaraishi, Nagatsu and coworkers (Bannerjee et al., 1992; Sasaoka et al., 1992), who have shown that ~5 kb of TH gene 5' flanking region is necessary to produce tissue-specific expression of the transgene. Since then, a number of other laboratories have produced TH-reporter transgenic mouse strains and the regions of the 5′ flanking region necessary for tissue-specific and ectopic expression have been further characterized (Banerjee *et al.*, 1994; Min *et al.*, 1996; Morgan *et al.*, 1996; Kim *et al.*, 1997; Liu *et al.*, 1997; Trocme *et al.*, 1998). These transgenic mice have also been used to study the response of the TH gene promoter to different stimuli (Bannerjee *et al.*, 1992; Min *et al.*, 1996; Osterhout *et al.*, 1997; Boundy *et al.*, 1998). This approach represents a powerful new way to address the question whether TH gene transcription rate is regulated by different types of stress in small brain nuclei, like the locus coeruleus.

Even though this transgenic mouse approach holds great promise for the study of TH gene regulation by stress, two major questions still need to be addressed: (1) Do changes in TH-reporter transgene expression correlate closely with changes in endogenous TH gene transcription rate? (2) Since most previous studies on stress have been performed in rats, does the TH gene in mice respond to stress similarly to rats? In this report we address some aspects of these two questions. Furthermore, recent studies have shown that stress induces TH gene expression at least partially via transcriptional mechanisms in the adrenal medulla (Nankova *et al.*, 1999) and that c-Fos expression correlates with this stress-induced stimulation of the TH gene (Nankova *et al.*, 1993). Hence, in this report we also address the question as to whether induction of c-Fos leads to stimulation of TH gene transcription rate, using rat pheochromocytoma PC18 cells as a model system.

Materials and Methods

Immobilization Studies

We based our protocol on the work of Kvetňanský, Sabban and coworkers (Kvetňanský *et al.*, 1970; Nankova *et al.*, 1993) in rats. Mice were immobilized by placing them on their stomachs. Their limbs were spread apart and taped to the bench using nonadhesive tape. Immobilizations were carried out for 2 hr each day for 1–7 days. Animals were anesthetized with an overdose of halothane prior to removal of tissues.

TH-hpAP Luminescent Assay

CSPD® (Disodium 3-(4-methoxyspiro [1,2-dioxetane-3,2′-(5′-chloro)tricyclo[3.3.1.13,7]decan]-4-yl) phenyl phosphate) is a 1,2 dioxetane substrate for alkaline phosphatase. The enzyme dephosphorylates CSPD® to yield a dioxetane anion. This anion then decomposes, emitting light in the process. The anion is relatively stable, therefore the decomposition is delayed and steady-state light emission is achieved (Figure 2). The hpAP assay was performed by a modification of the method of O'Connor (O'Connor and Culp, 1994). Tissues were homogenized on ice in a buffer containing 50 mM Tris (pH 7.5), 5 mM $MgCl_2$, 100 mM NaCl and 4% (w/v) CHAPS. The homogenate was centrifuged for 10 min at 10,000 rpm and an aliquot was removed for protein determination. The remaining homogenate

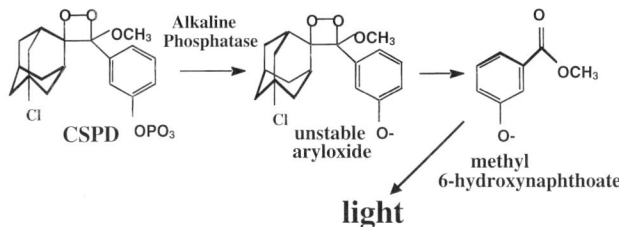

Fig. 2 TH-hpAP transgene. The top schematic is a diagram of the TH-human placental alkaline phosphatase (TH-hpAP) transgene. The bottom schematic depicts the chemical reaction by which hpAP activity was measured.

was heated to 65°C for 30 min to denature endogenous alkaline phosphatases and then clarified by centrifugation for 10 min at 10,000 rpm. The substrate buffer consisted of 0.25 mM CSPD®, 5 mM $MgCl_2$, 0.1 mM diethanolamine and 20 μL Emerald™ enhancer for a total volume of 200 μL. The substrate buffer was added to 20 μL of sample, incubated for 2–6 minutes and placed into a Turner luminometer. The luminescence was recorded with an interval of 5 seconds. Protein was determined by the Bradford method (Bradford, 1976). hpAP activity was expressed as relative light emission units (RLU)/mg protein.

Isolation of c-Fos-inducible PC18 Cell Lines

PC18 cells are a subclone of the rat pheochromocytoma PC12 cell line, in which numerous signaling pathways regulate the TH gene (Tank *et al.*, 1986). The tetracycline-inducible system described by Gossen and Bujard (1992) was used to produce stably-transfected cell lines in which c-Fos expression was under the control of the *tet operon*. PC18 cells were stably transfected sequentially with the pTet-tTAk and pTet-c-Fos or pTet-luciferase vectors, by slight modifications of the methods described by the manufacturer (GIBCO/BRL Life Technologies, Gaithersberg, MD). c-Fos-inducible cells were selected by their ability to express very low or negligible levels of c-Fos in the presence of tetracycline and very high levels in the absence of tetracycline. The luciferase-inducible cells expressed low levels of luciferase in the presence of tetracycline and high levels in its absence. These luciferase-inducible cells had low levels of c-Fos in the presence or absence of tetracycline and consequently were used as control cell lines.

For experiments, the c-Fos-inducible and luciferase-inducible cells were maintained in the presence of 0.5 ug/ml tetracycline. At the appropriate time the medium was removed and replaced with fresh medium lacking tetracycline. Cells were harvested at different time points after this removal and assayed for c-Fos levels and TH gene transcription rate.

Other Methods

TH enzyme activity was measured using the coupled decarboxylase assay (Osterhout et al., 1997); a saturating cofactor concentration (4 mM 6-methyl-5,6,7,8-tetrahydropterin) was used to measure Vmax activity. c-Fos was measured using standard western blotting procedures employing an antibody for c-Fos obtained from Santa Cruz Biotechnology, Inc. TH gene transcription rate was measured using nuclear run-on assays by a slight modification of the methods described by Fossom et al. (1992).

Results and Discussion

Regulation of TH Gene Expression by Stress in the Adrenal Medulla of Transgenic Mice

In a previous report (Osterhout et al., 1997) we used a transgenic mouse strain developed by Dona Chikaraishi's group (Bannerjee et al., 1992) to test whether stress activated the TH gene promoter in adrenal medulla. This transgenic line expressed a transgene encoding 4.5 kb of TH gene 5′ flanking region fused upstream of the chloramphenicol acetyltransferase (CAT) gene. Our results showed that TH-CAT gene expresson was activated by both cold stress and immobilization stress in the transgenic mouse adrenal glands. Furthermore, we found that endogenous adrenal TH activity and TH protein were induced by these stressful stimuli in a similar manner as observed in rats. Finally, we showed that repeated immobilization stress produced a more sustained induction of the TH-CAT transgene compared to that observed after a single immobilization stress. This latter finding was analogous to the sustained induction of TH mRNA in

rat adrenals after repeated immobilizations reported by Sabban and coworkers (Nankova et al., 1994). Three major conclusions were derived from this study: (1) both cold stress and immobilization stress induce adrenal TH at least partially via stimulation of the TH gene; (2) mouse adrenal TH gene expression responds to these stressors in a manner similar to that observed in rats; and (3) repeated immobilizations produce a sustained induction of TH mRNA at least partially because the transcription rate of the TH gene is stimulated for a sustained period of time. This third conclusion was recently confirmed in rat adrenals using nuclear run-on assays to measure TH gene transcription rate (Nankova et al., 1999).

Unfortunately, TH-CAT expression in brain nuclei, like the locus coeruleus or substantia nigra, of these transgenic mice was relatively low, and since a highly sensitive assay for CAT activity was not readily available, studies on the effect of stress on transgene expression in small brain nuclei from these mice were technically difficult. Hence, we initiated studies using a different line of transgenic mice that expressed a transgene encoding 4.5 kb of TH gene 5′ flanking region fused to the human placental alkaline phosphatase (hpAP) reporter gene. This transgenic line was produced by Gustincich et al. (1997) and expresses hpAP on the plasma membranes of central TH-expressing neurons, as well as adrenal chromaffin cells. We developed a sensitive luminescent assay for measuring hpAP activity in locus coeruleus and dopaminergic midbrain neurons, based upon the work of O'Connor and Culp (1994). This assay (depicted in Figure 2) converts a nonluminescent substrate into a luminescent product which can be measured using a luminometer. The assay was linear over time (Figure 3) and sensitive enough to measure TH-hpAP activity in brain nuclei or adrenal glands isolated from a single transgenic mouse.

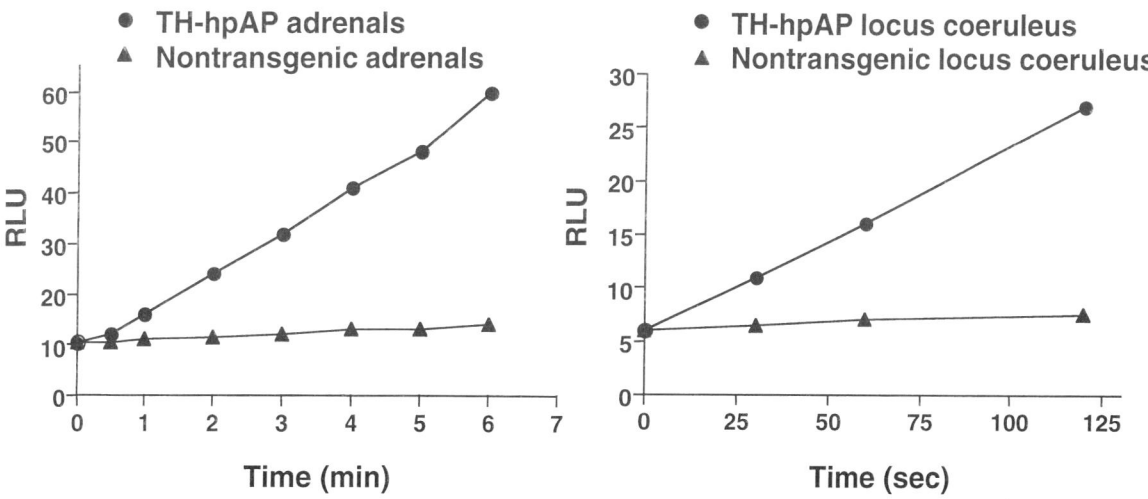

Fig. 3 Assay of TH-hpAP activity. Adrenal glands or locus coeruleus from a single mouse were assayed for TH-hpAP activity. The time courses for these assays are presented using either transgenic or nontransgenic mice. TH-hpAP activity is expressed as relative light emission units (RLU).

In our first set of studies we verified that the TH-hpAP mice reacted to immobilization stress in a manner similar to that observed using the TH-CAT mice. Mice were subjected to either one or three daily 2 hr immobilizations and adrenal glands were removed either immediately or 3 hr after the cessation of stress. TH-hpAP activity was measured in samples derived from a pair of adrenal glands. As presented in Figure 4, TH-hpAP activity increased when measured either 3 hr after a single immobilization or immediately after 3 immobilizations. This response was similar to that observed for TH-CAT expression in adrenal glands isolated from TH-CAT-expressing transgenic mice (Osterhout *et al.*, 1997) and provides confirmatory evidence that immobilization stress induces adrenal TH at least partially via stimulation of TH gene transcription rate.

Regulation of TH Gene Expression by Immobilization Stress in the Locus Coeruleus of Transgenic Mice

Numerous studies have shown that TH is induced by different types of stress in rat locus coeruleus (see Kvetňanský and Sabban (1993) for review). Recent work has shown that the response to immobilization stress in the rat locus coeruleus is accompanied by an initial induction of TH mRNA (Rusnak *et al.*, 1998). We have initiated studies to test whether endogenous TH expression is induced by immobilization in transgenic mice and whether this response is due to stimulation of the TH gene by measuring its effect on TH-hpAP activity in transgenic mouse locus coeruleus.

TH activity was measured using saturating pterin cofactor concentration and hence is an estimate of changes in TH protein levels. As presented in Figure 5,

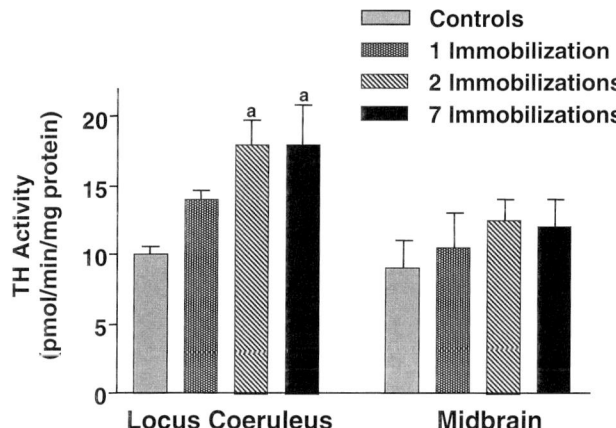

Fig. 5 Effect of immobilization stress on TH activity in mouse locus coeruleus and midbrain. Mice were immobilized for 2 hr either once, twice or seven times daily. Control mice were not manipulated. Brain regions were removed 24 hr after the initiation of stress for the single and twice-immobilized mice, and immediately after the stress for the mice immobilized 7 times. The results represent the means ± SEM from 3–5 mice. a: p < .05 compared to controls.

TH activity increased in the mouse locus coeruleus after 2 or 7 daily immobilizations, but not after a single immobilization. In contrast, midbrain TH activity was not increased by either single or repeated immobilization stress. TH activity also increased in the nerve terminal regions of the locus coeruleus after 7 immobilizations (Figure 6). These results are consistent with the hypothesis that stress induces TH expression in the mouse locus coeruleus as observed in the rat.

We also measured TH-hpAP activity in transgenic mouse locus coeruleus. TH-hpAP expression increased when measured immediately after or 3 hr after a single

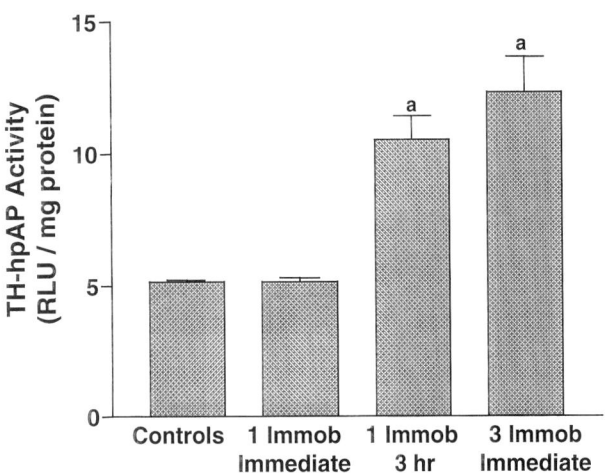

Fig. 4 Effect of immobilization stress on TH-hpAP transgene activity in mouse adrenal glands. Mice were immobilized for 2 hr once or three times daily. Control mice were not manipulated. Adrenals were removed under halothane anesthesia either immediately or 3 hr after the cessation of stress. The results represent the means ± SEM from 3 mice. a: p < .05 compared to controls.

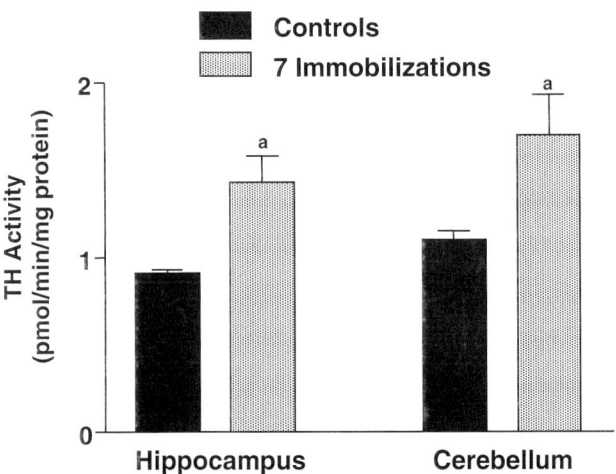

Fig. 6 Effect of immobilization stress on TH activity in hippocampus and cerebellum. Mice were immobilized for 2 hr on seven consecutive days. Control mice were not manipulated. Brain regions were removed immediately after the seventh immobilization. The results represent the means ± SEM from 5–6 mice. a: p < .05 compared to controls.

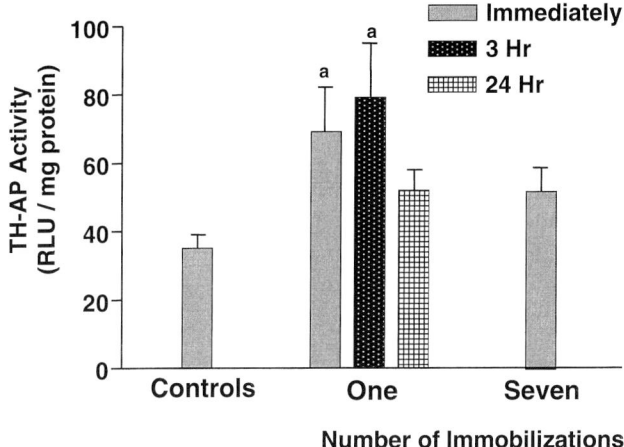

Fig. 7 Effect of immobilization stress on TH-hpAP expression in mouse locus coeruleus. Mice were immobilized for 2 hr either once or seven times daily. Control mice were not manipulated. Locus coeruleus was removed immediately, 3 hr or 24 hr after the cessation of stress. The results represent the means ± SEM from 3–5 mice. a: p < .05 compared to controls.

immobilizaton (Figure 7). However, this increase in TH-hpAP expression was transient, since its activity returned to control levels 24 hr after a single immobilization stress. These results provide evidence that the induction of TH in locus coeruleus is at least partially mediated via transcriptional mechanisms.

Perhaps our most interesting finding to date is that TH-hpAP expression was not apparently significantly elevated after 7 immobilizations (Figure 7). These results remain preliminary, but if confirmed by subsequent experiments, they suggest that repeated immobilization induces TH partially via post-transcriptional mechanisms in mouse locus coeruleus.

Over-expression of c-Fos in Rat Pheochromocytoma PC18 Cells Leads to Stimulation of TH Gene Transcription Rate

Nankova *et al.* (1993) showed that c-Fos is induced by immobilization stress in rat adrenal medulla. c-Fos is a transcription factor that dimerizes with members of the Jun family of transcription factors. These Fos/Jun dimers bind to AP1 sites in the promoter regions of responsive genes, leading to stimulation of transcription rate. The TH gene has a functional AP1 site in its proximal promoter at position –205 to –199. This site is responsive to phorbol esters, calcium, hypoxia, nerve growth factor and other stimuli, as well as being essential for optimal basal expression of the gene (for review see Kumer and Urana, 1996).

Since c-Fos was induced by immobilization in adrenal medulla, it is reasonable to speculate that it may participate in the regulation of the TH gene during stress. However, it is unclear whether induction of c-Fos actually regulates TH gene expression. Kinetic studies by Craviso and coworkers (1995) on the rates of induction of c-Fos and TH in bovine adrenal chromaffin

cells suggested that c-Fos did not participate in nicotinic receptor-mediated stimulation of the TH gene. However, in several other studies the induction of c-Fos correlated well with the induction of TH mRNA (Koistinaho *et al.*, 1993; Hiremagalur and Sabban, 1995; Norris and Millhorn, 1995). Equally confusing are several previous studies that measured the effects of over-expression of c-Fos on TH gene promoter activity in transiently-transfected cells (Gizang and Ziff, 1994; Ghee *et al.*, 1998). These studies found that c-Fos over-expression under these transient transfection conditions either activated or inhibited TH gene promoter activity. However, these studies were confounded by the long-term overexpression of c-Fos in the transiently-transfected cells. Since c-Fos was elevated for 1–2 days in these cells prior to measurement of TH gene promoter acivity and since c-Fos regulates numerous genes, it is difficult to assess whether the effects were directly due to c-Fos or to the induction/repression of downstream genes that regulate the TH gene indirectly.

In order to investigate this problem more precisely, we have developed stably-transfected rat pheochromocytoma PC18 cell lines in which c-Fos expression is inducible under the control of the *tet operon*. PC18 cells were stably transfected twice to produce these cell lines. The cells were first stably transfected with the expression vector for the tetracycline-inducible transactivator protein, tTAk. In the presence of tetracycline, this protein binds to *tet operon* sites and blocks transcription of downstream genes. However, when tetracycline is removed, it is no longer able to bind to the *tet operon*, and downstream genes are induced. PC18 cells that expressed large amounts of functional tTAk were then stably transfected a second time with an expression vector encoding the *c-fos* gene under the control of seven *tet operon* sites. Appropriate cell lines were chosen based on their ability to express very low amounts of c-Fos in the presence of tetracycline and very high amounts of c-Fos when tetracycline was removed. Control cells were stably transfected the second time with the luciferase gene under the control of the same *tet operon*. c-Fos expression in c-Fos-inducible and control luciferase-inducible PC18 cell lines is presented in Figure 8. Note that c-Fos expression under basal conditions in the presence of tetracycline is very low. In the absence of tetracyclin, c-Fos is induced more than 50-fold in the c-Fos-inducible cell line, but is not induced in the control cell line.

Nuclei were isolated from these cells at two different time points, 10 hr and 24 hr after tetracycline removal. At these time points c-Fos was induced ~10-fold and >40 fold, respectively, in the c-Fos-inducible cell line. The fold increase at the 10 hr time point was approximately equivalent to that observed after phorbol ester treatment of PC12 cells (data not shown). Nuclear run-on assays were performed on these nuclei. TH gene transcription rate was increased by ~3-fold and ~20-fold at the 10 hr and 24 hr time points, respectively, in the c-Fos-inducible cell line (Figure 9). In contrast, TH gene transcription rate was unaffected by tetracycline removal in the control cell line. These

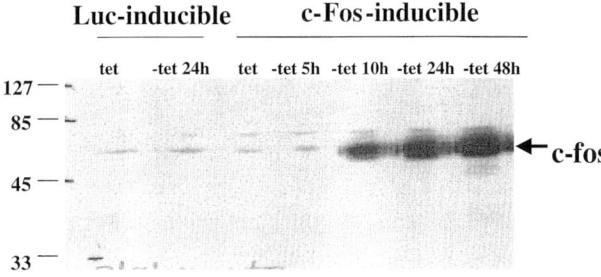

Fig. 8 Induction of c-Fos in the absence of tetracycline. The cells were maintained in the presence of 0.5 μg/ml tetracycline or the tetracycline-containing medium was removed and the cells were harvested at different time points after this removal as designated in the figure. Nuclear proteins were isolated and western analysis was performed to detect c-Fos protein. PC18 cells stably transfected with the tet operon-inducible luciferase gene (LUC-inducible) or the tet operon-inducible c-Fos gene (c-Fos-inducible) were used in these experiments.

results provide strong evidence that induction of c-Fos stimulates TH gene transcription rate. However, it remains to be determined whether this effect of c-Fos is due to its interaction with the TH AP1 site or whether it works via a less direct mechanism. Furthermore, these results only show that induction of c-Fos stimulates TH gene transcription rate; they do not prove that c-Fos induction is essential for the response of the TH gene to different stimuli.

Summary

Most of the work over the past 30 years have used rats as a model system to study the stress-initiated induc-

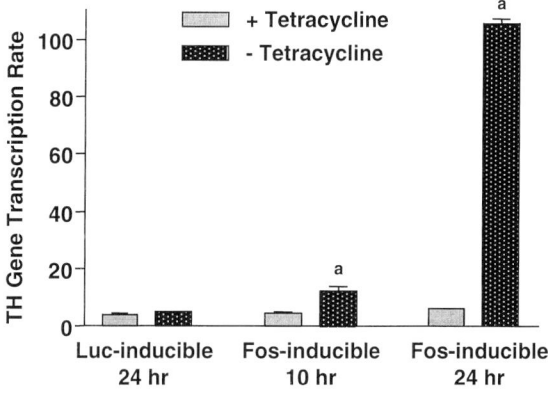

Fig. 9 Effect of c-Fos induction on TH gene transcription rate. PC18 cells that were stably transfected with either the tet-inducible luciferase gene or the tet-inducible *c-Fos* gene were incubated in the presence or absence of tetracycline. Cells were harvested at 10 or 24 hr after removal of the drug. Nuclear run-on assays were performed to measure TH gene transcription rate, which was expressed as the cpm TH divided by the cpm 28S; this quotient was multiplied by 1000 to obtain the y-axis values presented in the figure. The data represent the means ± SEM (range for N = 2) from 2–3 dishes. a: p < .01 compared to controls.

tion of TH. Recently, *in vivo* molecular approaches have been launched using transgenic mice to study the mechanisms involved in this response. The results from these studies provide evidence that the stress-initiated induction of TH in both the adrenal medulla and locus coeruleus is at least partially due to an initial stimulation of TH gene transcription rate. In addition, our results indicate that the stress-initiated response of the endogenous TH gene in the adrenal medulla and locus coeruleus is similar in transgenic mice as in rats. This latter finding suggests that transgenic mouse models will be useful in future studies to probe further the mechanisms responsible for this response. Finally, the results of our studies using cell culture models indicate that the induction of c-Fos is associated with stimulation of TH gene transcription rate. It remains to be determined whether this c-Fos-mediated response is due to direct interaction of this transcription factor with the TH gene promoter and whether it is essential for the response of the gene to stress or other stimuli.

Acknowledgements

This work was funded by NIDA grant 05014 and Smokeless Tobacco Research Council grant 0481 to AWT. CAO was partially supported by the Pharmacological Sciences Training Grant GM08427.

References

Banerjee, S.A., Roffler-Tarlov, S., Szabo, M., Frohman, L., and Chikaraishi, D.M. (1994). DNA regulatory sequences of the rat tyrosine hydroxylase gene direct correct catecholaminergic cell-type specificity of a human growth hormone reporter in the CNS of transgenic mice causing a dwarf phenotype. *Molecular Brain Research* **24**, 89–106.

Bannerjee, S.A., Hoppe, P., Brilliant, M., and Chikaraishi, D.M. (1992). 5′ Flanking sequences of the rat tyrosine hydroxylase gene target accurate tissue-specific, developmental, and transsynaptic expression in transgenic mice. *Journal of Neuroscience* **12**, 4460–4467.

Boundy, V.A., Gold, S.J., Messer, C.J., Chen, J., Son, J.H., Joh, T.H., and Nestler, E.J. (1998). Regulation of tyrosine hydroxylase promoter activity by chronic morphine in TH9. 0-LacZ transgenic mice. *Journal of Neuroscience* **18**, 9989–9995.

Bradford, M.M. (1976). A rapid and sensitive method for the quantitation of microgram quantities of protein utilizing the principle of protein-dye binding. *Analytical Biochemistry* **72**, 248–254.

Craviso, G.L., Hemelt, V.B., and Waymire, J.C. (1995). The transient nicotinic stimulation of tyrosine hydroxylase gene transcription in bovine adrenal chromaffin cells is independent of c-fos gene activation. *Molecular Brain Research* **29**, 233–244.

Fossom, L.H., Carlson, C.D., and Tank, A.W. (1991). Stimulation of tyrosine hydroxylase gene transcription rate by nicotine in rat adrenal medulla. *Molecular Pharmacology* **40**, 193–202.

Fossom, L.H., Sterling, C.R., and Tank, A.W. (1992). Regulation of tyrosine hydroxylase gene transcription rate and tyrosine hydroxylase mRNA stability by cyclic AMP and glucocorticoid. *Molecular Pharmacology* **42**, 898–908.

Ghee, M., Baker, H., Miller, J.C., and Ziff, E.B. (1998). AP-1, Creb and CBP transcription factors differentially regulate the tyrosine hydroxylase gene. *Molecular Brain Research* **55**, 101–114.

Gizang, G., and Ziff, E.B. (1994). Fos family members successively occupy the tyrosine hydroxylase AP-1 site after nerve growth factor or epidermal growth factor stimulation and can repress transcription. *Molecular Endocrinology* **8**, 249–262.

Gossen, M., and Bujard, H. (1992). Tight control of gene expression in mammalian cells by tetracycline-responsive promoters. *Proceedings of the National Academy of Sciences, USA* **89**, 5547–5551.

Gustincich, S., Feigenspan, A., Wu, D.K., Koopman, L.J., and Raviola, E. (1997). Control of dopamine release in the retina: a transgenic approach to neural networks. *Neuron* **18**, 723–736.

Hiremagalur, B., and Sabban, E.L. (1995). Nicotine elicits changes in expression of adrenal catecholamine biosynthetic enzymes, neuropeptide Y and immediate early genes by injection but not continuous administration. *Molecular Brain Research* **32**, 109–115.

Kim, S.J., Lee, J.W., Chun, H.S., Joh, T.H., and Son, J.H. (1997). Monitoring catecholamine differentiation in the embryonic brain and peripheral neurons using E. coli lacZ as a reporter gene. *Molecules and Cells* **7**, 394–398.

Koistinaho, J., Pelto-Huikko, M., Sagar, S.M., Dagerlind, A., Roivainen, R., and Hokfelt, T. (1993). Differential expression of immediate early genes in the superior cervical ganglion after nicotine treatment. *Neuroscience* **56**, 729–739.

Kumer, S.C., and Vrana, K.E. (1996). Intricate regulation of tyrosine hydroxylase activity and gene expression. *Journal of Neurochemistry* **67**, 443–462.

Kvetňanský, R., and Sabban, E.L. (1993). Stress-induced changes in tyrosine hydroxylase and other catecholamine biosynthetic enzymes, in *Tyrosine Hydroxylase: From Discovery to Cloning* (Naoi, M. and Parvez, S.H., eds), pp. 253–281. VSP Press, Utrecht.

Kvetňanský, R., Weise, V.K., and Kopin, I.J. (1970). Elevation of adrenal tyrosine hydroxylase and phenylethanolamine-N-methyltransferase by repeated immobilization of rats. *Endocrinology* **87**, 744–749.

Liu, J., Merlie, J.P., Todd, R.D., and O'Malley, K.L. (1997). Identification of cell type-specific promoter elements associated with the rat tyrosine hydroxylase gene using transgenic founder analysis. *Molecular Brain Research* **50**, 33–42.

Min, N., Joh, T.H., Corp, E.S., Baker, H., Cubells, J.F., and Son, J.H. (1996). A transgenic mouse model to study transsynaptic regulation of tyrosine hydroxylase gene expression. *Journal of Neurochemistry* **67**, 11–18.

Morgan, W.W., Walter, C.A., Windle, J.J., and Sharp, Z.D. (1996). 3.6 kb of the 5' flanking DNA activates the mouse tyrosine hydroxylase gene promoter without catecholaminergic-specific expression. *Journal of Neurochemistry* **66**, 20–25.

Nankova, B., Devlin, D., Kvetňanský, R., Kopin, I.J., and Sabban, E.L. (1993). Repeated immobilization stress increases the binding of c-Fos-like proteins to a rat dopamine beta-hydroxylase promoter enhancer sequence. *Journal of Neurochemistry* **61**, 776–779.

Nankova, B., Kvetňanský, R., McMahon, A., Viskupic, E., Hiremagalur, B., Frankle, G., Fukuhara, K., Kopin, I.J., and Sabban, E.L. (1994). Induction of tyrosine hydroxylase gene expression by a nonneuronal nonpituitary-mediated mechanism in immobilization stress. *Proceedings of the National Academy of Science USA* **91**, 5937–5941.

Nankova, B.B., Tank, A.W., and Sabban, E.L. (1999). Transient or sustained transcriptional activation of the genes encoding rat adrenomedullary biosynthetic enzymes by different durations of immobilization stress. *Neuroscience* **94**, 803–808.

Norris, M.L., and Millhorn, D.E. (1995). Hypoxia-induced protein binding to O_2-responsive sequences on the tyrosine hydroxylase gene. *Journal of Biological Chemistry* **270**, 23774–23779.

O'Connor, K.L., and Culp, L.A. (1994). Quantitation of two histochemical markers in the same extract using chemiluminescent substrates. *Biotechnology* **17**, 502–509.

Osterhout, C.A., Chikaraishi, D.M., and Tank, A.W. (1997). Induction of tyrosine hydroxylase protein and a transgene containing tyrosine hydroxylase 5' flanking sequences by stress in mouse adrenal gland. *Journal of Neurochemistry* **68**, 1071–1077.

Rusnak, M., Zorad, S., Buckendahl, P., Sabban, E.L., and Kvetňanský, R. (1998). Tyrosine hydroxylase mRNA levels in locus coeruleus of rats during adaptation to long-term immobilization stress exposure. *Molecular & Chemical Neuropathology* **33**, 249–258.

Sasaoka, T., Kobayashi, K., Nagatsu, I., Takahashi, R., Kimura, M., Yokoyama, M., Katsuki, M., and Nagatsu, T. (1992). Analysis of the human tyrosine hydroxylase promoter-chloramphenicol acetyltransferase chimeric gene expression in transgenic mice. *Molecular Brain Research* 274–286.

Serova, L.I., Nankova, B.B., Feng, Z., Hong, J.S., Hutt, M., and Sabban, E.L. (1999). Heightened transcription for enzymes involved in norepinephrine biosynthesis in the rat locus coeruleus by immobilization stress. *Biological Psychiatry* **45**, 853–862.

Sherman, T.G., and Moody, C.A. (1995). Alterations in tyrosine hydroxylase expression following partial lesions of the nigrostriatal bundle. *Molecular Brain Research* **29**, 285–296.

Tank, A.W., Ham, L., and Curella, P. (1986). Induction of tyrosine hydroxylase by cyclic AMP and glucocorticoids in a rat pheochromocytoma cell line: Effect of the inducing agents alone or in combination on the enzyme levels and rate of synthesis of tyrosine hydroxylase. *Molecular Pharmacology* **30**, 486–496.

Thoenen, H. (1970). Induction of tyrosine hydroxylase in peripheral and central adrenergic neurons by cold exposure. *Nature* **228**, 861–862.

Trocme, C., Sarkis, C., Hermel, J.M., Duchateau, R., Harrison, S., Simonneau, M., Alshawi, R., and Mallet, J. (1998). CRE and TRE sequences of the rat tyrosine hydroxylase promoter are required for TH basal expression in adult mice but not in the embryo. *European Journal of Neuroscience* **10**, 508–521.

20. Genes of Tetrahydrobiopterin-Synthesizing Enzymes in Relation to Stress

H. Ichinose, T. Ohye, T. Suzuki, H. Inagaki and T. Nagatsu

Introduction

Tyrosine hydroxylase (TH) is the first and regulatory enzyme in the biosynthetic pathway of catecholamines, i.e. tyrosine $\longrightarrow$ dopa $\longrightarrow$ dopamine $\longrightarrow$ norepinephrine $\longrightarrow$ epinephrine (Nagatsu et al., 1964; Levitt et al., 1965). Under stress, TH is acutely activated during sympathetic stimulation (Weiner et al., 1978). The acute activation of TH under stress in vivo is confirmed by increases in plasma dopa and dopamine (Goldstein et al., 1987; Kvetňanský et al., 1992). Under chronic stress, induction of catecholamine-synthesizing enzymes, especially TH, is an important regulatory mechanism (Kvetňanský et al., 1970).

TH is a pteridine-requiring monooxygenase, and its activity is partly regulated by the concentration of the pteridine cofactor (Nagatsu and Ichinose, 1999a, 1999b). Tetrahydrobiopterin (BH4) is the natural pteridine cofactor (Kaufman, 1963). The stereochemistry of BH4 is (6R)-L-erythro-BH4 (Matsuura et al., 1985). BH4 is synthesized from guanosine triphosphate (GTP) by the pathway: GTP $\xrightarrow{(1)}$ D-erythro-7,8-dihydroneopterin triphosphate (NH2P3) $\xrightarrow{(2)}$ 6-pyruvoyltetrahydropterin (PPH4) $\xrightarrow{(3)}$ (6R)-BH4 (Figures 1 and 2). There are three BH4-synthesizing enzymes: (1) GTP cyclohydrolase I (GCH1), (2) 6-pyruvoyltetrahydropterin synthase (PTPS), and (3) sepiapterin reductase (SPR). The first enzyme of BH4 biosynthesis, GCH1, is the regulatory enzyme (Nicol et al., 1985). Thus, the activity of GCH1 indirectly regulates TH activity via BH4 concentrations in catecholamine-producing cells (Nagatsu and Ichinose, 1999b).

Recent evidence indicates that under stress not only catecholamine levels, but also the levels of BH4 synthesized from GTP are increased (Hossain et al., 1992; Serova et al., 1997a; Serova et al., 1997b). We and other workers cloned cDNA and genomic DNA of the three BH4-synthesizing enzymes of humans and mice (Table 1). The structures of the genes and of the deduced enzyme proteins of the three BH4-synthesizing enzymes, especially those of GCH1, should be important in elucidating the molecular mechanism of stress.

GTP Cyclohydrolase I (GCH1) Gene

GCH1 (EC 3.5.4.16) is a 300-kDa decamer composed of identical 25 kDa subunits. The catalyzed reaction involves a ring expansion conductive to the formation of 7,8-dihydroneopterin triphosphate from GTP, the

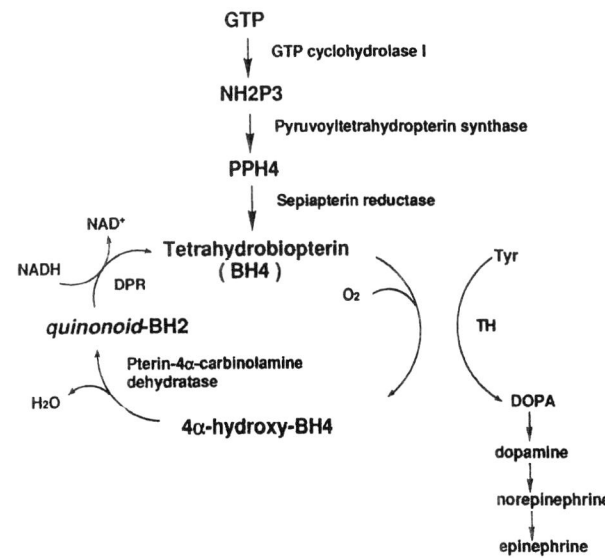

Fig. 1 Relation between the biosynthetic pathway of catecholamines mediated by tyrosine hydroxylase (TH) and the biosynthetic pathway of the cofactor of TH, tetrahydrobiopterin (BH4) from GTP. GTP, guanosine triphosphate; NH2P3, D-erythro-7, 8-dihydroneopterin triphosphate; PPH4, 6-pyruvoyltetrahydropterin; BH2, 7,8-dihydrobiopterin; and DPR, dihydropteridine reductase.

first step in BH4 biosynthesis (Figure 1, Figure 2). GCH1 cDNAs of humans (Togari et al., 1992), rats (Hatakeyama et al., 1991), and mice (Nomura et al., 1993), as well as the genomic DNAs of human and mouse GCH1 (Ichinose et al., 1995) were cloned, and their structures determined. The human GCH1 gene is mapped to chromosome 14q22.1-22.2 (Ichinose et al., 1994). Both mouse and human genes spanning about 30 kb consist of 6 exons interrupted by 5 introns (Figure 3).

The nucleotide sequences of the 5'-flanking region for the mouse and human genes were determined to about 600 bp upstream from the transcription start site. No obvious TATA box was observed in the promoter region. In the mouse gene, an AT-rich putative promoter motif (ATAAAAA) or a CCAAT box was present just upstream or 50 bp upstream from the transcriptional start site, respectively, and an H1-box, an IBP-1b, and a GT-2B binding consensus sequence were found. The AT-rich putative promoter motif and the CCAAT box found in the mouse gene were conserved in the human GCH1 gene. A consensus

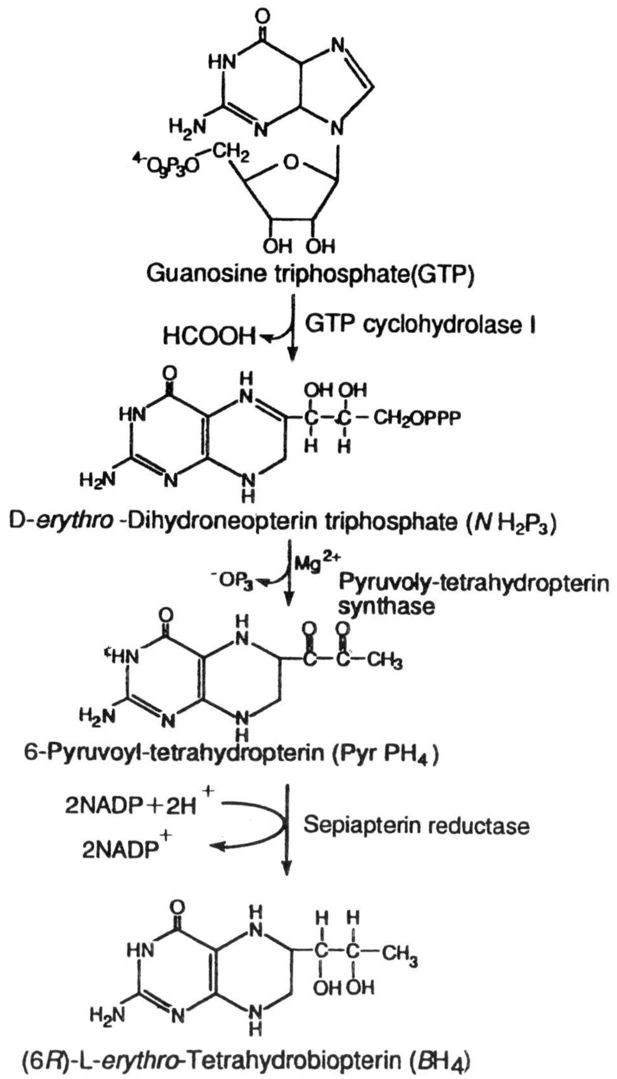

Fig. 2 Biosynthetic pathway of tetrahydrobiopterin (BH4).

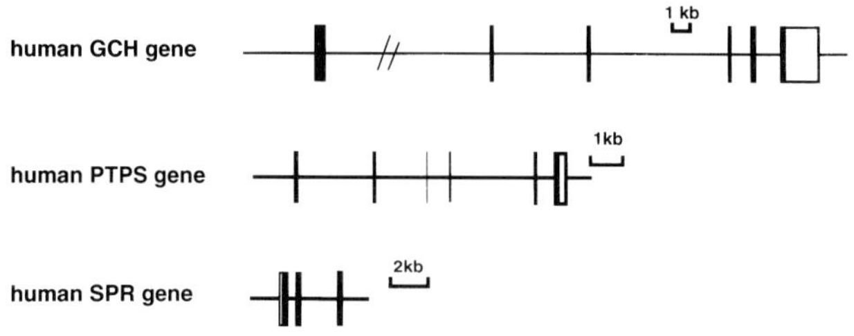

Fig. 3 Structures of human GTP cyclohydrolase I (GCH1) gene, human 6-pyruvoyltetrahydropterin; synthase (PTPS) gene, and human sepiapterin reductase (SPR) gene.

gene were not present in the human gene, whereas a leader binding protein LBP-1 binding sequence motif ((A/T)CTGG: positions −479 to −475), a T-antigen binding motif (positions −274 to −272), a TGGCA box (positions −675 to −671), and an SP-1 consensus sequence (positions −235 to −246) were present in the human gene (Ichinose et al., 1995).

We determined the nucleotide sequence of the 5′-flanking region of the human GCH1 gene up to 8kb, and analyzed the promoter activity. In PC12 cell line, the 8-kb, 5.4-kb, and 0.6 kb DNA fragments of the GCH1 gene had similar activity to express a chloramphenicol acetyltransferase (CAT) reporter gene. In NIE-115 cell line, however, a 0.6-kb fragment expressed CAT more predominantly than the longer fragments, suggesting the presence of some inhibitory elements between 5.4 and 0.6 kb.

6-Pyruvoyltetrahydropterin Synthase (PTPS) Gene

6-Pyruvoyltetrahydopterin synthase (PTPS, E.C4.6.1.10) is required for the second step of BH4 biosynthesis starting from GTP. PTPS converts 7,8-dihydroneopterin triphosphate to 6-pyruvoyltetrahydropterin. PTPS is a homohexamer which is formed by face-to-face association of the two trimers. The enzyme requires two metals for activity, Zn^{2+} and Mg^{2+} (Oppliger et al., 1995). The structure of human PTPS cDNA (Thöny et al., 1992;

sequence for the binding of GT-2B was found in the human gene as well as in the mouse gene. The IBP-1b and H1-box consensus sequences found in the mouse

Table 1 Human tetrahydrobiopterin-synthesizing enzymes.

Enzymes	Genes		Proteins	
	Chromosome	Exons	Mr (kDa × number of subunit)	Amino acid residues
Guanosine triphosphate (GTP) cyclohydrolase I (GCH1)	14q22.1–q22.2	6	30 × 10	250
Pyruvoyltetrahydropterin synthase (PTPS)	11q22.3–q23.3	6	17 × 6	145
Sepiapterin reductase (SPR)	2p13	3	25 × 2	261
Pterin-4α-carbinolamine dehydratase (PCBD)	10q22	4	11 × 2	104
Dihydropteridine reductase (QDPR)	4q15.31	(−)	25 × 2	244

Ashida *et al.*, 1993) and human genomic DNA (Kluge *et al.*, 1996) were elucidated. We also cloned the mouse and human PTPS genes (Figure 4). The PTPS gene is localized on chromosome 11q22.3-q22.3. The PTPS gene spans about 8kb and consists of 6 exons (Kluge *et al.*, 1996). The 5'-promoter regions of the PTPS gene contains the consensus sequences of SP-1, LBP-1, and T-antigen which also exist in the promoter region of the GCH1 gene.

Sepiapterin Reductase Gene

Sepiapterin reductase (SPR, EC 1.1.1.153) catalyzes the last step of BH4 biosynthesis, the NADPH-dependent reduction of 6-pyruvoyltetrahydropterin to BH4. SPR cDNA was cloned in rats (Citron *et al.*, 1990), mice (Ota *et al.*, 1995), and humans (Ichinose *et al.*, 1991). The complete amino acid sequence of the mature form of rat SPR was also elucidated (Oyama *et al.*, 1990). The protein begins with N-acetylmethionine at the N-terminus. The Mr of the subunit is 28,169 with 262 amino acid residues including the N-terminal acetyl group. The enzyme consists of two identical subunits. Therefore, the Mr of the native enzyme is 56,338. The human SPR gene (Figure 3) is composed of three exons spanning approximately 4 kb. There is no typical TATA-box within 300 bp from the transcription start point. There is a Sp1-binding consensus sequence in the 5'-flanking region. The human SPR gene is mapped to chromosome band 2p13 (Ohye *et al.*, 1998).

Mutations of GCH1 and PTPS Genes

Autosomal dominant hereditary progressive, dopa-responsive dystonia (Segawa's disease; HPD/DRD) that was first reported by Segawa *et al.* (1971) is caused by the mutations of one allele of the GCH1 gene (Ichinose *et al.*, 1994). The phenotype is dystonia with marked diurnal fluctuation, which is completely controlled by low doses of L-dopa. Autosomal recessive

GCH1 deficiency is caused by mutations of the two alleles of the GCH gene, resulting in nearly complete loss of GCH1 activity and BH4 levels. The phenotype of this recessive condition is complex: hyper-phenylalaninemia caused by low phenylalanine hydroxylase activity in the liver and severe neurological symptoms probably produced by both low TH activity (resulting in catecholamine deficiency) and low tryptophan hydroxylase activity (resulting in serotonin deficiency) in the brain. Reduction in NO production, as a result of decreased nitric oxide synthase may also contribute to the severe neurological phenotype (Nagatsu and Ichinose 1999a, 1999b; Nagatsu *et al.*, 1999).

PTPS mutations have only been found to be recessive. PTPS deficiency appears in three different phenotypes, a central, a peripheral, and a transient form, which may disappear during infancy. Patients with the central type of PTPS deficiency exhibit a general lack of BH4 in all organs and monoamine neurotransmitter shortage in the central nervous system, whereas patients with the peripheral form do not synthesize BH4 in peripheral organs but have normal BH4 and neurotransmitter levels in the central nervous system (Oppliger *et al.*, 1995).

In contrast to the GCH1 gene and the PTPS gene, SPR gene mutations have not been identified, although it is likely that an absence of SPR can lead to alterations in BH4 metabolism.

Regulation of the Genes for Tetrahydrobiopterin-Synthesizing Enzymes Under Stress

In murine neuroblastoma cell line NIE-115, twenty-four hour activation of this cell line with lipopolysaccharide resulted in statistically significant increases in the amounts of all three enzymes. However, the increase in GCH1 mRNA level was predominant. The results suggest that the GCH1 gene may be mainly influenced also under stress (Mori *et al.*, 1997).

The response of GCH1 to stress in acute phase is by increases in both Km and Vmax values (Hossain *et al.*, 1992). Chronic responses are reflected by the increase in GCH1 mRNA level. Serova *et al.* (1997b), reported that with immobilization stress GCH1 mRNA levels were increased 3–5 fold with an intact pituitary adrenal axis. It is speculated that the induction of TH and GCH1 mRNAs occur in parallel and is coordinated (Serova *et al.*, 1997b).

The common consensus sequences present in the upstream region of human GCH1 and PTPS genes are shown in Figure 5. However, the physiological relevance of these consensus sequences in relation to stress needs additional functional studies.

Conclusion

TH is a key enzyme of catecholamine systems in stress responses. TH activity may be regulated by GCH1 activity via changes in BH4 cofactor levels during

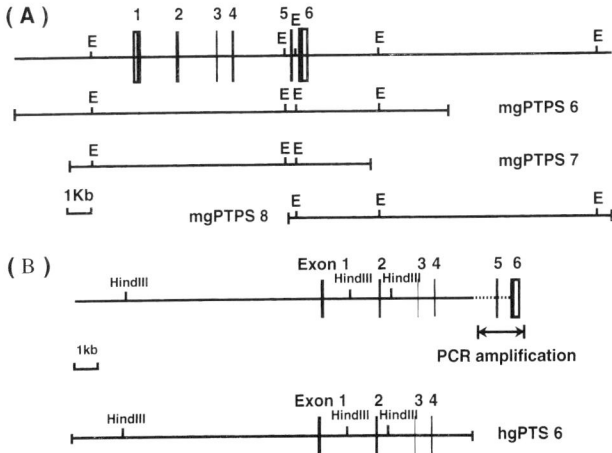

Fig. 4 Structures of mouse (A) and human (B) 6-pyruvoyltetrahydropterin synthase.

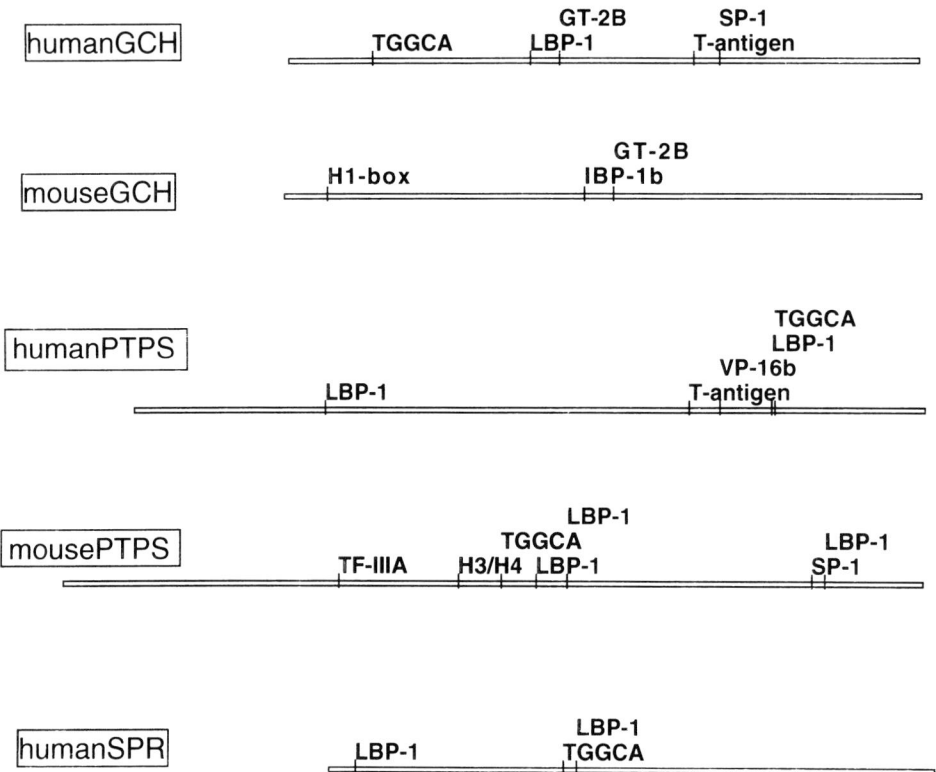

Fig. 5 Structures of the promoter regions of human and mouse GTP cyclohydrolase I (GCH1) genes, human and mouse 6-pyruvoyltetrahydropterin synthase (PTPS) genes, and human sepiapterin reductase (SPR) gene.

stress. Thus GCH1, the rate-limiting enzyme of BH4 cofactor biosynthesis, may also have an important regulatory role during stress. The structures of the genes of all three BH4-synthesizing enzymes will promote the elucidation of the molecular mechanism of the role of the BH4 system in relation to stress.

References

Ashida, A., Hatakeyama, K., and Kagamiyama, H. (1993). cDNA cloning, expression in *Escherichia coli* and purification of human pyruvoyltetrahydropterin synthase. *Biochemical and Biophysical Research Communications* **195**, 1386–1393.

Citron, B.A., Milstien, S., Gutierrez, J.C., Levine, R.A., Yanak, B.L., and Kaufman, S. (1990). Isolation and expression of rat liver sepiapterin reductase cDNA. *Proceedings of the National Academy of Sciences USA* **87**, 6436–6440.

Goldstein, D.S., Udelsman, R., Eisenhofer, G., Keiser, H.R., and Kopin, I.J. (1987). Neuronal source of plasma dihydroxyphenylalanine. *Journal of Clinical Endocrinology and Metabolism* **64**, 856–861.

Hatakeyama, K., Inoue, Y., Harada, T., and Kagamiyama, H. (1991). Cloning and sequencing of cDNA encoding rat GTP cyclohydrolase I. The first enzyme of the tetrahydrobiopterin biosynthetic pathway. *Journal of Biological Chemistry* **266**, 765–769.

Hossain, M.A., Masserano, J.M., and Weiner, N. (1992). Effects of electroconvulsive shock on tetrahydrobiopterin and GTP cyclohydrolase activity in the brain and adrenal gland of the rat. *Journal of Neurochemistry* **59**, 2237–2243.

Ichinose, H., Katoh, S., Sueoka, T., Titani, K., Fujita, K., and Nagatsu, T. (1991). Cloning and sequencing of cDNA encoding human sepiapterin reductase. An enzyme involved in tetrahydrobiopterin biosynthesis. *Biochemical and Biophysical Research Communications* **179**, 183–189.

Ichinose, H., Ohye, T., Matsuda, Y., Hori, T., Blau, N., Burlina, A., Rouse, B., Matalon, R., Fujita, K., and Nagatsu, T. (1995). Characterization of

mouse and human GTP cyclohydrolase I genes: Mutations in patients with GTP cyclohydrolase I deficiency. *Journal of Biological Chemistry* **270**, 10062–10071.

Ichinose, H., Ohye, T., Takahashi, E., Seki, N., Hori, T., Segawa, M., Nomura, Y., Endo, K., Tanaka, H., Tsuji, S., Fujita, K., and Nagatsu, T. (1994). Hereditary progressive dystonia with marked diurnal fluctuation caused by mutations in the GTP cyclohydroxylase I gene. *Nature Genetics* **8**, 236–242.

Kaufman, S. (1963). The structure of the phenylalanine hydroxylation cofactor. *Proceedings of the National Academy of Sciences, USA* **50**, 1085–1093.

Kluge, C., Brecevic, L., Heizmann, C.W. Blau, N., and Thöny, B. (1996). Chromosomal localization, genomic structure and characterization of the human gene and a retropseudogene for 6-pyruvoyltetrahydropterin synthase. *European Journal of Biochemistry* **240**, 477–484.

Kvetňanský, R., Fukuhara, K., Weise, V.K., Armando, I., Kopin, I.J., and Goldstein, D.S. (1992). Stress-induced changes in plasma dopa levels depend on sympathetic activity and tyrosine hydroxylation. In: R. Kvetňanský, R. McCarty, and J. Axelrod (Eds.), *Stress: Neuroendocrine and Molecular Approaches*, Volume 1, pp. 139–157. New York: Gordon and Breach Science Publishers.

Kvetňanský, R., Weise, V.K., and Kopin, I.J. (1970). Elevation of adrenal tyrosine hydroxylase and phenylethanolamine-N-methyltransferase by repeated immobilization of rats. *Endocrinology* **87**, 744–749.

Levitt, M., Spector, S., Sjoerdsma, A., and Udenfriend, S. (1965). Elicidation of the rate-limiting step in norepinephrine biosynthesis in the perfused guinea pig heart. *Journal of Pharmacology and Experimental Therapeutics* **148**, 1–8.

Matsuura, S., Sugimoto, T., Murata, S., Sugawara, Y., and Iwasaki, H. (1985). Stereochemistry of biopterin cofactor and facile methods for the determination of the stereochemistry of a biologically active 5,6,7,8-tetrahydrobiopterin. *Journal of Biochemistry* **98**, 1341–1348.

Mori, K., Nakashima, A., Nagatsu, T., and Ota, A. (1997). Effect of lipopolysaccharide on the gene expression of the enzymes involved in tetrahydrobiopterin de novo biosynthesis in murine neuroblastoma cell line NIE-115. *Neuroscience Letters* **238**, 21–24.

Nagatsu, T., and Ichinose, H. (1999a). Molecular biology of catecholamine-related enzymes in relation to Parkinson's disease. *Cellular and Molecular Neurobiology* **19**, 57–66.

Nagatsu, T., and Ichinose, H. (1999b). Regulation of pteridine-requiring enzymes by the cofactor tetrahydrobiopterin. *Molecular Neurobiology* **19**, 79–96.

Nagatsu, T., Ichinose, H., Mogi, M., and Togari, A. (1999). Neopterin and cytokines in hereditary dystonia and Parkinson's disease. *Pteridines* **10**, 5–13.

Nagatsu, T., Levitt, M., and Udenfriend, S. (1964). Tyrosine hydroxylase: The initial step in norepinephrine biosynthesis. *Journal of Biological Chemistry* **239**, 2910–2917.

Nichol, C.A., Smith, G.K., and Duch, D.S. (1985). Biosynthesis and metabolism of tetrahydrobiopterin and molybdopterin. *Annual Review of Biochemistry* **54**, 729–764.

Nomura, T., Ichinose, H., Sumi-Ichinose, C., Nomura, H., Hagino, Y., Fujita, K., and Nagatsu, T. (1993). Cloning and sequencing of cDNA encoding mouse GTP cyclohydrolase I. *Biochemical and Biophysical Research Communications* **191**, 523–527.

Nomura, T., Ohtsuki, M., Matsui, S., Sumi-Ichinose, C., Nomura, H., Hagino, Y., Iwase, K., Ichinose, H., Fujita, K., and Nagatsu, T. (1995). Isolation of a full-length cDNA clone for human GTP cyclohydrolase I type 1 from pheochromocytoma. *Journal of Neural Transmission* **101**, 237–242.

Ohye, T., Hori, T., Katoh, S., Nagatsu, T., and Ichinose, H. (1998). Genomic organization and chromosomal localization of the human sepiapterin reductase gene. *Biochemical and Biophysical Research Communications* **251**, 597–602.

Oppliger, T., Thöny, B., Nar, H., Bürgisser, D., Huber, R., Heizmann, C.W., and Blau, N. (1995). Structure and functional consequences of mutations in 6-pyruvoyltetrahydropterin synthase causing hyperphenylalanine-mia in humans: Phosphorylation is a requirement for in vivo activity. *Journal of Biological Chemistry* **270**, 29498–29506.

Ota, A., Ichinose, H., and Nagatsu, T. (1995). Mouse sepiapterin reductase: an enzyme involved in the final step of tetrahydrobiopterin biosynthesis. Primary structure deduced from the cDNA sequence. *Biochimica et Biophysica Acta* **1260**, 320–322.

Oyama, R., Katoh, S., Sueoka, T., Suzuki, M., Ichinose, H., Nagatsu, T., and Titani, K. (1990). The complete amino acid sequence of the mature form of rat sepiapterin reductase. *Biochemical and Biophysical Research Communications* **173**, 627–631.

Segawa, M., Ohmi, K., Itoh, S., Aoyama, M., and Hayakawa, H. (1971). Childhood basal ganglia disease with remarkable response to L-DOPA, hereditary basal ganglia disease with marked diurnal fluctuation. *Shinryo (Tokyo)* **24**, 667–672.

Serova, L.I., Nankova, B., Kvetňanský, R., and Sabban, E.L. (1997a). Glucocorticoid elevates GTP cyclohydrolase I mRNA levels in vivo and in PC12 cells. *Molecular Brain Research* **48**, 251–258.

Serova, L.I., Nankova, B., Kvetňanský, R., and Sabban, E.L. (1997b). Immobilization stress elevates GTP cyclohydrolase I mRNA levels in rat adrenals predominantly by hormonally mediated mechanisms. *Stress* **1**, 135–144.

Thöny, B., Leimbacher, W., Bürgisser, D., and Heizmann, C.W. (1992). Human 6-pyruvoyl-tetrahydropterin synthase: cDNA cloning and heterologous expression of the recombinant enzyme. *Biochemical and Biophysical Research Communications* **189**, 1437–1443.

Togari, A., Ichinose, H., Matsumoto, S., Fujita, K., and Nagatsu, T. (1992). Multiple mRNA forms of human GTP cyclohydrolase I. *Biochemical and Biophysical Research Communications* **187**, 359–365.

Weiner, N., Lee, F.L., Dreyer, E., and Barnes, E. (1978). The activation of tyrosine hydroxylase in noradrenergic neurons during acute nerve stimulation. *Life Sciences* **22**, 1197–1216.

21. Novel Stressors Exaggerate Tyrosine Hydroxylase Gene Expression in the Adrenal Medulla of Rats Exposed to Long-Term Cold Stress

R. Kvetňanský, J. Jeloková, M. Rusnák, S. Dronjak, L. Serova, B.B. Nankova, and E.L. Sabban

Introduction

It has been shown that animals chronically exposed to a particular stressor display an exaggerated response of the HPA axis to various novel stressors (Dhabhar *et al.*, 1997; Bhatnagar and Dallman, 1998). It has been also shown that prior chronic intermittent cold stress modifies cardiovascular function both under resting conditions and under stimulated conditions produced by restraint and formalin administration (Bhatnagar *et al.*, 1998). However, the activity of the sympathetic-adrenomedullary system in rats exposed to a novel, acute stressor after prior chronic stress exposure is poorly understood.

We have found a potentiation of sympathoadrenal system activity in chronically stressed rats exposed to novel stressors. Animals exposed to repeated immobilization stress displayed a much higher increase in plasma catecholamines in response to a novel stressor (e.g. trauma), when compared to the response of naive animals (Kvetňanský *et al.*, 1984). Animals acclimated to the long-term effects of a combination of cold and hypoxia at an altitude of 1350 m displayed significantly higher rises in plasma catecholamine levels and increased adrenal TH activity when subjected to a single immobilization stress compared to the age-matched animals kept at 150 m altitude (Balaz *et al.*, 1980). Rats exposed to chronic intermittent cold, foot shock or immobilization show enhanced peripheral catecholamine responses to a novel stressor (Kvetňanský *et al.*, 1971b; Stone and McCarty, 1983; McCarty and Stone, 1984; McCarty *et al.*, 1988; Konarska *et al.*, 1989).

Nisenbaum and Abercrombie (1992) and Nisenbaum *et al.* (1991) found that in rats exposed to chronic cold stress, *in vivo* tyrosine hydroxylation in the hippocampus did not differ from naive controls. However, when chronic cold stressed animals were exposed to a novel stressor, greater elevations were attained compared to naive rats. The exaggerated effect of novel stressors is an interesting phenomenon that has not been studied at the level of adrenomedullary gene expression of catecholamine biosynthetic enzymes.

The aim of the present study was to investigate changes in adrenomedullary tyrosine hydroxylase gene expression after short (one day) and long-term (28 days) cold exposure, as well as in rats adapted to long-term cold and then exposed to various novel stressors.

Effect of Long-term Cold Exposure on Tyrosine Hydroxylase mRNA Levels in the Rat Adrenal Medulla

Experiments were performed on male Sprague-Dawley rats weighing 400–450 g (Charles River, Wiga, Germany). The rats were divided into three groups. The first group was exposed to cold for one day only, the second group was exposed to long-term cold (28 days) and the third group was kept at room temperature (control). Rats were housed two per cage (metal cages without bedding) in a cold chamber at 4°C with a 12 hour light–12 hour dark cycle. Food and water were available *ad libitum*. Data are presented as means + SEM and were statistically evaluated by one-way ANOVA.

All procedures were in agreement with the standards for the care and use of laboratory animals as outlined in the NIH Guide for the Care and Use of Laboratory Animals.

We wanted to answer the question: "How does long-term cold exposure affect TH gene transcription and translation". TH mRNA levels were measured by Northern Blot as described previously (Kvetňanský *et al.*, 1996), TH immunoreactive protein levels by Western Blot (Nankova *et al.*, 1994) and TH activity by an adaptation of the micromethod (Waymire *et al.*, 1971).

TH mRNA levels in the adrenal medulla – one day cold exposure produced a significant elevation in TH mRNA levels. However, after 28 days of cold stress the levels were not significantly different from control rats kept at room temperature (Figure 1).

TH activity – showed a very similar picture as TH mRNA levels increased after the first day and no changes after the 28th day of cold exposure (Figure 2).

TH immunoreactive protein levels – showed a different picture; levels were not changed after one day of cold but were significantly elevated after 28 days of cold exposure (Figure 3).

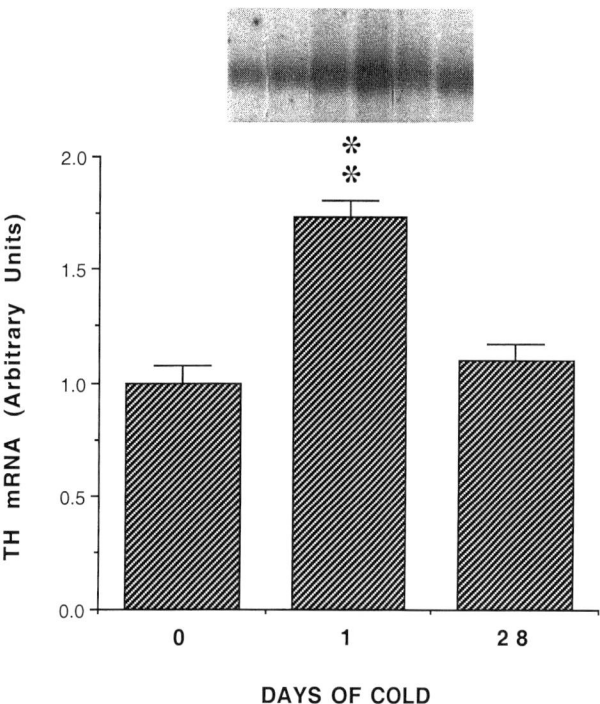

Fig. 1 Effect of short (1 day) and long-term cold exposure on tyrosine hydroxylase mRNA levels in the adrenal medulla of rats as measured by Northern blot. Mean values ± SEM for 7–8 rats per group are shown. Statistical significance calculated by one-factor ANOVA followed by Fisher's post-hoc test: (**) $p < 0.01$ vs. control or 28 day cold-exposed groups.

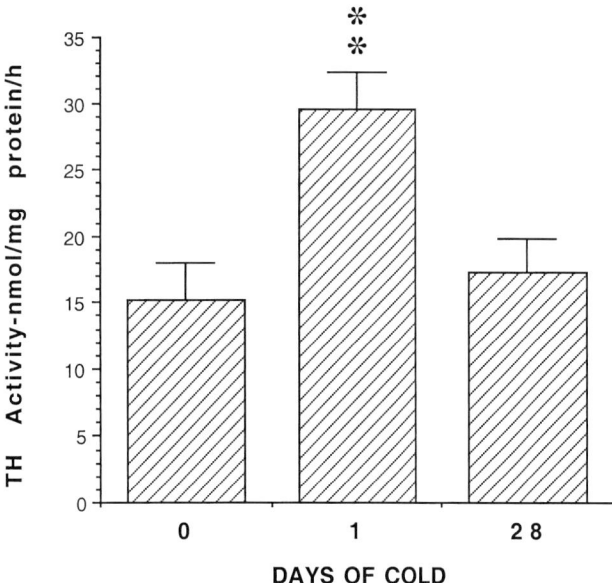

Fig. 2 Effect of short (1 day) and long-term cold exposure on tyrosine hydroxylase activity in the adrenal medulla of rats. Mean values ± SEM for 7–8 rats per group are shown. Statistical significance calculated by one-factor ANOVA followed by Fisher's post-hoc test: (**) $p < 0.01$ vs. control or 28 day cold-exposed groups.

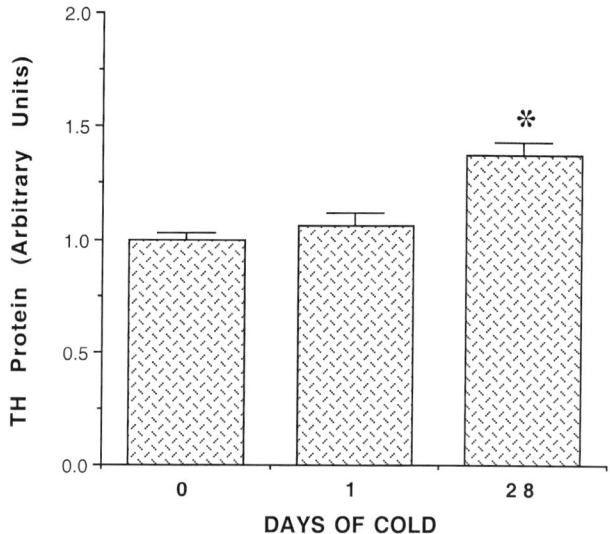

Fig. 3 Effect of short (1 day) and long-term cold exposure on levels of tyrosine hydroxylase immunoreactive protein in the adrenal medulla of rats as measured by Western blot. Mean values ± SEM for 7–8 rats per group. Statistical significance calculated by one-factor ANOVA followed by Fisher's post-hoc test: (*) $p < 0.05$ vs. control or one day cold-exposed groups.

Many researchers have found increased adrenal TH activity, TH immunoreactive protein and TH mRNA levels in rats exposed to cold stress (see reviews Kvetňanský and Sabban, 1993; 1998). Our present data have shown a significant increase in adrenal medullary TH activity after exposure of rats to short-term cold while TH activity returned almost to control levels in long-term cold exposed rats (28 days; "cold-adapted rats"). These data are in good agreement with our previous findings on adrenal TH activity which increased following up to 21 days of cold exposure but returned to near control values in rats exposed to cold for 28–42 days (Kvetňanský et al., 1971a). Another study also found significantly increased adrenal TH activity in rats kept in the cold for 1–3 weeks that did not return to basal levels (Fluharty et al., 1985a; Miner et al., 1992) but they did not study longer cold intervals. The differences with the present study might reflect the longer time of cold exposure or the different experimental design. The animals used were shaved but unshaved rats exposed to cold showed significantly lower adrenal TH activity (Fluharty et al., 1985). Our animals were not shaved and were housed two per metal cage without bedding. These factors may play a part in the almost complete adaptation of adrenal TH activity in rats exposed to long-term cold.

Cold stress is also accompanied by increased adrenal TH mRNA levels (see reviews by Kvetňanský and Sabban, 1993, 1998). Maximal TH mRNA increases were observed within 3–6 hours of cold exposure and remained elevated during longer cold exposure up to 7 days (Stachowiak et al., 1985; Richard et al., 1988; Baruchin et al., 1990; Miner et al., 1992). We have not found studies with a very long-term cold exposure. In the present work, we found a significant increase in adrenal TH mRNA levels after short-term cold exposure; however, in rats exposed to cold for 28 days, adrenal TH mRNA levels returned to near control values. Thus, the reduced TH mRNA levels correlate very well with the reduction of TH activity in the adrenal medulla of "cold-adapted rats" (Figure 2).

An increase in adrenal TH immunoreactive protein levels in response to cold exposure has also been reported (Hoeldtke et al., 1974; Richard et al., 1988; Baruchin et al., 1990). Data presented in this paper do not show any changes in TH protein levels during the first day of cold exposure; however, a significant increase after 28-day cold exposure was seen. How does one explain this discrepancy between increased TH protein, but not mRNA or enzyme activity? There is some evidence for a dissociation between TH activity and protein levels. In spite of the fast rise in TH mRNA levels, the maximal increase in TH protein levels was observed after three days of continuous cold, and TH activity reached maximal levels after 6–7 days of cold exposure (Baruchin et al., 1990; Miner et al., 1992). Thus, during cold, the rise in TH mRNA levels does not appear to be translated into immunoreactive protein until several days later, and there is a lag between the rise in TH protein and the elevation of TH activity. These authors showed that TH protein levels and TH enzyme activity can be differentially regulated. A more complicated situation may occur in the regulation of the three TH parameters in "cold-adapted rats". Andrews and coworkers (1993) examined the effects of prolonged cold exposure on TH protein and TH activity in sympathetic neurons of the superior cervical ganglion of rabbits and found that they were differentially regulated. TH protein levels rose while TH activity tended to decline. Decentralization of the ganglion abolished the cold-induced rise in TH protein but cold exposure resulted in an increase in TH activity. Thus, TH activity demonstrated a non-synaptic regulatory mechanism (Andrews et al., 1993). The factors which could influence TH activity, apart from synthesis of the enzyme, include phosphorylation of the enzyme to its active form, and availability of the substrate tyrosine or levels of the co-factor tetrahydrobiopterin (Massereno et al., 1989). Intracellular levels of tetrahydrobiopterin can be altered by stress and sympathetic nerve activity (Baruchin et al., 1990; Massereno et al., 1989) and in this way may affect TH activity without changing TH protein levels. Also, gene expression of the rate-limiting enzyme of tetrahydrobiopterin biosynthesis, GTP cyclohydrolase I, is increased with stress (Serova et al., 1997).

The dissociation between adrenal TH mRNA levels and TH protein levels after exposure to long-term cold might be explained by the differences in their half-life. Repeated immobilization-induced increases in levels of adrenal TH protein return to basal levels within 14 days with a half-life about 2.5 days (Kvetňanský et al., 1970). On the other hand, repeated IMO-induced increases in TH mRNA levels (5–10 fold)

return, after cessation of the stress stimuli, back to prestress levels faster, within 5–7 days (McMahon et al., 1992; Nankova et al., 1994). These findings might explain our data on reduced TH mRNA levels and increased TH protein levels found in long-term cold exposed rats. Thus, in cold adapted rats, adrenal TH mRNA is reduced but is prepared to be quickly elevated whenever it might be necessary, whereas TH protein (with a slow half-life) stays increased. This situation is documented by our data presented in the second part of this paper. Adrenal TH mRNA levels, which are in cold-adapted rats at control levels, are increased to a greater degree in response to a novel stressor compared to naive control rats (Figure 6), while TH protein levels are not changed (Figure 8). Thus, the observed differences in tyrosine hydroxylase mRNA levels in the adrenal medulla of rats exposed to short and long-term cold suggest an adaptation at the level of TH gene expression.

Plasma norepinephrine levels – How does long-term cold exposure affect plasma catecholamine levels? For blood sample collection, rats were cannulated in the tail artery to prevent an influence of unwanted stressors. Blood samples were taken through a polyethylene catheter which had been inserted into all animals under pentobarbital anaesthesia 24 h before the blood collection started. Cold adapted animals were after insertion of the catheters and waking up from anaesthesia transferred to their home cages in the cold room and the next morning blood was collected at intervals identical to animals exposed to cold for the first time. The animals exposed for the first time to cold were kept at room temperature and after baseline blood collection were carefully transferred within their home cages into the cold chamber and blood was collected at 30, 60, and 120 min of cold exposure. Plasma catecholamines were assayed by our modification of the radioenzymatic method (Peuler and Johnson, 1977). Data were statistically evaluated by two-way ANOVA. Results are presented as means $\pm$ SEM.

The effects of the first and long-term cold exposure on plasma norepinephrine (NE) levels are shown in Figure 4. The first cold exposure produced a significant increase in plasma NE levels with peak levels around 30 min of cold. Plasma NE stayed at these levels during the whole study period up to the 120 minute interval. In animals exposed to cold for 28 days, plasma NE levels were significantly elevated ($p < 0.01$). The adrenomedullary system was not noticeably affected by cold exposure and plasma EPI levels were not significantly changed (data not shown).

Thus, the elevated plasma levels of NE found in long-term cold exposed rats compared to the animals exposed to cold for the first time (Figure 4) are in good agreement with the finding of increased adrenal NE stores (Kvetňanský et al., 1971a) and increased levels of adrenal TH immunoreactive protein (Figure 3). Mechanism(s) of unchanged levels of TH mRNA and TH activity in the adrenal medulla of rats exposed to long-term cold are at present unknown.

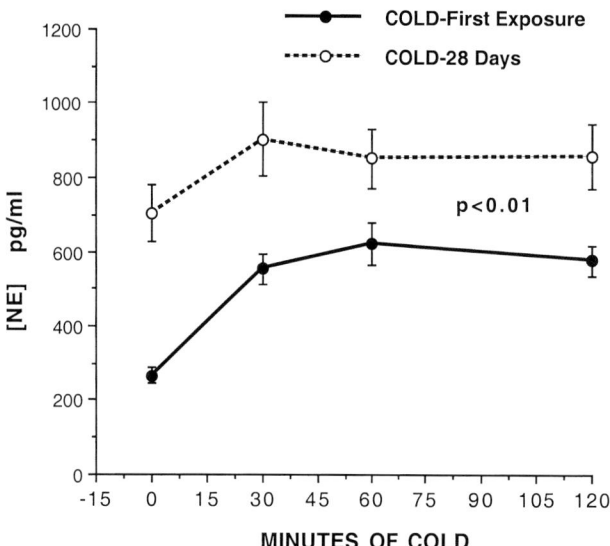

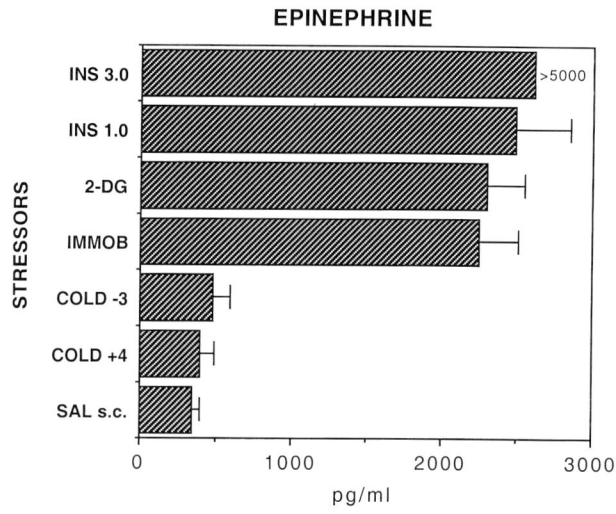

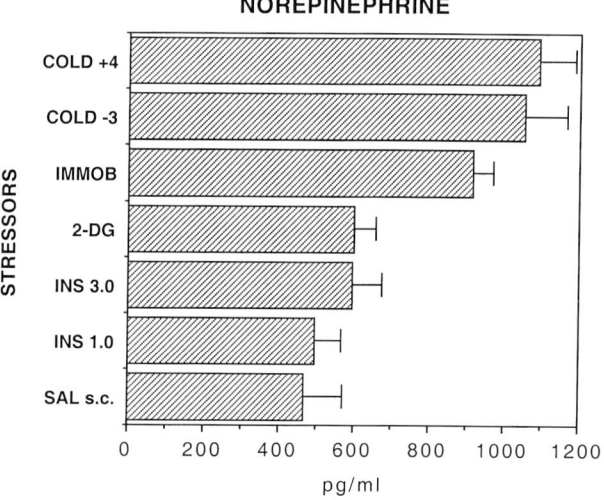

Fig. 4 Plasma norepinephrine (NE) levels in rats exposed to cold for the first time (for various intervals) or for 28 days. Blood was obtained repeatedly from the same animal via an indwelling tail arterial cannula. Mean values ± SEM for 8–9 rats per group are shown. Statistical significance calculated by two-factor ANOVA for repeated measures shows NE levels in all time intervals significantly increased ($p < 0.01$) in rats exposed to cold for the first time vs. control values (room temperature). Animals exposed to 28-day cold were permanently in cold room and no significant changes in NE levels were seen at various intervals of the 28th day. Time courses in rats exposed to cold for the first or 28th time are significantly different ($p < 0.01$).

Effect of Novel Acute Stressors on Tyrosine Hydroxylase mRNA Levels in the Adrenal Medulla of Rats Exposed to Long-term Cold

The question arose as to whether exposure of "cold-adapted animals" to a novel stressor would potentiate the usual changes in TH gene expression displayed in naive, previously unstressed animals. Experiments were basically performed on two groups of rats: those chronically exposed to cold (4°C) for 28 days (the interval at which TH mRNA levels and TH activity had returned to basal levels) and age-matched animals kept at room temperature. Animals of both groups were subjected to several novel stressors. The cold-adapted animals were kept in the cold even during exposure to novel stressors.

An important question arose concerning the type of novel stressors to be used in this study. Activation of the sympathoadrenal system (SAS) leads to outflow of catecholamines (CA) from the adrenal medulla and sympathetic nerve endings. Plasma levels of norepinephrine (NE) and epinephrine (EPI) are increased during sympathetic activation but some studies have demonstrated that the two branches of the SAS (sympathoneural and adrenomedullary) can be activated independently by various stressors of different quality or intensity (Kvetňanský *et al.*, 1998; Pacak *et al.*, 1998). As shown in Figure 5, a huge release of EPI from

Fig. 5 Plasma epinephrine and norepinephrine levels in rats exposed to different intensities of various stressors. The values represent maximal responses to stressors. Details concerning stressors are described in the text.

the adrenomedullary system is elicited by insulin-induced hypoglycemia (Goldstein *et al.*, 1996, Vietor *et al.*, 1996, Kvetňanský *et al.*, 1998), cellular glucoprivation caused by 2-deoxy-D-glucose (2DG) administration (Scheuring *et al.*, 1996), and by immobilization (Kvetňanský *et al.*, 1996). Cold exposure produces a specific activation of the sympathoneural system as evidenced by increased plasma NE levels (Figures 4 and 5) and NE metabolites (Fukuhara *et al.*, 1996). Immobilization also markedly activated the sympathoneural system (Figure 5).

Based on these data, immobilization, administration of insulin, and administration of 2-deoxy-glucose were chosen as "novel stressors":

Insulin-induced hypoglycemia (INSULIN) – porcine regular insulin (Actrapid, Novo Nordisk, Denmark) was injected i.p. at a dose of 5.0 IU/kg. Tissues were collected 5 hours after a single insulin administration.

2-deoxy-D-glucose-induced glucoprivation (2DG) – 2-deoxy-D-glucose (Sigma, St. Louis, USA) dissolved in saline was injected i.p.at a dose of 500 mg/kg. Tissues were collected 5 hours after a single 2DG administration.

Immobilization (IMO) – Immobilization stress was provided as described by Kvetňanský and Mikulaj (1970) for 120 min by taping each rat's limbs to a metal frame with hypoallergic tape. After the IMO interval, animals were transferred back to their home cages and decapitated 3 hours later.

TH mRNA levels in the adrenal medulla – A single immobilization, administration of insulin, or 2-deoxy-glucose, produced significant increases in TH mRNA levels in control animals kept at room temperature (Figure 6). Cold-adapted rats (28 days in cold) exposed once to these novel stressors showed not only significantly elevated adrenomedullary TH mRNA levels but displayed an exaggerated response compared to naive rats exposed to the same stressor at room temperature (Figure 6).

TH activity – There were no significant increases to the stressors in control rats. Changes in adrenomedullary TH activity of cold-adapted rats exposed to novel stressors were less pronounced than changes in TH mRNA levels but showed significant increases after insulin and 2DG administration (Figure 7).

TH immunoreactive protein levels – Levels of TH immunoreactive protein, as expected, did not show any changes in control rats. Protein levels were significantly increased in rats exposed to 28-day cold, but did not show any exaggerated responses after exposure to the novel stressors (Figure 8).

It is widely accepted that past experiences involving stressful events affect the subsequent response to a novel stressor. However, there is little knowledge concerning the effect of novel stressors on activity and protein levels of catecholamine-synthesizing enzymes in the adrenal medulla of rats previously exposed to repeated or chronic stress. We reported that when a group of rats acclimated to high-altitude conditions (one year stay at an altitude of 1,350 m) was exposed to a single IMO for 2 h and compared to a group of IMO rats which lived at 150 m above sea level, the adrenal response of the acclimated rats was more pronounced. These animals manifested a larger increase in adrenal TH activity and a larger reduction in adrenal catecholamine levels (Balaz *et al.*, 1980). Similar changes in adrenal medullary activity were seen after long-term space flight (18.5 days) on board Cosmos 1129 (Kvetňanský *et al.*, 1981). Although adrenal TH activity was unchanged in rats decapitated immediately after landing, compared to a control group, when these animals were subjected to a novel stressor, immobiliza-

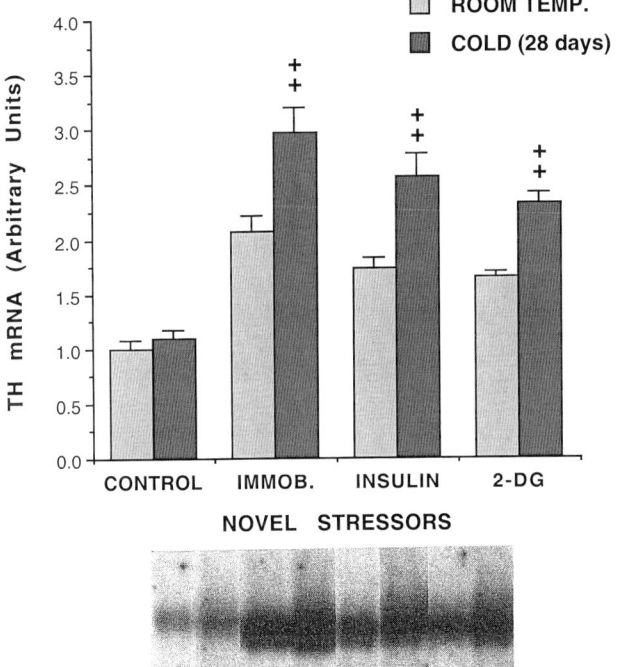

Fig. 6 Effects of novel stressors (immobilization, insulin or 2-deoxy-glucose, 2DG) on adrenomedullary tyrosine hydroxylase mRNA levels in rats exposed to long-term cold (28 days) vs. animals kept at room temperature. A representative Northern blot of TH mRNA levels in the adrenal medulla is shown. Details on novel stressors are described in the text. Mean values ± SEM for 7–8 rats per group are shown. Statistical significance calculated by one-factor ANOVA followed by Fisher's post-hoc test: (++) p < 0.01 vs. group of animals exposed to the same stressor at room temperature. Effects of novel stressors on TH mRNA levels in both long-term cold and room temperature exposed groups of rats were significantly increased (p < 0.01) vs. appropriate control groups.

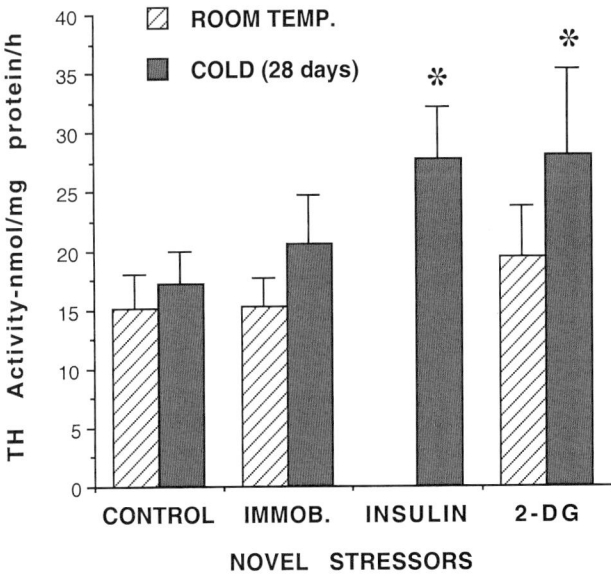

Fig. 7 Effects of novel stressors (immobilization, insulin or 2-deoxy-glucose, 2DG) on adrenomedullary tyrosine hydroxylase activity in rats exposed to long-term cold (28 days) vs. animals kept at room temperature. Details on novel stressors are described in the text. Mean values ± SEM for 7–8 rats per group are shown. Statistical significance calculated by one-factor ANOVA followed by Fisher's post-hoc test: (*) p < 0.05 vs. 28 day cold-exposed control group.

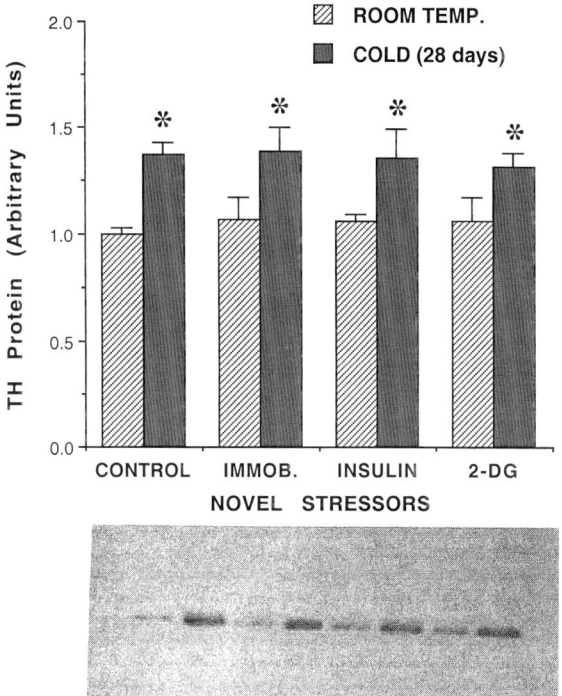

Fig. 8 Effects of novel stressors (immobilization, insulin or 2-deoxy-glucose, 2DG) on adrenomedullary tyrosine hydroxylase immunoreactive protein levels in rats exposed to long-term cold (28 days) vs. animals kept at room temperature. Details on novel stressors are described in the text. Mean values ± SEM for 7–8 rats per group are shown. Statistical significance calculated by one-factor ANOVA followed by Fisher's post-hoc test: (*) $p < 0.01$ vs. appropriate room temperature group.

tion (2.5 h daily for 5 days), an increase in TH activity was observed which was much greater than that observed with control rats (without space flight) that underwent a parallel IMO stress. Thus, although weightlessness and other factors associated with space flight do not change adrenal TH activity, they sensitize the adrenal medullary system for a greater response to a novel stressor (Kvetňanský et al., 1981). Bhatnagar et al. (1995) studied SAS function together with HPA function in rats exposed to chronic intermittent cold stress (4°C for 4 h a day for 21 days) and then exposed to a heterogenic stressor (20 min restraint). While plasma corticosterone levels were higher in chronically-cold stressed rats exposed to restraint, adrenal TH and PNMT activities as well as adrenal and plasma epinephrine and norepinephrine levels were not affected by the heterotypic stressor. These data are not consistent with our findings; however, the interval of novel stressor the authors used was too short to affect adrenal enzyme activities and CA concentrations. Plasma CA were in that study measured in blood of decapitated rats. In cannulated rats, the exaggerated peripheral CA responses to heterotypic stressors in rats chronically exposed to homotypic stressors were found (Balaz et al., 1980; Kvetňanský et al., 1984; McCarty et al., 1988; Dronjak et al., 2000).

Exposure to heterotypic stressors has been shown to produce greater increases in plasma corticosterone and ACTH levels (Marti et al., 1994; Bhatnagar and Dallman, 1998) as well as in plasma catecholamine levels (Balaz et al., 1980; Kvetňanský et al., 1984; McCarty et al., 1988; Dronjak et al., 2000) in chronically stressed rats compared to animals with no prior exposure to chronic stress. Prior chronic intermittent cold stress was found to modify cardiovascular function both under resting conditions and during stimulated conditions produced by novel stressors (Bhatnagar et al., 1998). Central mechanisms via paraventricular nucleus, parabrachial region and amygdaloid subnuclei may regulate changes in cardiovascular variables upon exposure to novel stressors (Bhatnagar et al., 1998).

Others supported these conclusions and extended them to several brain areas, where CA biosynthesis and TH activity were not altered by chronic stress, but a greater elevation was observed when the chronically stressed rats were exposed to a novel stressor (Nisenbaum and Abercrombie, 1992; Adell et al., 1988; Nisenbaum et al., 1991; Watanabe et al., 1995). Cuadra et al. (1999) found higher increases in frontal cortical dopamine release after exposure to a novel uncontrollable stressor (restraint) in chronically stressed rats compared with those without chronic stress exposure. Endogenous opiate mechanisms, presumably activated during chronic stress, have been proposed to be involved in the development of such a sensitization process (Cuadra et al., 1999). Our recent data on levels of TH mRNA in locus coeruleus have shown that long-term repeated stress can alter the sensitivity of LC neurons to novel stressors (Rusnak et al., 1999).

To examine the effect of a novel stressor, we exposed rats to cold at 4°C for 28 days. These conditions have been found to lead to a diminition of the initial increase in TH activity in the adrenal medulla (Figure 4). The question arose regarding how would long-term cold exposure affect TH gene expression and translation in rats exposed to novel stressors. In rats adapted to long-term cold, we found significantly exaggerated responses in adrenal TH mRNA levels to heterotypic novel stressors (immobilization, insulin-induced hypoglycemia, 2DG-induced glucoprivation) compared to naive rats kept at room temperature (Figure 6). Adrenal TH activity also shows exaggerated responses but to a much lesser extent (Figure 7). However, TH immunoreactive protein levels were not affected by novel stressors (Figure 8). The chronically-stressed animals were exposed to a novel stressor only once for 5 hours. This time interval is sufficient for increases in adrenal TH mRNA levels (Miner et al., 1992; McMahon et al., 1992; Nankova et al., 1994; Kvetňanský et al., 1996) and partly also for increases in TH activity (Kvetňanský et al., 1970), but is not long enough for producing changes in adrenal TH protein levels.

What are the mechanisms involved in the exaggerated responses of adrenal TH mRNA in chronically stressed rats to novel stressors? These responses might be affected by central regulatory mechanisms or by mechanisms acting directly at the level of transcriptional regulation in nuclei of adrenomedullary cells. Bhatnagar and Dallman (1998) proposed that increased activity in the parabrachial-posterior paraventricular thalamus-amygdala-parvocellular paraventricular hypothalamus underlies facilitation of the HPA axis to novel stress in chronically stressed rats. Since CRH may exhibit parallel actions at the level of pituitary and LC, and prior exposure to footshock selectively altered LC sensitivity to CRH (Curtis et al., 1999), we may expect an involvement of the proposed central mechanism also in the regulation of the SAS.

Alternatively, the increased sensitivity may be at the level of the adrenal medulla itself. Activation of transcription mechanisms may also play an important role in the stress-elicited induction of the genes encoding CA biosynthetic enzymes in rat adrenal medulla. Different signaling pathways and/or transcription factors are activated depending on the duration, frequency and specificity of the stressor (Sabban et al., 2000). It has been shown that several transcription factors are involved in the regulation of TH mRNA levels during stress (Kvetňanský and Sabban, 1998; Sabban et al., 2000). The 5' flanking region of the TH gene contains functional AP1 and perfect consensus CRE sites and a recently discovered Egr1 motif (Papanikolaou and Sabban, 1999). Increased binding of c-Jun/c-fos to the AP1 site of the TH gene was observed in the adrenal medulla of rats exposed to cold or to immobilization stress (Miner et al., 1992; Nankova et al., 1994). A recent study suggested, for the first time, that the Sp-1 and overlapping Egr motif may function in the regulation of TH transcription by stress (Papanikolaou and Sabban, 1999). Since acute and chronic immobilization stress increased Egr-1 mRNA levels in the adrenal medulla (Wong et al., 1997), Egr1 is a candidate, alone or together with c-fos, in mediating the transcriptional activation of TH in stress. Phosphorylation of CREB has been shown to occur rapidly with IMO stresss and can activate TH at the CRE element (Sabban et al., 1995). If the levels of CREB are elevated also in cold-adapted animals, then they could lend to an exaggerated response. Other promoter elements, that have not been investigated yet, may also participate in the response of the TH gene to different stressors. Thus, any of these promoter elements and/or transcription factors may be permanently changed in long-term cold exposed rats and be responsible for the exaggerated response of TH mRNA after exposure to a novel stressor. Such mechanisms could play a role in changes in human vulnerability to stress depending on prior experience.

Conclusions

- The observed differences in tyrosine hydroxylase mRNA levels in the adrenal medulla of rats exposed to a short and long-term cold stress suggest an adaptation of TH gene expression.
- Exposure of cold-adapted rats to a novel stressor induces exaggerated responses of TH gene expression,

moderate but significant increases in adrenal TH activity, but no changes in TH immunoreactive protein levels because of the relatively short period of the exposure.

- Thus, after exposure of long-term stressed rats to a novel stressor, there is a potentiation of the stress-elicited activation of TH gene expression which could reflect changes in central regulation and/or in adrenomedullary signalling pathway(s) leading to additional modifications or accumulation of transcription factors or synergistic effects. The precise mechanism(s) of this phenomenon remain to be elucidated.

Acknowledgments

The present work was supported by Fogarty International Research Collaboration Award, grant No. TW 00984, and by Slovak Grant Agency for Science (VEGA), Grant No 2-610999.

References

Adell, A., Garcia-Marquez, C., Armario, A., and Gelpi, D. (1988). Chronic stress increases serotonin and noradrenaline in rat brain and sensitizes their responses to a further acute stress. *Journal of Neurochemistry* **50**, 1678–1681.

Andrews, T., Lincoln, J., Milner, P., Burnstock, G., and Cowen, T. (1993). Differential regulation of tyrosine hydroxylase protein and activity in rabbit sympathetic neurones after long-term cold exposure: altered responses in ageing. *Brain Research* **624**, 69–74.

Balaz, V., Balazova, E., Blazicek, P., and Kvetňanský, R. (1980). The effect of one-year acclimatization of rats to maintain conditions on plasma catecholamines and dopamine-ß-hydroxylase activity. In: E. Usdin, R. Kvetňanský, and I.J. Kopin (Eds.), *Catecholamines and Stress: Recent Advances*, pp. 259–264. New York: Elsevier.

Baruchin, A.M., Weisberg, E.P., Miner, L.L., Ennis, D., Nisenbaum, L.K., Stricker, E., Zigmond, M.J., and Kaplan, B.B. (1990). Effect of cold exposure on rat adrenal tyrosine hydroxylase: an analysis of RNA, protein, enzyme activity and cofactor levels. *Journal of Neurochemistry* **54**, 1769–1775.

Bhatnagar, S., and Dallman, M.F. (1998). Neuroanatomical basis for facilitation of hypothalamic-pituitary-adrenal responses to a novel stressor after chronic stress. *Neuroscience* **84**, 1025–1039.

Bhatnagar, S., Dallman, M.F., Roderick, R.E., Basbaum, A.I., and Taylor, B.K. (1998). The effect of prior chronic stress on cardiovascular responses to acute restraint and formalin injection. *Brain Research* **797**, 313–320.

Bhatnagar, S., Mitchell, J.B., Betito, K., Boksa, P., and Meaney, M.J. (1995). Effects of chronic intermittent cold stress on pituitary adrenocortical and sympathetic adrenomedullary functioning. *Physiology and Behavior* **57**, 633–639.

Caudra, G., Zurita, A., Lacerra, C., and Molina, V. (1999). Chronic stress sensitizes frontal cortex dopamine release in response to a subsequent novel stressor: Reversal by naloxone. *Brain Research Bulletin* **48**, 303–308.

Curtis, A.L., Pavcovich, L.A., and Valentino, R.J. (1999). Long-term regulation of locus ceruleus sensitivity to corticotropin-releasing factor by swim stress. *Journal of Pharmacology and Experimental Therapeutics* **289**, 1211–1219.

Dhabhar, F.S., McEwen, B.S., and Spencer, R.L. (1997). Adaptation to prolonged or repeated stress – Comparison between rat strains showing intrinsic differences in reactivity to acute stress. *Neuroendocrinology* **65**, 360–368.

Dronjak, S., Ondriska, M., Svetlovska, D., Jezova, D., and Kvetňanský, R. (2001). Effect of novel stressors on plasma catecholamine levels in rats exposed to long-term cold. In: R. McCarty, G. Aguilera, E.L. Sabban and R. Kvetňanský (Eds.), *Stress: Neural, Endocrine and Molecular Studies*, New York: Gordon and Breach Sci. Publishers (pp. 83–89).

Fluharty, S.J., Snyder, G.L., Zigmond, M.J., and Stricker, E.M. (1985). Tyrosine hydroxylase activity and catecholamine biosynthesis in the adrenal medulla of rats during stress. *Journal of Pharmacology and Experimental Therapeutics* **233**, 32–38

Fukuhara, K., Kvetňanský, R., Weise, V.K., Ohara, H., Yoneda, R., Goldstein, D.S., and Kopin, I.J. (1996). Effects of continuous and intermittent cold (SART) stress on sympathoadrenal system activity in rats. *Journal of Neuroendocrinology* **8**, 65–72.

Goldstein, D.S., Pacak, K., and Kopin, I.J. (1996). Nonspecificity versus primitive specificity of stress responses. In: R. McCarty, G. Aguilera, E.L. Sabban, and R. Kvetňanský (Eds.) *Stress: Molecular Genetic and Neurobiological Advances*, pp. 3–22. New York: Gordon and Breach Sci. Publishers.

Hoeldtke, R., Lloyd, T., and Kaufman, S. (1974). An immunochemical study of the induction of tyrosine hydroxylase in rat adrenal glands. *Biochemical and Biophysical Research Communications* **57**, 1045–1053.

Konarska, M., Stewart, R.E., and McCarty, R. (1989). Habituation of sympathetic-adrenal medullary responses following exposure to chronic intermittent stress. *Physiology and Behavior* **45**, 255–261.

Kvetňanský, R., and Mikulaj, L. (1970). Adrenal and urinary catecholamines in rat during adaptation to repeated immobilization stress. *Endocrinology* **87**, 738–743.

Kvetňanský, R., and Sabban, E.L. (1993). Stress-induced changes in tyrosine hydroxylase and other catecholamine biosynthetic enzymes. In: M. Naoi, and S. H. Parvez (Eds.). *Tyrosine Hydroxylase: From Discovery To Cloning*, pp. 253–281. Utrecht: VSP Press.

Kvetňanský, R., and Sabban, E.L. (1998). Stress and molecular biology of neurotransmitter-related enzymes. In: P. Csermely (Ed.). *Stress of Life: From Molecules to man, Annals of the New York Academy of Sciences* **851**, 342–356.

Kvetňanský, R., Weise, V.K., and Kopin, I.J. (1970). Elevation of adrenal tyrosine hydroxylase and phenylethanolamine N-methyltransferase by repeated immobilization of rats. *Endocrinology* **87**, 744–749.

Kvetňanský, R., Gewirtz, G.P., Weise, V.K., and Kopin, I.J. (1971a). Catecholamine-synthesizing enzymes in the rat adrenal gland during exposure to cold. *American Journal of Physiology* **220**, 928–931.

Kvetňanský, R., Nankova, B., Hiremagalur, B., Viskupic, E., Vietor, I., Rusnak, M., McMahon, A., Kopin, I.J., and Sabban, E.L. (1996). Induction of adrenal tyrosine hydroxylase mRNA by single immobilization stress occurs even after splanchnic transection and in presence of cholinergic antagonists. *Journal of Neurochemistry* **66**, 138–146.

Kvetňanský, R., Nemeth, S., Vigas, M., Oprsalova, Z., and Jurcovicova, J. (1984). Plasma catecholamines in rats during adaptation to intermittent exposure to different stressors. In: E. Usdin, R. Kvetňanský, and J. Axelrod (Eds.), *Stress: The Role of Catecholamines and Other Neurotransmitters*, Vol. 1. pp. 537–562. New York: Gordon and Breach Sci. Publishers.

Kvetňanský, R., Pacak, K., Sabban, E.L., Kopin, I.J., and Goldstein, D.S. (1998). Stressor-specificity of peripheral catecholaminergic activation. In: D.S. Goldstein, G. Eisenhofer, and R. McCarty (Eds.), *Catecholamines: Bridging Basic Science with Clinical Medicine, Advances in Pharmacology* **42**, pp. 556–560. San Diego: Academic Press.

Kvetňanský, R., Silbergeld, S., Weise, V.K., and Kopin, I.J. (1971b). Effect of restraint on rat adrenomedullary response to 2-deoxy-D-glucose. *Psychopharmacologia (Berlin)* **20**, 22–31.

Kvetňanský, R., Torda, T., Macho, L., Tigranjan, R.A., Serova, L., and Genin, A.M. (1981). Effect of weightlessness on sympathetic-adrenomedullary activity of rats. *Acta Astronautica* **8**, 469–481.

Marti, O., Gavalda, A., Gomez, F., and Armario, A. (1994). Direct evidence for chronic stress-induced facilitation of the adrenocorticotropin response to a novel acute stressor. *Neuroendocrinology* **60**, 1–7.

Masserano, J.M., Vulliet, P.R., Tank, A.W., and Weiner, N. (1989). The role of tyrosine hydroxylase in the regulation of catecholamine synthesis. In: N. Weiner (Ed.). *Handbook of Experimental Pharmacology. Catecholamines 2*, Vol 90/II pp. 427–469. Berlin: Springer-Verlag.

McCarty, R., Horwatt, K., and Konarska, M. (1988). Chronic stress and sympathetic-adrenal medullary responsiveness. *Social Science and Medicine* **26**, 333–341.

McCarty, R., and Stone, E.A. (1984). Chronic stress and regulation of the sympathetic nervous system. In: E. Usdin, R. Kvetňanský, and J. Axelrod (Eds.), *Stress: The Role of Catecholamines and Other Neurotransmitters*, Vol. 1. pp. 563–576. New York: Gordon and Breach Sci. Publishers.

McMahon, A., Kvetňanský, R., Fukuhara, K., Weise, V.K., Kopin, I.J., and Sabban, E.L. (1992). Regulation of tyrosine hydroxylase and dopamine-β-hydroxylase mRNA levels in rat adrenals by a single and repeated immobilization stress. *Journal of Neurochemistry* **58**, 2124–2130.

Miner, L.L., Baruchin, A., and Kaplan B.B. (1992). Transsynaptic modulation of rat adrenal tyrosine hydroxylase gene expression during cold stress. In: R. Kvetňanský, R. McCarty, and J. Axelrod (Eds.), *Stress: Neuroendocrine and Molecular Approaches*, Vol. 1, pp. 313–324. New York: Gordon and Breach Sci. Publishers.

Nankova, B.B., Kvetňanský, R., McMahon, A., Viskupic, E., Hiremagalur, B., Frankle, W.G., Fukuhara, K., Kopin, I.J., and Sabban, E.L. (1994). Induction of tyrosine hydroxylase gene expression by a non-neuronal non-pituitary mediated mechanism in immobilization stress. *Proceedings of the National Academy of Sciences, USA* **91**, 5937–5941.

Nisenbaum, L.K., and Abercrombie, E.D. (1992). Enhanced tyrosine hydroxylation in hippocampus of chronically stressed rats upon exposure to a novel stressor. *Journal of Neurochemistry* **58**, 276–281.

Nisenbaum, L.K., Zigmond, M.J., Sved, A.F., and Abercrombie, D. (1991). Prior exposure to chronic stress results in enhanced synthesis and release of hippocampal norepinephrine in response to a novel stressor. *Journal of Neuroscience* **11**, 1478–1484.

Pacak, K., Palkovits, M., Yadid, G., Kvetňanský, R., Kopin, I.J., and Goldstein, D.S. (1998). Heterogenous neurochemical responses to different stressors: a test of Selye's doctrine of nonspecificity. *American Journal of Physiology* **275**, R1247–R1255.

Papanikolaou, N.A., and Sabban, E.L. (1999). Sp1/Egr1 motif: a new candidate in the regulation of rat tyrosine hydroxylase gene transcription by immobilization stress. *Journal of Neurochemistry* **73**, 433–436.

Peuler, J.D., and Johnson, G.A. (1977). Simultaneous single isotope radioenzymatic assay of plasma norepinephrine, epinephrine and dopamine. *Life Sciences* **21**, 625–636.

Richard, F., Faucon-Biguet, N., Labatut, R., Rollet, D., Mallet, J., and Buda, M. (1988). Modulation of tyrosine hydroxylase gene expression in rat brain and adrenals by exposure to cold. *Journal of Neuroscience Research* **20**, 32–37.

Rusnak, M., Kvetňanský, R., Jelokova, J., Eisenhofer, G., and Palkovits, M. (1999). Effect of various stressors on the activity of locus coeruleus neurons in long-term repeatedly immobilized rats. *Neuroscience Abstracts* **25**, Part 1, p. 75.

Sabban, E.L., Hiremagalur, B., Nankova, B., and Kvetňanský, R. (1995). Molecular biology of stress-elicited induction of catecholamine biosynthetic enzymes. *Annals of the New York Academy of Sciences* **771**, 327–338.

Sabban, E.L., Nankova, B.B., Serova, L.I., and Kvetňanský, R. (2001). Transcriptional mechanisms of stress-triggered activation of neurotransmitter gene expression. In: R. McCarty, G. Aguilera, E.L. Sabban and R. Kvetňanský (Eds.), *Stress: Neural, Endocrine and Molecular Studies*, New York: Gordon and Breach Sci. Publishers, pp. 99–105.

Scheurink, A.J.W., DeBoer, S.F., Van Dijk, G., and Steffens, A.B. (1996). Central and peripheral mechanisms involved in the regulation of sympathoadrenal outflow. In: R. McCarty, G. Aguilera, E.L. Sabban, and R. Kvetňanský (Eds.) *Stress: Molecular Genetic and Neurobiological Advances*, pp. 227–241. New York: Gordon and Breach Sci. Publishers.

Serova, L., Nankova, B., Kvetňanský, R., and Sabban, E.L. (1997). Immobilization stress elevates GTP cyclohydrolase I mRNA levels in rat adrenals predominantly by hormonally mediated mechanisms. *Stress* **1**, 135–144.

Stachowiak, M., Sebbane, R., Stricker, E.M., Zigmond, M.J., and Kaplan B.B. (1985). Effect of chronic cold exposure on tyrosine hydroxylase mRNA in rat adrenal gland. *Brain Research* **359**, 356–359.

Stone, E.A., and McCarty, R. (1983). Adaptation to stress: tyrosine hydroxylase activity and catecholamine release. *Neuroscience and Biobehavioral Review* **7**, 29–34.

Vietor, I., Rusnak, M., Viskupic, E., Blazicek, P., Sabban, E.L., and Kvetňanský, R. (1996). Glucoprivation by insulin leads to trans-synaptic increase in rat adrenal tyrosine hydroxylase mRNA levels. *European Journal of Pharmacology* **313**, 119–127.

Watanabe, Y., McKittrick, C.R., Blanchard, D.C., Blanchard, R.J., McEwen, B.S., and Sakai, R.R. (1995). Effects of chronic social stress on tyrosine hydroxylase mRNA and protein levels. *Molecular Brain Research* **32**, 176–180.

Waymire, J.C., Bjur, R., and Weiner, N. (1971). Assay of tyrosine hydroxylase by coupled decarboxylation of DOPA formed from L-(1-^{14}C) tyrosine. *Analytical Biochemistry* **43**, 588–600.

Wong, D.L., Siddall, B.J., Lindley, S.E., and Kvetňanský, R. (1997). Regulation of phenylethanolamine N-methyltransferase gene expression and epinephrine biosynthesis during stress. *Abstracts Stress of Life (Budapest)*, No. B5–4, p. 77.

22. Stress-Induced Expression of Phenylethanolamine N-Methyltransferase: Normal and Knock out Animals

D.L. Wong, S. Her, T.C. Tai, R.A. Bell, M. Rusnák, R. Farkas, R. Kvetňanský, and J.C. Shih

Neuropsychiatric disorders are classic examples of illnesses where stress is a major contributor. The hallmark of these diseases is the dysfunction of the hypothalamic-pituitary-adrenal (HPA or stress) axis. Normally, in response to a stressor, whether environmental, physiological or psychological, the HPA axis is activated, initiating a cascade that eventuates in the synthesis and release of glucocorticoids from the adrenal cortex and epinephrine from the adrenal medulla. In response to this surge in glucocorticoids and epinephrine, hormonal and neural inhibitory feedback pathways are activated, suppressing further corticosteroid and epinephrine synthesis and release.

The consequences of HPA axis dysfunction in the case of psychiatric patients includes the following: 1) hypercortisolemia, i.e. patients show elevated basal cortisol levels, 2) glucocorticoid loading, as evidenced from the dexamethasone suppression test (DST), at best results in blunted feedback inhibition of endogenous corticosteroid production, suggesting that hormonal inhibitory feedback pathways are malfunctioning, and 3) adrenal medullary levels of epinephrine are also reportedly elevated, indicating that neural inhibitory pathways are malfunctioning as well. Relating these apparent disparate findings is the fact that glucocorticoids control epinephrine through its biosynthetic enyzme, phenylethanolamine N-methyltransferase (PNMT).

Glucocorticoids regulate PNMT in two ways, post-transcriptionally and transcriptionally (Figure 1). Post-transcriptionally, glucocorticoids control PNMT through its methyl donor and co-substrate, S-adenosylmethionine (SAM). Specifically, they regulate the SAM metabolic enzymes, methionine adenosyltransferase (MAT) and S-adenosylhomocysteine hydrolase (SAHHase), thereby sustaining levels of SAM so that more is available for enzymatic activity. The binding of SAM to PNMT then simply protects it against proteolytic degradation. Glucocorticoids also control PNMT gene expression. For every species for which the PNMT gene has been cloned, human (Baetge et al., 1988), cow (Baetge et al., 1986; Batter et al., 1988; Kaneda et al., 1988), mouse (Kaneda et al., 1988), and rat (Ross et al., 1990), at least one glucocorticoid response element (GRE) has been identified. Moreover, the GRE for the

rat PNMT gene has been demonstrated to be functional in vitro (Ross et al., 1990), although this response element shows weak and variable sensitivity to corticosteroid activation.

Originally, glucocorticoids were thought to provide the singular cue for the initiation of PNMT gene expression and hence, adrenergic differentiation. Evidence to date, in fact, points to the contrary. For example, in vivo, glucocorticoid administration to the developing rat does not elicit the premature expression of PNMT prior to its normal time of onset, embryonic day 14.5 (E14.5) based on mRNA (Hemmick et al., 1995) and E16.5 based on enzyme activity (Bohn et al., 1981). Similarly, in vitro, it is difficult to elicit PNMT expression in PC12 cells through glucocorticoid exposure. PC12 cells are derived from a rat adrenal medullary tumor (Greene and Tischler 1976), and in the normal rat adrenal gland 85–90% of the chromaffin cells express the adrenergic phenotype (Verhofstad et al., 1979). If glucocorticoids could independently elicit PNMT expression, then a very robust induction of this enzyme should occur. However, at best, the enzyme is only weakly activated and extremely high concentrations of corticosteroids are required (30 μM DEX).

These findings, in aggregate, have led to the hypothesis that the glucocorticoid-activated glucocorticoid receptor must function in conjunction with another transcription factor(s) to induce the PNMT gene. We have previously demonstrated that the glucocorticoid receptor interacts cooperatively with Egr-1, a member of the immediate early gene family of transcriptional activators (Wong et al., 1998), and AP-2, a developmental factor (Ebert et al., 1998) important for the differentiation of neural crest derived tissues such as the adrenal medulla. The interaction of Egr-1 and the glucocorticoid receptor is being further investigated using two in vivo animal models, the normal rat subjected to acute and chronic immobilization stress (IMMO) and the monoamine oxidase B knock out (MAOB KO) mouse subjected to acute and chronic immobilization stress as well.

Studies in rats were initiated by examining the effects of immobilization stress on adrenal PNMT expression, varying the length of immobilization, the number of immobilizations, and the time to sacrifice in order to

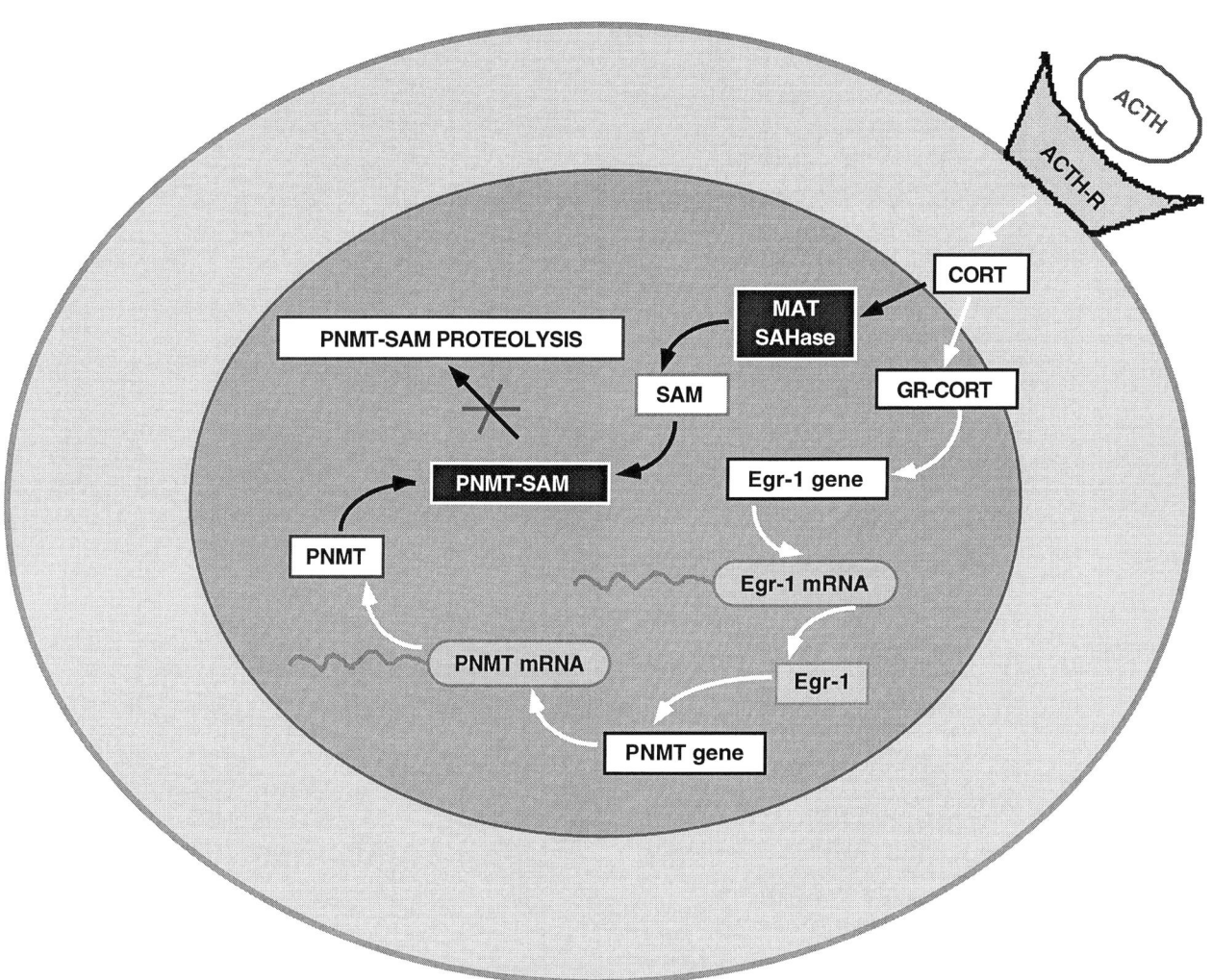

Fig. 1 Hormonal control of PNMT gene transcription and PNMT enzyme stability.

ascertain conditions maximizing changes in PNMT. For acute stress, animals were immobilized for 30 min and sacrificed immediately or immobilized for 120 min and sacrificed immediately or 3 hr later. In the case of chronic IMMO stress, rats were immobilized and sacrificed identically, but IMMO was repeated once daily for 7 days prior to sacrifice. Following termination, adrenal glands were immediately removed from the animals, quick frozen on dry ice, and segregated into left and right adrenals for measurement of Egr-1 and PNMT mRNA and enzyme activity, respectively. Levels of Egr-1 and PNMT message expression were quantified by ribonuclease protection assay using ^{32}P-labelled Egr-1 and PNMT cRNA transcripts generated as described earlier (Wong *et al.*, 1992), and a rat GADPD cRNA as a normalization control. PNMT enzymatic activity was measured by radioenzymatic assay with ^{3}H-S-adenosylmethionine providing the radiolabelled methyl group (Wong *et al.*, 1982) while relative amounts of PNMT protein were semi-quantitatively measured by ECL western analysis. Finally, total cellular protein was determined by

the Bradford method, and corticosterone (CORT) quantified by radioimmunoassay (Wong *et al.*, 1995). At least 6 animals were included in each experimental group.

As shown in Figure 2, Egr-1 mRNA increased 25-fold in response to acute IMMO stress, with the greatest increase apparent with 30 min of IMMO and immediate sacrifice. Prolongation of IMMO for an additional 90 min attenuated the magnitude of rise in Egr-1 mRNA. If the length of IMMO was extended as well as the time to sacrifice (3 hr), no significant differences in Egr-1 mRNA levels were observed by comparison to control. A very similar pattern of expression in Egr-1 mRNA was apparent when animals were chronically IMMO-stressed except that the incremental rise was not as great, perhaps reflecting adaptation to the stress. When PNMT mRNA responses were determined (Figure 2), a rise in expression was also observed, except that the time course was protracted so that PNMT message levels were most elevated with 120 min of IMMO and sacrifice 3 hr later (~6-fold control), whether the animals were singly or repeatedly stressed.

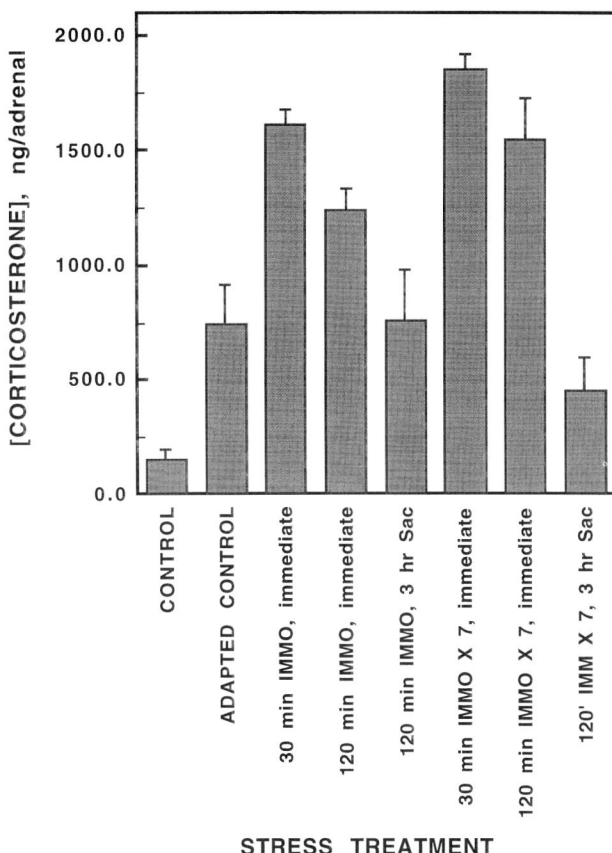

Fig. 2 Stress-induced changes in adrenal corticosterone in Sprague-Dawley rats.

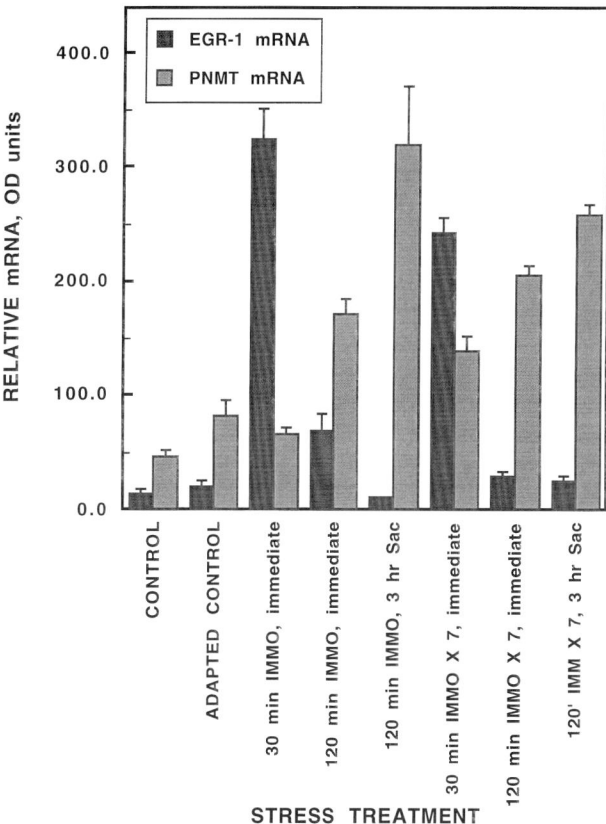

Fig. 3 Stress-induced changes in adrenal Egr-1 and PNMT mRNA in Sprague-Dawley rats.

Assessment of adrenal CORT demonstrated that both the acute and chronic IMMO paradigms elicited the desired stress response, rapidly and markedly elevating CORT (Figure 3). An approximately 11-fold stimulation in CORT was apparent in animals stressed for 30 min and sacrificed immediately. CORT was also elevated (8.1-fold basal), but significantly lower, following 2 hr of IMMO with immediate sacrifice. If sacrifice was delayed for 3 hr, after 120 min of IMMO, CORT levels were still lower (4.9-fold basal). Chronic stress showed a similar pattern of change in CORT, with elevation above control values of 12.1-fold, 10.1-fold and 2.9-fold, respectively.

Finally, no significant differences between experimental and control animals were observed in PNMT activity and protein. Furthermore, total cellular protein remained unchanged irrespective of treatment.

In aggregate, the observed changes in Egr-1 and PNMT mRNA elicited in response to IMMO stress and CORT induction likely reflect the time course of changes in the various indices examined rather than their absolute incremental changes. As an immediate early gene transcription factor, Egr-1 is very rapidly stimulated, with peak levels in mRNA achieved between 30 min to 1 hr. In contrast, PNMT mRNA does not show peak induction until 6 to 8 hr later. PNMT protein and activity changes require additional

time to reach maximum stimulated levels, approximately 18–24 hr. Finally, CORT shows the same rapid response as Egr-1 with a rapid rise (30 min to 1 hr), followed by a decline to basal levels.

To identify times for examining maximum stress-induced changes in the indices of interest, a second study was executed in which rats were IMMO stressed for 30 min and sacrificed immediately or 6 or 24 hr later. Both the effects of a single stressor and repeated stress were again examined, and changes in adrenal Egr-1 and PNMT mRNA and PNMT activity determined. As shown in Figure 4, Egr-1 mRNA shows peak levels when the animals are immediately sacrificed after stress while PNMT mRNA is not maximally induced until 6 hr later. Both indices are restored to basal levels by 24 hr post-stress. Again, however, no differences in PNMT enzymatic activity were apparent, even at the 24 hr time point, where we have seen very robust changes in enzyme expression in response to both hormonal and neural stimulation.

It has been suggested that longer times of IMMO (120 min) are required in order to evoke changes in adrenal PNMT expression by comparison to other catecholamine biosynthetic enzymes, e.g. tyrosine hydroxylase and dopamine β-hydroxylase. To examine this possibility, rats were subjected to acute and chronic IMMO for 120 min and as previously, sacrificed immediately or 6 or 24 hr later. In this case, PNMT

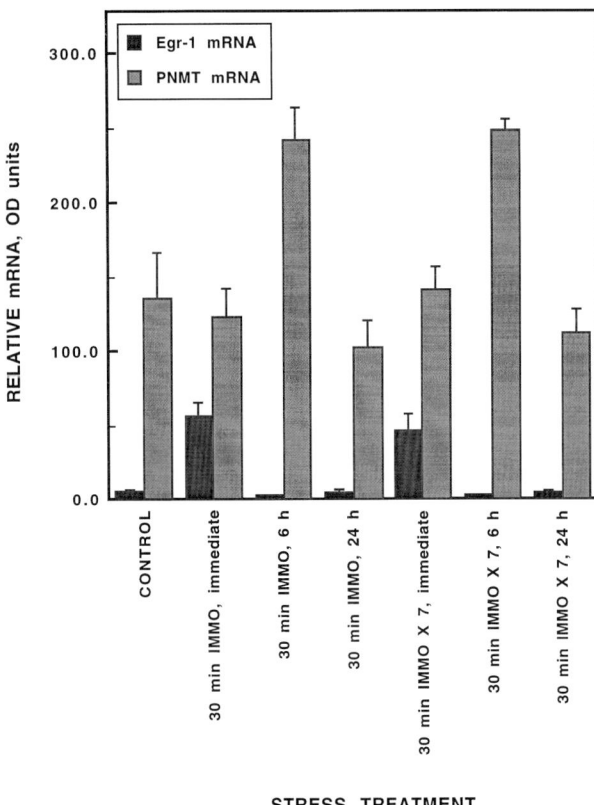

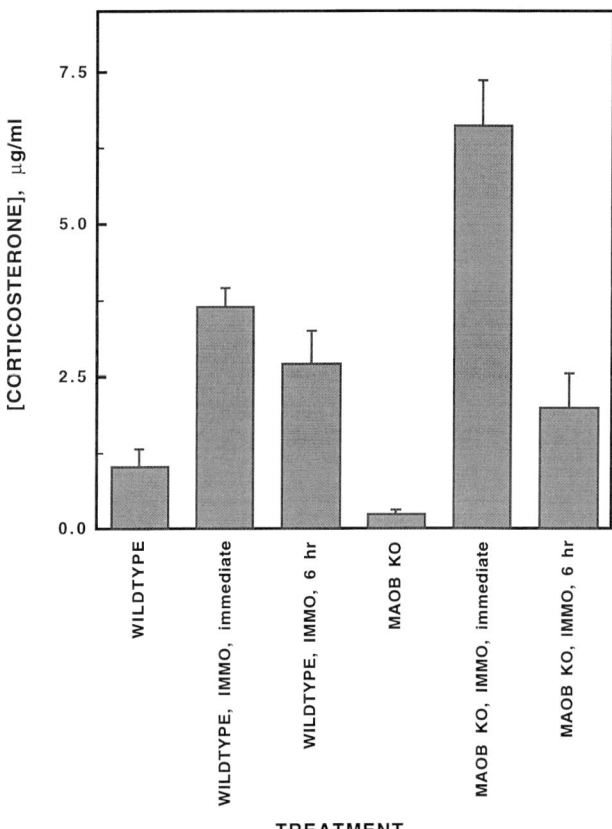

Fig. 4 Temporal pattern for stress-induced changes in adrenal Egr-1 and PNMT mRNA in Sprague-Dawley rats.

Fig. 5 Stress-induced changes in adrenal corticosterone in MAOB KO mice.

mRNA was significantly elevated when sacrifice was delayed until 6 or 24 hr in the case of acute stress while it was significantly elevated irrespective of sacrifice time in the case of chronic stress. However, the greatest induction of PNMT mRNA was observed when animals were repeatedly stressed and sacrificed 6 hr after the final IMMO stress (~5-fold). In contrast, PNMT activity appeared to be attenuated, rather than induced. While the time course for changes in PNMT protein responses to IMMO may differ from that observed previously, it is more likely that post-transcriptional regulatory controls may be compensating for transcriptional induction. The latter would not be surprising, given that down-regulation of stress-induced rises in corticosteroids and catecholamines is critical for the organism's survival.

The monoamine oxidase B knock out (MAOB KO) mouse is the second model that we are using to examine stress-induced changes in PNMT gene expression. In contrast to its counterpart, the monoamine oxidase A knock out mouse, this animal shows normal neurochemistry with the exception that urinary phenylethylamine levels are 8-fold elevated (Grimbsy *et al.*, 1997). While phenylethylamine reportedly affects behavior, the MAOB KO mouse did not show any abnormalities in the open field test, which assesses exploratory behavior or stereotypy. In addition, it did not display any untoward anxiety in novel fear-inducing environments (plus maze test). However, the

MAOB KO mice did exhibit increased mobility in the Porsolt forced swim test, suggestive of greater vigilance to the inescapable stressor. We have hypothesized that this KO animal may provide a model with a hyper-responsive HPA axis so that the mouse is not incapacitated by stress and further that the HPA axis may also have a broader range of responsiveness. As a first step to investigate this possibility, wildtype (129V founder) and KO mice were subjected to acute IMMO stress for 30 min, sacrificed immediately, 6 or 18 hr later and CORT, Egr-1 and PNMT mRNA, and PNMT activity determined. Interestingly, basal CORT levels (Figure 5) in the KO mice were markedly lower by comparison to control (0.22-fold of control) and stimulated to much greater levels (29.2-fold basal vs. 2.6-fold basal, sacrifice immediately after stress). Despite these changes in CORT, only the wildtype mice showed an elevation in Egr-1 mRNA (Figure 6, 2-fold, immediate sacrifice), and no induction of PNMT mRNA or activity was observed in either the wildtype or KO mice. As in the case of the rat, we are now examining whether repeated IMMO or prolonged IMMO (120 min) may be required to evoke changes in PNMT expression. However, preliminary findings do not indicate that 30 min of repeated stress significantly alters PNMT mRNA.

As described earlier, glucocorticoids stimulate PNMT transcription through GREs located with the

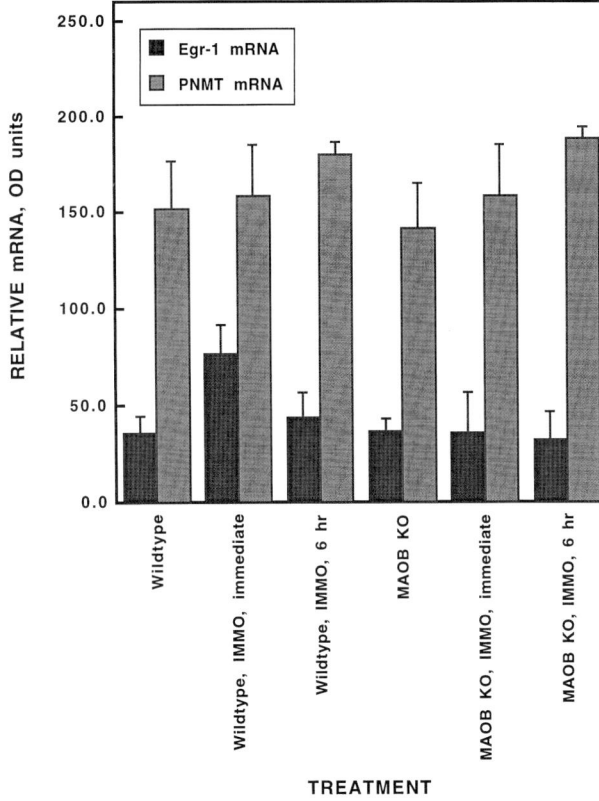

Fig. 6 Stress-induced changes in adrenal Egr-1 and PNMT mRNA in MAOB KO mice.

5' upstream promoter/regulatory sequences of the PNMT gene. While a GRE was originally identified at –513 bp (–539 bp correcting for misalignment) upstream of the site of transcription initiation in the rat gene, the variable responsiveness and sensitivity of this GRE has been puzzling (Figure 7). We have recently identified two new GREs upstream of the original that show very robust glucocorticoid responses (Her *et al.*, submitted). These GREs were identified when we subcloned the proximal –893 bp of upstream PNMT promoter regulatory sequences into another plasmid vector pGL3basic (Promega, Madison, WI). A series of

nested deletion constructs were generated from the full length construct, transient transfection assays performed in the PC12-derived RS1 cells, and the transfected cells maintained in the absence and presence of 1 μM dexamethasone (DEX). As shown in Figure 8, constructs harboring >–745 bp of upstream promoter sequence showed marked induction of luciferase reporter gene expression in response to DEX (5.5 to 6.5-fold), suggesting that the glucocorticoid sensitive sequences lay between –745 and –798 bp. When the nucleotides spanning this region where matched to the 15 bp consensus GRE, two response elements were identified with a 1 bp overlap. Based on the most conserved portion of the consensus GRE, the 3' hexanucleotide TGTTCT, the 3' extent of these GREs was fixed at –759 and –773 bp.

Site-directed mutagenesis was used to demonstrate the functionality of the GREs. Mutant GRE PNMT promoter-reporter gene constructs were generated by introducing point mutations into each individual GRE or both GREs. Transient transfection assays showed that mutation of the distal GRE reduced DEX-stimulated luciferase activation by 18% while mutation of the proximal GRE reduced reporter gene expression by 56%, indicating that the latter was the more important GRE functionally. However, mutation of both GREs was necessary to nearly completely eliminate glucocorticoid activation of the PNMT promoter.

We have further demonstrated that activated glucocorticoid receptors bound to these GREs can interact with GRIP1, a facilitory protein of the p160 family (Hong *et al.*, 1996), to greatly enhance glucocorticoid stimulated PNMT promoter activity. When a series of nested deletion constructs and a GRIP1 expression construct were cotransfected into RS1 cells, luciferase activity rose as much as 3-fold above levels observed with DEX alone in constructs harboring greater than –893 bp of upstream promoter sequence (Table 1). Since the presence of GRIP1 did not incrementally induce luciferase expression from the –893 bp PNMT promoter-luciferase reporter gene construct beyond levels seen with DEX alone, it appears that GRIP1 and the GR/GRE interact with an additional transcription factor(s)

PNMT PROMOTER/REGULATORY SEQUENCES

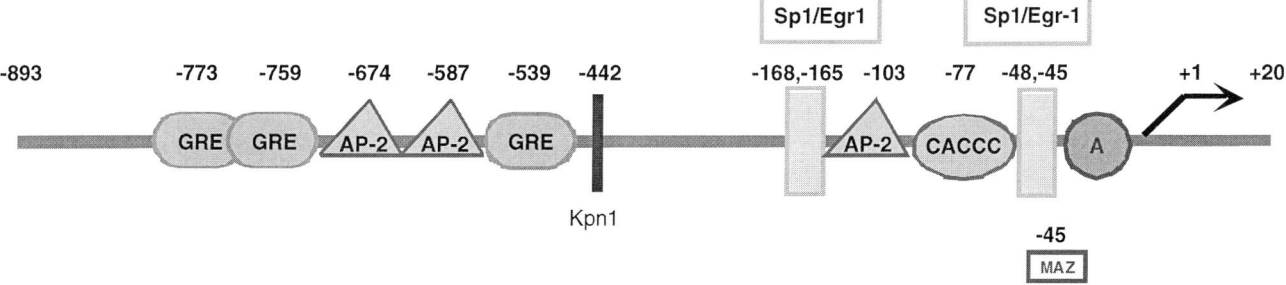

Fig. 7 Transcription factor binding elements in proximal PNMT promoter sequences.

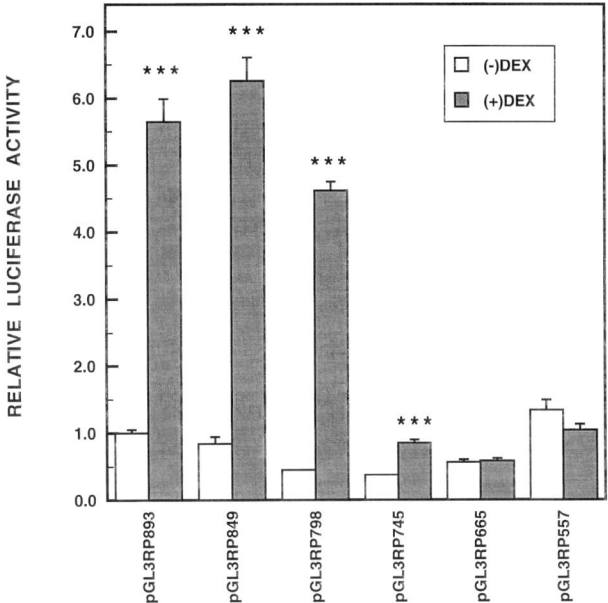

Fig. 8 Upstream glucocorticoid responsive sequences in proximal PNMT promoter.

bound to more distal upstream sequences to synergistically activate the PNMT promoter. Likely the binding site for this unknown transcriptional co-activator(s) lies between −893 bp and −3.0 kb since no incremental activation of luciferase is apparent with constructs containing greater than −3.0 kb of upstream sequences.

In summary, the immediate early gene transcription factor Egr-1 appears to participate in the stress-induced rise in PNMT gene expression in the normal rat. While the MAOB KO mouse seems to be a potential model for HPA axis dysregulation, no similar association between Egr-1 and PNMT mRNA appears to be evoked by stress. Finally, we have recently identified two new GREs in the rat PNMT promoter which are very responsive to glucocorticoid receptor activation in contrast to the original GRE identified at −513 bp. These GREs are located at −759 and −773 bp and overlap each other by 1 bp. In addition, they can apparently interact with the facilitory protein GRIP1 and an as yet unidentified transcription factor(s) to

synergistically stimulate PNMT promoter activity, a mechanism perhaps important for stress-induced changes in PNMT expression.

Acknowledgement

This work was supported by the NIDDK, National Institute of Health, the Marianne Gerschel Charitable Account, the Spunk Fund, Inc., and the Denise Sobel Research Support Fund.

References

Baetge, E.E., Behringer, R.R., Messing, A., Brinster, R.L., and Palmiter, R.D. (1988). Transgenic mice express the human phenylethanolamine N-methyltransferase gene in adrenal medulla and retina. *Proceedings of the National Academy of Sciences USA* **85**, 3648–3652.

Baetge, E.E., Suh, Y.H., and Joh, T.H. (1986). Complete nucleotide and deduced amino acid sequence of bovine phenylethanolamine N-methyltransferase: Partial amino acid homology with rat tyrosine hydroxylase. *Proceedings of the National Academy of Sciences USA* **83**, 5454–5458.

Batter, D.K., D'Mello, S.R., Turzai, L.M., Hughes, H.B., III, Gioio, A.E., and Kaplan, B.B. (1988). The complete nucleotide sequence and structure of the gene encoding bovine phenylethanolamine N-methyltransferase. *Journal of Neuroscience Research* **19**, 367–376.

Bohn, M.C., Goldstein, M., and Black, I.B. (1981). Role of glucocorticoids in the adrenergic phenotype in rat embryonic adrenal gland. *Developmental Biology* **82**, 1–10.

Ebert, S.N., Ficklin, M.B., Her, S., Siddall, B.J., Bell, R.A., Morita, K., Ganguly, K., and Wong, D.L. (1998). Glucocorticoid-dependent action of neural crest factor AP-2: stimulation of phenylethanolamine N-methyltransferase gene expression. *Journal of Neurochemistry* **70**, 2286–2295.

Greene, L.A., and Tischler, A.S. (1976). Establishment of a noradrenergic clonal line of rat adrenal pheochromocytoma cells which respond to nerve growth factor. *Proceedings of the National Academy of Sciences USA* **73**, 2424–2428.

Grimbsy, J., Toth, M., Chen, K., Kumazawa, T., Klaidman, L., Adams, J.D., Karoum, F., Gal, J., and Shih, J.C. (1997). Increased stress response and β-phenylethylamine in MAOB-deficient mice. *Nature Genetics* **17**, 206–210.

Hemmick, L.M., Ross, M.E., and Evinger, M.J. (1995). Regulation of PNMT gene promoter constructs transfected into the TE671 medulloblastoma. *Neuroscience Letters* **201**, 77–80.

Her, S., Bloom, A.K., Tai, T.C., Bell, R.A., and Wong, D.L. (submitted). Glucocorticoid responsiveness of the rat phenylethanolamine N-methyltransferase gene.

Hong, H., Kohli, K., Trivedi, A., Johnson, D.L., and Stallcup, M.R. (1996). GRIP1, a novel mouse protein that serves as a transcriptional coactivator in yeast for the hormone binding domains of steroid receptors. *Proceedings of the National Academy of Sciences USA* **93**, 4948–4952.

Kaneda, N., Ichinose, H., Kobayashi, K., Oka, K., Kishi, F., Nakazawa, A., Kurosawa, Y., Fujita, K., and Nagatsu, T. (1988). Molecular cloning of

Table 1 Luciferase activity in nested deletion constructs.

Construct	Relative Luciferase Activity 1 µM DEX	Relative Luciferase Activity 1 µM DEX, (+)GRIP1
pGL3RP8.4	1.12 ± 0.24	2.28 ± 0.31
pGL3RP4.2	0.58 ± 0.02	2.11 ± 0.20
pGL3RP3.0	1.19 ± 0.11	2.08 ± 0.15
pGL3RP893	1.42 ± 0.01	1.23 ± 0.06
pGL3RP392	0.30 ± 0.02	0.14 ± 0.02
pGL3RP60	0.11 ± 0.02	0.03 ± 0.00

cDNA and chromosomal assignment of the gene for human phenylethanolamine N-methyltransferase, the enzyme for epinephrine biosynthesis. *Journal of Biological Chemistry* **263**, 7672–7677.

Ross, M.E., Evinger, M.J., Hyman, S.E., Carroll, J.M., Mucke, L., Comb, M., Reis, D.J., Joh, T.H., and Goodman, H.M. (1990). Identification of a functional glucocorticoid response element in the phenylethanolamine N-methyltransferase promoter using fusion genes introduced into chromaffin cells in primary culture. *Journal of Neuroscience* **10**, 520–530.

Verhofstad, A.A.J., Hokfelt, T., Goldstein, M., Steinbusch, H.W.M., and Joosten, H.W.J. (1979). Appearence of tyrosine hydroxylase, aromatic amino-acid decarboxylase, dopamine beta-hydroxylase and phenylethanolamine N-methyltransferase during ontogenesis of the adrenal medulla. *Cellular Tissue Research* **200**, 1–13.

Wong, D.L., Lesage, A., White, S., and Siddall, B. (1992). Adrenergic expression in the rat adrenal gland: Multiple developmental regulatory mechanisms. *Developmental Brain Research* **67**, 229–236.

Wong, D.L., Siddall, B., and Wang, W. (1995). Hormonal control of rat adrenal phenylethanolamine N-methyltransferase: enzyme activity, the final critical pathway. *Neuropsychopharmacology* **13**, 223–234.

Wong, D.L., Siddall, B.J., Ebert, S.N., Bell, R.A., and Her, S. (1998). Phenylethanolamine N-methyltransferase gene expression: Synergistic activation by Egr-1, AP-2 and the glucocorticoid receptor. *Molecular Brain Research* **61**, 154–161.

Wong, D.L., Zager, E.L., and Ciaranello, R.D. (1982). Effects of hypophysectomy and dexamethasone administration on central and peripheral S-adenosylmethionine levels. *Journal of Neuroscience* **2**, 758–764.

23. Reacting to Stress in the Absence of Norepinephrine: Lessons from Dopamine β-Hydroxylase- and Tyrosine Hydroxylase-Deficient Mice

S.A. Thomas

Introduction

Many different approaches have been used to study the importance of adrenergic signaling in the response to various stressors. These approaches include surgical ablation, chemical ablation, pharmacological manipulation, and physiological recording and stimulation. Much has been learned about the importance of adrenergic signaling with respect to the regulation of cardiovascular and metabolic functions. To further our understanding of adrenergic signaling and its importance in the response to stress, a new model of adrenergic function has recently been created. In this model, the adrenergic ligands norepinephrine (NE) and epinephrine (EPI) have been eliminated by the genetic ablation of the enzyme responsible for their synthesis, dopamine β-hydroxylase (DBH). One advantage of this approach is that it should be more specific than surgical and chemical ablation, where cotransmitters are also lost when the cells or their terminals are lesioned. This approach is also potentially more specific than pharmacological manipulations that may include effects beyond those impinging on the adrenergic system. The mutant mouse model ($dbh^{-/-}$) has permitted the identification of several periods during development when adrenergic signaling is critical. Furthermore, it has allowed us to examine the response to specific stressors in the complete absence of NE and EPI in adults as described below.

Methods

The creation of $dbh^{-/-}$ mice was achieved by standard gene-targeting methods (Thomas *et al.*, 1995). Approximately 1 kb of proximal promoter, and the first exon and intron were replaced by a neomycin resistance cassette. Gene targeting was done in AB1 129/SvCPJ embryonic stem cells. Chimeric mice created with these targeted cells were bred to C57BL/6J females and the resulting 129 × B6 F1 $dbh^{+/-}$ mice were inbred to maintain the mutation on this hybrid background for all subsequent generations. Mice were housed in either conventional or specific pathogen free areas at the University of Washington in Seattle.

The $dbh^{-/-}$ mutation resulted in a level of mRNA expression from the dbh gene that was at least 1000-fold lower in $dbh^{-/-}$ fetuses compared to $dbh^{+/+}$ fetuses as detected by PCR (Thomas *et al.*, 1995). Furthermore, in adult $dbh^{-/-}$ mice NE and EPI are below the limits of detection when assayed for by HPLC with electrochemical detection (Thomas and Palmiter, 1997b; Thomas *et al.*, 1998). $Dbh^{+/-}$ mice are routinely used as controls because their NE and EPI levels do not differ from those of $dbh^{+/+}$ mice and no phenotypic differences have been noted between heterozygous and wild type mice (Thomas *et al.*, 1998). Studies of adolescents and adults have been performed on mice that were treated prenatally with 1 mg/ml L-threo- 3,4-dihydroxyphenylserine (DOPS) in the maternal drinking water from embryonic day (E) 9.5 until birth. No treatment was administered after birth. Mating was performed between $dbh^{+/-}$ females and $dbh^{-/-}$ males treated daily with a subcutaneous injection of 10 mg of DOPS. Sex-matched littermate controls were used for all studies.

Results and Discussion

Fetus

The first critical period for adrenergic signaling as defined by an observable phenotype associated with the $dbh^{-/-}$ mouse is during midgestation (Thomas *et al.*, 1995). During development, NE is first detected at significant levels in wild type fetuses at E10.5, while EPI is not detected at significant levels until after E13.5. Interestingly, the $dbh^{-/-}$ fetuses begin dying around E11, shortly after the appearance of NE. If NE is partly restored in the $dbh^{-/-}$ fetuses by providing 1 mg/ml DOPS in the maternal drinking water, the mutant fetuses all survive to birth. Histological examination of fetuses between E9.5 and E15.5 has not revealed any major morphological differences between $dbh^{-/-}$ and control fetuses. However, subtle alterations in the organization and size of the cardiomyocytes are apparent in the majority of $dbh^{-/-}$ fetuses that survive beyond E11. In addition, cardiovascular function is compromised in the $dbh^{-/-}$ fetuses under hypoxic conditions. Mutant fetuses excised with their yolk sac and circulation intact have a heart rate that is ~25%

lower than that for control fetuses at 37°C in non-oxygenated phosphate-buffered saline. Hypoxia is a known stimulus for the release of catecholamines in the fetus (Jones, 1980; Slotkin and Seidler, 1988). One hypothesis for the role of adrenergic stimulation at midgestation has to do with the regulation of cardio-vascular function in a fetus that is rapidly growing at this stage. NE may be required to maintain adequate cardiac output and thus tissue perfusion as the fetus grows. Future studies will examine whether $dbh^{-/-}$ fetuses are under hypoxic stress prior to their demise.

As the precursor of NE, dopamine (DA) is present in the $dbh^{-/-}$ fetuses at ~5 times its normal level, and is about half the concentration of NE in control fetuses (Thomas et al., 1995). The excess DA does not seem to have a large impact on the phenotype, however, because mice lacking tyrosine hydroxylase ($th^{-/-}$) and thus both NE and DA exhibit an almost identical phenotype with respect to the timing of lethality and histological changes (Zhou et al., 1995). In the absence of NE fetuses appear to suffer cardiovascular failure as suggested by dilated right atria and blood vessels in some cases. However, it is not clear whether the cardiovascular changes observed in these mutants are due to a primary deficit in cardiovascular function or secondary changes indicative of a dying fetus. Future studies will attempt to test this possibility directly by "restoring" adrenergic signaling specifically in the heart. For example, this could be achieved by expressing a DA receptor that signals through the appropriate second messenger system using a heart specific promoter that is expressed in the fetus (for example, that for the atrial natriuretic peptide (Cameron et al., 1996).

Neonate

The second critical period for adrenergic signaling as defined by an observable phenotype associated with the $dbh^{-/-}$ mouse is shortly after birth. About 10% of the $dbh^{-/-}$ fetuses survive to term without DOPS in the maternal drinking water (Thomas et al., 1995). This is probably due to the small amounts of NE and EPI that reach the fetus from the maternal circulation. Interestingly, about half of these naturally born $dbh^{-/-}$ fetuses die within two days of birth. Partial restoration of NE by prenatal DOPS treatment results in over 90% of $dbh^{-/-}$ neonates surviving into adolescence. This is probably due to additional NE and EPI present in the prenatally treated $dbh^{-/-}$ neonates during the first day of postnatal life.

Most insights into the pathophysiology of adrenergic deficient mutant mice during the neonatal period have come from studies on the $th^{-/-}$ mice reported by Nagatsu's laboratory (Kobayashi et al., 1995). Similar to our fetal data, they found ~30% decrease in the heart rate of E18.5 fetuses delivered by caesarean section. In addition, there was an elevation in the S-T segment of the electrocardiogram. They also examined fluid resorption in the lungs and plasma glucose levels but no differences were observed between mutant and control mice. This was interesting in light of previous

pharmacological data indicating an important role for adrenergic signaling in regulating metabolic and pulmonary function shortly after birth (Walters and Olver, 1978; Jones, 1980; Slotkin and Seidler, 1988). As for the fetus, however, it is not clear whether differences in cardiac function account for the increased mortality of neonates that lack NE and EPI. For example, adrenergic signaling is also thought to be critical for olfactory learning in the neonate (Sullivan et al., 1989). It is possible that some of the mutant neonates succumb because they are unable to locate the nipple to nurse. Studies in adult $dbh^{-/-}$ mice indicate that olfactory learning that depends on either the main or accessory olfactory systems is intact. Whether the same is true for $dbh^{-/-}$ neonates remains to be tested.

Adolescent

About one third of the $dbh^{-/-}$ mice die in their third, fourth and fifth postnatal weeks when raised in a conventional facility known to contain murine pathogens such as mouse hepatitis virus. However, when these mice are raised in a specific pathogen free (SPF) environment that excludes most known murine pathogens, only ~3% of the mutant mice die during adolescence. This observation led us to test whether the $dbh^{-/-}$ mice have impaired immunity. We found that some $dbh^{-/-}$ adolescents raised in a conventional facility displayed thymic involution and a relative loss of CD4$^+$/CD8$^+$ thymocytes (Alaniz et al., 1999). To study immune function under more controlled conditions, we examined the immune system of SPF-raised mice. The mutants had normal numbers of blood granulocytes, monocytes and lymphocytes, as well as normal numbers and subpopulations of thymocytes and splenocytes. Thymocyte and splenocyte proliferation in response to mitogens was normal, as was splenocyte production of IFN-γ and IL-4. These results suggest that intrinsic immunity is normal in the $dbh^{-/-}$ mice.

To examine the response to infection, SPF-raised mice were injected IP with 10^5 Listeria monocytogenes colony-forming units. Bacterial burdens in both the liver and the spleen were at least 10-fold greater in the mutant as compared to the control mice 5 days after injection (Figure 1). By 8 days 5/23 mutant mice had died while only 1/19 control mice had died. When this experiment was repeated to examine secondary infection with L. monocytogenes, similar results were obtained for primary infection. Following secondary infection, the bacterial burden in the liver was 100-fold greater in the mutant as compared to the control mice.

Thymocytes harvested from uninfected mutant SPF mice were normal in total number and distribution of CD4 and CD8 subpopulations. In contrast, thymocytes harvested from mutant mice 5 days after primary infection with L. monocytogenes were relatively small and yielded a total number of thymocytes that was ~10% of those from infected controls (Figure 2). Furthermore, there was a pronounced loss of CD4$^+$/CD8$^+$ thymocytes, a scenario reminiscent of results obtained from some of the mutant adolescents housed

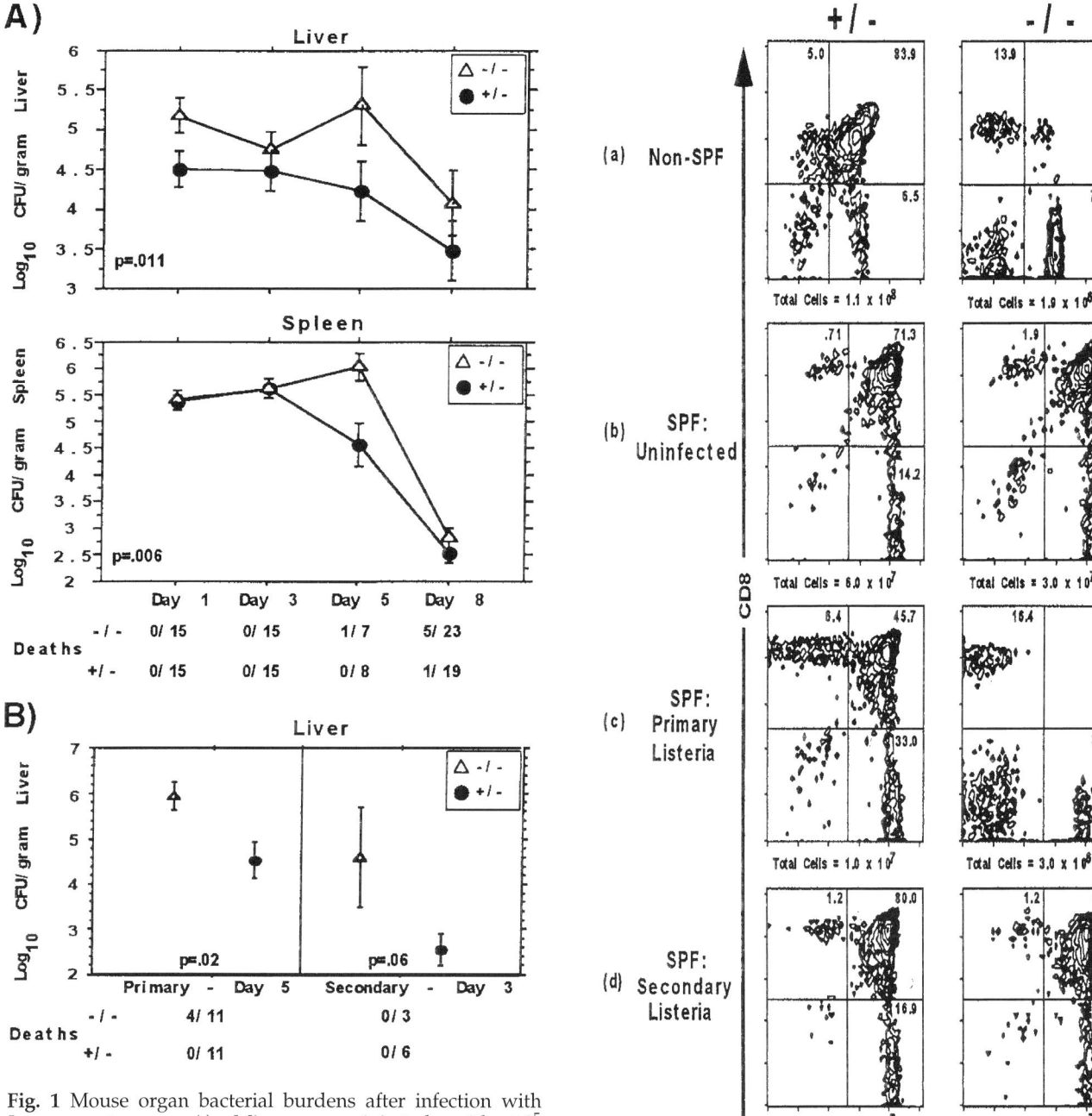

Fig. 1 Mouse organ bacterial burdens after infection with *L. monocytogenes*. A) Mice were injected with 10⁵ *L. monocytogenes* CFU and sacrificed on the indicated days. B) Mice were injected with 10⁵ *L. monocytogenes* CFU and sacrificed on either day 5 (primary) or day 24 (secondary), 3 days after a second injection of 10⁵ *L. monocytogenes* CFU on day 21. The table below each panel shows the frequency of deaths for the total number of mice for each day. [After Alaniz *et al.*, 1999]

Fig. 2 Thymic cellularity in *dbh⁻/⁻* and *dbh⁺/⁻* mice as determined by flow cytometry. A) Thymocytes were derived from mice housed in a conventional facility. B) Thymocytes from uninfected mice housed under SPF conditions. C) Thymocytes from mice after 5 days of primary *L. monocytogenes* infection. D) Thymocytes from mice after 3 days of secondary *L. monocytogenes* infection. [After Alaniz *et al.*, 1999]

in the conventional facility. Interestingly, secondary infection did not result in any differences in thymocyte number or distribution. Splenocytes harvested from mutant mice 5 days after primary infection were normal in number and proliferation in response to anti-CD3 antibody. However splenocyte cytokine production in response to anti-CD3 was altered. IFN-γ, TNF-α and IL-10 were all substantially reduced

during both primary and secondary infection in the mutants (Figure 3).

Because the above findings suggested that the *dbh⁻/⁻* mice may have impaired type 1 helper T cell (Th1)

A)   Splenocytes from Primary Listeria

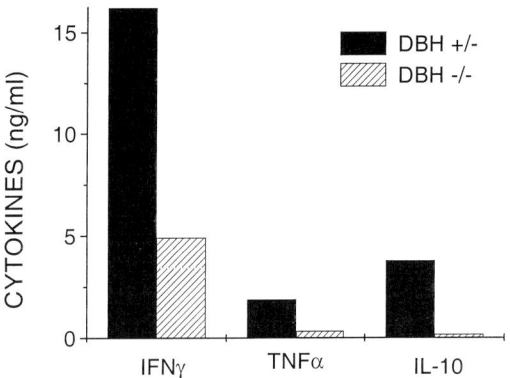

B)   Splenocytes from Secondary Listeria

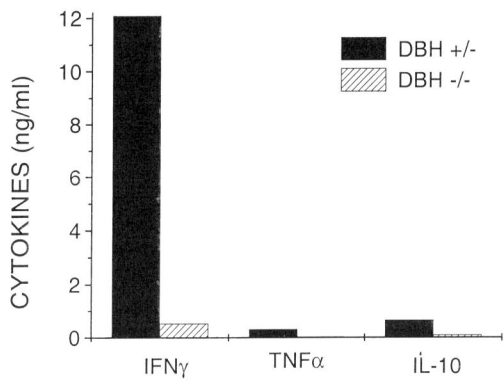

Fig. 3 Cytokine production and proliferation by splenocytes from mice infected with *L. monocytogenes*. A) 5 days after primary infection. B) 3 days after secondary infection. [Adapted from Alaniz *et al.*, 1999]

function, these mice were also infected with *M. tuberculosis*, another intracellular pathogen that elicits an immune response dependent on Th1 processes. IFN-λb and TNF-α were reduced in splenocytes from infected mice that were stimulated with either anti-CD3 or culture filtrate proteins from *M. tuberculosis* (Alaniz *et al.*, 1999). To further assess T cell function, mice were immunized with trinitrophenol-keyhole limpet hemocyanin (TNP-KLH). $dbh^{-/-}$ mice had 10-fold lower titers of IFN-γ dependent IgG2a anti-TNP antibodies, but normal titers of IgG1 and IgM anti-TNP antibodies. All of these results are consistent with a specific impairment of Th1 function in the mutant mice. The phenotype may present during adolescence in a conventional facility because the mice are weaning themselves from the dam at this time and passive immunity should decline as a result.

Adult

Cold Stress

The response to cold stress is multifaceted. Two of the most important elements include heat conservation and heat production, both of which are influenced by the

sympathetic nervous system. Heat conservation is achieved in several ways, including peripheral vaso-constriction, piloerection, and a reduction in exposed body surface. Thermogenesis is achieved by increasing physical (shivering, ambulating) and metabolic activity. To test the extent to which adrenergic signaling is critical for adaptation to cold stress, mice were placed at 4°C for up to 2 hours (Thomas and Palmiter, 1997c). The core body temperature of the mutant mice dropped ~15°C in the first hour while that for control mice dropped 2°C in the first 30 minutes and then remained stable (Figure 4A). Restoration of NE in the mutant mice 5 hours after injection of DOPS at 1 mg/g resulted in a 3-fold lower rate of drop in core body temperature.

Heat conservation was impaired in the mutant mice in at least two ways. The mutants failed to piloerect, and peripheral vasoconstriction was impaired. The latter was assessed by monitoring blood flow to the skin before and after the transition to 4°C using laser Doppler. The mutants reduced their blood flow by 35% while the controls did so by 85% (Figure 4C). Thermogenesis was also impaired in the mutant mice. Rodents depend on the stimulation of metabolism in brown fat for much of their adaptive thermogenesis (Nicholls and Locke, 1984). Thermogenesis in brown fat is modulated by controlling the expression of the brown fat-specific uncoupling protein, UCP1 (Klaus *et al.*, 1991). Heat is generated when the proton gradient is dissipated by UCP1 after it is inserted into the inner mitochondrial membrane. $Dbh^{-/-}$ mice express less than 10% of the normal level of UCP1 mRNA in their brown fat, both at 22°C and after 2 hours at 4°C (Figure 4B). Consistent with the idea that their brown fat is less active metabolically, the lipid vacuoles of the mutant brown adipocytes are much larger than normal. These experiments confirm the idea that adrenergic signaling is critical for adaptation to cold stress.

Fasting

Adrenergic stimulation has been implicated in the mobilization of energy stores, specifically glucose and fatty acids (Landsberg and Young, 1992). The $dbh^{-/-}$ mice have normal energy stores as assessed by the combined weight of several fat pads relative to body weight. To assess the importance of adrenergic stimulation for energy substrate mobilization during a fast, food was withheld from mutant and control mice for up to 3 days. All mice survived 24 and 48 hours without food, but 2/21 control mice had died by 72 hours of fasting while 0/19 mutant mice died. At each of the 3 time points, the mutant mice had lost significantly less total body weight and percent body weight than controls. Preliminary observations showed that fat pads could still be found in the mutant but not the control mice at 72 hours. These observations suggest that fatty acid mobilization may be impaired during a fast in the mutant mice. Fasting fat pad weights, plasma glycerol and free fatty acids will be measured in future studies to test this idea.

Plasma glucose levels were also measured prior to and during a fast. Plasma was collected after asphyxia-

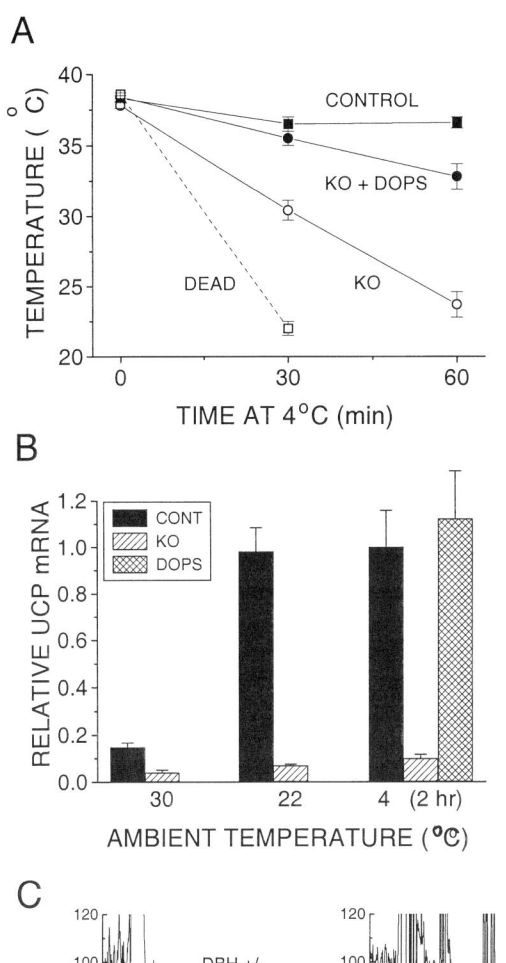

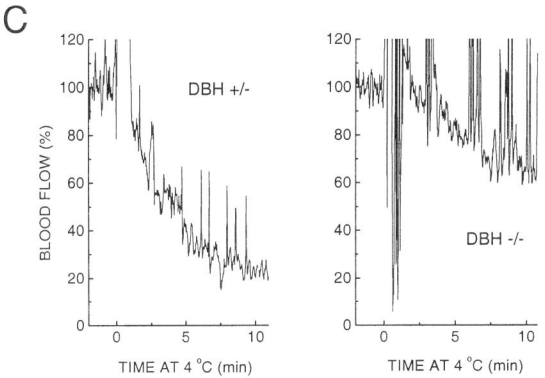

Fig. 4 Thermoregulation during cold stress. A) Core body temperature of control mice, $dbh^{-/-}$ mice, $dbh^{-/-}$ mice injected with DOPS for 5 days, and mice sacrificed just prior to being placed at 4°C (DEAD). B) UCPI mRNA levels in BAT after two weeks at 30°C, chronically at 22°C, after 2 hr at 4°C, and after 3 injections of DOPS in $dbh^{-/-}$ mice. UCP1 mRNA levels were normalized to 1.0 for $dbh^{+/-}$ mice (CONT) kept at 4°C for 2 hr. C) Peripheral blood flow in the tail of a $dbh^{+/-}$ mouse and $dbh^{-/-}$ mouse prior to and after placement at 4°C from 22°C. Time = 0 denotes when the transition into the cold was made. The mean blood flow at 22°C was normalized to 100%. Spikes in blood flow are movement artifacts. [Adapted from Thomas and Palmiter, 1997c]

tion of the mice with CO_2. Therefore the measurements represent hypoxia-induced release of glucose into the blood. Hypoxic plasma glucose levels (~300 mg/dl) were about 10% lower in the mutant mice prior to

fasting. However, there was no difference in these levels after one or three days of fasting, and only a slight difference after two days of fasting. $Dbh^{-/-}$ mice had normal plasma glucagon and insulin levels at all time points. These results suggest that adrenergic stimulation does not have a substantial role in the surge of blood glucose levels in response to hypoxia.

Aversive Conditioning

Learning aversively motivated paradigms provides one means of examining the behavioral response to stress. Pharmacological studies have suggested that adrenergic signaling in the CNS and the periphery may be important for learning such paradigms (Barnes and Pompeiano, 1991). To test this possibility further, the $dbh^{-/-}$ mice were trained with a simple passive avoidance paradigm in which they learn to associate the dark side of a two-compartment chamber with an electrical shock. The mice were tested at low, intermediate and high intensity shock levels over a period beginning 1 hour after the training trial and extending up to 31 days later (Thomas and Palmiter, 1997a). Remarkably, no difference in avoidance was observed between the mutants and controls at any shock intensity or duration after testing. Because DA is packaged in and released from the adrenergic terminals of the mutant mice, and DA is at least a weak agonist at some of the adrenergic receptors, it is possible that not all adrenergic signaling has been ablated in these mice. If this were true, then some phenotypes that would be apparent in the absence of NE may be masked in the $dbh^{-/-}$ mice. This will be tested in future studies with mice that have neither DA nor NE in their adrenergic terminals.

The mice were also tested in three more complex paradigms that required trials over many days (Thomas and Palmiter, 1997a). The mutant mice learned an active avoidance paradigm more slowly than controls. In this paradigm the mice were required to move to the other compartment of the chamber during a 10-second tone. If they failed to move in this time, they received a shock.

In a second paradigm, the mice were required to start running when a rod on which they were standing began to rotate, otherwise they would fall to the ground. About 20% of the mutant mice repeatedly failed to begin moving when the rod began to rotate. Results from these two paradigms suggest the possibility that the mutant mice may be impaired in their motor learning abilities. Additional motor learning paradigms will be used in the future to test this idea.

In a third paradigm, the mice were required to learn the location of a hidden platform while swimming in an opaque pool of water (Morris water maze). The mutant mice learned this task as well as controls, but consolidation of the memory for the location of the hidden platform was reduced in the mutants. How well the mutant mice perform in reward based tasks remains to be tested.

Glucocorticoids

As a final topic for consideration, we have begun to examine the activity of the hypothalamic-pituitary-adrenal axis (HPA) Activation of the HPA and elevated glucocorticoids represent another major response to stress. Therefore it seemed reasonable to examine the possibility of cross talk between the adrenergic and glucocorticoid systems. As a first step, we have measured plasma levels of corticosterone and adreno-corticotropic hormone (ACTH) under basal conditions and following immobilization-induced stress. Stress was induced by placing the mice in a 50-ml plastic centrifuge tube for 30 minutes. Trunk blood was collected after decapitation. Interestingly, the mutant mice had significantly higher basal levels of ACTH and corticosterone. However, there was no difference in these hormone levels between genotypes following immobilization stress. Further studies need to be performed to determine whether the loss of adrenergic signaling, or perhaps the excess DA present in the $dbh^{-/-}$ mice, is responsible for the alteration in basal glucocorticoid regulation.

Conclusions

Use of the $dbh^{-/-}$ mice as a model for adrenergic function has confirmed the idea that adrenergic signaling is critical for the regulation of the cardiovascular system (see also Cho *et al.*, 1999) and in particular its adaptation to stress. Studies from the mutant mice have demonstrated that adrenergic regulation of cardiovascular function begins at midgestation and continues into adulthood. Furthermore, the $dbh^{-/-}$ mice have provided convincing evidence for the importance of catecholamines in controlling infection-related illness. Studies using these mice also support the hypothesis that adrenergic stimulation has a role in learning some aversive conditioning paradigms. Future studies with these and other mutant mice should help define the extent to which adrenergic signaling is critical for the reaction to various stressors and the mechanisms by which the adrenergic system effects adaptation.

Acknowledgments

I would like to thank Sumitomo Pharmaceuticals, Ltd of Osaka, Japan for their generous gifts of DOPS; A. Matsumoto for the catecholamine measurements; J. Licinio for the measurements of plasma ACTH and corticosterone; G. Taborsky for the measurements of plasma glucagon and insulin; and Richard Palmiter whose generous scientific support made this work possible.

References

Alaniz, R.C., Thomas, S.A., Perez-Melgosa, M., Mueller, K., Farr, A.G., Palmiter, R.D., and Wilson, C.B. (1999). Dopamine beta-hydroxylase deficiency impairs cellular immunity. *Proceedings of the National Academy of Sciences USA* **96**, 2274–8.

Barnes, C.D., and Pompeiano, O. (1991). *Neurobiology of the locus coeruleus* New York, NY, USA: Elsevier.

Cameron, V.A., Aitken, G.D., Ellmers, L.J., Kennedy, M.A., and Espiner, E.A. (1996). The sites of gene expression of atrial, brain, and C-type natriuretic peptides in mouse fetal development: temporal changes in embryos and placenta. *Endocrinology* **137**, 817–24.

Cho, M.C., Rao, M., Koch, W.J., Thomas, S.A., Palmiter, R.D., and Rockman, H.A. (1999). Enhanced contractility and decreased beta-adrenergic receptor kinase-1 in mice lacking endogenous norepinephrine and epinephrine. *Circulation* **99**, 2702–7.

Jones, C.T. (1980). Circulating catecholamines in the fetus, their origin, actions and significance. In: H. Parvez and S. Parvez (Eds.), *Biogenic Amines in Development*, pp. 63–86. Amsterdam: Elsevier/North Holland Biomedical Press.

Klaus, S., Casteilla, L., Bouillaud, F., and Ricquier, D. (1991). The uncoupling protein UCP: a membranous mitochondrial ion carrier exclusively expressed in brown adipose tissue. *International Journal of Biochemistry* **23**, 791–801.

Kobayashi, K., Morita, S., Sawada, H., Mizuguchi, T., Yamada, K., Nagatsu, I., Hata, T., Watanabe, Y., Fujita, K., and Nagatsu, T. (1995). Targeted disruption of the tyrosine hydroxylase locus results in severe catecholamine depletion and perinatal lethality in mice. *Journal of Biological Chemistry* **270**, 27235–43.

Landsberg, L., and Young, J.B. (1992). Catecholamines and the adrenal medulla. In: R.H. Williams, J.D., Wilson, and D.W. Foster (Eds.), *Williams' textbook of endocrinology*, pp. 621–705. Philadelphia: W.B. Saunders Co.

Nicholls, D.G., and Locke, R.M. (1984). Thermogenic mechanisms in brown fat. *Physiological Reviews* **64**, 1–64.

Slotkin, T.A., and Seidler, F.J. (1988). Adrenomedullary catecholamine release in the fetus and newborn: secretory mechanisms and their role in stress and survival. *Journal of Developmental Physiology* **10**, 1–16.

Sullivan, R.M., Wilson, D.A., and Leon, M. (1989). Norepinephrine and learning-induced plasticity in infant rat olfactory system. *Journal of Neuroscience* **9**, 3998–4006.

Thomas, S.A., Marck, B.T., Palmiter, R.D., and Matsumoto, A.M. (1998). Restoration of norepinephrine and reversal of phenotypes in mice lacking dopamine beta-hydroxylase. *Journal of Neurochemistry* **70**, 2468–76.

Thomas, S.A., Matsumoto, A.M., and Palmiter, R.D. (1995). Noradrenaline is essential for mouse fetal development. *Nature* **374**, 643–6.

Thomas, S.A., and Palmiter, R.D. (1997a). Disruption of the dopamine beta-hydroxylase gene in mice suggests roles for norepinephrine in motor function, learning, and memory. *Behavioral Neuroscience* **111**, 579–89.

Thomas, S.A., and Palmiter, R.D. (1997b). Impaired maternal behavior in mice lacking norepinephrine and epinephrine. *Cell* **91**, 583–92.

Thomas, S.A., and Palmiter, R.D. (1997c). Thermoregulatory and metabolic phenotypes of mice lacking noradrenaline and adrenaline. *Nature* **387**, 94–7.

Walters, D.V., and Olver, R.E. (1978). The role of catecholamines in lung liquid absorption at birth. *Pediatric Research* **12**, 239–42.

Zhou, Q.Y., Quaife, C.J., and Palmiter, R.D. (1995). Targeted disruption of the tyrosine hydroxylase gene reveals that catecholamines are required for mouse fetal development. *Nature* **374**, 640–3.

24. Role of CRH-Binding Protein in the Mammalian Stress Response

A.F. Seasholtz, S.J. McClennen, D.N. Cortright, H.L. Burrows, D.B. Speert and S.A. Camper

Introduction

Corticotropin-releasing hormone (CRH) is widely recognized as the major hypophysiotropic hormone in the mammalian stress response. Neural input to the hypothalamus in response to stress causes an increase in the synthesis of CRH in the paraventricular nucleus and release of CRH from the median eminence. CRH acts on pituitary corticotropes to stimulate POMC synthesis and ACTH release, resulting in increased adrenal glucocorticoid production and secretion. Glucocorticoids mediate a variety of metabolic changes, allowing the body to respond to the stressful stimulus. Glucocorticoids also feed back at multiple levels in the hypothalamic-pituitary-adrenal (HPA) axis to negatively regulate CRH and POMC expression, thereby maintaining homeostasis within the system. Outside of the HPA axis, CRH is thought to act as a neurotransmitter in the CNS to mediate the autonomic, behavioral and immunological responses to stress. Aberrant regulation of CRH activity has been implicated in the pathophysiology of a number of clinical disorders including depression, anxiety disorders, and anorexia (reviewed in Owens et al., 1991).

To mediate its biological effects, CRH binds to two classes of CRH receptors. Both classes of receptors are seven transmembrane domain G_s-protein coupled receptors, and they show 70% amino acid identity (reviewed in Chalmers et al., 1996). The type I CRH receptor (CRH-R1) is expressed in anterior pituitary corticotropes and in the intermediate lobe of the pituitary. The many CNS sites of CRH-R1 expression include the neo-, olfactory, and hippocampal cortices, cerebellum, septum, amygdala, and brainstem sensory relay structures (Potter et al., 1994). The second CRH receptor, CRH-R2, has been isolated in two alternatively spliced forms in the rat and three alternatively spliced forms in human. CRH-R2 receptors have a distinct mRNA expression profile from CRH-R1. In the rodent, CRH-R2α is expressed primarily in brain, including olfactory bulb, lateral septum, medial amygdala, inferior colliculus, and supraoptic, ventromedial, and paraventricular nuclei of the hypothalamus (Chalmers et al., 1996). CRH-R2β in rodents is found largely in the periphery (heart, skeletal muscle, intestine, lung, kidney, and epididymis, but also choroid plexus and cerebral arterioles), while human CRH-R2β is found predominantly in the CNS. None of the forms of CRH-R2 are expressed in the pituitary, suggesting that CRH-R1 is the key receptor for CRH in the pituitary.

A CRH-binding protein (CRH-BP) distinct from the CRH receptors has also been isolated and characterized from human plasma (Orth et al., 1987; Behan et al., 1989). The CRH-BP is a 37 kDa secreted glycoprotein that binds CRH with an equal or greater affinity than the CRH-receptor (Orth et al., 1987; Behan et al., 1989). CRH-BP also binds with very high affinity to the recently described CRH-like peptide, urocortin (Vaughan et al., 1995). The binding protein is expressed at many sites in the brain, colocalizing to several sites of CRH expression or CRH target sites (Potter et al., 1991; Potter et al., 1992). These sites of coexpression suggest that the CRH-BP could modulate the endocrine and synaptic activity of CRH in the brain and periphery. Of particular interest is the demonstration of CRH-BP mRNA and immunoreactivity in anterior pituitary corticotropes, the CRH target site in the stress axis (Potter et al., 1992). In vitro studies have shown that CRH-BP can block CRH-mediated ACTH secretion from anterior pituitary cells (Potter et al., 1991; Cortright et al., 1995), leading us to hypothesize that CRH-BP may act as a negative regulator of CRH activity of the HPA axis in vivo.

In an effort to determine the in vivo role of CRH-BP in pituitary corticotropes and its potential modulatory role in the stress axis, we examined the in vivo regulation of CRH-BP expression in the pituitary by stress and glucocorticoids. Results from these in vivo studies are consistent with in vitro analyses of the transcriptional regulation of CRH-BP gene expression using transfection assays. Finally, we examined the in vivo consequences of pituitary CRH-BP overexpression using a transgenic mouse model.

In vivo Regulation of CRH-BP mRNA Levels in Response to Stress or Adrenalectomy

The overall effect of acute restraint stress on the hypothalamic-pituitary-adrenal axis is well-documented (Owens et al., 1991). CRH, ACTH, and glucocorticoid secretion all increase significantly during the stressor period but return to basal levels within an hour after stress. CRH and POMC transcription also increase in response to stressors. Experiments were conducted to determine the effects of acute restraint stress on pituitary CRH-BP steady-state mRNA levels

(McClennen *et al.*, 1998). Rats were sacrificed either directly before (control), directly after (30), or 60, 120, or 240 minutes after the start of a 30 minute restraint stress. Total RNA was isolated from each pituitary and used for RNase protection assays with a 252 base CRH-BP cRNA probe and a 114 base cyclophilin cRNA internal control. The effect of restraint stress on pituitary CRH-BP mRNA levels is shown in Figure 1. Figure 1a shows a representative RNase protection assay with the CRH-BP protected hybrids in the top panels and the cyclophilin internal control hybrids in the bottom panels. Each lane is representative of RNA isolated from a single rat pituitary, and the intensity of the CRH-BP hybrid was normalized to the intensity of the cyclophilin hybrid. The quantitation of the RNase protection assay is shown in Figure 1b. Immediately following the restraint stress (30 min time point), the normalized levels of CRH-BP mRNA are 2.1 times greater than basal levels (p = 0.09 compared to control). At the 60 min time point, steady-state CRH-BP mRNA levels are 3.1 times basal levels (p = 0.009 compared to

control), decreasing to 2.1 times basal level (p = 0.08 compared to control) at the 120 min time point and basal levels by 240 minutes after the start of the restraint period. These results demonstrate that acute restraint stress significantly increases pituitary CRH-BP steady-state mRNA levels. The rapid time course profiles for CRH, ACTH, and glucocorticoid release compared with the time course of CRH-BP mRNA changes suggest that both CRH and glucocorticoids may play important roles in the observed increase in pituitary steady-state CRH-BP mRNA levels in response to acute restraint stress.

Glucocorticoids maintain homeostasis in the HPA axis and have previously been shown to down-regulate the stress response at many levels. To examine the effect of glucocorticoids on pituitary CRH-BP mRNA levels *in vivo*, we compared steady-state CRH-BP mRNA levels in unstressed adrenalectomized (Adx) and sham-adrenalectomized (sham-Adx) rats (McClennen *et al.*, 1998). Figure 2 depicts the effect of adrenalectomy on CRH-BP steady-state

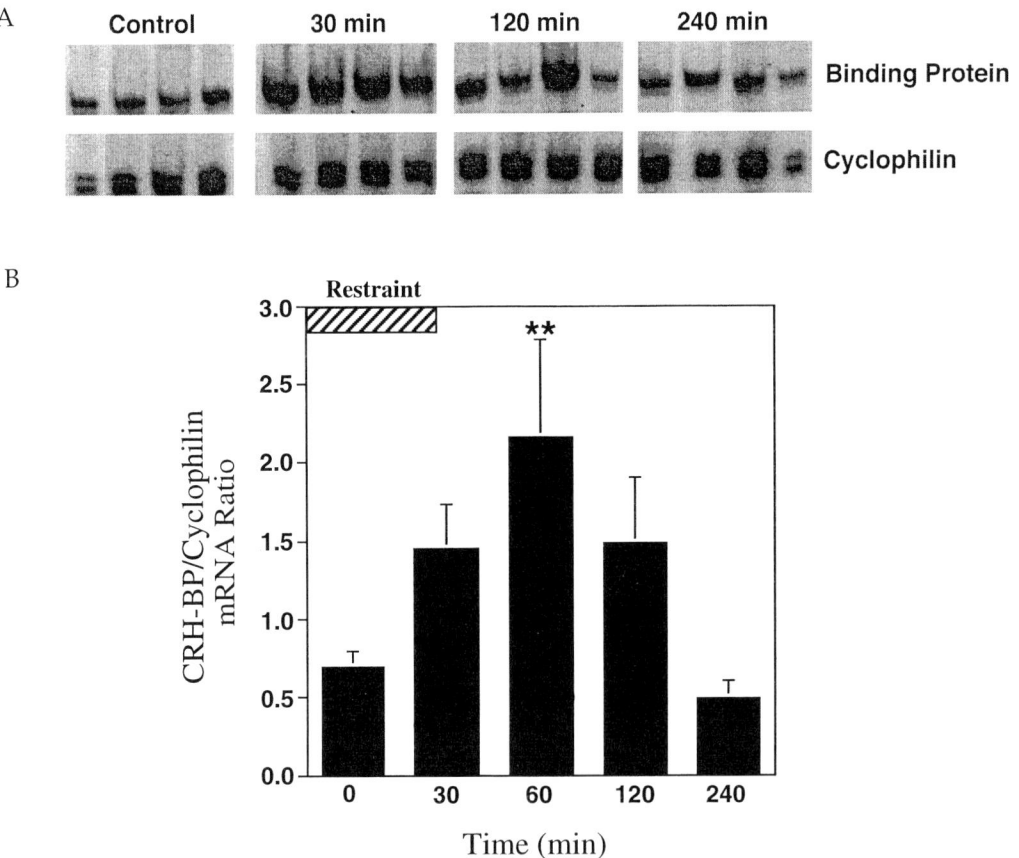

Fig. 1 Acute restraint stress increases pituitary CRH-BP steady-state mRNA levels. Individual pituitaries were isolated, harvested for total RNA, and subjected to RNase protection assays. **A.** A representative RNase protection assay depicting both protected CRH-BP hybrids and internal control cyclophilin hybrids for the time points: control, 30, 120, and 240 minutes after initiation of restraint stress. **B.** Quantitation of all restraint stress RNase protection assays. Data are presented as CRH-BP/cyclophilin mRNA ratios. ** represents p < 0.05 vs. 0 or 240 minute samples when data were analyzed by ANOVA with Fisher's post-hoc analysis (n = 11 for 0 min; n = 12 for 30 and 120 min; and n = 6 for 60 and 240 min). This figure is taken from McClennen *et al., Endocrinology* **139**: 4435–4441 (1998).

A

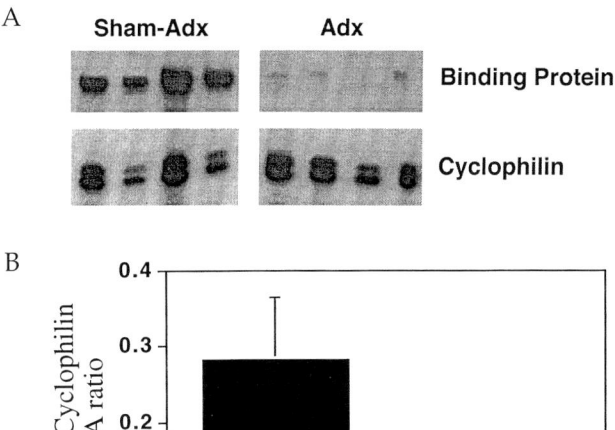

B

Fig. 2 **Adrenalectomy decreases pituitary CRH-BP steady-state mRNA levels in unstressed animals. A.** Representative RNase protection assay depicting differences in CRH-BP mRNA levels in sham-Adx and Adx rats as compared to internal cyclophilin controls. **B.** Quantitation of Adx RNase protection assays. Data are presented as CRH-BP/cyclophilin mRNA ratios. ** represents p < 0.05. This figure is taken from McClennen et al., *Endocrinology* **139**: 4435–4441 (1998).

mRNA levels. Figure 2a shows an RNase protection assay with representative CRH-BP hybrids in the top panels and the corresponding cyclophilin internal control hybrids in the bottom panels. Figure 2b represents the quantitation of the RNase protection assay shown in Figure 2a. The data are presented as CRH-BP/cyclophilin mRNA ratios. The CRH-BP mRNA levels at 7 days post-ADX are approximately 8% of the sham-Adx control levels (p = 0.01, n = 6), demonstrating a significant decrease in CRH-BP steady-state mRNA levels after adrenalectomy in unstressed animals. This decrease suggests that glucocorticoids positively regulate CRH-BP gene expression in the normal pituitary and confirms that CRH-BP has a role in the mammalian HPA axis during normal axis activity.

Together, these data suggest an additional role for CRH and/or glucocorticoids in HPA axis regulation by rapidly increasing the expression of CRH-BP which can then bind and sequester free CRH peptide, thus inhibiting its ACTH-releasing activity and attenuating the stress response. These results suggest that CRH-BP is directly involved in the mammalian stress response and may be important in HPA axis homeostasis. Additional studies on the impact of stress and/or glucocorticoid levels on CRH-BP protein expression and binding activity will further address the *in vivo* role of CRH-BP in HPA axis regulation.

Molecular Mechanisms of Transcriptional Regulation of CRH-BP Gene Expression

While *in vivo* studies demonstrate that CRH-BP gene expression is highly regulated, very little is known about the molecular mechanisms controlling CRH-BP mRNA expression. To further define the molecular mechanisms regulating CRH-BP gene expression, we characterized the rat CRH-BP gene structure (Cortright et al., 1997). In addition, we utilized transient transfection experiments with CRH-BP reporter constructs to identify functional CRH-BP promoter sequences.

Comparison of the 5′ flanking and 5′ untranslated DNA sequences of the rat (Cortright et al., 1997), human (Behan et al., 1993), and mouse (A. Seasholtz, unpublished data) CRH-BP genes indicates high sequence identity (85 to 94%) between the three genes from nucleotides –215 to +77 bp (Figure 3). Promoter sequence analysis reveals multiple potential transcription factor binding sites that are conserved in the rat, human, and mouse CRH-BP 5′ flanking DNAs based on homology to consensus binding sites. These potential transcription factor binding sites include a CREB/ATF site (–127 to –123), one OTX site (–93 to –88), one AP-2 site (–121 to –114), two Sp1 sites (–147 to –139 and –169 to –162), and two AP-1 sites (–177 to –171, –196 to –190). Other potential binding sites found in the rat CRH-BP gene sequence include two TATA sites (–221 to –217 and –478 to –474), a Pit-1 site (–297 to –290), an NF-kB site (–550 to –541), an additional AP-1 site (–238 to –232) and an additional AP-2 site (–188 to –182) (Faisst et al., 1992).

The identification of potential CREB/ATF and AP-2 sequences in the rat CRH-BP promoter sequence suggested that promoter activity might be transcriptionally regulated by cAMP. COS-1 cells transfected with the 3500BP-CAT construct (3500 bp of rat CRH-BP 5′ flanking DNA and 66 bp of 5′ untranslated sequences fused to the bacterial chloramphenicol acetyltransferase (CAT) gene) and treated with forskolin/IBMX (For/I) for 6 or 24 hours demonstrate significant inductions in CAT activity (Figure 4A). Since CRH receptors are G_s-protein coupled and therefore activate adenylate cyclase, the positive regulation of CRH-BP promoter activity by cyclic AMP suggested that CRH might also alter CRH-BP transcription. To assay for CRH regulation of CRH-BP promoter activity, αTSH cells were transiently co-transfected with 3500BP-CAT and the CRH receptor expression construct, CMV-CRH-R1. Alternatively, cells were co-transfected with 3500BP-CAT and the CMVneo vector construct as a negative control. As shown in Figure 4B, a 12 hour treatment of transfected αTSH cells with 10 μM forskolin (For) results in an 8-fold induction in CAT activity relative to control in the absence or presence of the CRH receptor. In contrast, a 12 hour treatment of transfected αTSH cells with 20 nM ovine CRH (oCRH) results in increased activity of the 3500BP-CAT construct relative to control levels, but only in the presence of CRH receptor. CRH regulation of CRH-BP promoter activity is also observed in AtT-20 cells which express endogenous CRH

Fig. 3 Nucleotide sequence homology of the 5′ flanking DNA of the rat (R), human (H), and mouse (M) CRH-BP genes. The rat CRH-BP 5′ flanking and 5′ untranslated DNA are shown in the top line. Homologous nucleotides in the human and mouse genes are shown by dashes with only divergent nucleotides shown. Potential transcription factor binding sites are underlined and labeled. Also indicated are three TATA sequences and several restriction endonuclease sites. The transcription initiation site of the CRH-BP gene in human liver is indicated with an asterisk (*). The human CRH-BP gene sequence was taken from (Behan *et al.*, 1993) This figure is modified from Cortright *et al.*, (1997) *Endocrinology* **138**: 2098–2108.

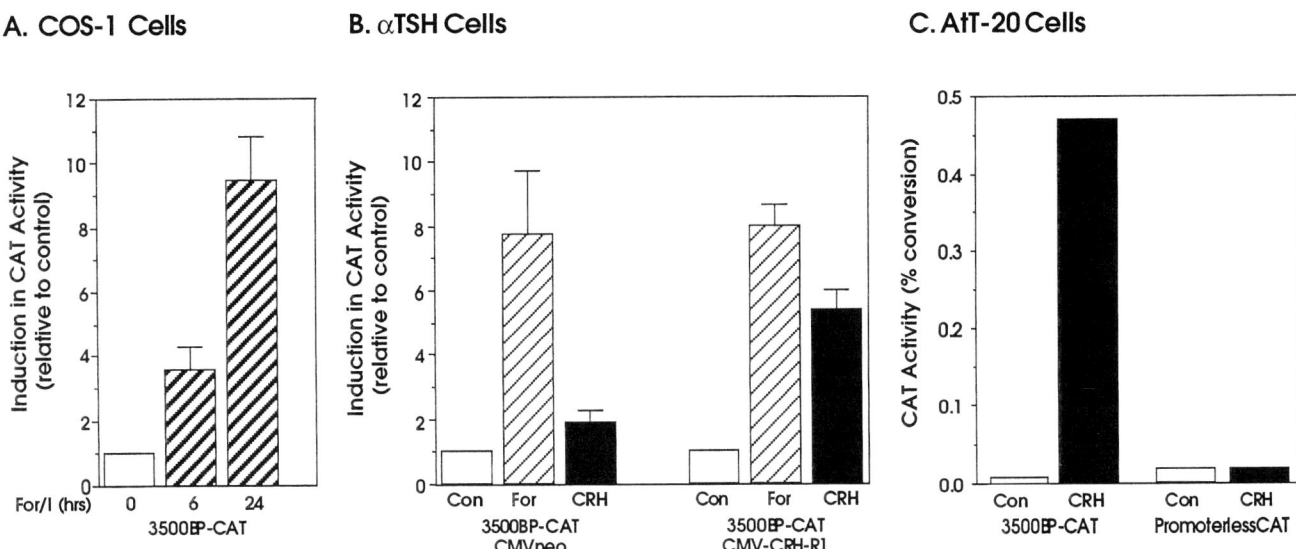

Fig. 4 CRH-BP promoter activity is induced by cAMP and CRH. (A) COS-1 cells were transiently transfected with the 3500BP-CAT construct and cultured in the absence (0 hrs) or presence of 10 μM forskolin/0.25 mM IBMX for 6 or 24 hours before harvest. (B) TSH cells were co-transfected with 3500BP-CAT/CMVneo or 3500BP-CAT/CMV-CRH-R1 and cultured in the absence (Con) or presence of either 10 μM forskolin (For) or 20 nM ovine CRH (CRH) for 12 hours prior to harvest. The inductions in CAT activity shown in A and B represent the induced CAT activity divided by the control CAT activity for each construct. The data represent the average of 3 or more experiments ± SEM. (C) AtT-20 cells were transfected with 3500BP-CAT or the promoterless CAT construct (pGSVOCAT). Transfected cells were cultured in the absence (Con) or presence of 20 nM CRH (CRH) for 15 hours prior to harvest. CAT activity is shown as % conversion of ^{14}C-chloramphenicol to acetylated products. The experiment was repeated multiple times with similar results; data shown are from a representative experiment. This figure is taken from Cortright *et al.*, (1997) *Endocrinology* **138**: 2098–2108.

receptors. Transfection of AtT-20 cells with the 3500BP-CAT or promoterless CAT construct results in barely detectable levels of basal CAT activity. However, treatment of 3500BP-CAT-transfected cells with 20 nM oCRH for 15 hours dramatically increases CAT activity, while no CRH-dependent increase in CAT activity is detected in cells transfected with the promoterless CAT construct. Together, these results demonstrate that DNA sequences within the 5' flanking region of the CRH-BP gene mediate positive cAMP and CRH regulation of CRH-BP promoter activity.

The observation that CRH mediates positive regulation of CRH-BP-reporter activity via CRH receptors suggests that CRH-BP expression *in vivo* may be increased in CRH target cells after exposure to CRH. Thus, the increased hypothalamic CRH secretion after acute restraint stress may play a role in the increased pituitary CRH-BP mRNA levels observed in Figure 1. In light of the ability of CRH-BP to attenuate CRH activity *in vitro*, these results further support the hypothesis that the CRH-BP may play an important role as a negative regulator of CRH activity.

The identification of potential CREB/ATF and AP-2 transcription factor binding sites within this region suggested that these CRH-BP DNA sequences might be important for cAMP and/or CRH regulation. The potential cAMP response element (CRE) sequence CGTCA (–127 to –123 bp) is a truncated version of the palindromic consensus CRE sequence TGACGTCA which was previously shown to be important in the

cAMP regulation of many genes. To determine the role of this potential transcription factor binding site in CRH-BP promoter regulation, the CGTCA sequence in the CRH-BP 5' flanking DNA was mutated to CTCGA using site-directed mutagenesis to create 341(ΔCRE)BP-CAT. This construct is identical to a deletion construct, 341BP-CAT, except for the mutation of the putative CRE sequences within the 341 bp of 5' flanking DNA. When the 341BP-CAT plasmid is transfected into COS-1 cells, the For/I-mediated induction in CAT activity is 19-fold. However, transfection of the 341(ΔCRE)BP-CAT construct results in only a 1.2 fold increase in CAT activity in the presence of For/I (Figure 5A). Similar decreases in forskolin- and CRH-mediated inductions are observed upon transfection of the 341(ΔCRE)BP-CAT into αTSH cells co-transfected with the CRH receptor (Figure 5B). The small cAMP and CRH inductions remaining in αTSH cells after mutation of the CRE sequence are comparable to the inductions in the 88BP-CAT construct (contains only 88 bp of CRH-BP 5' flanking DNA). These results demonstrate that the CRE-like DNA sequence at –127 to –123 bp in the CRH-BP promoter mediates, in large part, the cAMP and CRH inducibility of the CRH-BP promoter in transfected cells. In addition, these results suggest that the CREB/ATF family of transcription factors that binds to CRE sequences may be important mediators of the cAMP and CRH regulation of CRH-BP promoter activity.

Transcription of the pro-opiomelanocortin (POMC) gene in anterior pituitary corticotropes is also positively

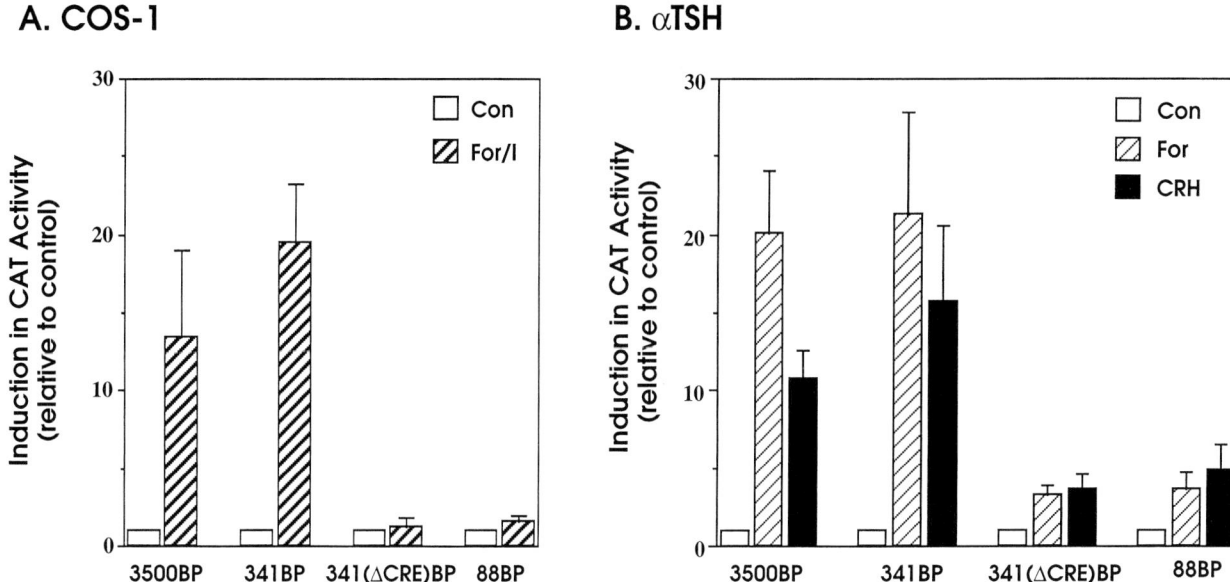

Fig. 5 Localization of 5′ CRH-BP DNA sequences involved in cAMP and CRH regulation. (A) COS-1 cells were transiently transfected with the 3500BP-CAT, 341BP-CAT, 341(▶CRE)BP-CAT or 88BP-CAT constructs and cultured in the absence (Con) or presence of 10 μM forskolin/0.25 mM IBMX (For/I) for 24 hours before harvest. (B) αTSH cells were co-transfected with the 3500BP-CAT, 341BP-CAT, 341(▶CRE)BP-CAT or 88BP-CAT constructs in addition to the CMV-CRH-R1 plasmid and cultured in the absence (Con) or presence of either 10 μM forskolin (For) or 20 nM ovine CRH (CRH) for 24 hours prior to harvest. The induction in CAT activity represents the induced CAT activity divided by the control CAT activity for each construct. The data represent the average of 3 or more experiments ± SEM. This figure is taken from Cortright *et al.*, (1997) *Endocrinology* **138**: 2098–2108.

regulated by CRH. While the mechanisms and transcription factors that mediate this response are not completely understood, the POMC promoter does not contain a classic CRE-like sequence. CRH regulation of the POMC promoter is thought to be mediated, in part, by AP-1 transcription factors cFos and cJun via a conserved AP-1 site in exon 1 (Boutillier *et al.*, 1995) and by other AP-1-independent pathways, possibly through the actions of the transcription factor AP-2 or the CRH-responsive element-binding protein (Bishop *et al.*, 1993; Jin *et al.*, 1994). The CRH-BP promoter, like POMC, contains consensus binding sites for AP-1 and AP-2 transcription factors. However, deletion of all the putative AP-1 sites in the rat CRH-BP promoter (169BP-CAT) does not block the positive CRH regulation of CRH-BP reporter constructs (D. Speert, unpublished data). Thus, the molecular mechanisms for CRH regulation of CRH-BP transcription and POMC transcription appear to be distinct.

Overexpression of CRH-BP in Pituitary of Transgenic Mice Alters HPA Axis Activity

To further investigate the *in vivo* function of CRH-BP in modulation of hypothalamic CRH activity in the HPA axis, transgenic mice were generated that constitutively express elevated levels of CRH-BP in the anterior pituitary gland (Burrows *et al.*, 1998). We chose to express the mouse CRH-BP cDNA under the control of the mouse pituitary glycoprotein hormone α-subunit (α-GSU) promoter. A 4.6 kb fragment of the α-GSU

5′ flanking-sequence was linked to the mouse CRH-BP cDNA (Figure 6, αCRH-BP) to direct high levels of transgene expression in adult pituitary gonadotropes and thyrotropes with varying low levels of transgene expression in a few sites within the CNS (Kendall *et al.*, 1994). Since the αGSU promoter is not regulated by corticosterone levels, the expression of the αCRH-BP transgene will be unaffected by any alterations in the HPA axis. Thus, the CRH-BP produced by this

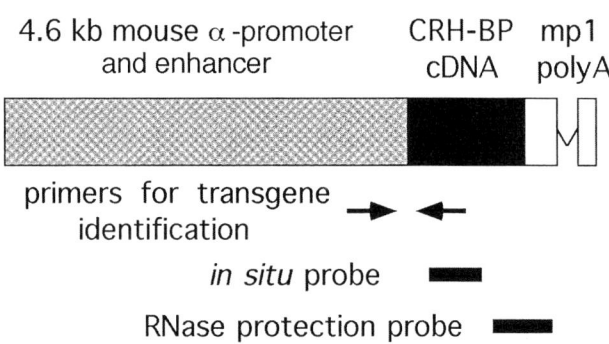

Fig. 6 Transgene Construct. The promoter consists of 4.6 kb of the a-GSU 5′-flanking region, attached to 1.3 kb of the mCRH-BP cDNA and 0.5 kb of the mouse protamine 1 intron and polyadenylation signal. PCR primers used to identify transgenic mice are indicated by arrows. The locations of riboprobes used for *in situ* hybridization and RNase protection are also indicated. This figure is taken from Burrows *et al.*, (1998) *J. Clin. Invest.* **101**: 1439–1447.

transgene should be highly expressed by the gonado-tropes and thyrotropes and secreted into the heterogeneous pituitary where it could potentially interact with CRH in the extracellular regions around the corticotropes.

Several lines of αCRH-BP transgenic mice were generated, and RNase protection assays, in situ hybridization histochemistry, and ^{125}I-hCRH crosslinking assays were used to confirm the expression of the transgene (Burrows *et al.*, 1998). These assays demonstrated significantly elevated levels of CRH-BP mRNA and protein in the pituitaries of several lines of transgenic mice (compared to wild-type littermates). This transgenic mouse model therefore tests the role of chronic elevation of secreted CRH-BP specifically in the HPA axis.

Progeny from the two transgenic lines with highest levels of pituitary CRH-BP activity were analyzed to determine the effects of chronic excess pituitary CRH-BP. There were no significant differences in basal corticosterone or ACTH plasma levels between transgenic and nontransgenic animals of either sex (Burrows *et al.*, 1998). In addition, adrenal tissue from transgenic mice exhibited normal morphology (data not shown). A restraint stress protocol was used to assess the ability of the transgenic mice to increase serum ACTH and corticosterone levels in response to stress. In a study of 20 transgenic male mice and their littermates, the transgenic mice exhibited a stress response virtually identical to the nontransgenic mice (Burrows *et al.*, 1998). Female transgenic mice also demonstrated a normal increase in corticosterone following restraint stress (H. Burrows, unpublished data).

These results could be interpreted to suggest that CRH-BP has no role in the HPA axis or that CRH-BP must be expressed in corticotropes to mediate its effect. Alternatively, the normal basal and stressed corticosterone and ACTH levels in mice with chronic excess CRH-BP could be explained by the existence of homeostatic mechanisms designed to maintain balance within the HPA axis. For example, a compensatory increase in CRH levels could result in a normal level of "free" CRH. Alternatively, an increase in another ACTH secretagogue such as arginine-vasopressin (AVP) that potentiates the activity of CRH could normalize ACTH production. To investigate these possibilities, brain sections from unstressed animals were examined by quantitative *in situ* hybridization with CRH and AVP riboprobes. Hypothalamic (PVN) CRH mRNA levels were elevated by 82% in transgenic mice ($P < 0.01$) (Figure 7a) and AVP expression was increased by approximately 35% ($P < 0.05$) in the transgenic mice (Figure 7b) (Burrows *et al.*, 1998). These results indicate that both of these hypothalamic factors are involved in maintaining homeostasis in response to the challenge of chronically elevated pituitary CRH-BP.

Homeostatic mechanisms are important for the survival and well-being of all animals, and the HPA axis is tightly regulated to avoid hyper- and hypoactivity. In response to restraint stress, rodents increase pituitary CRH-BP levels, providing an additional

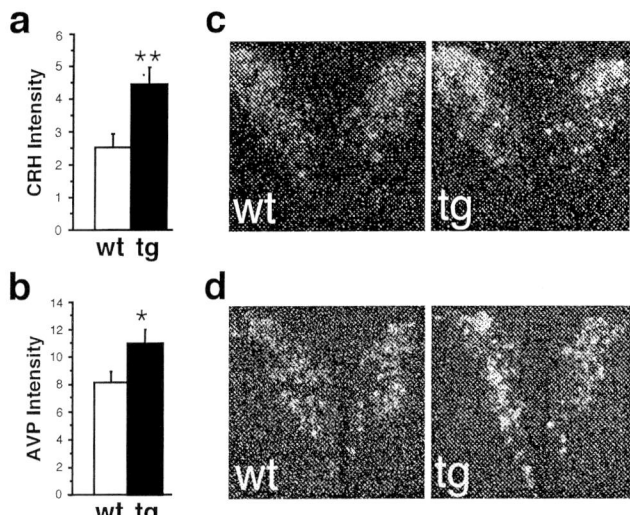

Fig. 7 Increased Basal CRH and AVP Transcripts in Transgenic Mice May Compensate For Excess CRH-BP. (a) Average integrated signal intensity of CRH mRNA is increased by 82% in the PVN of transgenic mice. (b) AVP mRNA levels are also increased by 35% in transgenic mice. (c) Representative photomicrographs of slides hybridized for CRH mRNA illustrate the increased signal present in transgenic brain. (d) Representative photomicrographs of slides demonstrate the increase in AVP expression. (n = 10 slides per genotype, line 6; * $P < 0.05$; ** $P < 0.01$) This figure is taken from Burrows *et al.*, (1998) *J. Clin. Invest.* **101**: 1439–1447.

mechanism for binding CRH and returning the system to homeostasis. In the αCRH-BP transgenic mice, the normal ACTH and corticosterone levels along with elevated CRH and AVP expression demonstrate the ability of the HPA axis to maintain appropriate levels of corticosterone despite the chronic challenge of elevated CRH-BP. The compensatory increases in CRH and AVP in the CRH-BP transgenic mice support the hypothesis that CRH-BP normally acts to blunt the effects of CRH upon the corticotropes. The complexity of the control mechanisms of the HPA axis and the redundancies of the system demonstrate the importance of an intact HPA axis for the animal. The CRH-BP may play an important role in the pituitary to help maintain homeostasis in the mammalian stress axis.

References:

Behan, D.P., Linton, E.A., and Lowry, P.J. (1989). Isolation of the human plasma corticotrophin-releasing factor-binding protein. *Journal of Endocrinology* **122**, 23–31.

Behan, D.P., Potter, E., Lewis, K.A., Jenkins, N.A., Copeland, N., Lowry, P.J., and Vale, W.W. (1993). Cloning and structure of the human corticotrophin releasing factor-binding protein gene (CRHBP) [published erratum appears in *Genomics* 1994 Jan 1;19(1), 198]. *Genomics* **16**, 63–8.

Behan, D.P., Potter, E., Lewis, K.A., Jenkins, N.A., Copeland, N., Lowry, P.J., and Vale, W.W. (1993). Cloning and structure of the human corticotrophin releasing factor-binding protein gene (CRHBP). *Genomics* **16**, 63–68.

Bishop, J.F., and Mouradian, M.M. (1993). Characterization of a corticotropin releasing hormone responsive region in the murine proopiomelanocortin gene. *Molecular and Cellular Endocrinology* **97**, 165–71.

Boutillier, A.L., Monnier, D., Lorang, D., Lundblad, J.R., Roberts, J.L., and Loeffler, J.P. (1995). Corticotropin-releasing hormone stimulates proopiomelanocortin transcription by cFos-dependent and -independent pathways: characterization of an AP1 site in exon 1. *Molecular Endocrinology* **9**, 745–55.

Burrows, H.L., Nakajima, M., Lesh, J.S., Goosens, K.A., Samuelson, L.C., Inui, A., Camper, S.A., and Seasholtz, A.F. (1998). Excess corticotropin-releasing hormone-binding protein in the hypothalamic-pituitary-adrenal axis in transgenic mice. *Journal of Clinical Investigation* **101**, 1–9.

Chalmers, D.T., Lovenberg, T.W., Grigoriadis, D.E., Behan, D.P., and De Souza, E.B. (1996). Corticotrophin-releasing factor receptors: from molecular biology to drug design. *Trends in Pharmacological Science* **17**, 166–172.

Cortright, D.N., Goosens, K.A., Lesh, J.S., and Seasholtz, A.F. (1997). Isolation and characterization of the rat corticotropin-releasing hormone-binding protein gene: transcriptional regulation by cyclic adenosine monophosphate and CRH. *Endocrinology* **138**, 2098–2108.

Cortright, D.N., Nicoletti, A., and Seasholtz, A. (1995). Molecular and biochemical characterization of the mouse brain corticotropin-releasing hormone-binding protein. *Molecular and Cellular Endocrinology* **111**, 147–57.

Faisst, S., and Meyer, S. (1992). Compilation of vertebrate-encoded transcription factors. *Nucleic Acids Research* **20**, 3–26.

Jin, W.D., Boutillier, A.L., Glucksman, M.J., Salton, S.R.J., Loeffler, J.P., and Roberts, J.L. (1994). Characterization of a CRH-responsive element in the rat POMC gene promoter and molecular cloning of its binding protein. *Molecular Endocrinology* **8**, 1377–1388.

Kendall, S.K., Gordon, D.F., Birkmeier, T.S., Petrey, D., Sarapura, V.D., O'Shea, K.S., Wood, W.M., Lloyd, R.V., Ridgway, E.C., and Camper, S.A. (1994). Enhancer-mediated high level expression of mouse pituitary glycoprotein hormone alpha-subunit transgene in thyrotropes, gonadotropes, and developing pituitary gland. *Molecular Endocrinology* **8**, 1420–1433.

McClennen, S., Cortright, D., and Seasholtz, A. (1998). Regulation of pituitary corticotropin-releasing hormone-binding protein messenger ribonucleic acid levels by restraint stress and adrenalectomy. *Endocrinology* **139**, 4435–4441.

Orth, D.N., and Mount, C.D. (1987). Specific high-affinity binding protein for human corticotropin-releasing hormone in normal human plasma. *Biochemical and Biophysical Research Communications* **143**, 411–417.

Owens, M.J., and Nemeroff, C.B. (1991). Physiology and pharmacology of corticotropin- releasing factor. *Pharmacological Review* **43**, 425–73.

Potter, E., Behan, D.P., Fischer, W.H., Linton, E.A., Lowry, P.J., and Vale, W.W. (1991). Cloning and characterization of the cDNAs for human and rat corticotropin releasing factor-binding proteins. *Nature* **349**, 423–426.

Potter, E., Behan, D.P., Linton, E.A., Lowry, P.J., Sawchenko, P.E., and Vale, W.W. (1992). The central distribution of a corticotropin-releasing factor (CRF)- binding protein predicts multiple sites and modes of interaction with CRF. *Proceedings of the National Academy of Sciences USA* **89**, 4192–6.

Potter, E., Sutton, S., Donaldson, C., Chen, R., Perrin, M., Lewis, K., Sawchenko, P.E., and Vale, W. (1994). Distribution of corticotropin-releasing factor receptor mRNA expression in the rat brain and pituitary. *Proceedings of the National Academy of Sciences USA* **91**, 8777–8781.

Vaughan, J., Donaldson, C., Bittencourt, J., Perrin, M.H., Lewis, K., Sutton, S., Chan, R., Turnbull, A.V., Lovejoy, D., Rivier, C., Rivier, J., Sawchenko, P.E., and Vale, W. (1995). Urocortin, a mammalian neuropeptide related to fish urotensin I and to corticotropin-releasing factor. *Nature* **378**, 287–292.

25. Gene Expression of the Cardiac Renin-Angiotensin Components and its Modulation by Stress

O. Križanová, D. Jurkovičová and R. Kvetňanský

Introduction

The renin-angiotensin system (RAS) is a metabolic pathway producing a biologically active angiotensin II (Figure 1). Angiotensin II (AII), as a potent modulator of many physiological functions, is produced not only in the circulation, but also locally in some tissues, e.g. kidney, heart, brain, etc. (Saavedra 1992, Phillips *et al.*, 1993). Vasoconstrictive effects of the circulating angiotensin II are very strong, since within 20 min. after intravenous administration it increases arterial pressure by direct vasoconstriction of arterioles (Schmermund *et al.*, 1999). Local angiotensin II exerts both autocrine and paracrine effects. In the heart, angiotensin II is considered to be an important factor in cardiovascular pathology, such as cardiac left ventricular hypertrophy and fibrosis, vascular media hypertrophy and structural alterations of the heart, such as postinfarct remodeling (for review see Krizanova, 1998). A local RAS in the heart is often invoked to explain the beneficial effects of ACE inhibitors in heart failure (Danser *et al.*, 1997). In addition, increased renin mRNA expression in the border zone of the infarcted left ventricle suggested a role for intracardiac angiotensin II in infarct healing (Passier *et al.*, 1996).

Renin is a key enzyme of the renin-angiotensin system and changes in the biosynthesis and renal secretion of renin are normally the prime and apparently the sole determinants of a change in plasma AII formation. Renin cleaves the substrate angiotensinogen to the decapeptide angiotensin I, which is subsequently cleaved by angiotensin converting enzyme (ACE) to biologically active angiotensin II (Figure 1). Angiotensin II acts through AT-receptors, either AT_1, or AT_2 types. AT_1 receptor mediates many important cardiovascular responses, including inotropic effect, vasoconstriction, vascular and cardiac remodeling (e.g. proliferation, hypertrophy, etc.) and cell survival/cell death (Sadoshima *et al.*, 1993). AT_2 receptor is probably involved in cellular growth, differentiation and adhesion (Chung *et al.*, 1998). AT_2 receptors can be reexpressed under pathophysiological conditions involving tissue remodeling or repair, such as in vascular neointima formation, postmyocardial infarction, as well as apoptosis (Yamada *et al.*, 1996).

Several stimuli are known to regulate the production of angiotensin II. Among them, stress is one of the most important ones, since it is assumed to be involved in the development of some diseases, especially of the cardiovascular system. Numerous studies have demonstrated differential neuroendocrine responses during exposure to different stressors (Pacak *et al.*, 1998). Moreover, during exposure to different stressors, sympathetic responses exhibit pronounced heterogeneity among different organs (Victor *et al.*, 1989), and the response patterns are dependent on the duration and intensity of the stressor (Morita and Vatner, 1985; Schadt and Ludbrook, 1991). Immobilization (Kvetňanský and Mikulaj, 1970) is known to be one of the most potent stressors, since it activates both pathways of the sympathoadrenal system. We have already shown that immobilization stress increased both plasma (Figure 2A) and kidney (Figure 2B) renin activities. When animals were immobilized for 2 hours, plasma renin activity (PRA) showed a peak increase after 30 min. of immobilization, while kidney renin activity (KRA) peaked after 2 hours of immobilization. These results are in a good agreement with those previously published by Jindra and Kvetňanský (1982) and might point to different patterns of regulation of

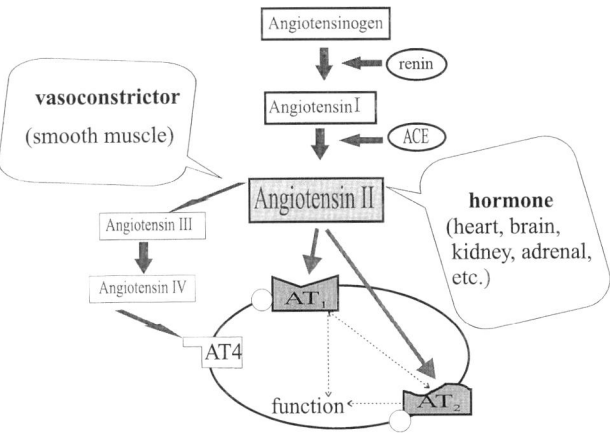

Fig. 1 The renin-angiotensin pathway. Angiotensin II is formed from its substrate angiotensinogen by two enzymes – renin and angiotensin converting enzyme (ACE). Its signaling is performed through the AT receptors – AT_1 receptor (AT_1) and AT_2-receptor (AT_2). Angiotensin II is degraded to smaller peptides, angiotensin III and IV, which have their own receptors (e.g. AT4).

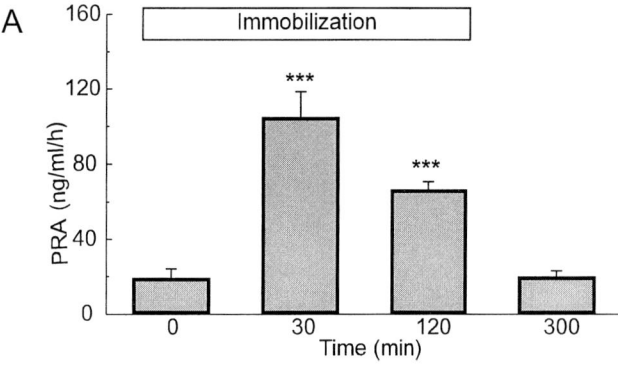

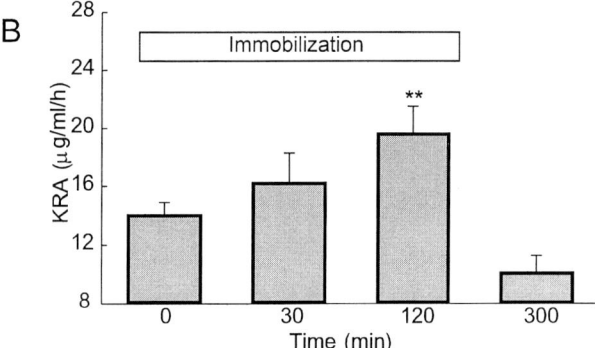

Fig. 2 The effect of immobilization stress on plasma renin activity (PRA; A) and kidney renin activity (KRA; B). The peak of PRA (A) appeared after 30 min of immobilization with subsequent decline in PRA after 120 min of immobilization stress and return to the control levels three hours after the end of stimulus. Single immobilization revealed a maximal increase in KRA (B) after 120 min of immobilization, although an indication of the increase appeared after 30 min of immobilization. Each value is expressed as a mean ± S.E.M. and represents an average of 7–8 measurements. Significance was calculated by ANOVA (*** – p < 0.001, ** – p < 0.01).

the same system by the same stimulus in different tissues. Increases in PRA were also observed during the chronic physical-psychological stress (Aguilera *et al.*, 1995), which was, at least in part, due to increased renin expression in the kidney as a result of stress-induced sympathetic stimulation. Since during immobilization stress PRA was increased much more quickly than KRA, we propose that besides renal renin, there might also be additional factors involved in PRA increases due to immobilization stress.

Cardiac RAS and Immobilization Stress

Different opinions about the existence of cardiac renin appear in the literature. Renin gene expression has been documented in the heart and brain of several species, including rodents and primates (Jin *et al.*, 1988, Chernin *et al.*, 1990). Cardiac expression of renin and angiotensinogen genes suggests that RAS may be involved in growth and development of the myocardium (Chernin *et al.*, 1990). Renin and angiotensinogen gene expression

was detected in cardiac atria and ventricles by Northern blot and hybridization technique (Linz *et al.*, 1989). On the other hand, Lou and coworkers (1993) stated that ventricular tissue does not express the renin gene in normal or pathological states.

Up to now, it is quite difficult to measure the components of the cardiac RAS, because of the contamination with circulating RAS. The presence of mRNA for the appropriate protein is the first, but not the sole requirement for its presence in certain tissues. Since mRNA content for the individual RAS components in cardiac tissue seems to be a good marker to distinguish between circulating and local RAS, using reverse transcription and subsequent polymerase chain reaction we amplified the fragments for renin, angiotensinogen, ACE and AT$_1$ receptors (Figure 3). For renin, we used kidney (KR) as a positive control, since kidney produces renin for the circulation. We have found the signal for all RAS components tested in the whole heart (HR), right and left atria (RA, LA), right and left ventricles (RV, LV) and also in isolated myocytes (MC). As a negative control was performed the PCR from myocyte RNA (MCC), proving that the signal was not an artifact from DNA contamination. These results point to the real existence of the cardiac RAS and support a number of previously published papers on this topic (Chernin *et al.*, 1990; Paul and Schunkert, 1992; Studer *et al.*, 1994; Sawa *et al.*, 1992; Jin *et al.*, 1988).

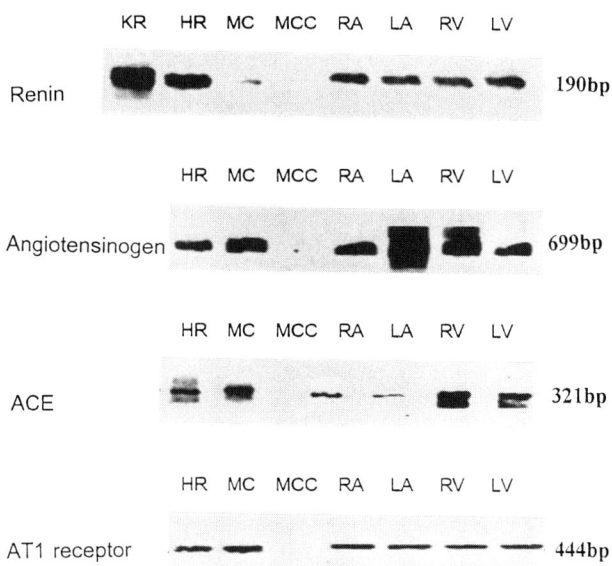

Fig. 3 mRNA amounts of renin, angiotensinogen, ACE and AT$_1$ receptors in the whole heart (HR), isolated and purified myocytes (MC), left (LV) and right (RV) ventricles, left (LA) and right (RA) atria of Sprague-Dawley rats. For renin, mRNA levels in kidney (KR) were used as a positive control. In order to determine that the signal in myocytes was due to mRNA expression of RAS compounds and not due to genomic DNA contamination, a negative control (MCC) was performed, omitting the reverse transcription. Quantification was done relative to β-actin and GAPDH (not shown).

As was already mentioned, immobilization stress affects plasma and kidney renin activities (Figure 2). From these observations the question arises, whether the components of cardiac RAS are vulnerable to immobilization stress at the level of mRNA. The comparative evaluation of gene expression of individual components relative to an appropriate housekeeper seems to be the most suitable tool to measure changes in this system due to different stimuli. Using reverse transcription with subsequent polymerase chain reaction (RT-PCR) we were able to study the effect of immobilization stress on mRNA levels of angiotensinogen, renin and AT_1 receptors in rat heart. A single (1^*IMO) and repeated immobilizations for seven (7^*IMO) and 42 times (42^*IMO) were performed daily for two hours on normotensive Sprague-Dawley rats (Figure 4). Animals were sacrificed either immediately or 24 hours after the last immobilization. A significant increase in mRNA levels for angiotensinogen, renin and AT_1 receptors was observed after a single immobilization (Figure 4). However, after 24 hours rest a decrease to original levels was observed in all RAS components in the heart. After repeated immobilization for seven and forty-two times, a similar upregulation in these components occurred, with no return to control levels after 24 hours. The mRNA levels for housekeepers' β-actin and GAPDH were not changed by immobilization stress. Although results obtained by RT-PCR cannot give a quantitative estimation of individual mRNA levels, they are suitable enough for comparative evaluation between control and immobilized groups. Therefore, this result might point to the adaptation of the cardiac RAS to immobilization stress, on the contrary to the PRA, where no adaptation was found even after 42 immobilizations. Regression analysis between the mRNA levels of renin and AT_1 receptors showed a significant positive correlation (Figure 5; $R = 0.88$, $p = 0.035$). No positive correlation was found between renin and angiotensinogen mRNAs ($p = 0.100$), or AT_1 receptors and angiotensinogen mRNAs ($p = 0.537$). Thus, renin and AT_1 receptor mRNAs in the heart are probably similarly affected by immobilization stress.

Gene Expression of RAS Components in Genetically Hypertensive Rat Strains.

Angiotensin II and hypertension are intimately connected. However, the pathophysiology of blood pressure includes not only increases in systolic and diastolic blood pressure, but also hypertrophic structural and functional changes in the heart and other tissues (Quadri et al., 1998). Therefore, we compared the effect of immobilization stress on the mRNA levels of RAS components in hypertensive animals.

Spontaneously hypertensive rats (SHR) are the most frequently used animal model of essential hypertension. It was already demonstrated that generation of circulating angiotensin II and expression of AT recep-

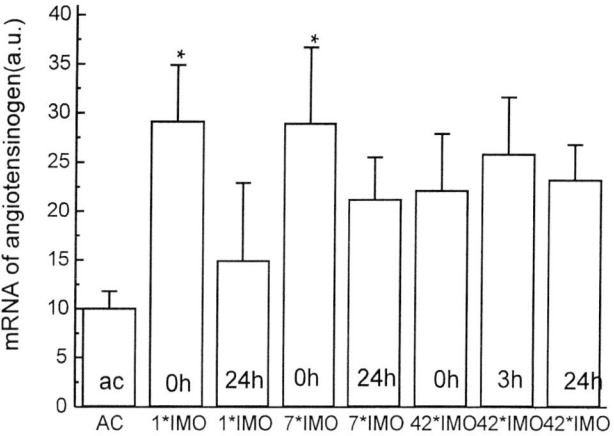

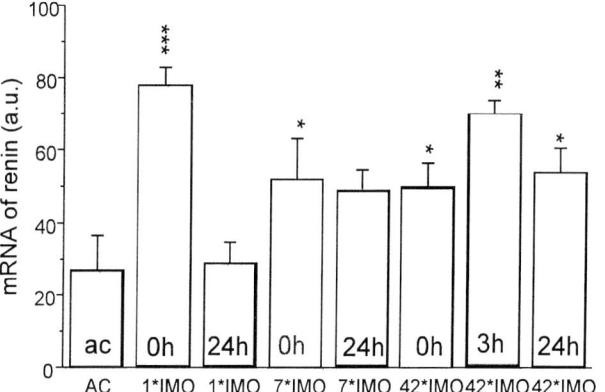

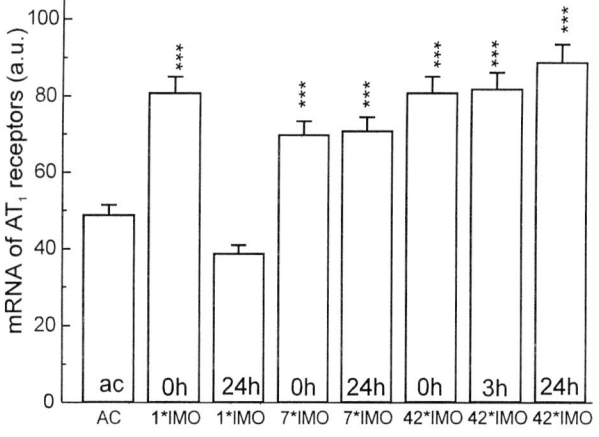

Fig. 4 The effect of immobilization stress on cardiac angiotensinogen, renin and AT_1 receptor gene expression in Sprague-Dawley rats. The mRNA of these components was detected by RT-PCR and quantified relatively to two housekeeping genes, β-actin and GAPDH. Each value is expressed as a mean ± S.E.M. and represents an average of 7–8 measurements. Significance was calculated by ANOVA ($^{***} - p < 0.001$; $^{**} - p < 0.01$; $^* - p < 0.05$). Immobilization upregulated mRNA for all these components after a single immobilization (1^*IMO) compared to the control group (AC), with subsequent decrease to control levels after a 24-hour rest. After 7 (7^*IMO) and 42 immobilizations (42^*IMO), mRNA for angiotensinogen, renin and AT_1 receptors remained increased even after a 24 hour rest.

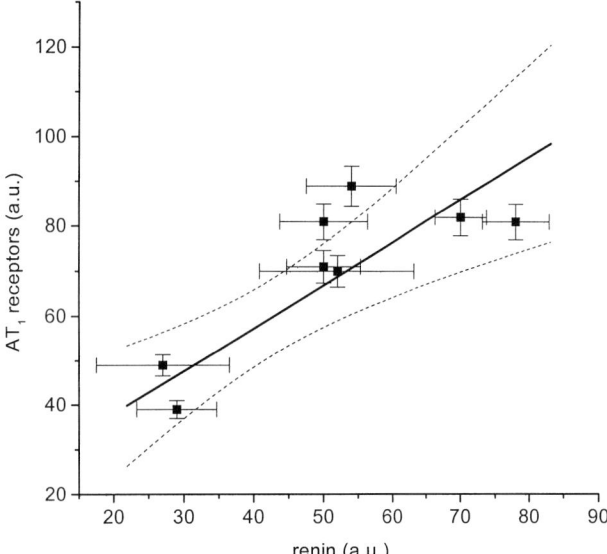

Fig. 5 Regression analysis between renin and AT_1 receptor mRNA levels. This analysis shows a significant positive correlation (R = 0.880, p = 0.035).

tors and angiotensinogen is higher in SHR than in genetically related normotensive rats (Yongue *et al.*, 1991). Hereditary triglyceridemic (HTG) rats were originally developed as a genetic model of hypertriglyceridemia. Nevertheless, they also exhibit insulin resistance, glucose intolerance and hypertension (Klimes *et al.*, 1995).

It is clear that hypertension is known to be tightly related to some cardiac diseases, e.g. left ventricular hypertrophy. The development of hypertensive left ventricular hypertrophy is an early and important finding that may have direct pathophysiological implications on the progression from early hypertension to cardiovascular death (Kahan, 1998). Comparison of the components of the cardiac renin angiotensin system between normotensive and hypertensive rats may contribute to an understanding of mechanisms leading to development of the pathophysiological state of the heart.

Our results clearly demonstrated that expression of cardiac renin is significantly higher in both hypertensive (SHR, HTG) strains when compared to normotensive WKY and LEW controls (Figure 6). No significant differences in gene expression of cardiac angiotensinogen between hypertensive and normotensive strains studied were observed. Increased mRNA for renin in ventricles of SHR rats compared to WKY was also observed by Sano *et al.* (1998), who suggested a role for cardiac RAS in collagen accumulation during cardiac hypertrophy. This is in agreement with the proposal that not the circulating, but the activated tissue RAS can have a detrimental effect on cardiac function (Pinto *et al.*, 1996). mRNA for ACE was significantly increased only in HTG, but not in SHR hearts, compared to corresponding controls (Figure 6). Increased ACE activity and mRNA expression were observed in rat hearts with pressure overload, in infarcted rat hearts

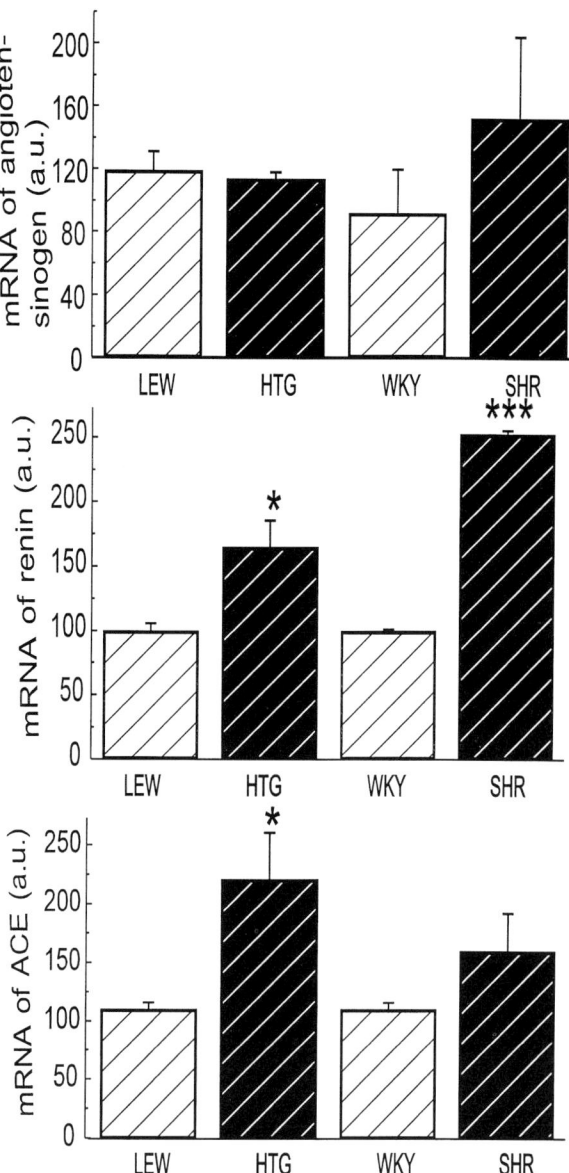

Fig. 6 Comparison of mRNA levels for cardiac angiotensinogen, renin and AT_1 receptors in normotensive (WKY and LEW) and genetically hypertensive (HTG and SHR) rats. mRNA of these components was detected by RT-PCR and quantified relative to two housekeepers, β-actin and GAPDH. Each value is expressed as a mean ± S.E.M. and represents an average of 7–8 measurements. Significance was calculated by ANOVA (*** – p < 0.001; * – p < 0.05). mRNA for the cardiac angiotensinogen did not show any significant differences among normotensive and hypertensive strains. mRNA for the cardiac renin was significantly upregulated in both hypertensive strains compared to the normotensive ones. mRNA for ACE was significantly increased in HTG hearts, but not in SHR, compared to their normotensive controls.

and in volume overload induced hypertrophy (for review see Lijnen and Petrov, 1999).

Although immobilization significantly upregulates mRNA for renin, angiotensinogen and AT1 receptors in hearts of normotensive rats, no response in any of these compounds was observed in the hearts of spontaneously hypertensive animals (Figure 7). This result

points to the fact that regulation of the cardiac renin angiotensin system in hypertensive animals differs from that in normotensive animals

Conclusion

Physiological relevance of increased mRNA levels of cardiac RAS during immobilization stress remains to be elucidated. However, this upregulation might result in increased levels of angiotensin II in the rat heart. Angiotensin II is known to increase intracellular calcium concentration in the heart, in rat portal vein myocytes and in many other cells. Therefore, increased angiotensin II production might result in elevated levels of intracellular calcium in the myocardium. Nevertheless, this proposal remains to be tested empirically.

Acknowledgement

This work was supported by grant VEGA 2/4128. The authors wish to thank to Dr Zahradnikova for helpful comments and Mrs. Zacikova for help with manuscript preparation.

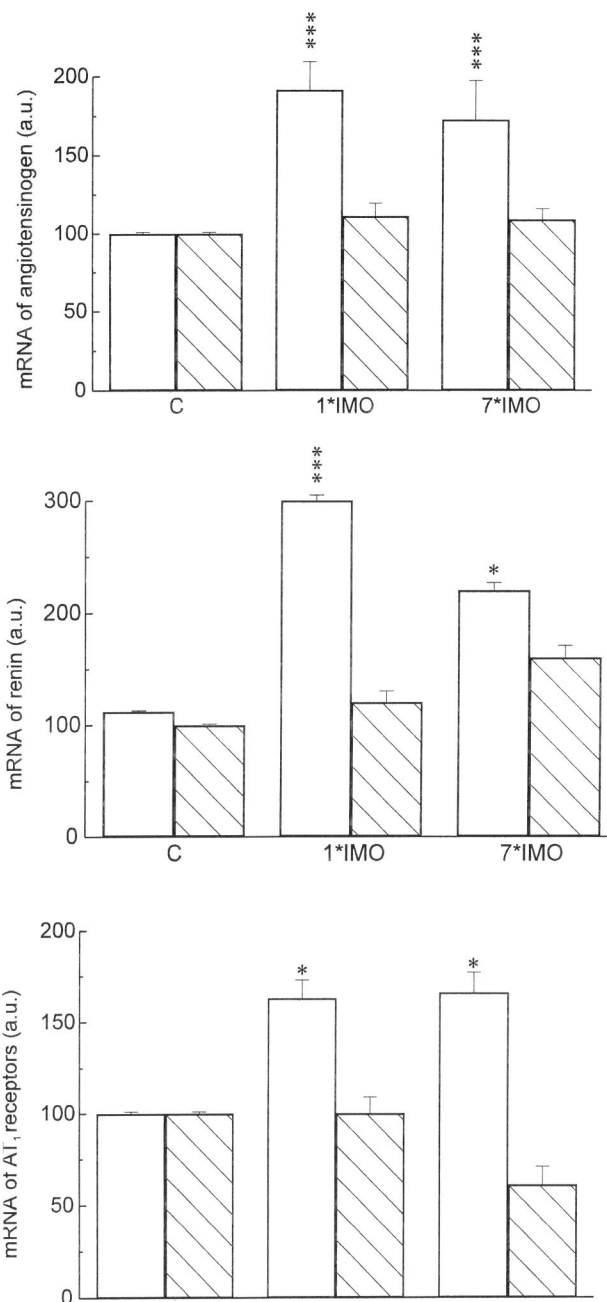

Fig. 7 The effect of immobilization stress on mRNA levels of cardiac angiotensinogen, renin and AT_1 receptor in normotensive (empty columns) and hypertensive (hatched columns) rats. mRNA of these components was detected by RT-PCR and quantified relatively to two housekeepers, β-actin and GAPDH. Each value is expressed as a mean ± S.E.M. and represents an average of 7–8 measurements. Significance was calculated by ANOVA (*** – $p < 0.001$; * – $p < 0.05$). Although immobilization stimulus upregulated mRNA for all these components after the single (1*IMO) and repeated (7*IMO) immobilization compared to the control group (C), no effect was observed in SHR rats.

References

Aguilera, G., Kiss, A., Luo, X. and Akbasak, B.S. (1995). The renin angiotensin system and the stress response. *Annals of the New York Academy of Sciences* **771**, 173–186.

Chernin, M.I., Candia, A.F., Stark, L.L., Aceto, J.F. and Baker, K.M. (1990). Fetal expression of renin, angiotensinogen, and atriopeptin genes in chick heart. *Clinical and Experimental Hypertension [A]* **12**, 617–29.

Chung, O., Kuhl, H., Stoll, M., and Unger, T. (1998). Physiological and pharmacological implications of AT1 versus AT2 receptors. *Kidney International* **67**, S95–S99.

Danser, A.H., van Kats, J.P., Verdouw, P.D. and Schalekamp, M.A. (1997). Evidence for the existence of a functional cardiac renin-angiotensin system in humans. *Circulation* **96**, 3795–3796.

Jin, M., Wilhelm, M.J., Lang, R.E., Unger, T., Lindpaintner, K. and Ganten, D. (1988). Endogenous tissue renin-angiotensin systems. From molecular biology to therapy. *American Journal of Medicine* **84**, 28–36.

Jindra, A. and Kvetňanský, R. (1982). Stress-induced activation of inactive renin. *Journal of Biological Chemistry* **257**, 5997–5999.

Kahan, T. (1998). The importance of left ventricular hypertrophy in human hypertension. *Journal of Hypertension* **16**, S23–S29.

Klimes, I., Sebokova, E., Vrana, A., Stolba, P., Kazdova, L., Kunes, J., Bohov, P., Fickova, M., Zicha, J., Raucinova, D. and Kren, V. (1995). The hereditary hypertriglyceridemic rat, a new animal model of the plurimetabolic syndrome. *in: Lessons from animal diabetes V*. Ed. E. Shafrir, 271–283.

Krizanova, O. (1998). Renin-angiotensin system and its role in myocardial diseases. *Experimental Clinical Cardiology* **3**, 144–150.

Kvetňanský, R. and Mikulaj, L. (1970). Adrenal and urinary catecholamines in rats during adaptation to repeated immobilization stress. *Endocrinology* **87**, 738–743.

Lijnen, P. and Petrov, V. (1999). Renin-angiotensin system, hypertrophy and gene expression in cardiac myocytes. *Journal of Molecular and Cellular Cardiology* **31**, 949–970.

Linz, W., Scholkens, B., Lindpaintner, K. and Ganten, D. (1989). Cardiac renin-angiotensin system. *American Journal of Hypertension* **2**, 307–10.

Lou, Y.K., Robinson, B.G. and Morris, B.J. (1993). Renin messenger RNA, detected by polymerase chain reaction, can be switched on in rat atrium. *Journal of Hypertension* **11**, 237–243.

Morita, H. and Vatner, S.F. (1985). Effects of volume expansion on renal nerve activity, renal blood flow and sodium and water excretion in conscious dogs. *Circulation Research* **57**, 788–793.

Pacak, K., Palkovits, M., Yadid, G., Kvetňanský, R., Kopin, I.J. and Goldstein, D.S. (1998). Heterogeneous neurochemical responses to different stressors: a test of Selye's doctrine of nonspecificity. *American Journal of Physiology* **275**, R-1247–1255.

Passier, R.C., Smits, J.F., Verluyten, M.J. and Daemen, M.J. (1996). Expression and localization of renin and angiotensinogen in rat heart after myocardial infarction. *American Journal of Physiology* **271**, H1040–H1048.

Paul, M. and Schunkert, H. (1992). Detection of angiotensin converting enzyme mRNA in the rat heart by use of the polymerase chain reaction (PCR). *Agents and Actions Suppl.* **38**, 384–391.

Phillips, M.I., Speakman, E.A. and Kimura, B. (1993). Levels of angiotensin and molecular biology of the tissue renin angiotensin systems. *Regulatory Peptides* **43**, 1–20.

Pinto Y.M., Buikema H., van Gilst W.H. and Lie K.I. (1996). Activated tissue renin-angiotensin systems add to the progression of heart failure. *Basic Research in Cardiology* **91**, 85–90.

Quadri, L., Gobbini, M. and Monti, L. (1998). Recent advances in antihypertensive therapy. *Current Pharmaceutical Design* **4**, 489–512.

Saavedra, J.M. (1992). Brain and pituitary angiotensin. *Endocrine Review* **13**, 329–80.

Sadoshima, J. and Izumo, S. (1993). Molecular characterization of angiotensin II–induced hypertrophy of cardiac myocytes and hyperplasia of cardiac fibroblasts. Critical role of the AT1 receptor subtype. *Circulation Research* **73**, 413–23.

Sano, H., Okamoto, H., Kitabatake, A., Iizuka, K., Murakami, T. and Kawaguchi, H. (1998). Increased mRNA expression of cardiac renin-angiotensin system and collagen synthesis in spontaneously hypertensive rats. *Molecular and Cellular Biochemistry* **178**, 51–8.

Sawa, H., Tokuchi, F., Mochizuki, N., Endo, Y., Furuta, Y., Shinohara, T., Takada, A., Kawaguchi, H., Yasuda, H. and Nagashima, K. (1992). Expression of the angiotensinogen gene and localization of its protein in the human heart. *Circulation* **86**, 138–46.

Schadt, J.C. and Ludbrook, J. (1991). Hemodynamic and neurohormonal responses to acute hypovolemia in conscious mammals. *American Journal of Physiology* **260**, H305–H318.

Schmermund, A., Lerman, L.O., Ritman, E.L., Rumberger, J.A. (1999). Cardiac production of angiotensin II and its pharmacologic inhibition: effects on the coronary circulation. *Mayo Clinical Proceedings* **74**, 503–13.

Studer, R., Reinecke, H., Muller, B., Holtz, J., Just, H., Drexler, H. (1994). Increased angiotensin I converting enzyme gene expression in the failing human heart. Quantification by competitive RNA polymerase chain reaction. *Journal of Clinical Investigation* **94**, 301–310.

Victor, R.G., Thoren, P., Morgan, D.A., Mark, A.L. (1989). Differential control of adrenal and renal sympathetic nerve activity during hemorrhagic hypotension in rats. *Circulation Research* **64**, 686–694.

Yamada, T., Horiuchi, M., Dzau, V.J. (1996). Angiotensin II type 2 receptor mediates programmed cell death. *Proceedings of the National Academy of Sciences USA* **93**, 156–60.

Yongue, B.G., Angulo, J.A., McEwen, B.S., Myers, M.M. (1991). Brain and liver angiotensinogen messenger RNA in genetic hypertensive and normotensive rats. *Hypertension* **17**, 485–91.

26. Association Between Brain COMT Gene Expression and Aggressive Experience in Daily Agonistic Confrontations in Male Mice

M.L. Filipenko, A.G. Beilina, O.V. Alekseyenko, O.A. Timofeeva, D.F. Avgustinovich and N.N. Kudryavtseva

Introduction

An association between catecholaminergic systems and aggressive behavior is widely accepted. Pharmacological data (Poshivalov, 1986; Bell and Hepper, 1987; Navarro *et al.*, 1993), concerning inhibition of catecholaminergic activity, has revealed its suppressing effects on aggression in different models and in various species: injection of the synthesis enzyme blocker of catecholamines or antagonists of dopamine receptors decreased aggression in animals. Catecholamine enhancing drugs (precursors, receptor agonists in low doses), increased the tone of catecholaminergic systems and facilitated natural mechanisms that regulate aggression. It has also been reported that the execution of aggressive behavior in resident mice that attacked and threatened an intruder results in increases of dopamine (DA) turnover in the nucleus accumbens (Haney *et al.*, 1990). In current *in vivo* microdialysis studies on aggressive rats, large and long-lasting increases in DA have been measured in the nucleus accumbens immediately before, during, and after episodes of threat and attack toward an intruding opponent (Miczek and Tornatzky, 1996). Our previous data showed that chronic experience of aggression in daily agonistic confrontations is accompanied by total activation of dopaminergic metabolism through formation of DOPAC in different brain areas of aggressive winners (Kudryavtseva and Bakshtanovskaya, 1991).

Catechol-O-methyltransferase (COMT) is involved in the metabolic degradation of catecholamines by O-methylation. Considering the role of COMT in dopaminergic metabolism and the role of dopaminergic pathways in the mechanisms of aggression, it was suggested that COMT is involved in the regulation of agonistic behavior. Moreover, COMT enzyme activity is associated with violent behavior in patients with schizophrenia and schizoaffective disorder (Strous *et al.*, 1997; Lachman, 1998).

The aim of this study was to examine levels of COMT mRNA in the midbrain with pons (containing the bulk of monoaminergic cell bodies) in male mice with repeated experience of aggression in daily agonistic confrontations (winners). For comparison, COMT mRNA levels were determined in male mice with experience of social defeats (losers) and in controls (5 days of individual housing) with no experience with aggression.

Materials and Methods

Animals

Adult male CBA/Lac mice, maintained at the Institute of Cytology and Genetics, were used. The animals were housed under standard vivarium conditions and a natural light regimen, food and water were available *ad libitum*. Males were weaned at the age of one month and housed in one-litter groups of 8–10 individuals in plastic $36 \times 23 \times 12$ cm cages. Mice used in the experiments were 10–12 weeks of age.

Generation of Aggressive and Submissive Types of Behavior in Male Mice

To generate aggressive and submissive behavior in male mice, the sensory contact model (Kudryavtseva, 1991) was used. Males were weighed and caged individually for 5 days to remove group effects. Animals of approximately the same weight were then placed in pairs in steel cages ($28 \times 14 \times 10$ cm) divided into halves by a perforated transparent partition permitting them to see, hear and smell their neighbor, while preventing physical contact. After two days of adaptation to the housing conditions and sensory contact, testing commenced. Every afternoon (14:00–17:00 p.m.), the steel cover of the cage was replaced by a transparent one and, 5 min later (the period necessary for individuals' activation) the partition was removed for 10 min to allow agonistic interaction. Undoubted superiority of one of the partners was evident within 2 or 3 tests in daily social encounters with the same opponent. One member of each pair was seen to attack, bite, and chase the other, who displayed only defensive behavior (sideways, upright postures, and also 'on the back' or 'freezing') during the test. As a rule, in our experiments, aggressive confrontations between males were discontinued by lowering the partition if the aggression lasted more then 3 minutes. Then, every day after the test, each defeated male of 1 pair was paired with a winning member of another pair behind the partition in an unfamiliar cage. The aggressive males remained in their own compartments. This procedure resulted in equal numbers of winners and losers. Three experimental groups were studied:

Winners – aggressive males that were victorious in 10 daily agonistic confrontations;

Losers – submissive males with repeated experience of defeats in 10 daily agonistic confrontations;

Control – males housed individually for 5 days. They were thought to be best as intact controls, because, in this case, the submissiveness of grouped males would be removed, and the repeated experience of aggression would not yet be acquired (Kudryavtseva, 1991).

In the 24 hours after the last agonistic confrontation, the animals were decapitated; their brains were removed and chilled rapidly on ice. The midbrains with pons, which contain monoaminergic neurons synthesizing DA, NA and 5-HT and their key enzymes of synthesis and catabolism, were dissected, rapidly frozen in liquid nitrogen and stored at –70°C until assayed.

DNA Competitors Used for RT-Competitive PCR

Synthetic DNA competitors were prepared by PCR amplification of a few DNA fragments from unrelated template DNA by including a single initial cycle with a low stringency annealing step using COMT and β-actine primers as described (Carrasco et al., 1997). Briefly, for each gene three PCR were performed using T4 phage DNA as a template. Of these three reactions, two were performed with the primer U or R alone and in the third one both primers were added (for β-actine – ACT-U: CGGAACCGCTCATTGCC and ACT-R: CGGAACCGCTCATTGCC; or COMT – COMT-U: TCCACAACCTGCTCATGGGT and COMT-R: ACATCGTACTTCTTCTTCAGCTGG).

The temperature cycles were (95°C, 1 min; 37°C, 3 min; 72°C, 1 min) and then 35 × (94°C, 1 min; 57°C, 0.5 min for β-actine or 60°C for COMT; 72°C, 1 min). After gel electrophoresis, the DNA fragments with appropriate size amplified from two primers were selected as DNA competitors. DNA fragments were re-amplified, gel purified and diluted for subsequent PCR. Appropriate dilution of DNA competitors was determined by a test titration with a wide range of the competitor amounts for rough estimation of the competition equivalence point.

Determination of Brain mRNA Level

Total RNA was prepared from pooled samples (3-5 midbrains) for each experimental group as described (Chattopadhyay et al., 1993). The RNA was re-suspended in RNAse-free water and treated with RNAse-free DNAse. The samples were then treated with phenol/chloroform (1:1) and precipitated. The quality of RNA was evaluated by electrophoresis on a 1% agarose gel. The amount of RNA in each sample was determined spectrophotometrically by absorption at 260 nm. Samples were stored at –70°C until used.

1 μg or 300 ng of total RNA per reaction were converted to cDNA at 42°C for 1 h using an oligo-dT primer and AMV reverse transcriptase in 20-μl reaction volume. The PCR reaction mixture consisted of 1 μl cDNA, 1.4 μl of DNA competitor, 1 μM of each primer, 0.2 units of Taq-pol, 0.2 mM dNTPs in 10 μl of buffer containing 67 mM Tris-HCl (pH 8.9), 16 mM $(NH4)_2SO_4$, 1.5 mM $MgCl_2$, 0.05% Tween-20. The number of PCR cycles was optimized for each primer set. Aliquots (1 μl) of cDNA were subjected to varying numbers of amplification cycles. Amplification was exponential up to 29 cycles for the β-actine primers and up to 36 cycles for COMT primers. The amplification products were resolved on a 2% agarose gel, stained with ethidium bromide, and photographed. For each RNA sample two reverse transcription reactions were accomplished. Subsequently all RT reactions went through repeated competitive PCR assays.

Statistics

Negative films of ethidium bromide-stained gels were digitized into gray-scale images. Analysis was performed using the public domain Scion Image program (Web Site: HYPERLINK http://www.scioncorp.com). The competitor/target band density ratio of the competitor was plotted versus the initial amount of the competitor. The equivalence point (log ratio = 0), where the initial amount of target cDNA corresponds to the initial amount of the competitor, was detected using linear regression (STATISTICA program). For each RNA sample the mean COMT RNA level was calculated in four separate assays. The COMT mRNA levels for the winners and losers were expressed in relation to the control, which was set at 100%.

Results

Since RT-competitive PCR allows the quantification of very low levels of mRNA, requiring only minute amounts of starting material (Auboeuf and Hubert, 1997), it is a valid method for investigating the regulation of gene expression, especially in brain tissue where only small samples are available. Figure 1 shows a typical RT-competitive PCR assay of target RNA. The logarithm of the band density ratio plotted versus the logarithm of the initial amount of the competitor gave a linear titration curve (r > 0.95). The variation of β-actine and COMT, determined in repeated measurements of the same RNA sample, was below 20%, confirming that our quantitative PCR was accurate and reliable.

In three independent experiments there was variability in COMT mRNA levels between samples within the same experimental group which is shown in Table 1. Nevertheless a significant decrease in COMT mRNA content (6.7 time – in first experiment, 3.25 time – in second experiment and 2.3 time – in third experiment) was found in the midbrain of the winners in comparison with the control level. The mean value in the three experiments was 30.4% for the winners. The COMT mRNA level in the losers was similar to controls in

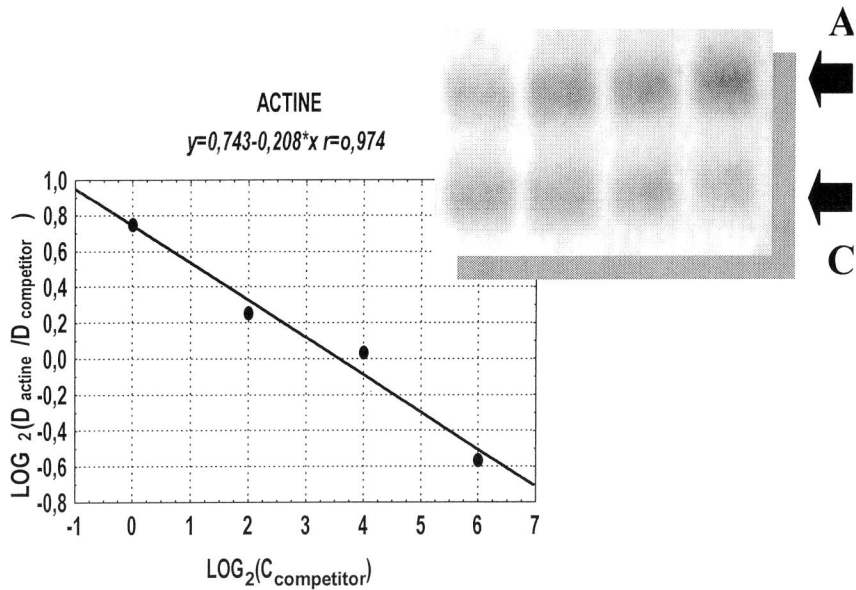

ACTINE

$y=0,743-0,208*x$ $r=0,974$

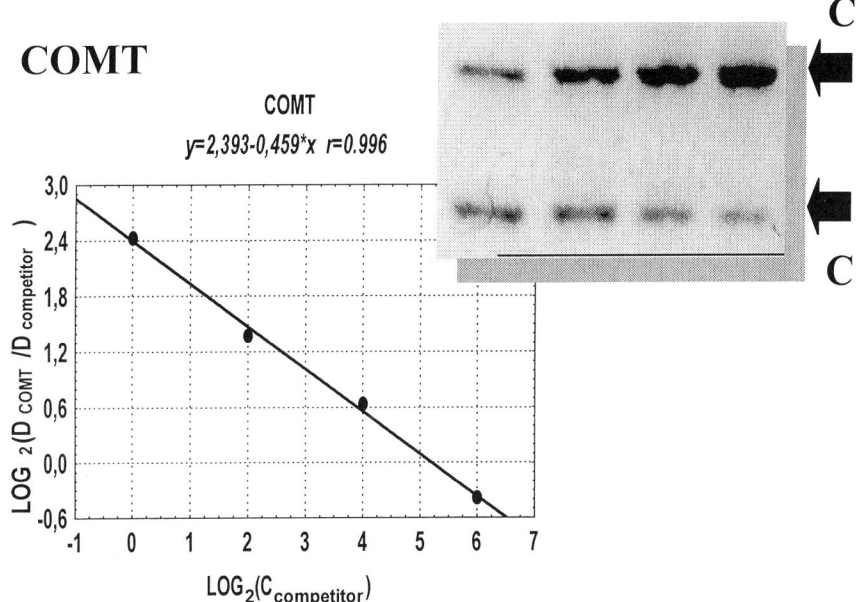

COMT

COMT

$y=2,393-0,459*x$ $r=0.996$

Fig. 1 Quantitative analysis of COMT and β-actine mRNA levels in mouse midbrain.
A – b-actine DNA fragment (250 bp), C_A – β-actine competitor DNA (195 bp). C – COMT DNA fragment (408 bp), C_C – COMT competitor DNA (220 bp).
The competitor concentrations are expressed in arbitrary units.

experiments 2 and 3, and was 43.5% in experiment 1. The mean value of COMT mRNA level in the midbrain of the losers over the three experiments was 80.4%.

Discussion

Earlier it was shown that repeated experience of aggression in 10 daily agonistic confrontations is accompanied by total activation of DA catabolism which leads to an increase in DOPAC formation in the midbrain with pons, amygdala, nucleus accumbens, hippocampus and olfactory bulbs in male mice of the CBA/Lac strain (Kudryavtseva and Bakshtanovskaya, 1991). Similar data were shown for the winners of C57BL/6J strain in another laboratory (Devoino, 1998). The increase of DOPAC levels in the brain areas may be a consequence of activation of DA synthesis. On the other hand, an increase of DOPAC levels may be a result of inhibited activity of CA inactivating enzymes – MAO or COMT. The present study revealed intriguing

Table 1 COMT mRNA levels in mouse midbrain. Data for three independent experiments – 1, 2, 3 are shown. The results of each experiment were obtained from four repeated measurements. COMT mRNA levels for the winners and losers were expressed in relation to controls which were set at 100% and presented as MEAN ± SD (standard deviation).

Experimental groups	Experiments			Mean Values for 3 experiments	Confidential limits of mean (p = 0.95)
	1	2	3		
Controls	100	100	100	100	
Winners	14.9 ± 8.8	30.7 ± 4.5	45.5 ± 5.3	30.37	17.3
Losers	43.5 ± 4.5	109.5 ± 6.9	88.2 ± 6.6	80.4	38.1

experimental results – a three-fold decrease of COMT mRNA levels in the midbrain of the aggressive winners in comparison with control levels in male mice. Data obtained testify that repeated experiences of aggression are accompanied by suppression of COMT gene expression in catecholaminergic neurons of the midbrain. It also suggests that increased DOPAC levels in the brain areas in the winners may result from decreased COMT enzyme activity. In the losers, repeated experiences of defeat in daily agonistic confrontations were accompanied by changes in brain serotonergic activity (Kudryavtseva and Bakshtanovskaya, 1991). MAO enzyme is involved in brain serotonin degradation (Chaouloff, 1993). The absence of systematic changes in brain COMT mRNA levels in the midbrain of the losers suggests an involvement of COMT expression, specifically in the mechanisms of aggressive behavior. The intrinsic molecular mechanisms associated with aggressive experience, COMT gene expression and neurochemical events developing in the brains of the winners require further study.

Some data confirm a possible association of COMT deficiency and aggressive behavior in animals and humans. COMT deficient heterozygous null mutant male mice exhibited increased aggressive behavior (Gogos et al., 1998). In humans, high activity and low activity COMT alleles were recently described (Lachman et al., 1996). A significant association between COMT genotype and history of violent behavior was shown in humans: schizophrenic patients who were homozygous for the low activity allele were judged by their psychiatrists to be at higher risk for aggressive and dangerous behavior that those who were homozygous for the high activity allele (Lachman et al., 1998).

Catechol-O-methyltransferase (COMT) is considered a candidate for several psychiatric disorders, including schizophrenia (Lachman et al., 1998; Kunugi et al., 1997), bipolar affective disorder (Mynett-Johnson et al., 1998; Kirov et al., 1998), and unipolar depression (Ohara et al., 1998). However, there are numerous limitations associated with techniques for examining central nervous system functions in humans. Our experimental approach could provide a framework for understanding COMT gene regulation in the brain under influences of environmental factors, behavior and physiology.

Acknowledgement

This work was supported by Russian Foundation for Basic Research (grant 97-04-49688)

References

Auboeuf, D., and Hubert, V. (1997). The use of the reverse transcription-competitive polymerase chain reaction to investigate the *in vivo* regulation of gene expression in small tissue samples. *Analytical Biochemistry* **245**, 141–148.

Bell, R., and Hepper, P.G. (1987). Catecholamines and aggression in animals. *Behavioural Brain Research* **23**, 1–21.

Carrasco, C., Esteban, O., and Domingo, A. (1997). *In vitro* generation of competitor DNA segments for the quantitation of any nucleic acid of known sequence by competitive polymerase chain reaction. *Analytical Biochemistry* **244**, 406–407.

Chaouloff, F. (1993). Psychopharmacological interactions between stress hormones and central serotonergic systems. *Brain Research Reviews* **18**, 1–32.

Chattopadhyay, N., Kher, R., and Godbole, M. (1993). Inexpensive SDS/phenol method for RNA extraction from tissues. *Biotechniques* **15**, 24–26.

Devoino, L.V., Idova, G.V., Alperina, E.L., and Cheido, M.A. (1998). Neurochemical set-up of the brain is an extraimmune mechanism of psychoneuro immunomodulation. *Vestnik Rossijskoj Academii meditsinskih nauk* **9**, 19–24. (in Russian)

Gogos, J.A., Morgan, M., Luine, V., Santha, M., Ogawa, S., Pfaff, D., and Karayiorgou, M. (1998). Catechol-O-methyltransferase-deficient mice exhibit sexually dimorphic changes in catecholamine levels and behavior. *Proceedings of the National Academy of Sciences USA* **95**, 9991–9996.

Haney, M., Noda, K., Kream, R., and Miczek, K.A. (1990). Regional 5-HT and dopamine activity: Sensitivity to amphetamine and aggressive behavior in mice. *Aggressive Behavior* **16**, 259–270.

Kirov, G., Murphy, K.C., Arranz, M.J., Jones, I., McCandles, F., Kunugi, H., Murray, R.M., McGuffin, P., Collier, D.A., Owen, M.J., and Craddock, N. (1998). Low activity allele of catechol-O-methyltransferase gene associated with rapid cycling bipolar disorder. *Molecular Psychiatry* **3**, 342–345.

Kudryavtseva, N.N., and Bakshtanovskaya, I.V. (1991). Neurochemical control of aggression and submission. *Zhurnal Visshey Nervnoy Deyatelnosti* **41**, 459–466. (in Russian)

Kudryavtseva, N.N. (1991). The sensory contact model for the study of aggressive and submissive behavior in male mice. *Aggressive Behavior* **17**, 285–291.

Kunugi, H., Vallada, H.P., Sham, P.C., Hoda, F., Arranz, M.J., Li,T, Nanko, S., Murray, R.M., McGuffin, P., Owen, M., Gill, M., and Collier, D.A. (1997). Catechol-O-methyltransferase polymorphisms and schizophrenia: a transmission disequilibrium study in multiply affected families. *Psychiatric Genetics* **7**, 97–101.

Lachman, H.M., Papolos, D.F., Saito, T., Yu, Y.-M., Szumlanski, C.L., and Weinshilboum, R.M. (1996). Human catechol-O-methyltransferase pharmacogenetics: description of a functional polymorphism and its potential application to neuropsychiatric disorders. *Pharmacogenetics* **6**, 243–250.

Lachman, H.M., Nolan, K.A., Mohr, P., Saito, T., and Volavka, J. (1998). Association between catechol O-methyltransferase genotype and violence in schizophrenia and schizoaffective disorder. *American Journal of Psychiatry* **155**, 835–837.

Miczek, K.A., and Tornatzky, W. (1996). Ethopharmacology of aggression: impact on autonomic and mesocorticolimbic activity. In: C. Ferris and T. Grisso (Eds.), *Understanding aggressive behavior in children, Annals of the New York Academy of Sciences*, **794**, pp. 60–77.

Mynett-Johnson, L.A., Murphy, V.E., Claffey, E., Shields D.C., and McKeon, P. (1998). Preliminary evidence of an association between bipolar disorder in females and the catechol-O-methyltransferase gene. *Psychiatric Genetics* **8**, 221–225.

Navarro, J.F., Minarro, J., and Simon, V.M. (1993). Antiaggressive and motor effects of haloperidol show different temporal patterns in the development of tolerance. *Physiology and Behavior* **53**, 1055–1059.

Ohara, K., Nagai, M., Suzuki, Y., and Ohara, K. (1998). Low activity allele of catechol-o-methyltransferase gene and Japanese unipolar depression. *Neuroreport* **9**, 1305–1308.

Poshivalov, V.P. (1986). *Experimental psychopharmacology of aggressive behavior*. Leningrad: Nauka. (in Russian)

Strous, R.D., Bark, N., Parsia, S.S., Volavka, J., and Lachman, H.M. (1997). Analysis of a functional catechol-O-methyltransferase gene polymorphism in schizophrenia: evidence for association with aggressive and antisocial behavior. *Psychiatry Research* **69**, 71–77.

27. Cytokines and Stress: Neurochemistry, Endocrinology and Behavior

A.J. Dunn and A.H. Swiergiel

Introduction

Administration of interleukin-1 (IL-1) to mice and rats mimics many of the responses characteristic of commonly studied stressors. IL-1 is a potent stimulator of the hypothalamo-pituitary-adrenocortical (HPA) axis (Besedovsky *et al.*, 1986, Sapolsky *et al.*, 1986). It also induces a modest activation of the adrenal medulla and sympathetic nervous system (Berkenbosch *et al.*, 1989). In the CNS, there are increases in the turnover of noradrenaline (NA) and serotonin (5-HT) (Dunn, 1988; Kabiersch *et al.*, 1988). Body temperature is also affected, but in the mouse we observed relatively small effects that were not clearly related to the neurochemical and endocrine changes (Wang *et al.*, 1997). Certain other cytokines share some of these properties. For example, IL-6 activates the HPA axis, although it is considerably less potent than IL-1, and the activation is quite short (Wang and Dunn, 1998). IL-6 also elevates brain tryptophan and 5-hydroxyindoleacetic acid (5-HIAA), but fails to affect brain NA (Wang and Dunn, 1998). TNFα has similar effects on the HPA axis, but only alters tryptophan and NA metabolism at high doses (Ando and Dunn, 1999). Mouse interferon α administration had none of these activities in mice. A summary of the HPA and neurochemical properties of administration of homologous cytokines in the mouse is shown in Table 1.

IL-1 is behaviorally active. Peripheral or central administration can decrease locomotor activity, feeding, exploration, and sexual activity, and increase slow-wave sleep (Kent *et al.*, 1992). Similar responses have been observed following stressors and have been associated with depression in human (Weiss *et al.*, 1985) (see Table 2). The behavioral responses associated

with illness and IL-1 or endotoxin (LPS) administration have been characterized as an adaptive "sickness behavior" that can enable an animal to cope with the stress of tissue damage or infection and allow pathogen rejection and healing to occur (Kent *et al.*, 1992; Hart, 1988).

Association of the Neurochemical and Endocrine Responses

Our interest was to identify the function of the neurochemical responses to IL-1. First we addressed the relationship between the neurochemical responses and the HPA activation. The noradrenergic response to intraperitoneal (ip) injection of mice with IL-1β, as indicated by increases in the catabolite, 3-methoxy,4-hydroxyphenylethyleneglycol (MHPG) starts early, reaches a peak around 2 h, and disappears within 8 h (Dunn, 1988, 1992). The increases in brain tryptophan and serotonin metabolism (5-HIAA) are significantly slower and do not reach a peak until around 4 h (Dunn, 1988, 1992). The time course of the HPA response, indicated by increases in plasma concentrations of ACTH and corticosterone, follows quite closely the noradrenergic response (Dunn, 1988). Moreover, in individual animals, the correlation between the increases in plasma corticosterone and hypothalamic MHPG is almost perfect. This correlation is indicated even more clearly in a study in which the release of NA from rat hypothalamus was assessed by microdialysis while simultaneously sampling plasma for corticosterone, following ip or intravenous injection of IL-1β (Smagin *et al.*, 1994). The only dissociation of the two responses occurred in mice treated with cyclooxygen-

Table 1 Neurochemical responses to cytokine administration in mice.

Cytokine	Corticosterone	NE	DA	Tryptophan	5-HT
IL-1α/IL-1β	+	+	0	+	+
IL-2	0	+	+	nd	0
IL-6	+	0	0	+	+
TNFα	+	(+)	0	(+)	0
IFNα	0	0	0	0	0

+ increased; ++ large increase; 0 no change; (+) indicates increases only at the highest doses of TNFα (1 μg or more); nd not determined.

Table 2 Behavioral Responses to Interleukin-1.

Hyperthermia (absent or very slight in mice)

Hypomotility

Anorexia

Decreased Interest in Exploring the Environment

Decreased Libido

Increased Time spent in Slow-Wave Sleep

Decreased Learning

Taste aversion

ase (COX) inhibitors (such as indomethacin), which prevents the MHPG response to ip IL-1β, while the plasma corticosterone response is unimpaired (Dunn and Chuluyan, 1992). When the noradrenergic input to the hypothalamus is reduced by injecting the selective NA neurotoxin, 6-hydroxydopamine (6-OHDA) directly into the paraventricular nucleus (PVN) or the ventral ascending noradrenergic bundle (VNAB) of rats, the plasma corticosterone response to ip IL-1 was markedly impaired (Chuluyan et al., 1992). However, treatment with α- and β-adrenergic antagonists had relatively little effect on the HPA response to IL-1 in rats (Rivier, 1996) and mice (Dunn et al., 1999). Also, very large depletions of mouse brain NA have little or no effect on the IL-1-induced elevation of plasma corticosterone (Swierqiel et al., 1996). Most likely the regulation of HPA axis activation involves not only NA, but neuropeptides, such as substance P, neuropeptide Y or galanin.

Association of the Neurochemical and Behavioral Responses

An interesting question is the relationship between the neurochemical and the behavioral responses. We have studied the hypophagia commonly associated with illness and which is readily apparent after acute administration of IL-1 or LPS (Weingarleu, 1996; Swiergiel et al., 1997a). Changes in feeding have been associated with alterations of central NA and 5-HT, so this was an appropriate target. The measurement of ingestive behavior was twofold. We provided mice access to sweetened condensed milk (diluted 4-fold with water) for a period of 30 minutes every day in the mid morning (Swiergiel et al., 1997a,b). After exposure to the milk for 3–5 days, most mice rapidly drank 1.5–3 ml of milk in the 30 min. The mice were allowed free access to food pellets and water. Food pellet intake over the subsequent 22 h and body weight were also monitored. The milk paradigm was set up because it was anticipated that the duration of the effects of IL-1 and other cytokines would be limited, as has proven to be the case.

Our studies have shown that ip administration of mouse IL-1α or IL-1β induces a dose-dependent reduction in milk intake that is greatest around 30 min to 2 h after injection and recovers by around 3 h (Swiergiel et al., 1997a). Food pellet intake was also reduced slightly, although this effect was not always statistically significant. Body weight was not consistently affected. No significant changes were observed in milk or food pellet intake on subsequent days. Ip injection of LPS produced similar responses in milk intake but perhaps slightly slower with a nadir around 2–2.5 h. LPS reliably reduced food pellet intake over the following 22 h, and body weight 24 h later. Recovery was not always complete the day after LPS administration. Administration of mouse IL-6 in doses up to 1 μg per mouse failed to alter milk or food pellet intake (Swiergiel et al., 1997b). Mouse TNFα had some effect at a dose of 2.5 μg per mouse, but none at 1 μg or lower (Swiergiel et al., 1997b).

In an extensive series of experiments, we studied the effects of α- or β-adrenergic receptor antagonists, either alone or in combination. No attenuation whatsoever of the reductions in milk or food pellet intake induced by IL-1β (100 ng/mouse ip) and LPS (1 μg/mouse ip) were observed. Moreover, treatment of mice with 6-OHDA which depleted brain NA by 98% failed to alter the responses to IL-1β or LPS (Figure 1) (Swiergiel et al., 1999). Similar results were obtained with DSP-4 which depletes forebrain NA. We also studied a variety of other possible mediators, including histamine, acetylcholine, dopamine, as well as endorphins and nitric oxide (NO). Histamine depletion with α-fluoromethylhistidine, antagonists of H_1, H_2, and H_3 receptors, scopolamine, haloperidol, naloxone and the NO

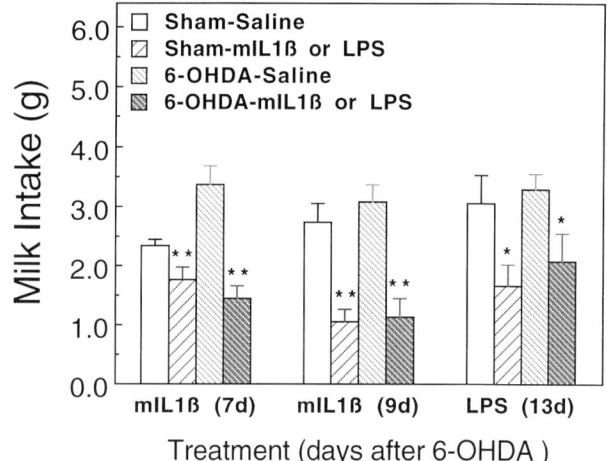

Fig. 1 Effect of 6-OHDA pretreatment on mIL-1β- and LPS-induced milk intake. Mice were pretreated intracerebroventricularly (icv) with 6-OHDA (100 μm icv) and injected 7 or 9 days later with mIL-1β (100 ng ip), and 13 days later with LPS (1 μg ip). Mice injected with mIL-1β on Day 7 were injected with saline on Day 9 and vice versa. Milk intake was assessed 2 h after mIL-1β or LPS. N = 8. Significantly different from the corresponding control group (*p < 0.05 or **p < 0.01, respectively). Reproduced from (Swiergiel et al., 1999)

Table 3 Treatments that failed to alter IL-1 β-induced reductions in milk intake.

Prazosin (α_1-adrenergic receptor antagonist)

Phentolamine (α-adrenergic receptor antagonist)

Propranolol (β-adrenergic receptor antagonist)

6-Hydroxydopamine (NA neurotoxin)

Scopolamine (muscarinic receptor antagonist)

Haloperidol (dopamine receptor antagonist)

α-Fluoromethylhistidine (histamine synthesis inhibitor)

Pyrilamine (H_1 receptor antagonist)

Cimetidine (H_2 receptor antagonist)

Thioperamide (H_3 receptor antagonist)

8-OH-DPAT and Ipsapirone (5-HT$_{1A}$ receptor agonists)

WAY100135 (5-HT$_{1A}$ receptor antagonist)

GR127935 (5-HT$_{1B}$ receptor antagonist)

Metergoline, Methysergide, Ritanserin, Ketanserin
    (5-HT$_2$ receptor antagonists)

RS102221 (5-HT$_{2C}$ receptor antagonist)

SB206553 (5-HT$_{2B/2C}$ receptor antagonist)

Tropisetron (5-HT$_3$ receptor antagonist)

5,7-dihydroxytryptamine (serotonin neurotoxin)

Naloxone (opiate receptor antagonist)

L-NAME (NO synthase inhibitor)

BIBP3226 (Y$_1$ receptor antagonist)

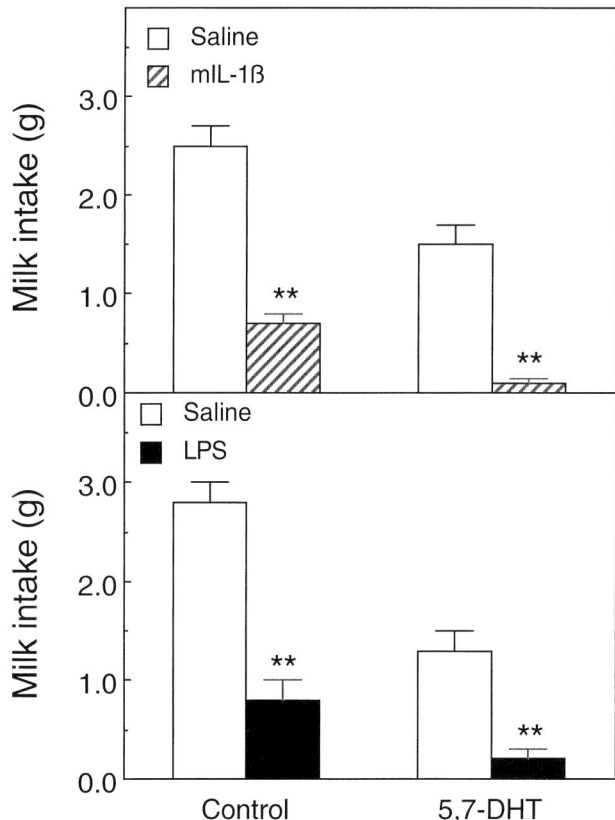

Fig. 2 Effect of mIL-1β and LPS on milk intake in 5,7-DHT-treated mice. Mice were pretreated with 5,7-DHT (80 μg icv) and injected 20 days later with mIL-1β or 22 days later with LPS. Milk intake was assessed 90 min after mIL-1β or 2 h after LPS. N = 8. **Significantly different from the corresponding control group (p < 0.001). Reproduced from (Swiergiel and Dunn, 2000)

synthase inhibitor, L-NAME, all failed to alter the hypophagic responses to IL-1β or LPS (Swiergiel et al., 1999). A summary of the treatments tested to date is shown in Table 3.

Because of the well known relationship between serotonin and anorexia (e.g., the hypophagic effects of 5-HT-releasing drugs, such as fenfluramine), we focussed on 5-HT-receptor antagonists and treatments that deplete brain 5-HT. Antagonists of 5-HT$_{1A}$, 5-HT$_{1B}$, 5-HT$_{2A}$, 5-HT$_{2B}$, 5-HT$_{2C}$ and 5-HT$_3$ receptors, as well as 5-HT$_{1A}$ agonists all failed to induce any attenuation of IL-1-induced hypophagia (Swiergiel and Dunn, 2000). Also, depletion of whole brain 5-HT by treatment with 5,7-dihydroxytryptamine (5,7-DHT) failed to alter the responses to IL-1β and LPS, although the neurotoxin treatment itself reduced milk intake chronically (Figure 2).

We have also tested the involvement of certain peptides. Antagonists of Y$_1$ receptors failed to alter the IL-1-induced reduction in milk intake, and CRF knockout mice showed normal milk intake responses to IL-1β and LPS (Swiergiel and Dunn, 1999).

The only treatment we have tested that reduces the hypophagic responses to IL-1 and LPS is the use of COX inhibitors. Such treatments largely (but not completely) prevented the responses to IL-1, and partially attenuated the responses to LPS (Swiergiel et al., 1997a). Our recent results indicate the effect may be mediated by the COX1 form of the enzyme, because COX1-selective inhibitors, such as piroxicam, were more effective than COX2-selective inhibitors, such as nimesulide and NS-398.

Acknowledgements

The technical assistance of Charles Dempsey and Galina Mikhaylova is greatly appreciated as well as the clerical assistance of Sharon Farrar. This research was supported by grants from the U.S. National Institute of Mental Health (MH46261) and the U.S. National Institute of Neurological Diseases and Stroke (NS25370).

References

Ando, T., and Dunn, A.J. (1999). Mouse tumor necrosis factor α increases brain tryptophan concentrations and norepinephrine metabolism while activating the HPA axis in mice. *Neuroimmunomodulation* **6**, 319–329.

Berkenbosch, F., De Goeij, D.E.C., del Rey, A., and Besedovsky, H.O. (1989). Neuroendocrine, sympathetic and metabolic responses induced by interleukin-1. *Neuroendocrinology* **50**, 570–576.

Besedovsky, H.O., del Rey, A., Sorkin, E., and Dinarello, C.A. (1986). Immunoregulatory feedback between interleukin-1 and glucocorticoid hormones. *Science* **233**, 652–654.

Chuluyan, H., Saphier, D., Rohn, W.M., and Dunn, A.J. (1992). Noradrenergic innervation of the hypothalamus participates in the adrenocortical responses to interleukin-1. *Neuroendocrinology* **56**, 106–111.

Dunn, A.J. (1988). Systemic interleukin-1 administration stimulates hypothalamic norepinephrine metabolism paralleling the increased plasma corticosterone. *Life Sciences* **43**, 429–435.

Dunn, A.J. (1992). Endotoxin-induced activation of cerebral catecholamine and serotonin metabolism: comparison with interleukin-1. *Journal of Pharmacology and Experimental Therapeutics* **261**, 964–969.

Dunn, A.J., and Chuluyan, H. (1992). The role of cyclo-oxygenase and lipoxygenase in the interleukin-1-induced activation of the HPA axis: dependence on the route of injection. *Life Sciences* **51**, 219–225.

Dunn, A.J., Wang, J.-P., and Ando, T. (1999). Effects of cytokines on central neurotransmission: Comparison with the effects of stress. In *Cytokines, Stress and Depression* Dantzer, B. Leonard and R. Yirmiya, Eds. Plenum Press. London, England.

Hart, B.L. (1988). Biological basis of the behavior of sick animals. *Neuroscience and Biobehavioral Reviews* **12**, 123–137.

Kabiersch, A., del Rey, A., Honegger, C.G., and Besedovsky, H.O. (1988). Interleukin-1 induces changes in norepinephrine metabolism in the rat brain. *Brain Behavior and Immunity* **2**, 267–274.

Kent, S., Bluthé, R.-M., Kelley, K.W., and Dantzer, R. (1992). Sickness behavior as a new target for drug development. *Trends in Pharmacological Science* **13**, 24–28.

Rivier, C. (1995). Influence of immune signals on the hypothalamic-pituitary axis of the rodent. *Frontiers in Neuroendocrinology* **16**, 151–182.

Sapolsky, R., Rivier, C., Yamamoto, G., Plotsky, P., and Vale, W. (1987). Interleukin-1 stimulates the secretion of hypothalamic corticotropin-releasing factor. *Science* **238**, 522–524.

Smagin, G.N., Swiergiel, A.H., and Dunn, A.J. (1996). Peripheral administration of interleukin-1 increases extracellular concentrations of norepinephrine in rat hypothalamus: comparison with plasma corticosterone. *Psychoneuroendocrinology* **21**, 83–93.

Swiergiel, A.H., Dunn, A.J., and Stone, E.A. (1996). The role of cerebral noradrenergic systems in the Fos response to interleukin-1. *Brain Research Bulletin* **41**, 61–64.

Swiergiel, A.H., Smagin, G.N., and Dunn, A.J. (1997a). Influenza virus infection of mice induces anorexia: comparison with endotoxin and interleukin-1 and the effects of indomethacin. *Pharmacology Biochemistry and Behavior* **57**, 389–396.

Swiergiel, A.H., Smagin, G.N., Johnson, L.J., and Dunn, A.J. (1997b). The role of cytokines in the behavioral responses to endotoxin and influenza virus infection in mice: effects of acute and chronic administration of the interleukin-1-receptor antagonist (IL-1ra). *Brain Research* **776**, 96–104.

Swiergiel, A.H., Burunda, T., Patterson, B., and Dunn, A.J. (1999). Endotoxin- and interleukin-1-induced hypophagia are not affected by noradrenergic, dopaminergic, histaminergic and muscarinic antagonists. *Pharmacology Biochemistry and Behavior* **63**, 629–637.

Swiergiel, A.H., and Dunn, A.J. (1999). CRF-deficient mice respond like wild type mice to hypophagic stimuli. *Pharmacology Biochemistry and Behavior* **64**, 59–64.

Swiergiel, A.H., and Dunn, A.J. (2000). Lack of evidence for a role of serotonin in interleukin-1-induced hypophagia. *Pharmacology, Biochemistry and Behavior*, **65**, 531–537.

Wang, J.P., Ando, T., and Dunn, A.J. (1997). Effect of homologous interleukin-1, interleukin-6 and tumor necrosis factor α on the core body temperature of mice. *NeuroImmunomodulation* **4**, 230–236.

Wang, J.P., and Dunn, A.J. (1998). Mouse interleukin-6 stimulates the HPA axis and increases brain tryptophan and serotonin metabolism. *Neurochemistry International* **33**, 143–154.

Weingarten, H.P. (1996). Cytokines and food intake: the relevance of the immune system to the student of ingestive behavior. *Neuroscience and Biobehavioral Reviews* **20**, 163–170.

Weiss, J.M., Simson, P.G., Ambrose, M.J., Webster, A., and Hoffman, L.J. (1985). Neurochemical basis of behavioral depression. *Advances in Behavioral Medicine* **1**, 233–275.

28. Central Neurochemical Effects of Cytokines: Parallels with Stressor Effects

H. Anisman, S. Hayley and Z. Merali

Introduction

In addition to its other functions, the immune system may serve in a sensory capacity informing the brain of antigenic challenges (Blalock, 1994). In this respect, it was suggested that cytokines released from activated macrophages serve as signaling molecules between the immune system and the CNS (Dunn, 1995), and the CNS may interpret immune activation as if it were a stressor (Dunn, 1990; Merali *et al.*, 1997). As in the case of other stressors, systemic endotoxin administration, as well as cytokines released from activated immune cells, may stimulate the release of hypothalamic monoamines (Hayley *et al.*, 1999; Shintani *et al.*, 1993), as well as CRH release from the paraventricular nucleus (PVN) of the hypothalamus, hence stimulating pituitary ACTH release and that of adrenal glucocorticoids (Dunn, 1992; Rivier, 1993). It appears, however, that in several respects the neural circuitry associated with cytokine treatments is not entirely congruent with that associated with more traditional stressors. For instance, although both processive stressors (i.e., those involving higher order sensory processing, such as conditioned fear cues, exposure to a novel environment or predators) and systemic (metabolic) stressors give rise to hypothalamic-pituitary-adrenal (HPA) activation, processive stressors may involve activation of limbic forebrain regions, which in turn, may stimulate GABAergic neurons at an intervening synapse (e.g., bed nucleus of the stria terminalis) before acting upon corticotropic PVN cells. In contrast, HPA alterations elicited by a systemic stressor (e.g., cytokines) may reflect the stimulation of brainstem nuclei which directly innervate the PVN (Herman and Cullinan, 1997).

Despite the possibility that psychogenic stressors and cytokines influence CNS functioning through different circuitry, the cytokines might be part of a regulatory loop that, by virtue of their effects on the CNS, contribute to the symptoms of behavioral pathologies, including mood and anxiety-related disorders (Anisman and Merali, 1999; Maes, 1995). Depressive illness is associated with elevated levels of IL-1β, IL-2, IL-6, and soluble cytokine receptors (Maes, 1995). Moreover, acute systemic IL-2 administration may provoke anhedonia, as reflected by disturbed responding for rewarding stimulation in rats (Anisman *et al.*, 1998), and antidepressants attenuate the anorexic effects of endotoxin treatment (Yirimiya, 1996). Further, humans undergoing immunotherapy (e.g., with IL-2) display numerous adverse neuropsychological, neurologic and psychiatric disturbances, including depression (Capuron *et al.*, 1998; Denicoff *et al.*, 1987).

Several explanations have been offered as to how immune activation may come to affect CNS processes. For instance, it is possible that peripheral cytokines target the vagus nerve, hence affecting central neurotransmitter activity (Dantzer *et al.*, 1998; Maier and Watkins, 1998). Alternatively, several macrophage-derived cytokines, such as interleukin (IL)-1β, IL-6, and tumor necrosis factor-α (TNFα), may act as mediators (immunotransmitters) in this respect. These cytokines and their receptors are endogenous to the brain, may be affected by psychological factors (e.g., stressors) and by physical insults (concussive injury, seizure), as well as neurodegenerative processes (Rothwell, 1999).

Neurochemical Effects of Interleukin-1β

While a number of studies reported the effects of cytokine or endotoxin manipulations on hypothalamic monoamine activity, there have been relatively few that examined the effects of these treatments at extrahypothalamic sites. Of course, in order to determine how cytokines come to provoke behavioral changes, particularly those related to mood disturbances, it is necessary to identify the neurochemical consequences of such challenges within extrahypothalamic regions, particularly limbic sites. Given the view that cytokines may elicit stressor-like effects, our studies evaluated the behavior alterations in paradigms where stressors ordinarily influence behavior (e.g., in tests of anxiety, exploratory behaviors and habituation, consumption of highly palatable foods, and responding for rewarding brain stimulation). These experiments were coupled with the analysis of the central neurochemical alterations (both in vivo and postmortem tissue) elicited by cytokines or bacterial endotoxins, notably those commonly induced by stressors of either a psychogenic or a neurogenic nature.

Like other investigators, we observed that the bacterial endotoxin, lipopolysaccharide (LPS), and recombinant human IL-1β dose-dependently increased the release of ACTH and corticosterone (Lacosta *et al.*, 1998, 1999), and provoked region-specific variations of norepinephrine (NE), dopamine (DA) and serotonin (5-HT) activity. For instance, as seen in Figure 1, acute

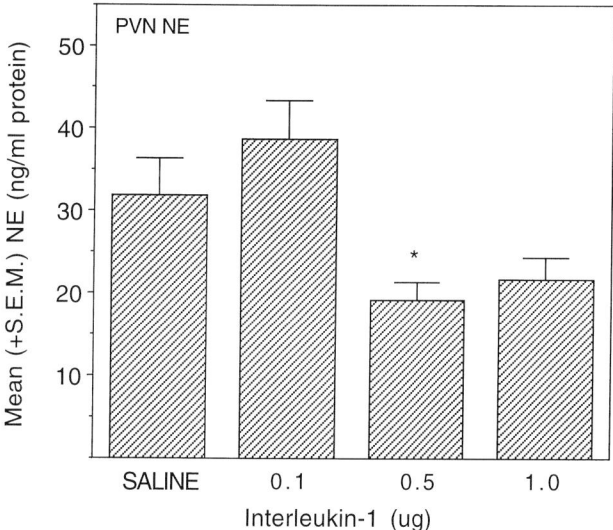

Fig. 1 Mean (±SEM) NE concentrations in the paraventricular nucleus of the hypothalamus (PVN) 50 min following treatment with several doses of IL-1β (* P < 0.05 relative to saline treated animals). From Lacosta et al. (1998).

administration of this cytokine dose-dependently reduced the concentrations of NE levels within the PVN. Furthermore, as seen in Figure 2, within the locus coeruleus, arcuate nucleus plus median eminence, and the medial prefrontal cortex (mPFC) IL-1β increased the utilization of NE, as reflected by elevated MHPG accumulation. These effects, it will be recognized, were reminiscent of those frequently associated with neurogenic or psychogenic stressors. However, while such stressors ordinarily influence DA activity within the mPFC and the nucleus accumbens (Deutch and Roth, 1990), DA turnover was unaffected by the cytokine in these regions. Thus, while there are similarities between the effects of neurogenic stressors and IL-1β, their effects are distinguishable from one another (Lacosta et al., 1998).

Given the uncertain fate of amine metabolites over lengthy periods following cytokine treatment, amine utilization may be more clearly reflected by in vivo

analyses than in postmortem tissues. Accordingly, several experiments were conducted in rats, to assess the in vivo release of monoamines within the nucleus accumbens, mPFC and hippocampus (Merali et al., 1997; Song et al., 1998). Paralleling the postmortem findings, IL-1β (1.0 μg) did not influence extracellular DA concentrations within these brain regions. However, within both the nucleus accumbens and the hippocampus (see also Linthorst et al., 1995) the release of 5-HT (in this instance reflected by increased extracellular 5-HIAA) was appreciably enhanced by IL-1β, an effect that was augmented by an air-puff stressor (see Figure 3). Evidently, although IL-1β elicited some stressor-like effects, they were not entirely like those elicited by stressors, at least with respect to mesolimbic DA variations.

When animals are exposed to antigenic challenge, the cytokine changes are not restricted to IL-1β, but involve variations of a constellation of cytokines, acute phase proteins, and hormones (IL-1β may promote other cytokine alterations which could elicit behavioral and neurochemical changes; Wang and Dunn, 1999). Thus, the absence of mesolimbic DA variations in response to IL-1β should not be interpreted to mean that immune insults do not engender such an effect, but rather that this particular cytokine, when administered alone, does not affect DA functioning. Indeed, Figure 4 shows that LPS markedly influenced in vivo accumbal DA and 5-HIAA release (Borowski et al., 1998). As in the case of IL-1β, the 5-HIAA variations commenced soon after LPS administration, and persisted over the ensuing 4 hr of testing. However, unlike the effects of IL-1β, the endotoxin also provoked DA release within 90 min of LPS administration and persisted over 90 min before returning to basal levels. Why the time course for the two amines was so different is unclear; however, it is apparent that immune activation, like a processive stressor, will affect mesolimbic DA release. At the doses used in our studies, both LPS and IL-1β, provoked a moderate degree of sickness (ptosis, ruffled fur, inactivity), and thus the DA variations induced by LPS were not likely secondary to malaise.

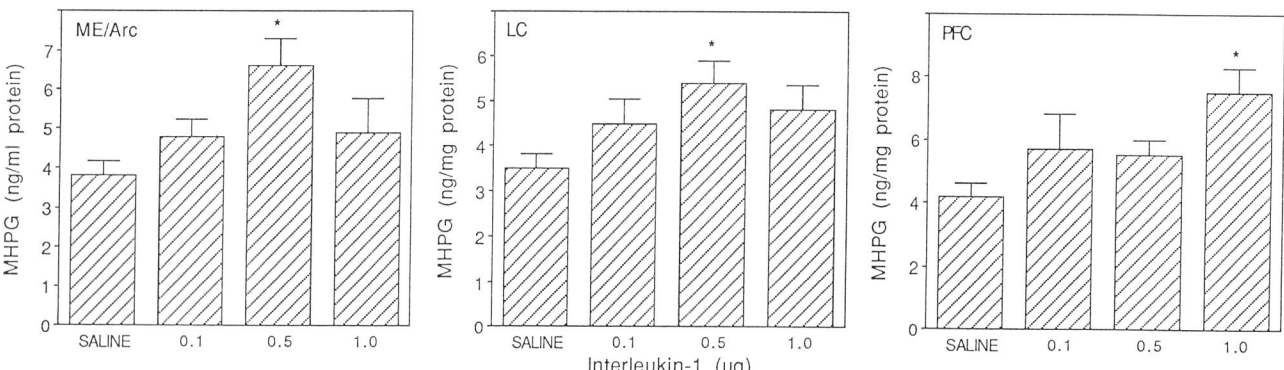

Fig. 2 Mean (±SEM) MHPG accumulation in the locus coeruleus (LC), arcuate nucleus plus median eminence (ARC/ME) and the medial prefrontal cortex (PFC) 50 min following treatment with several doses of IL-1β (* p < .05 relative to saline treated animals). From Lacosta et al. (1998).

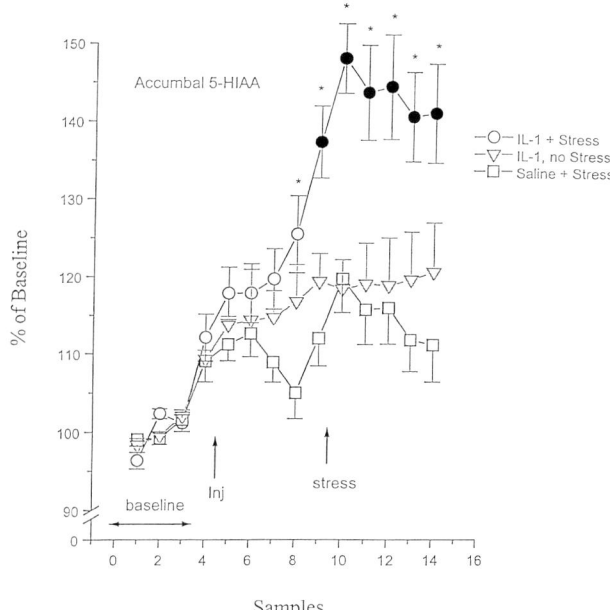

Fig. 3 Extracellular 5HIAA in the nucleus accumbens, expressed as a percent of baseline, over 30 min dialysate samples. Following 4 baseline samples, rats were injected with either saline (squares) or IL-1β (circles and triangles) and 5 samples collected at 30 min intervals. Thereafter an air-puff stress (5 puffs of 1 sec duration) was delivered (circles and squares) and dialysates collected over 5 additional periods. Filled-in symbols denote significant differences form the average of the baseline period, asterisks denote significant differences from saline treated rats. Similar effects were also obtained from the hippocampus. From Merali *et al.*, 1997.

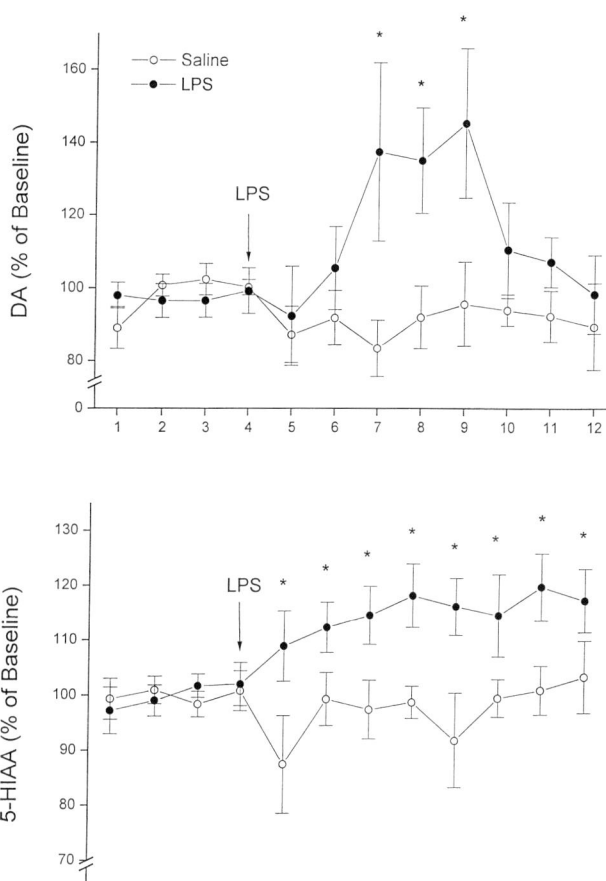

Fig. 4 Mean (+SEM) interstitial DA (upper panel) and 5-HIAA (lower panel) at the nucleus accumbens as a percentage of baseline (four samples) and after i.p. administration of either saline or LPS (100 μg). Each point represents dialysates collected over a 30 min period. *$P < 0.05$ relative to saline-treated rats. From Borowski *et al.* (1998).

In our postmortem studies only modest NE changes were apparent in the central amygdala following IL-1β treatment. This contrasts with the effects of other stressors which reliably influence monoamine activity in the different amygdaloid nuclei (McIntyre *et al.*, 1999). Thus, the finding that IL-1β did not impact on amygdala NE activity in postmortem tissue was somewhat surprising. Accordingly, we assessed whether amygdala NE variations would be detected *in vivo*, which as indicated earlier, yields a better view of the effects stemming from endogenous or exogenous challenges. Given the marked interindividual differences reported concerning the impact of stressors, in experiments conducted in collaboration with D. McIntyre, we assessed the effects of systemic IL-1β on amine variations in two rat strains selectively bred for differences of their amygdala excitability (realized by either fast or slow development of kindled epileptic seizures elicited by focal amygdala stimulation) (McIntyre, Kelly and Dufresne, 1999). These strains, termed FAST and SLOW, also exhibit marked differences in emotionality, and have been found to display different behavioral profiles in response to stressors. The SLOW rats tend to be more fearful, while the FAST are somewhat hyper-reactive and show disturbed habituation (Anisman *et al.*, 1997; Mohapel and McIntyre,

1998). Figure 5 shows that in response to 1.0 μg of IL-1β (i.p.) the emotional SLOW rats displayed a pronounced *in vivo* release of NE from the central amygdala, whereas virtually no change from baseline was seen in the less emotional FAST rats. Thus, administration of the cytokine indeed has marked effects on NE activity within the central amygdala, but that these effects appear to be strain dependent. Whether the observed effects were related to amygdala excitability (consider that elevated NE activity within the amygdala is inhibitory to seizure provocation) or to emotionality differences, remains to be determined.

Sensitization of Cytokine-induced Neurochemical Changes

It is well established that in addition to immediate neurochemical alterations, stressors may have marked proactive effects. Although stressor-induced neuro-

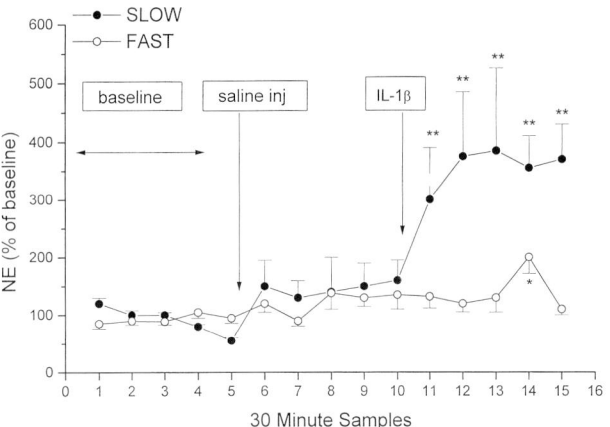

Fig. 5 Mean ± SEM interstitial NE as a percentage of baseline at the central amygdala among FAST (high seizure susceptible) and SLOW (low seizure susceptible) rats at baseline (first 5 samples), following i.p. administration of saline (5 samples) or subsequent administration of IL-1β (1.0 μg) (5 samples).

chemical changes are transient, reexposure even to a relatively modest insult may promote marked mono-amine variations in several brain regions (Kalivas and Stewart, 1991). Such a sensitization effect is also seen when the reexposure treatment differs from the initial stressor experience (cross-sensitization) (Nissenbaum *et al.*, 1991). It has likewise been observed (Bartanusz *et al.*, 1993; Tilders *et al.*, 1993) that acute administration of IL-1β provoked a progressive, time-dependent change with respect to HPA functioning (Schmidt *et al.*, 1995, 1996; Tilders and Schmidt, 1998). Specifically, with the passage of time a phenotypic change occurred within CRH neurons of the PVN terminating in the external zone of the median eminence, such that colocalization of AVP and CRH was greatly increased. Since CRH and AVP synergistically stimulate pituitary ACTH secretion, the elevated levels of the two peptides might account for the increased HPA responsivity observed upon reexposure to a stressor. These findings are important as they indicate that (1) stressors and cytokine treatments may proactively influence the response to later stressors, (2) the development of the CRH and AVP colocalization may be time-dependent, and (3) the colocalization was relatively long-lasting and might thus have proactive repercussions long after the initial insult. Indeed, endotoxin challenge early in life may influence HPA functioning in response to stressors encountered months later (Shanks *et al.*, 1995).

Given the sensitization effects seen with respect to IL-1β on HPA functioning, it was of interest to determine whether a comparable outcome would be evident with respect to TNF-α, particularly since immune activation is also associated with the release of this macrophage-derived proinflammatory cytokine. Moreover, it was of interest to establish whether this cytokine would also elicit a sensitization effect with respect to central monoamine functioning, just as more traditional

stressors were previously found to provoke such effects (Anisman *et al.*, 1991; Anisman and Merali, 1999). Indeed, treatment with TNF-α was not only found to elicit immediate behavioral and neurochemical consequences, but also provoked profound time-dependent sensitization effects. In several preliminary studies (Anisman and Merali, 1999; Brebner *et al.*, 1998; Hayley *et al.*, 1999) we observed that systemic administration of recombinant human TNF-α (1.0, 2.0 and 4.0 μg) dose-dependently influenced sickness behaviors, as reflected by reduced consumption of a highly favored solution (chocolate milk), social interaction, locomotor activity, and general appearance (e.g., lethargy, ptosis, ruffled fur) (see Figure 6). As depicted in the lower panel of Figure 6, the behavioral changes were also accompanied by elevated plasma corticosterone.

In an ensuing experiment (Hayley *et al.*, 1999) we evaluated behavior, plasma corticosterone and central monoamine activity in mice that had initially been exposed to TNF-α (4.0 μg), and then at various times later (1, 7, 14 or 28 days) were reexposed to a lower dose (1.0 μg) of the cytokine. Additional groups of mice received either vehicle on 2 occasions, only the initial TNF-α treatment or only the reexposure treatment.

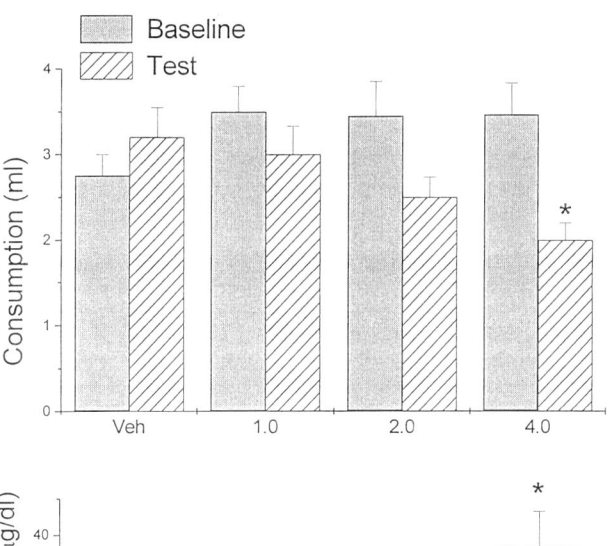

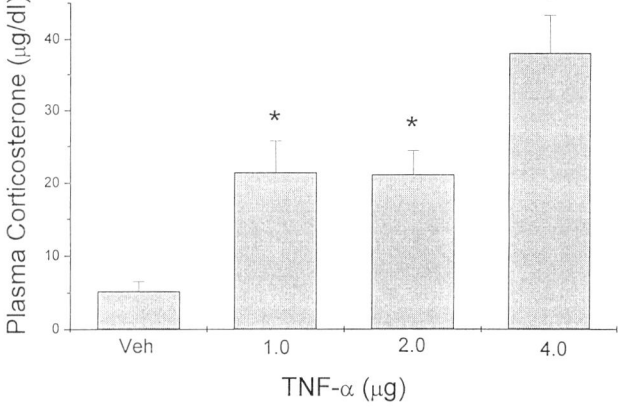

Fig. 6 Consumption of a highly favored food (mean ± SEM chocolate milk intake) (upper panel) and plasma corticosterone levels (lower panel) in mice treated with different doses of either vehicle or TNF-α. *P < 0.05 relative to vehicle.

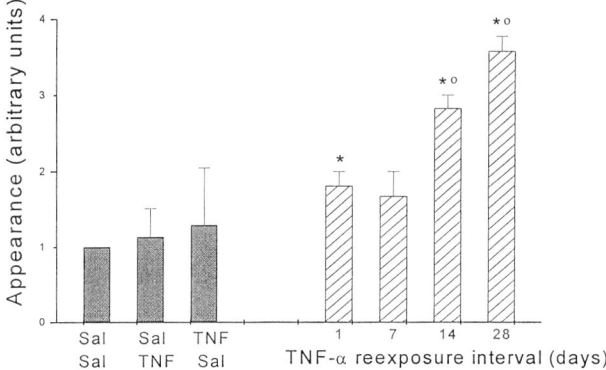

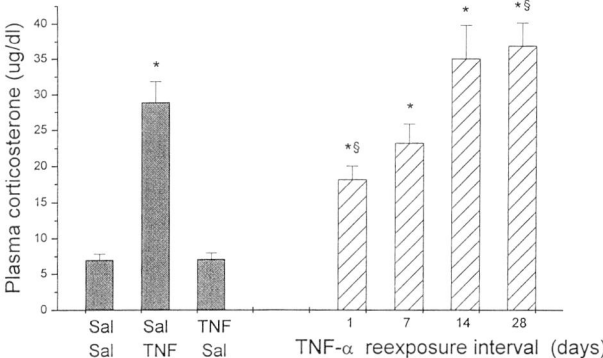

Fig. 7 Overall sickness profile (upper panel) and plasma corticosterone levels (x ± SEM) (lower panel) among mice exposed to various TNF-α regimens. All mice received two ip injections; the three groups to the left (solid bars) received either saline only, saline followed two weeks later by a low dose of TNF-α (1.0 µg) or a high dose of TNF-α (4.0 µg) followed two weeks later by saline treatment. The four groups on the right (hatched bars) received two injections of TNF-α; an initial 4.0 µg dose followed by a second 1.0 µg dose either 1, 7, 14 or 28 days later. *P < 0.05 relative to saline treated mice, § P < 0.05 relative to mice that received acute TNF-α (modified from Hayley *et al.*, 1999)

Figure 7 shows sickness behavior as reflected by general appearance of the animals (lethargy, ptosis, ruffled fur) rated on a 4-point scale. These data were paralleled by similar effects with respect to locomotor functioning and social interaction (data not shown). When acutely administered, TNF-α (1.0 µg) had very modest effects on sickness behaviors measured 1 hr later. Likewise acute administration of a higher dose (4.0 µg) given 2 weeks earlier had no effect on the various behaviors. However, in mice treated with the 4.0 µg dose, subsequent reexposure to the lower dose of the cytokine had marked time-dependent effects on sickness. At the 1 and 7 day reexposure intervals there was little evidence of sickness, whereas at the 14 and 28 day reexposure intervals mice displayed marked illness.

As seen in the lower panel of Figure 7, plasma corticosterone levels in response to TNF-α were likewise subject to a sensitization effect that was dependent on the passage of time. Acute administration of 1.0 µg of TNF-α increased plasma corticosterone concentra-

tions measured 1 h later. TNF-α at a higher dose (4.0 µg) had no carryover effects, such that plasma corticosterone levels 14 days later were comparable to those of saline treated animals. However, in mice that had received TNF-α (4.0 µg) followed subsequently by a low dose of the cytokine (1.0 µg), time-dependent biphasic variations of corticosterone were observed. Specifically, a desensitization effect was evident in animals that had been reexposed to the cytokine 1 day after initial treatment (i.e., corticosterone levels were significantly lower than those seen 1 hr after acute cytokine administration). With progressively longer intervals the plasma corticosterone response increased, such that when a 28-day interval elapsed between the treatments corticosterone levels were higher than those observed after a single injection. Parenthetically, these effects were not unique to the human recombinant form of this cytokine, as recent studies assessing the effects of murine TNF-α at the 28-day reexposure interval yielded essentially identical results with respect to both behavior and plasma corticosterone variations (Hayley *et al.*, 1999). Thus, the sensitization effects were not due to immune changes related to cross-species cytokine effects (i.e., the use of the human recombinant cytokine in the mouse). Moreover, these data suggest that the outcome was not a reflection of effects unique to the murine form of the cytokine, including activation of both the p55 and the p75 receptor, as opposed to just the p55 receptor activation induced by human TNF-α.

As in the case of the corticosterone changes, TNF-α influenced central monoamine functioning in response to cytokine reexposure. However, the nature of the time-dependent effects differed with the specific amine and brain region being assessed. To a considerable extent, NE activity (as reflected by MHPG accumulation) within the PVN was subject to a sensitization effect that paralleled the corticosterone alterations (see Figure 8). Specifically, the metabolite accumulation became more pronounced when TNF-α was administered at longer intervals following initial cytokine treatment. Analysis of the MHPG/NE ratio also indicated that at the 1 day reexposure interval NE turnover was reduced relative to that induced by acute TNF-α. Inasmuch as NE may play a fundamental role in the provocation of CRH release in the PVN, which then comes to affect adrenal glucocorticoid release, it is possible that the sensitization with respect to corticosterone changes were related to the NE alterations.

In contrast to the PVN, a different profile was apparent within the mPFC, central amygdala, and hippocampus. As shown in Figure 9, a single administration of TNF-α did not influence NE activity within the central amygdala, and produced a modest increase of NE utilization within the mPFC. However, within both regions, a marked sensitization was evident among mice reexposed to the cytokine 1 day later. At longer intervals, TNF-α increased NE utilization, but the sensitization was entirely absent. Interestingly, however, the accumulation of 5-HIAA within the mPFC increased over the course of the reexposure interval, peaking at 2 weeks following initial treatment.

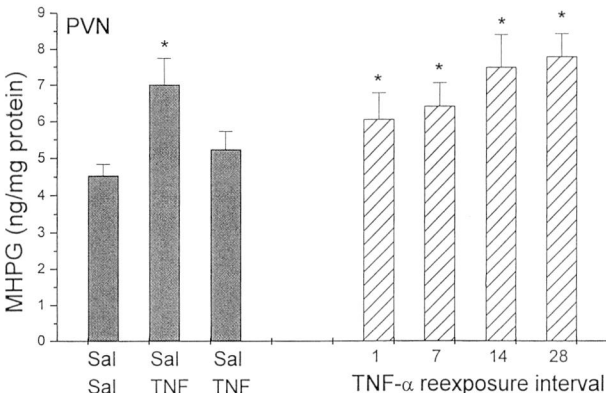

Fig. 8 Mean (±SEM) concentrations of MHPG within the PVN as a function of the initial and reexposure TNF-α treatments. The three groups on the left (solid bars) received either saline on two occasions, saline followed two weeks later by the low dose of TNF-α (1.0 μg), or a high dose of TNF-α (4.0 μg) followed two weeks later by saline. The four groups on the right (hatched bars) received an initial 4.0 μg dose of TNF-α followed by a second 1.0 μg dose either 1, 7, 14 or 28 days later. P < 0.05 relative to saline treated mice (From Hayley et al., 1999).

It is clear that administration of TNF-α may induce a sensitization effect, such that the aminergic changes exerted by the cytokine are augmented upon cytokine reexposure, but the time dependent profile is readily distinguishable from that seen in the hypothalamus, and is also dependent upon the specific transmitter being determined.

Summary

Both IL-1β and TNF-α, which had previously been shown to induce sickness-like behaviors and variations of anxiety, also induced marked neuroendocrine and central monoamine changes. The latter effects, significantly, were not restricted to the hypothalamus but occurred in several mesolimbic sites, and in several respects were reminiscent of the actions elicited by psychogenic stressors. Furthermore, it appears that a cytokine challenge may engender a sensitization effect, wherein the response to subsequent cytokine challenge is augmented. As well, it has been reported that cytokine administration may influence the response to subsequently applied stressor experiences (Tilders and Schmidt, 1998).

In several respects, the actions of cytokines and endotoxin challenge are reminiscent of the effects ordinarily elicited by stressors. Yet, it needs to be considered that a preparation comprising administration of a single cytokine, as a bolus injection, does not fully represent the effects of metabolic stressors. After all, viral and bacterial insults are sustained over protracted periods, and it is obviously relevant to assess the effects of continuous infusion of cytokines or endotoxins (e.g., Ilyin and Plata-Salaman, 1996). In the

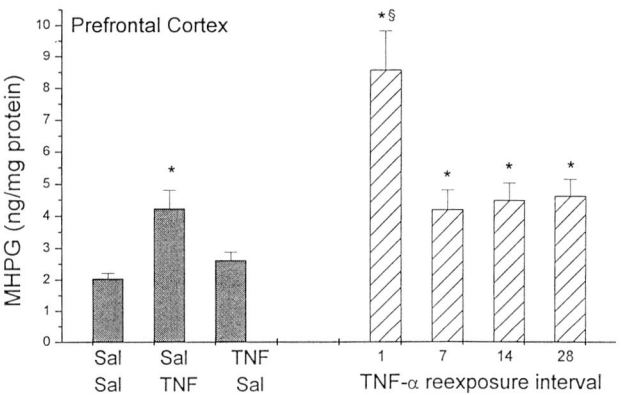

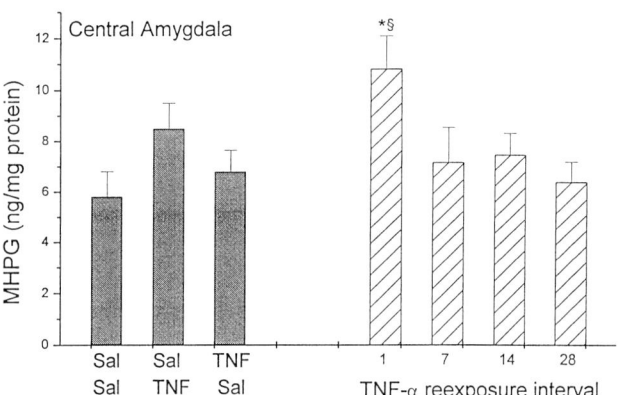

Fig. 9 Mean (±SEM) concentrations of MHPG (top) and 5-HIAA (bottom) within the central amygdala as a function of the TNF-α treatments mice received. The three groups on the left (solid bars) received ip administration of either saline on two occasions, saline followed two weeks later by the low dose of TNF-α (1.0 μg), or a high dose of TNF-α (4.0 μg) followed two weeks later by saline. The four groups on the right (hatched bars) received an initial 4.0 μg dose of TNF-α followed by a second 1.0 μg dose either 1, 7, 14 or 28 days later. *P < 0.05 relative to saline treated mice (From Hayley et al., 1999).

case of processive stressors, it is known that chronic insults promote a compensatory increase of amine synthesis, such that stores of the transmitter will not decline (e.g., Anisman et al., 1991). While this may serve in an adaptive capacity to ensure adequate supplies of the transmitter, it was suggested that the wear and tear induced by attempts to adapt to a chronic stressor (allostatic load), particularly with respect to CRH functioning in the amygdala, may have adverse behavioral repercussions (Schulkin et al., 1998). At this juncture, it is unclear what behavioral and central neurochemical alterations are introduced by sustained cytokine administration; however, analyses regarding immune-brain interactions and their ramifications for psychopathology, need to consider the impact of continuous and sustained insults.

Acknowledgements

Supported by the Medical Research Council of Canada. H.A. is a Senior Research Fellow of the Ontario Mental Health Foundation.

References

Anisman, H., and Merali, Z. (1999). Cytokines and stress in relation to anxiety and anhedonia. In: R. Dantzer, E. E. Wollmann and R. Yirmiya (Eds.), *Cytokines, Stress and Depression* London, Plenum Press.

Anisman, H. Kokkinidis, L. Borowski, T., and Merali, Z. (1998). Differential effects of IL-1, IL-2 and IL-6 on responding for rewarding lateral hypothalamic stimulation. *Brain Research* **779**, 177–187.

Anisman, H., Zalcman, S., Shanks, N., and Zacharko, R.M., (1991). Multisystem regulation of performance deficits induced by stressors: An animal model of depression. In: A. Boulton, G. Baker and M. Martin-Iverson (Eds.), *Neuromethods, vol. 19: Animal Models of Psychiatry, II*, pp. 1–59. New Jersey: Humana Press.

Anisman, H. Lu, Z.W. Song, C. Kent, P. McIntyre, D.C., and Merali, Z. (1997). Influence of psychogenic and neurogenic stressors on endocrine and immune activity: differential effects in fast and slow seizing rat strains. *Brain, Behavior and Immunity* **11**, 63–74.

Bartanusz, V., Jezova, D., Bertini, L.T., Tilders, F.J., Aubry, J.M., and Kiss J.Z. (1993). Stress-induced increase in vasopressin and corticotropin-releasing factor expression in hypophysiotrophic paraventricular neurons. *Endocrinology* **132**, 895–902.

Blalock, J.E. (1994). The syntax of immune-endocrine communication. *Immunology Today* **15**, 504–544.

Borowski, T., Kokkinidis, L., Merali, Z., and Anisman, H. (1998). Lipopolysaccharide, central in vivo biogenic amine variations, and anhedonia. *NeuroReport* **9**, 3797–3802.

Brebner, K., Hayley, S., Merali, Z., and Anisman, H. (1998). Synergistic actions of interleukin-1β, interleukin-6 and TNF-α: Central neurochemical, neuroendocrine and behavioral alterations. *Neuroscience Abstracts* **24**, 2075.

Capuron, L., Ravaud, A., Radat, F., Dantzer, R., and Goodall, G. (1998). Affects of interleukin-2 and alpha-interferon cytokine immunotherapy on the mood and cognitive performance of cancer patients. *Neuroimmunomodulation*, **5**, 9.

Dantzer, R., Bluthe, R.M., Laye, S., Bret-Dibat, J.L., Parnet, P., and Kelley, K.W. (1998). Cytokines and sickness behavior. *Annals of the New York Academy of Sciences* **840**, 586–590.

Denicoff, K.D., Rubinow, D.R., Papa, M.Z., Simpson, L., Seipp, L.A., Lotze, M.T., Chang, A.E., Rosenstein, D., and Rosenberg, S.A., (1987). The neuropsychiatric effects of treatment with interleukin-2 and lymphokine-activated killer cells. *Annals of Internal Medicine* **107**, 293–300.

Deutch, A.Y., and Roth, R.H. (1990). The determinants of stress-induced activation of the prefrontal cortical dopamine system. In: H.B.M. Uylings, C.G. Van Eden, J.P.C. De Bruin, M.A. Corner, and M.G.P. Feenstra (Eds.), *Progress in Brain Research* Vol. 85., pp. 367–403, New York, Elsevier.

Dunn, A.J. (1990). Interleukin-1 as a stimulator of hormone secretion. *Progress in NeuroEndocrinImmunology* **3**, 26–34.

Dunn, A.J. (1992). The role of interleukin-1 and tumor necrosis factor in the neurochemical and neuroendocrine responses to endotoxin. *Brain Research Bulletin* **29**, 807–812.

Hayley, S., Brebner, K., Lacosta, S., Merali, Z., and Anisman, H. (1999). Sensitization to the effects of tumor necrosis factor-α: Neuroendocrine, central monoamine and behavioral variations. *Journal of Neuroscience* **19**, 5654–5665.

Herman, J.P., and Cullinan, W.E. (1997). Neurocircuitry of stress: Central control of hypothalamo-pituitary-adrenocortical axis. *Trends in Neuroscience* **20**, 78–84.

Ilyin, S.E., and Plata-Salaman, C.R. (1996). An approach to study molecular mechanisms involved in cytokine-induced anorexia. *Journal of Neuroscience Methods* **70**, 33–8.

Kalivas, P.W., and Stewart, J. (1991). Dopamine transmission in the initiation and expression of drug- and stress-induced sensitization of motor activity. *Brain Research Reviews* **16**, 223–244.

Lacosta, S., Merali, Z., and Anisman, H. (1999). Behavioral and neurochemical consequences of lipopolysaccharide in mice: anxiogenic-like effects. *Brain Research* **818**, 291–303.

Lacosta, S., Merali, Z., and Anisman, H. (1998). Influence of interleukin-1 on exploratory behaviors, plasma ACTH and cortisol, and central biogenic amines in mice. *Psychopharmacology* **137**, 351–361.

Linthorst, A.C.E., Flachskamm, C., Muller-Preuss, P., Holsboer, F., and Reul, J.M.H.M., (1995). Effect of bacterial endotoxin and interleukin-1β on hippocampal serotonergic neurotransmission, behavioral activity, and free corticosterone levels: An in vivo microdialysis study. *Journal of Neuroscience* **15**, 2920–2934.

Maes M. (1995). Evidence for an immune response in major depression: A review and hypothesis. *Progress in Neuropsychopharmacology and Biological Psychiatry* **19**, 11–38.

Maier S.F., and Watkins L.R. (1998). Cytokines for psychologists: Implications of bidirectional immune-to-brain communication for understanding behavior, mood, and cognition. *Psychological Review* **105**, 83–107.

McIntyre, D.C. Kelly, M.E., and Dufresne, C. (1999). FAST and SLOW amygdala kindling rat strains: Comparison of amygdala, hippocampal, piriform and perirhinal cortex kindling. *Epilepsy Research*, **35**, 197–209.

McIntyre, D.C., Kent, P., Hayley, S., Merali, Z., and Anisman, H. (1999). Influence of psychogenic and neurogenic stressors on neuroendocrine and central monoamine activity in fast and slow kindling rats. *Brain Research*, **840**, 65–74.

Merali, Z., Lacosta, S., and Anisman, H. (1997). Effects of interleukin-1β and mild stress on central monoamine alterations: A regional microdialysis study. *Brain Research* **761**, 225–235.

Mohapel, P., and McIntyre, D.C. (1998). Amygdala kindling-resistant (SLOW) or-prone (FAST) rats strains show differential fear responses. *Behavioral Neuroscience* **112**, 1402–1413.

Nisenbaum, L.K., Zigmond, M.J., Sved, A.F., and Abercrombie, E.D. (1991). Prior exposure to chronic stress results in enhanced synthesis and release of hippocampal norepinephrine in response to a novel stressor. *Journal of Neuroscience* **11**, 1478–1484.

Rivier, C. (1993). Effect of peripheral and central cytokines on the hypothalamic-pituitary-adrenal axis of the rat. *Annals of the New York Academy of Sciences* **697**, 97–105.

Rothwell, N.J. (1999) Cytokines-killers in the brain. *Journal of Physiology* **514**, 3–17.

Schmidt, E.D., Janszen, A.W.J.W., Wouterlood, F.G., Tilders, F.J.H., (1995). Interleukin-1 induced long-lasting changes in hypothalamic corticotropin-releasing hormone (CRH) neurons and hyperresponsiveness of the hypothalamic-pituitary-adrenal axis. *Journal of Neuroscience* **15**, 7417–7426.

Schmidt E.D., Binnekade R., Janszen A.W., Tilders F.J.H. (1996). Short stressor induced long-lasting increases of vasopressin stores in hypothalamic corticotropin-releasing hormone (CRH) neurons in adult rat. *Journal of Neuroendocrinology* **8**, 703–712.

Schulkin, J., Gold, P.W., and McEwen, B.S. (1998). Induction of corticotropin-releasing hormone gene expression by glucocorticoids: Implication for understanding the states of fear and anxiety and allostatic load. *Psychoneuroendocrinology* **23**, 219–243.

Shanks, N., Larocque, S. & Meaney, N. (1995). Neonatal endotoxin exposure alters the development of the hypothalamic-pituitary-adrenal axis: Early illness and later responsivity to stress. *Journal of Neuroscience* **15**, 376–384.

Shintani, F., Kanba, S., Nakaki, T., Nibuya, M., Kinoshita, N., Suzuki, E., Yagi, G., Kato, R. and Asai M. (1993). Interleukin-1β augments release of norepinephrine, dopamine and serotonin in the rat anterior hypothalamus. *Journal of Neuroscience* **13**, 3574–3581.

Song, C., Merali, Z., and Anisman, H. (1998). Variations of nucleus accumbens dopamine and serotonin following systemic interleukin-1, interleukin-2 or interleukin-6 treatment. *Neuroscience* **88**, 823–836.

Tilders F.J.H., and Schmidt E.D. (1998) Interleukin-1-induced plasticity of hypothalamic CRH neurons and long-term stress hyperresponsiveness. *Annals of the New York Academy of Sciences* **840**, 65–73.

Tilders, F.J.H., Schmidt, E.D., and De Goeij D.C.E. (1993). Phenotypic plasticity of CRF neurons during stress. *Annals of the New York Academy of Sciences* **697**, 39–52.

Wang, J. & Dunn, A.J. (1999). The role of interleukin-6 in the activation of the hypothalamo-pituitary-adrenocortical axis and brain indoleamines by endotoxin and interleukin-1β. *Brain Research* **815**, 337–348.

Yirmiya R. (1996). Endotoxin produces a depressive-like episode in rats. *Brain Research* **711**, 163–174.

29. Neuropeptides in Immune Tissues: Mediators of the Stress Response

A.J. Fulford and D.S. Jessop

Introduction

Exposure of the body to stress results in a complex and integrated homeostatic response involving activation of the central nervous system, autonomic nervous system and immune systems. A range of stressful stimuli, either psychological or physical, results in activation of the hypothalamic-pituitary adrenal (HPA) axis to release ACTH from the anterior pituitary (Buckingham *et al.*, 1997). The end result of this activation is the secretion of glucocorticoids from the adrenal cortex into the blood. An inability to mount a robust response to stress is potentially pathophysiological and can be associated with serious illness, including disorders of the cardiovascular system, gastrointestinal tract and psychiatric illness.

The primary HPA axis neuropeptides stimulated by stress, CRH, AVP and the POMC peptide products ACTH and β-endorphin, are also found in the immune system. CRH immunoreactivity and mRNA has been identified in all cell types of the immune system including macrophages, fibroblasts, T and B lymphocytes, leucocytes and in the primary lymphoid organs, the spleen and thymus (Stephanou *et al.*, 1990; Aird *et al.*, 1993; Jessop *et al.*, 1994). Immune cells appear to contain the structural equivalent of CRH found in neuroendocrine cells. Additionally, T lymphocytes and thymocytes appear to synthesize novel forms of the CRH peptide (Chowdrey *et al.*, 1994). The fact that immune cells express CRH mRNA and CRH receptors indicates that these immune cells synthesize and respond to CRH (Carr, 1992). The physiological role of immune-derived CRH remains to be fully established however, it may play an important role in regulating or mediating inflammation. Immune CRH was first reported in human lymphocytes (Stephanou *et al.*, 1990), followed by a demonstration that CRH expression was increased in inflamed synovial tissue (Karalis *et al.*, 1991). Evidence exists that, in contrast to the anti-inflammatory effects of hypothalamic CRH, peripherally synthesized CRH can be pro-inflammatory (Crofford *et al.*, 1995; Jessop *et al.*, 1997a).

Immune cells and lymphoid organs synthesize and secrete POMC-derived peptides equivalent to those found in pituitary cells. In addition to full length ACTH, they produce truncated transcripts of the POMC gene (Galin *et al.*, 1991) which may be responsible for the isolation of multiple fragments of ACTH in immune cells (Jessop *et al.*, 1994; Lyons and Blalock, 1995, 1997). Multiple isoforms of immunoneuropeptides may have different functional effects on immune function. Early studies demonstrated that activated leucocytes contain increased POMC mRNA expression (Galin *et al.*, 1991; Lolait *et al.*, 1986). Other cell types such as splenic macrophages constitutively express the POMC gene (Lyons and Blalock, 1995).

POMC products could exert local paracrine or autocrine influences over immune cells by interacting with their cell-surface receptors. In the early 1980's ACTH was identified in human leucocytes and macrophages (Blalock and Smith, 1980). It appears that stimulation of T and B lymphocytes is required for expression of ACTH, whereas macrophages express ACTH only when activated. The POMC gene in immune cells appears to respond to salient stimuli in a comparable manner to the POMC gene in neuroendocrine tissues. Mitogens, superantigens as well as neuropeptides such as CRF, augment expression of the POMC gene.

In addition to ACTH, another POMC encoded peptide, β-endorphin, is found in immune cells (Heijnen *et al.*, 1991; Panerai and Sacerdote, 1997). β-endorphin contents of the spleen and thymus can be altered by the chronic inflammatory stress of adjuvant-induced arthritis in the rat (Jessop *et al.*, 1995). Both ACTH and β-endorphin have been shown to increase lymphocyte proliferation responses (Gilman *et al.*, 1982) and increase production of the inflammatory cytokine, tumor necrosis factor alpha (Harbour *et al.*, 1991). In addition, β-endorphin increases stimulation of calcium mobilization in immune cells (Shahabi *et al.*, 1996).

Opioid Peptides and Immune Function

Radioligand binding and mRNA expression studies have identified all classes of opioid receptors and their encoding mRNAs in immune tissues (Sibinga and Goldstein, 1988; Carr *et al.*, 1996; Madden *et al.*, 1998). This is suggestive of an important role for the opioids in the regulation of immune function and inflammatory responses. Morphine, the prototypical μ-opiate agonist, and its effects on immune function have been the focus of research for many years. A major side effect of morphine use clinically is the drug's propensity to cause immunosuppression (Roy *et al.*, 1997; Hall *et al.*, 1998). Morphine's effects on natural killer cell activity and lymphocyte proliferation responses to polyclonal mitogens, and its more complex cellular effects in immune cells, have been attributed to a central site of action (Eisenstein and Hilburger, 1998). However, the isolation of μ-opiate receptors on peripheral immune cells implicates a peripheral site of action in addition to a central modulatory effect. Morphine, despite having high affinity for the μ-receptor, also activates other opioid receptors. Thus, it is unclear which of the effects of morphine are due to activation of μ or other opioid receptor subtypes. The recent identification of endomorphins as likely candidates for endogenous μ-selective agonists (Zadina et al., 1997), will enable the delineation of specific μ-mediated effects on immune cells. We recently identified endomorphins in primary lymphoid organs suggesting that these opioids may modulate immune cell function through paracrine mechanisms (Jessop et al., 1999).

Another important family of endogenous opioid peptides are the enkephalins. Enkephalin peptides and their prohormone precursor, proenkephalin A (PEA) are found in immune tissues (Przewlocki et al., 1992) including all types of immune cell; T and B lymphocytes (Zurawski et al., 1986; Rosen et al., 1989), thymocytes (Linner et al., 1991), macrophages and mast cells (Martin et al., 1987). PEA is proteolytically cleaved to form six copies of met-enkephalin and one copy of leu-enkephalin. These peptides have highest affinity for the d-opioid receptor and are putatively involved in local inflammation and autocrine/ paracrine control of immune function. For instance, met-enkephalin and PEA are found in immune cells at sites of inflammation (Przewlocki et al., 1992). Enkephalins can modulate several aspects of immune status including B cell production of antibodies (Jankovic and Maric, 1987), natural killer (NK) cell activity (Faith et al., 1987), macrophage-induced phagocytosis and mitogen-induced proliferation of lymphocytes (Plotnikoff and Miller, 1983).

The precise functional role of the various opioid peptides at inflamed sites requires further investigation but they could modulate inflammation by activation of opioid receptors on peripheral sensory afferent nerves in inflamed sites to suppress neurogenic pain (Chang et al., 1989; Stein et al., 1990, 1993). Opioid peptides appear to have multiple effects on immune function depending on the cell type studied, dose of the peptide and whether its effects are assessed in vitro or in vivo. Opioids exert potent effects in the CNS including regulation of mood, reinforcement and analgesia. It is likely that the paracrine effects of opioids at immune sites may be obscured by the effects of opioids acting centrally.

Neuropeptides, Stress and Immune Activation-Methodological Perspectives

There is widespread interest in the role of the HPA axis in the etiology of inflammatory disorders as it has been demonstrated that stress can influence disease outcome. Attention has focused on the influence of various stressors on neuropeptide expression and immunoreactivity in immune tissues. Many studies have demonstrated that stress can influence immune function. The ability of various stressors to modulate immunoneuropeptide contents in immune tissues has led to the suggestion that the effects of stress on immune functions may be mediated through immunoneuropeptides, as well as through the actions of glucocorticoids.

Studies of the influence of stress on HPA axis activation in rodents have utilized either acute, repeated or chronic stressors. We have widely employed the model of adjuvant-induced arthritis (AA) in the rat which is associated with the development of chronic inflammatory disease, release of pituitary hormones and subsequent elevation of levels of circulating corticosteroids. AA is a T-lymphocyte-dependent autoimmune disease which is induced in specific strains of rat by injection of a heat-killed suspension of Mycobacterium butyricum in paraffin oil intradermally at the tail-base. Hindpaw inflammation develops 12–14 days later (Sarlis et al., 1992). We have observed changes in CRF, ACTH and β-endorphin contents in immune tissues in response to acute restraint stress or cytokine injection (Jessop et al., 1997b) and the chronic inflammatory stressors of AA (Jessop et al., 1995) and experimental allergic encephalomyelitis (EAE). EAE in rats was associated with changes in CRF, ACTH and β-endorphin contents of the thymus (Figure 1). These results extend our previous reports that tissue contents of neuropeptides in the spleen and thymus can be altered by the chronic inflammatory stress associated with autoimmune disease (Jessop et al., 1995). In the EAE experiment there was a consistent pattern showing decreases in CRF, ACTH and β-endorphin contents in the thymus at day 3 of the preclinical phase and an increase at day 11 following recovery, whereas no changes in peptide contents were observed in the day 7 group, which exhibited the symptoms of weight loss, hind-limb hypotonia and hind limb paralysis associated with EAE. Decreased expression of these neuropeptides in the period preceding the onset of inflammation, coupled with enhanced expression during the recovery phase, is consistent with a net protective role for CRF, ACTH and β-endorphin in the inflammatory autoimmune disease of EAE.

Assessments of Exogenous or Endogenous Immunoneuropeptide Effects

Most studies investigating immunoneuropeptides have studied the effects of exogenously applied neuropeptides on indices of immune function, either in vitro or in vivo. These studies have demonstrated how synthetic peptides can modulate immune cell function, presumably via interactions with specific receptors on immune cells. However, several studies report conflicting results with either immunoenhancing or immunosuppressive effects. Given the reports of structural variants of some immunoneuropeptides such as ACTH and CRF (Jessop

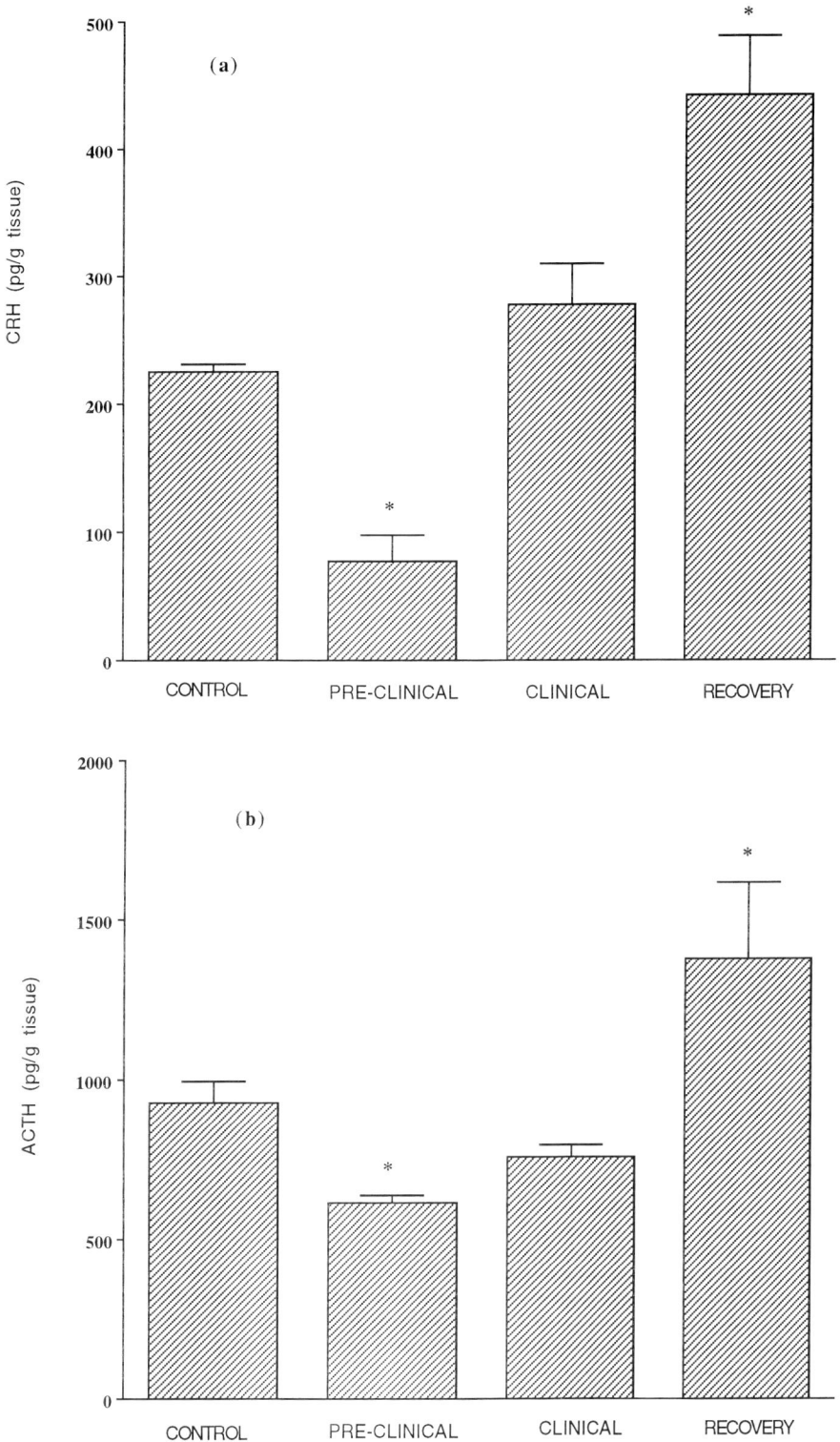

Fig. 1 Effect of EAE on (a) CRH, (b) ACTH and (c) β-endorphin contents in thymus extracts from EAE rats in control, preclinical, clinical and recovery groups. Values are mean ± SEM, n = 6–8 per group. *P < 0.05 treatment compared to control group, one-way ANOVA followed by Fisher PLSD test. EAE was induced in female Lewis rats by immunization of the hind foot with complete Freund's adjuvant (50 μl) containing heat-killed *Mycobacterium butyricum* and myelin basic protein. Eleven days after immunization, spleens were removed and splenocytes cultured for three days. After culture, cells were injected intraperitoneally into naive recipients, groups of which were decapitated on days 3 (preclinical), 7 (clinical) and 11 (recovery). The clinical course of adoptively transferred EAE was scored according to the method previously described (Leonard *et al.*, 1990). Spleens and thymuses were removed after decapitation and frozen on dry ice. Tissues were then weighed and homogenized in 2 ml ice-cold acetic acid (0.1 M) containing 2-mercaptoethanol (0.01 M). After centrifugation, supernatants were heated for 15 min at 90°C, dried by vacuum centrifuge and reconstituted in buffer for radioimmunoassay.

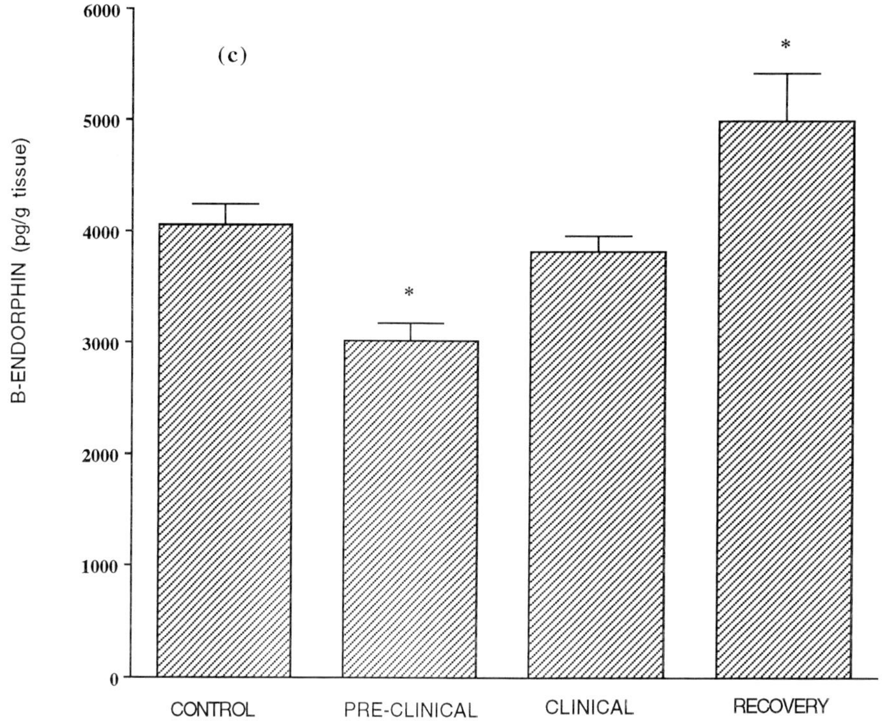

Fig. 1 (*continued*)

et al., 1994; Chowdrey *et al.*, 1994), in addition to the existence of several POMC transcripts and the ineffectiveness of receptor antagonists in reversing some peptide effects, it is likely that application of synthetic peptide equivalents may not be the most suitable strategy for targeting immunoneuropeptide systems.

As an alternative approach we have employed antisense technology to target endogenous immuno-neuropeptide systems. We have constructed 18 base length oligodeoxynucleotide antisense and nonsense probes which are sequences of DNA complementary to target sequences of mRNA of the gene of interest. Antisense probes hybridize to host mRNA molecules and this is proposed to prevent mRNA translation into protein in the cytoplasm of the cell. We routinely use phosphodiester probes as these have demonstrated specificity and efficacy *in vitro* without the unwanted toxic effects associated with modified oligonucleotides, in particular those with phosphothioate moieties. We have used antisense probes which hybridize to the CRF (Jessop *et al.*, 1997b), POMC and PEA (Fulford *et al.*, 1999) mRNAs in rat primary immunocyte cultures and found reliable and reproducible effects on mitogen-induced cell proliferation and on neuropeptide immunoreactivity. Incubation of rat splenocytes with a POMC antisense probe for 72 h results in inhibition of cell proliferation (Figure 2). Conversely, incubation with the PEA antisense probe significantly increased splenocyte proliferation to concanavalin A (Figure 3).These findings suggest a proinflammatory role for

POMC-derived peptides and an antiinflammatory role for PEA-derived peptides in immune cells. This opioid-mediated immunomodulation may represent an important regulatory mechanism in immune cells and is potentially important in inflammation.

Conclusions

There is now ample evidence for the expression of CRF, ACTH, β-endorphin and the enkephalins and their receptors within immune tissues. Opioid peptides are found in many types of immune cell and cloning of their receptors has revealed these to be equivalent to those found centrally. The incidence of opportunistic infections in drug misusers and AIDS patients is a major public health concern and this demonstrates the impact of opioids on resistance to infection. Significant advances in opioid research have been made but more research is essential to advance our appreciation of the opioids and other immunoneuropeptides in the regulation of immune homeostasis.

Low tissue concentrations of immunoneuropeptides suggest that they are paracrine/autocrine mediators of immune function. One possible mechanism is that immunoneuropeptides play a critical role in the way the brain communicates with the immune system by controlling release of cytokines from lymphocytes in response to stress activation of the sympathetic nervous system or HPA axis. Now that many of the cellular components for neuropeptide/immune interactions

POMC

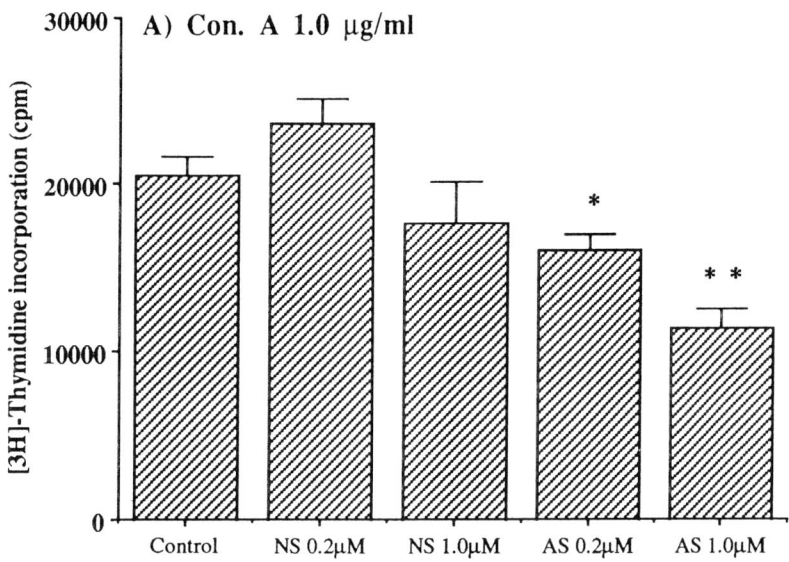

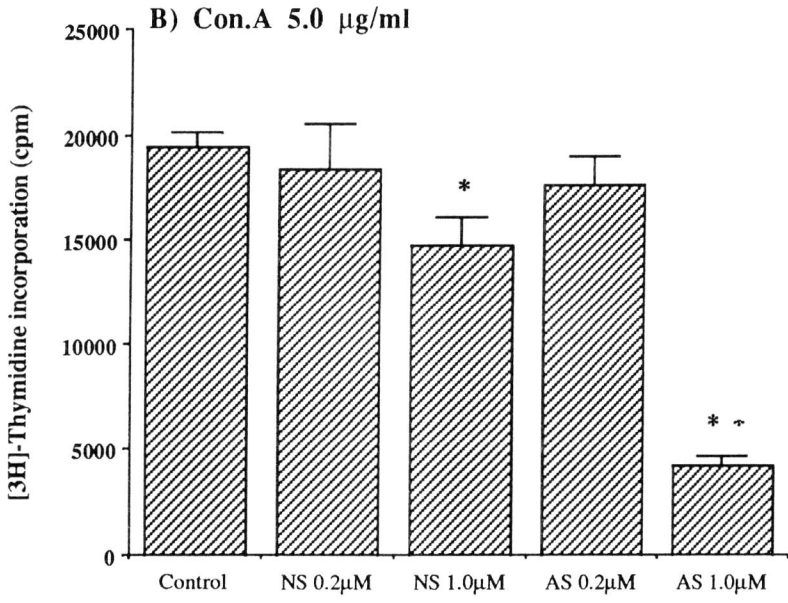

Fig. 2 Effect of proopiomelanocortin (POMC) antisense (AS) and nonsense (NS) oligodeoxynucleotides (0.2 and 1.0 μM) on proliferation of rat splenocytes to concanavalin A 1.0 μg/ml (panel A) and 5.0 μg/ml (panel B). Values are means $\pm$ SEM (n = 6–12) and are representative data from five experiments. *P < 0.05, **P < 0.01 treatment versus control (no probe), one-way ANOVA followed by posthoc Fisher PLSD test. Cells from the spleens of male Sprague-Dawley rats (220–280 g) were dispersed in Hanks solution, erythrocytes lysed and splenocytes suspended in RPMI-1640 medium supplemented with streptomycin/penicillin (1%), sodium pyruvate (0.1%), 2-mercaptoethanol (0.1%) and fetal calf serum (10%). Viable cells were seeded in flat-bottomed 96 well plates at a density of 5×10^5 cells/200 μl well and incubated at 37°C in 95% O_2/ 5% CO_2 in the presence of POMC probes (0.2 or 1.0 μM). 24 h later cells were activated with concanavalin A and incubated for a further 48 h. Cells were then incubated overnight in the presence of [^{3}H]-thymidine (0.25 μCi/ well) and incorporation subsequently measured to give an index of cell [^{3}H] proliferation.

have been identified, the real challenge is to test the physiological relevance of these interactions by experimentally disrupting immunoneuropeptide synthesis and establishing causal links with immunological dysfunction. The reward of meeting this challenge is that novel neuroendocrine mechanisms may be revealed through which the immune system responds to stress, disease and infection.

PREPROENKEPHALIN

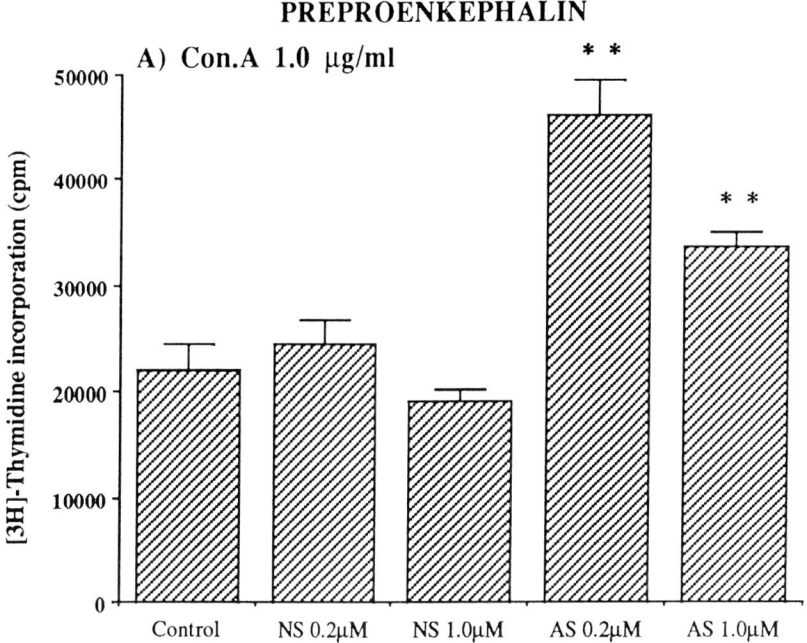

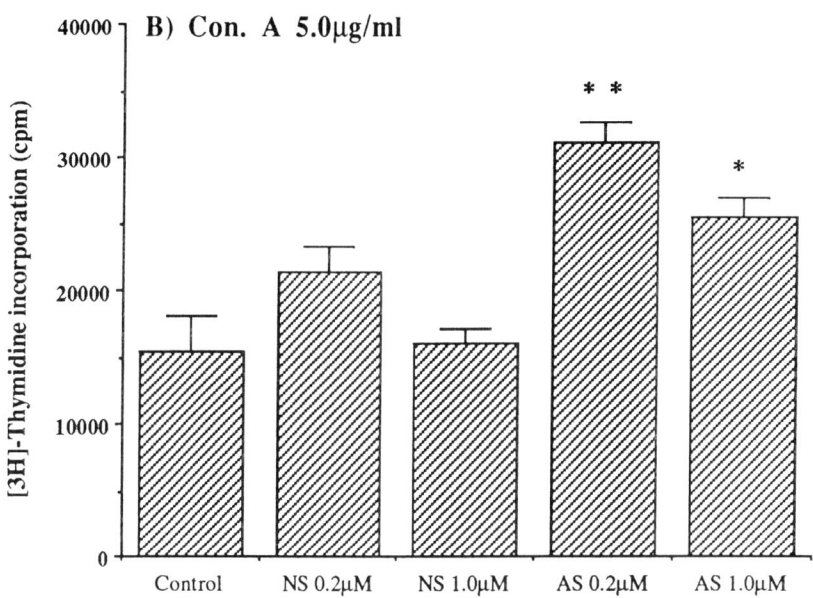

Fig. 3 Effect of proenkephalin A (PEA) antisense (AS) and nonsense (NS) oligodeoxynucleotides on proliferation of rat splenocytes to concanavalin A 1.0 μg/ml (panel A) and 5.0 μg/ml (panel B). Values are expressed as means ± SEM (n = 6–12) and are representative results from five independent experiments. *P < 0.05, **P < 0.01 treatment versus control (no probe), one-way ANOVA with posthoc Fisher PLSD test. Experiments were performed according to the method described in the legend to Figure 2. In all cases oligos were phosphodiester probes, synthesized using a PerSeptive Biosystems Expedite 8909 Nucleic Acid Synthesizer and purifed by ethanol precipitation. Antisense probes were 18 mer complementary to the AUG initiation codon regions of the mRNA encoding the rat POMC and PEA genes. Nonsense probes had an identical base composition to antisense probes but their sequences were scrambled.

References

Aird, F., Clevenger, C.V., Prystowsky, M., and Redei, E. (1993). Corticotropin-releasing factor mRNA in rat thymus and spleen. *Proceedings of the National Academy of Sciences* U.S.A. **90**, 7104–7108.

Blalock, J.E., and Smith, E.M. (1980). Human leukocyte interferon: structural and biological relatedness to adrenocortical hormone and endorphins. *Proceedings of the National Academy of Sciences* U.S.A. **77**, 5972–5974.

Buckingham, J.C., Cowell, A.-M., Gillies, G.E., Herbison, A.E., and Steel, J.H. (1997). The neuroendocrine system: anatomy, physiology and responses to stress. In: J.C. Buckingham, A-M. Cowell and G.E. Gillies (Eds.), *Stress, Stress Hormones and the Immune System*, pp. 9–47. John Wiley and Sons, Chichester.

Carr, D.J.J. (1992). Neuroendocrine peptide receptors on cells of the immune system. In: *Neuroimmunoendocrinology*, Second Rev. Ed. *Chemical Immunology*, Karger, Basel, Vol. 52, pp. 84–105.

Carr, D.J.J., Rogers, T.J., and Weber, R.J. (1996). The relevance of opioid receptors on immunocompetence and immune homeostasis. *Proceedings of the Society for Experimental Biology and Medicine* **213**, 248–257.

Chang, H.M., Berde, C.B., Holz, G.G., Steward, G.F., and Kream, R.M. (1989). Sufentanil, morphine, met-enkephalin and k-agonist (U-50,488H) inhibit substance P release from primary sensory neurons: a model for presynaptic signal opioid actions. *Anaesthesiology* **70**, 672–677.

Chowdrey, H.S., Lightman, S.L., Harbuz, M.S., Larsen, P.J., and Jessop, D.S. (1994). Contents of corticotropin-releasing hormone and arginine vasopressin immunoreactivity in the spleen and thymus during a chronic inflammatory stress. *Journal of Neuroimmunology* **53**, 17–21.

Crofford, L.J., Sano, H., Karalis, K., Webster, E., Friedman, T.C., Chrousos, G., and Wilder, R. (1995). Local expression of corticotropin-releasing hormone in inflammatory arthritis. *Annals of the New York Academy of Sciences* **771**, 459–471.

Eisenstein, T.K., and Hilburger, M.E. (1998). Opioid modulation of immune responses: effects on phagocytic and lymphoid cell populations. *Journal of Neuroimmunology* **83** 36–44.

Faith, R.E., Jurgo, A.J., Clinkscales, C.W., and Plotnikoff, N.P. (1987). Enhancement of host resistance to viral and tumour challenge with methionine-enkephalin. *Annals of the New York Academy of Sciences* **496**, 137–145.

Fulford, A.J., Harbuz, M.S., and Jessop, D.S. (1999). Antisense inhibition of opioid peptide synthesis alters rat immunocyte function *in vitro*. *Neuroimmunomodulation* **6**, 219.

Galin, F.S., LeBoeuf, R.D., and Blalock, J.E. (1991). Corticotropin-releasing factor upregulates expression of two truncated pro-opiomelanocortin transcripts in murine lymphocytes. *Journal of Neuroimmunology* **31**, 51–58.

Gilman, S.J., Schwartz, J.M., Miner, R.J., Bloom, F.E., and Feldman, J.D. (1982). b-Endorphin enhances lymphocyte proliferative responses. *Proceedings of the National Academy of Sciences U.S.A.* **79**, 4226–4230.

Hall, D.M., Suo, J.-L., and Weber, R.J. (1998). Opioid mediated effects on the immune system: sympathetic nervous system involvement. *Journal of Neuroimmunology* **83**, 29–35.

Harbour, D.V., Galin, F., Hughes, T., Smith, E.M., and Blalock, J.E. (1991). Role of leukocyte-derived pro-opiomelanocortin peptides in endotoxic shock. *Circulatory Shock* **35**, 181–191.

Heijnen, C.J., Kavelaars, A., and Ballieux, R.E. (1991). b-Endorphin: cytokine and neuropeptide. *Immunological Reviews* **119**, 41–63.

Jankovic, B.D., and Maric, D. (1987). Enkephalins and immunity: *in vivo* suppression and potentiation of humoral immune response. *Annals of the New York Academy of Sciences* **496**, 115–125.

Jessop, D.S., Lightman, S.L., and Chowdrey, H.S. (1994). Effects of a chronic inflammatory stress on levels of pro-opiomelanocortin-derived peptides in the rat spleen and thymus. *Journal of Neuroimmunology* **49**, 197–203.

Jessop, D.S., Renshaw, D., Lightman, S.L., and Harbuz, M.S. (1995). Changes in ACTH and β-endorphin immunoreactivity in immune tissues during a chronic inflammatory stress are not correlated with changes in corticotropin-releasing hormone and arginine vasopressin. *Journal of Neuroimmunology* **60**, 29–35.

Jessop, D.S., Harbuz, M.S., Snelson, C.L., Dayan, C.M., and Lightman, S.L. (1997a). An antisense oligodeoxynucleotide complementary to corticotropin-releasing hormone mRNA inhibits rat splenocyte proliferation *in vitro*. *Journal of Neuroimmunology* **75**, 135–140.

Jessop, D.S., Douthwaite, J.A., Conde, G.L., Lightman, S.L., Dayan, C.M., and Harbuz, M.S. (1997b). Effects of acute stress or centrally injected interleukin-1b on neuropeptide expression in the immune system. *Stress* **2**, 133–144.

Jessop, D.S., Kaye, S., Coventry, T., Fulford, A.J., and Harbuz, M.S. (1999). Novel opioid peptides endomorphin-1 and endomorphin-2 are located in rat spleen and thymus and circulate in the blood. *Neuroimmunomodulation* **6**, 227.

Karalis, K., Sano, H., Redwine, J., Listwak, J., Wilder, R., and Chrousos, G. (1991). Autocrine or paracrine inflammatory actions of corticotropin-releasing hormone *in vivo*. *Science* **254**, 421–423.

Leonard, J.P., MacKenzie, F.J., Patel, H.A., and Cuzner, M.L. (1990). Splenic noradrenergic and adrenocortical responses during the preclinical and clinical stages of adoptively transferred experimental autoimmune encephalomyelitis (EAE). *Journal of Neuroimmunology* **26**, 183–186.

Linner, K.M., Beyer, H.S., and Sharp, B.M. (1991). Induction of the messenger ribonucleic acid for proenkephalin A in cultured murine CD4-positive thymocytes. *Endocrinology* **128**, 717–724.

Lolait, S., Clements, J.A., Markwick, A., Cheng, C., McNulty, M., Smith, A.I., and Funder, J.W. (1986). Pro-opiomelanocortin messenger ribonucleic acid and posttranslational processing of beta-endorphin in spleen macrophages. *Journal of Clinical Investigation* **77**, 1776–1779.

Lyons, P.D., and Blalock, J.E. (1995). The kinetics of ACTH expression in rat leukocyte subpopulations. *Journal of Neuroimmunology* **63**, 103–112.

Lyons, P.D., and Blalock, J.E. (1997). Pro-opiomelanocortin gene expression and protein processing in rat mononuclear leukocytes. *Journal of Neuroimmunology* **78**, 47–56.

Madden, J.J., Whaley, W.L., and Ketelson, D. (1998). Opiate binding sites in the cellular immune system: expression and regulation. *Journal of Neuroimmunology* **83**, 57–62.

Martin, J., Prystowsky, M.B., and Angeletti, R.H. (1987). Preproenkephalin mRNA in T cells, macrophages and mast cells. *Journal of Neuroscience Research* **18**, 82–87.

Panerai, A.E., and Sacerdote, P. (1997). β-Endorphin in the immune system: a role at last? *Immunology Today* **18**, 317–319.

Plotnikoff, N.P., and Miller, G.C. (1983). Enkephalins as immunomodulators. *International Journal of Immunopharmacology* **5**, 437–441.

Przewlocki, R., Hassan, A.H.S., Lason, W., Epplen, C., Herz, A., and Stein, C. (1992). Gene expression and localisation of opioid peptides in immune cells of inflamed tissue. Functional role in antinociception. *Neuroscience* **48**, 491–500.

Rosen, H., Behar, O., Abramsky, O., and Ovadia, H. (1989). Regulated expression of proenkephalin A in normal lymphocytes. *Journal of Immunology* **143**, 3703–3707.

Roy, S., Chapin, R., Cain, K., Charboneau, R., Ramakrishnan, S., and Barke, R.A. (1997). Morphine inhibits transcriptional regulation of Il-2 synthesis in thymocytes. *Cell Immunology* **179**, 1–9.

Sarlis, N.J., Chowdrey, H.S., Stephanou, A., and Lightman, S.L. (1992). Chronic activation of the hypothalamo-pituitary-adrenal axis and loss of circadian rhthym during adjuvant-induced arthritis in the rat. *Endocrinology* **130**, 1775–1779.

Shahabi, N.A., Heagy, W., and Sharp, B.M. (1996). β-Endorphin enhances concanavalin A-stimulated calcium mobilisation by murine splenic T cells. *Endocrinology* **137**, 3386–3393.

Sibinga, N.E.S., and Goldstein, A. (1988). Opioid peptides and opioid receptors in cells of the immune system. *Annual Review of Immunology* **6**, 219–249.

Stein, C., Hassan, A.H.S., Pzewlocki, R., Gramsch, C., and Herz, A. (1990). Opioids from immunocytes interact with sensory nerves to inhibit nociception in inflammatory cells. *Proceedings of National Academy of Sciences U.S.A.* **87**, 5935–5939.

Stein, C., Hassan, A.H.S., Lehrberger, K., Giefing, J., and Yassourdis, A. (1993). Local analgesic effects of endogenous opioid peptides. *Lancet* **342**, 321–324.

Stephanou, A., Jessop, D.S., Knight, R.A., and Lightman, S.L. (1990). Corticotropin-releasing factor-like immunoreactivity and mRNA in human leukocytes. *Brain Behaviour and Immunity* **4**, 67–73.

Zadina, J.E., Hackler, L., Ge, L.-J., and Kastin, A.J. (1997). A potent and selective endogenous agonist for the opiate receptor. *Nature* **386**, 499–502.

Zurawski, G., Benedik, G.M., Kamb, B.J., Abrams, J.S., Zurawski, S.M., and Lee, F.D. (1986). Activation of mouse T-helper cells induces abundant preproenkephalin mRNA synthesis. *Science* **232**, 772–775.

30. The Ontogeny of the Neural Responses to Stress: The Influence of Maternal Factors

S. Levine, G. Dent, M.A. Smith and E.R. de Kloet

Introduction

The rodent Hypothalamic-Pituitary-Adrenal (HPA) axis of the neonate is in many ways very different from the adult. Specifically, beginning at postnatal day 4 and until approximately postnatal day 12, rat pups respond weakly to a variety of acute challenges, which include handling, electric shock, restraint, ether and surgery. These stimuli would normally result in a rapid and profound increase in circulating levels of corticosterone (CORT). It was originally thought that this diminished stress response occurred at all levels of the HPA axis (for review see de Kloet et al., 1988). However, most of the data which supported the concept of a stress hyporesponsive period (SHRP) was based primarily on the failure to observe a significant increase in CORT secretion. However, the pituitary ACTH response has been shown to be time and stressor specific and in general the ACTH response is somewhat blunted compared to the adult (Walker et al., 1991). Further, whereas in the adult increases in CRF gene expression are frequently observed, the literature contains very few indications that the neonatal CRF system is capable of increasing CRF mRNA following stress. The purpose of this paper is to examine more closely the central components of the neonatal stress response.

All of available information supports the view that during the SHRP the adrenal response to stress is severely inhibited and there is general agreement that there exists an adrenal hyporesponsive period. Although there appear to be rate-limiting factors that act developmentally to limit the secretion of CORT in the neonate, there is evidence that indicates that the adrenal is actively suppressed during the SHRP. In recent years it has been extensively documented that certain aspects of rodent maternal behavior play an important role in regulating the neonate's HPA axis. In particular, two specific components of the dam's care giving activities seem to be critical: licking/stroking and feeding. In numerous studies (Levine, 1994) it has been demonstrated that feeding is in part responsible for the down regulation of the pups' capacity to both secrete and to clear CORT from the circulation. Thus removing the mother from the litter for 24 hours results in a significantly higher basal level and a further increase in the secretion of CORT following stress or administration of ACTH. Providing the pup with milk during the 24-hr deprivation period completely restores the low basal levels and prevents CORT increases following stress (Suchecki et al., 1993).

The question we will address in this paper is whether the concept of the SHRP is valid when applied to other components of the HPA axis. In order to confront this question we will examine the development of the regulation of ACTH and the neural responses to stress. These include changes in the immediate early gene c-fos and alterations in the CRF primary transcript (hRNA) and CRF gene expression.

ACTH

The concept of an absolute SHRP with regard to the response of the pituitary following stress in the neonate is much more problematic. Whether the pituitary can show an increase in ACTH in response to stress is dependent on numerous factors. Amongst these are the age of the neonate, the type of stress imposed and once again maternal factors. The early findings concerning the stress response of the pituitary suggested that there was a deficiency in the neonate's capacity to synthesize ACTH. Thus as a result the pup should exhibit a reduction in the magnitude of the ACTH stress response. However, there are sufficient data that indicate that the pituitary of the neonate does have the capacity to synthesize and release ACTH that very much resembles the adult response. What seems to discriminate the neonate from the adult is that for the pup the response of the pituitary is much more stimulus dependent. Further, the ability to terminate the stress response is also not fully developed and does not mature until quite late in development.

Perhaps the earliest demonstration that the neonate can indeed mount an ACTH response to at least some types of challenges was a study, which challenged neonates with an injection of endotoxin throughout the period from birth to weaning (Witek-Januchek 1988). At all ages the neonate exhibited a significant elevation of ACTH which, beginning at day 5, was equivalent to that of the adult. Of interest is that although there was a robust ACTH response, the CORT response was markedly reduced from day 5 and did not begin to approach adult values until about day 15. It has been reported that administration of IL-1β and adrenalectomy do elicit an ACTH response in pups early in development (Levine et al., 1994 and van Oers et al., 1998a). The peak of the response followed a similar time course to that of the adult, although the magnitude of the response was lower early in development. The reduced response in the neonates cannot, however, be interpreted as a reduction in the neonate's capacity to produce ACTH. Three hours following ADX a robust increase in ACTH occurs as early as day 5, presumably due to the absence of a CORT negative feedback signal. This magnitude of the ACTH response is as great as that seen in older neonates at day 18, which are well out of the SHRP.

It has been reported that the neonate does show a significant increase in ACTH in response to a variety of different stimuli (Walker *et al.*, 1991). It was concluded that the neonate can respond in an "adult-like manner". It is noteworthy that for each stimulus examined there appears to be an idiosyncratic time course that is dependent on age and the type of stimulus. Regardless, it is apparent that the capacity for a pituitary response is present early in development. Under some circumstances the pup can show a greater ACTH response early in development than later. Following treatment with kainic acid the ACTH response of day 12 pups exceeded that of day 6 and day 18 neonates. Brief periods of maternal separation, exposure to novelty, injections of isotonic saline, and restraint for 30 min all fail to elicit an ACTH response in normally reared pups until they escape from the SHRP.

Why do neonates appear to discriminate between different classes of stimuli, whereas older pups, which have escaped from the SHRP and adults, appear to respond in a similar manner regardless of the stress-inducing stimulus? There are several hypotheses that could account for this phenomenon. First, it could simply be a matter of stimulus intensity. Thus the neonate may be less responsive to stimuli of lower intensities and therefore require a more intense stressor to activate the neuroendocrine cascade that eventually leads to a release of ACTH. Second, it has been well documented that different stimuli activate distinct neural pathways that lead to the release of CRF and thus ACTH (Sawchenko, 1991). It is conceivable that the neural pathways that regulate the response to different classes of stimuli mature differentially and thus if a particular stimulus activates a pathway, which matures early in development, then it is likely that a pituitary response will be manifest. However, if the regulating pathways are developing more slowly these stimuli may not be able to be processed neurally and produce the neuroendocrine cascade required to activate the pituitary.

There is a third factor that appears to contribute to the reduced capacity of the mother-reared pup to respond to milder stress inducing procedures. In the previous section of this paper we discussed the role of the mother's care giving activities on the developing adrenal. There is evidence that maternal factors can also actively inhibit the release of ACTH. Pups deprived of maternal care for 24 hr showed an increase in ACTH following an injection of saline as early as day 6. The effects of maternal deprivation were even more apparent at days 9 and 12. Although in subsequent

experiments the response at day 6 was not reliable, significant increases in ACTH were reliably reproduced in 9 and 12-day old neonates (Suchecki *et al.*, 1993b). These increases are also observed following 30 min of restraint. In contrast non-deprived pups failed to show an acute release of ACTH following stress. Whereas in order to reduce the sensitivity of the adrenal, feeding was required, anogenital stroking can reverse the increased ACTH secretion following deprivation (Suchecki *et al.*, 1993a; Van Oers *et al.*, 1998b). Thus, different components of the mother's behavior appear to be involved in regulating different components of the endocrine stress response. In pups that were stroked and fed, both ACTH and CORT are suppressed. In pups that received only stroking, ACTH was down regulated but CORT was still elevated. These data would suggest that the dam's behavior was actively inhibiting the neuroendocrine cascade that ultimately results in the peripheral endocrine responses to stress. Thus, the capacity to respond is present early in development but is only observable if the maternal inhibitory factors are not present. It is important to note, however, that although the pup during the SHRP can be induced to show an endocrine response to stress, maternal inhibition is not the only rate-limiting factor. If one examines the body of data on the ACTH responses what emerges is that even when the infant during the SHRP responds to mild stress the magnitude of the response is always considerably lower in the SHRP than that of the older pups (day 18) and adults. Since we have previously concluded that the reduction of ACTH is not due to a reduction in pituitary ACTH, we are forced to conclude that during the SHRP the neural pathways that regulate the response to these milder stimuli are not yet mature, or that deficiency may exist in the communication between the neuroendocrine events and the ultimate action of the neuropeptides on the pituitary. These could occur in a number of sites of action amongst which are the production of the secretagogue, the transport of the critical neuropeptides, the release of these peptides from the nerve terminals of the median eminence, and/or some deficiency in developing CRF receptors in the pituitary.

Maternal deprivation also modifies the responses to stimuli that are capable of activating the HPA axis in non-deprived pups. Maternal deprivation altered the ACTH response to IL-1β (Kent *et al.*, 1997). The changes were, however, age-dependent. At day 6 deprived pups showed a more robust ACTH response. At day 12 the deprived pups exhibited a more rapid response but at the end of two hours both deprived and non-deprived pups achieved similar elevated levels. Paradoxically at day 18 maternal deprivation resulted in marked suppression of ACTH following the injection of this cytokine. Although we have been emphasizing the phenomena of increased responsivity in maternal deprivation there are several studies that reveal that the effects of maternal deprivation are indeed age-dependent. At older ages (days 18 and 20) removal of the dam accompanied by fasting results in down regulation of ACTH secretion and a more rapid return to basal levels which indicates a more efficient negative feedback regulation (Smith *et al.*, 1997)

It is difficult to view the body of data on the ontogeny of the pituitary response to stress and conclude there is an absolute SHRP. We do not agree that the neonate's pituitary response to stress is "adult-like". The neonate is different in many ways that we have already discussed. First, there is the issue of stimulus specificity. Second, in general the levels of ACTH secreted tend to be reduced compared to older pups and adults. Third, there is a major deficiency in the ability of the infant to terminate the response once it is initiated. The pup is "adult-like" in only one regard: it can under certain conditions clearly mount an ACTH response during the so-called SHRP.

CRF

The brain is clearly a stress sensitive organ. It is not the purpose of this paper to review all of the changes in the brain that have been shown to occur in response to stress. For this review the focus will be on a specific neural marker that has a direct effect on the regulation of the pituitary. In their 1991 review, Rosenfeld *et al.* concluded that "the low basal and stress induced levels of CRF may be due to the immaturity (especially of those neural pathways that provide stimulatory input to the hypophysiotropic cells) and perhaps to chronic maternal inhibition. The mechanism(s) underlying this inhibition has (have) not yet been examined. The ontogenetic changes in stimulus-specificity of the stress response may reflect concurrent maturation in the specific pathways that mediate the effects of each particular stimulus." Thus the general consensus based on several studies is that early in development there is a deficiency in the neonate's capacity to increase gene transcription and peptide expression of ACTH secreta-gogues even though there is a striking increase in ACTH release. The hyporesponsiveness of CRF has been demonstrated using several distinct and potent stimuli which all induce changes in the peripheral manifestations of increased HPA activity. Hypoglyce-mia failed to induce changes in CRF mRNA in day 8 pups, in contrast to day 25 animals that did increase CRF mRNA (Grino *et al.*, 1989). However, a sub-population of CRF neurons that co-express AVP did respond at day 8. It was argued that different pathways regulate different populations of cells. Pups exposed to "maximal tolerated cold" showed a significant increase in CORT secretion (Yi, *et al.*, 1994). However, at day 6 no changes in CRF mRNA could be detected. By day 8 increases in CRF message did occur. In this experiment CRF antiserum was administered to pups at all ages and surprisingly the antiserum was able to diminish the CORT response in day 6 pups. Thus, although CRF mRNA does not appear to be altered by the severe exposure to cold, the peripheral response appears to be CRF-dependent. No compensatory changes in CRF and AVP gene expression were observed following ADX although there were marked increases in circulating

levels of ACTH which were observed in older pups (day 14–19). Thus, during the first 10 days of life, within the adrenal SHRP, hypothalamic CRF and AVP neurons appear not to be sensitive to glucocorticoid feedback and that basal ACTH secretion appears to be relatively independent of hypothalamic input.

On the basis of existing data the conclusion that CRF is also not involved in the stress response would also be warranted. However, there is recent evidence that indicates that although the neonates CRF system is very different from the adult, it may indeed be hyper rather than hyporesponsive. Dent *et al.* (2000) examined the primary CRF transcript as well as total CRF message in mother reared and maternally deprived pups at 6, 12, and 18 days of age. It was hypothesized that although CRF message was reduced in maternally deprived pups at 12 and 19 days of age, the hRNA would more likely reflect the increased stress responsiveness of the maternally deprived neonates. Surprisingly, both hRNA and CRF mRNA were down regulated by maternal deprivation. However, in mother-reared pups there was a robust response to stress in both the primary transcript and total message which was unusual in two respects. First, the response was far more rapid than has previously been observed in the adult. It was expected that hRNA would increase within 15 min of the onset of the stress, However, CRF mRNA was also significantly increased within 15 min following stress and at 60 min was below basal levels. In the adult the general consensus is that it requires between 2–4 hrs to detect significant increases in CRF mRNA. The dynamics of CRF gene expression in the neonate are clearly distinctly different from the adult. Since in the prior studies which examined CRF message in the neonate the assumption was made that the temporal aspects of gene expression would be similar to the adult, the time selected for measuring CRF mRNA was similar to the adult, 24 hrs. The data obtained by Dent *et al.* suggest that any changes in the neonate would be no longer present at these times. Further it has been reported (Lightman and Harbuz, 1993) that injection of isotonic saline failed to elicit an increase in hRNA or CRF mRNA. In general, it appears that the stimuli which result in changes in CRF message in the adult are relatively severe. The neonate at all ages tested showed a rapid and robust increase in CRF gene expression to the mild stimulus of a saline injection. Thus, these findings indicate that the pup is more responsive with regard to the ability to activate the CRF gene (Figures 1 and 2).

It appears as though early in development there is a dissociation between the extent of CRF gene activation and the peripheral response to stress. In contrast, by day 18 the pituitary stress response seems to correspond with CRF gene activation both in deprived and non-deprived pups.

It is difficult to account for the paradoxical finding that although ACTH and CORT are elevated during stress in maternally deprived pups CRH gene expression is suppressed. At postnatal day 12 CRH mRNA is significantly lower in deprived pups (Smith *et al.*, 1997;

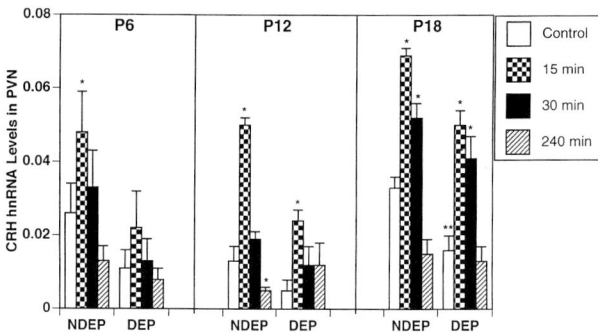

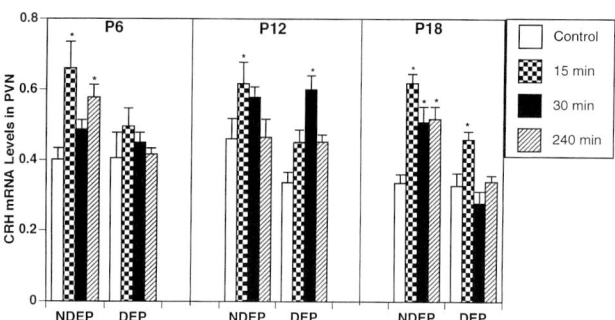

Fig. 1 Effects of stress and maternal deprivation on CRH hnRNA expression in the PVN. Rats (maternally deprived (DEP) and non-deprived (NDEP) at 6, 12 and 18 days of age) were stressed by a saline injection and sacrificed at 15, 30 and 240 min later. Non-injected pups served as controls. *, Stressed group (within each age) is significantly different from corresponding control (non-injected) group at $P < 0.05$. **, Basal CRH hnRNA expression in deprived pups is significantly different from their non-deprived counterparts.

Fig. 2 Effects of stress and maternal deprivation on CRH mRNA expression in the PVN. Rats (maternally deprived (DEP) and non-deprived (NDEP) at 6, 12 and 18 days of age) were stressed by a saline injection and sacrificed at 15, 30 and 240 min later. Non-injected pups served as controls. *, Stressed group (within each age) is significantly different from corresponding control (non-injected) group at $P < 0.05$.

van Oers *et al.*, 1998a). It is tempting to account for these data by invoking a negative feedback explanation insofar as circulating CORT levels are elevated in the deprived pups as early as 8 hr following the onset of deprivation. However, anogenital stroking which does not reduce CORT does reverse these reductions in CRF gene expression. Although there does not appear to be a simple explanation for these results it is apparent that as far as the regulation of the CRF system is concerned, the neonate cannot be described as stress hyporesponsive (Figure 3).

Immediate Early Genes

One of the techniques that is being used to investigate the response to stress has been to examine changes in

the expression of some of the immediate early genes (IEG). In particular c-fos has been used as a marker of increased neuronal activity in response to stress. It is important to note that increases in these stress responsive genes should not be interpreted as an indication of the expression of the neuropeptides. The early literature on changes in c-fos in the brain of the neonatal rat indicated that there were little or no changes in the immediate early genes to stress. However, probably due to methodological advances in recent studies (Smith *et al.*, 1997) both c-fos and NGF1-B were found to be significantly elevated in the PVN of the neonate following the mild stress of a saline injection. Maternally deprived pups show a significant enhancement of the expression of these two IEGs. Once again, stroking normalized the response of the immediate early genes (van Oers *et al.*, 1998b). In all of the

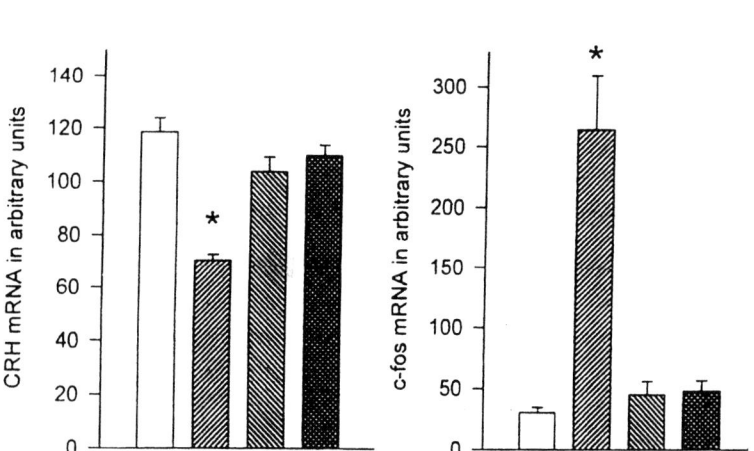

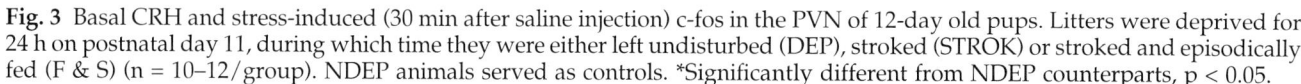

Fig. 3 Basal CRH and stress-induced (30 min after saline injection) c-fos in the PVN of 12-day old pups. Litters were deprived for 24 h on postnatal day 11, during which time they were either left undisturbed (DEP), stroked (STROK) or stroked and episodically fed (F & S) (n = 10–12/group). NDEP animals served as controls. *Significantly different from NDEP counterparts, p < 0.05.

experiments that have examined the response of the immediate early genes following stress, the magnitude of the expression of these genes in the PVN is more reflective of the ACTH response than CRF gene activation.

Conclusions

Is there an SHRP? The concept of the SHRP in its most traditional form described a period in development when all aspects of the HPA axis appeared to be down regulated, resulting in an organism which either failed to respond or at best in which the response was markedly reduced compared to older pups and adults. This may still be applicable to the response of the adrenal if we take total CORT as the index of the stress response. The adrenocortical secretion of CORT is strongly reduced during the SHRP. This reduction in adrenal function is a combination of maturational processes and is further down regulated by maternal factors, such as feeding and stroking.

The evidence for an SHRP with regards to the secretion of ACTH from the pituitary is even more problematic. Robust ACTH responses are observed under many conditions. The assumption that there is a reduction in the synthesis of ACTH in the neonate does not appear to be justified. Under appropriate conditions the pup can and does release quantities of ACTH that are as great and perhaps even greater than in the adult. The reduction of ACTH during the SHRP does not appear to be related to the capacity of the pituitary to produce ACTH. What is unique about this period is that the neonate exhibits greater selectivity in terms of the stimulus which will elicit an increase in ACTH secretion. One possible explanation for this stimulus specificity is that the different pathways that regulate the pituitary may be maturing differentially. It has also been shown that maternal factors inhibit the ACTH response under some stress conditions and can, early in development, enhance the response to stimuli that lead to an ACTH response during the SHRP in mother-reared pups.

The data regarding the neuronal responses to stress are sparse. Recent studies have demonstrated that a significant increase in PVN c-fos mRNA can be seen following a mild stress in non-deprived pups. Maternal deprivation results in an augmented c-fos response. Further other regions of the brain also show marked increases in c-fos expression following stress. Thus, in at least one aspect the brain is clearly stress responsive. There are further data which indicate that increases in PVN CRF mRNA can occur in neonates during the SHRP but do so much more rapidly than in the adult.

In summary, the existing information on the ontogeny of the HPA axis does not support the concept of an SHRP. This is not meant to imply that the HPA system of the neonate is similar to the adult. There are many aspects of the developing HPA axis which are idiosyn-cratic to the neonate. We do not believe, however, that a description of the developing organism as stress hyporesponsive accurately characterizes the ontogeny of the HPA axis.

Acknowledgement

This work was supported by a grant from the National Institutes of Mental Health, MH-45006 (to S.L.), a NATO collaborative Research Grant, CRG 97.0477 and the Netherlands Organization for Scientific Research (NWO), 554–545.

References

De Kloet, E.R., Rosenfeld, P., Van Eekelen, J.A.M., Sutanto, W., and Levine, S. (1998). Stress, glucocorticoids and development. *Progress in Brain Research* 101–120, Amsterdam: Elsevier Science Publishers.

Dent, G.W., Smith, M.A., and Levine, S. (2000). Rapid induction of corticotropin-releasing hormone gene transcription in the paraventricular nucleus of the developing rat. *Endocrinology* **141**, 1593–1598.

Grino, M., Young, W.S., and Burgunder, J.M. (1989). Ontogeny of expression of the corticotropin-releasing factor gene in the hypothalamic paraventricular nucleus and of the proopiomelanocortin gene in rat pituitary. *Endocrinology* **124**, 60–68.

Kent, S., Kernahan, S.D., and Levine, S. (1996). Effects of excitatory amino acids on the hypothalamic-pituitary-adrenal axis of the neonatal rat. *Developmental Brain Research* **96**, 1–13.

Levine, S. (1994). The ontogeny of the hypothalamic-pituitary-adrenal axis: the influence of maternal factors. In: E.R. de Kloet, E.C. Azmitia, P.W. Landfield (eds.), Brain Corticosteroid Receptors: studies on the mechanism, function and neurotoxicity of corticosteroid action, *Annals of the New York Academy of Sciences* **740**, 275–288.

Levine, S., Berkenbosch, F., Suchecki, D., and Tilders, F.J.H. (1994). Pituitary-adrenal and interleukin-6 responses to recombinant interleukin-1 in neonatal rats. *Psychoneuroendocrinology* **19**, 143–153.

Lightman, S.L., and Harbuz, M.S. (1993). Expression of corticotropin-releasing factor mRNA in response to stress. In: *CIBA Foundation Symposium*, Chichester: Wiley, **172**, 173–198.

Smith, M.A., Kim, S.Y., Van Oers, H.J.J., and Levine, S. (1997). Maternal deprivation and stress induce immediate early genes in the infant rat brain. *Endocrinology* **138**, 4622–4628.

Sawchenko, P.E. (1991). The final common pathway: issues concerning the organization and central mechanisms controlling corticotropin secretion. In: M.E. Brown, G.F. Koob and C. River (Eds.), *Stress: Neurobiology and Neuroendocrinology*, pp. 55–71. New York: Marcel Dekker Inc.

Suchecki, D., Mozaffarian, D., Gross, G., Rosenfeld, P., and Levine, S. (1993a). Effects of maternal deprivation on the ACTH stress response in the infant rat. *Neuroendocrinology* **57**, 204–212.

Suchecki, D., Rosenfeld, P., and Levine, S. (1993b). Maternal regulation of the hypothalamic-pituitary-adrenal axis in the infant rat: the roles of feeding and stroking. *Development Brain Research* **75**, 185–192.

Van Oers, H.J.J., De Kloet, E.R., Li, C., and Levine, S. (1998a). The ontogeny of glucocorticoid negative feedback: influence of maternal deprivation. *Endocrinology* **139**, 2838–2846.

Van Oers, H.J.J., De Kloet, E.R., Whelan, T., and Levine, S. (1998b). Maternal deprivation; effect on the infant's neural stress markers is reversed by tactile stimulation and feeding but not by suppressing corticosterone. *Journal of Neuroscience* **18**, 10171–10179.

Walker, C.D., Scribner, K.A., Cascio, C.S., and Dallman, M.F. (1991). The pituitary-adrenocortical system of rats is responsive to stress throughout development in a time-dependent and stressor specific fashion. *Endocrinology* **128**, 1385–1395.

Witek-Janusek, L. (1998). Pituitary-adrenal response to bacterial endotoxin in developing rats. *American Journal of Physiology* **255**, E525–E530.

Yi, S.-J., Masters, J.N., and Baram, T.Z. (1993). Effects of specific glucocorticoid receptor antagonist on corticotropin releasing hormone gene expression in the paraventricular nucleus of the neonatal rat. *Developmental Brain Research* **73**, 253–259.

31. Age and Stress: Differential Effects on Neurotransmitter Gene Expression in Adrenal Medulla and Brain Catecholaminergic Neurons

L.I. Serova, B.B. Nankova, K. Pacak, O. Tjurmina, G. Chrousos, E.L. Sabban and R. Kvetňanský

Introduction

The catecholaminergic (CA) nervous systems have been implicated in many of the metabolic, cardiovascular and behavioral changes which accompany advancing age. Basal plasma levels of norepinephrine (NE) and its metabolites dihydroxyphenylglycol and methoxyhydroxyphenylglycol increase with age in humans and in experimental animals (Morrow et al., 1987; Milakofsky et al., 1993; Cizza et al., 1995) parallel to an increase in the prevalence of hypertension with age (Folkow and Svanborg, 1993; Scarpace and Lowenthal, 1994; Goldstein et al., 1996). Basal plasma levels of epinephrine (EPI) are similar (McCarty, 1984, 1986; Mabry et al., 1995a; Milakofsky et al., 1993) or higher (Cizza et al., 1995) in aged rats. Despite elevated plasma NE levels, NE stores decline significantly in the heart, atria, ventricle, kidney and caudal artery with age in male Fischer 344/N rats (Dawson and Meldrum, 1992). The concentrations of both NE and EPI in the adrenal gland are maximal by 20 months of age (Kvetňanský et al., 1978; Ivanisevic-Milovanovic et al., 1997).

In the central nervous system, age-related changes in CA metabolism are observed in nigrostriatal, mesocortical and coeruleohippocampal systems (Goudsmit et al., 1990). In the dopaminergic system, several characteristic changes are found with age: dopamine (DA) concentrations are lower in striatum of aged humans, monkeys and rodents (Woods and Druse, 1996; Morgan and Finch, 1988; Gozlan et al., 1990; Irwin et al., 1994). Declines in DA levels and reduction in the D1 and D2 receptors are observed in neostriatum (Woods et al., 1995). DA content is also reduced in the ventral tegmental area (VTA) (Woods and Druse, 1996). In addition, the functional dopamine transporters are down-regulated with age in nigrostriatal and mesolimbic systems (Hebert et al., 1999). Locus coeruleus (LC) neurons have been shown to be far more vulnerable to damage in the old versus young animals (Date et al., 1990; Unnerstall and Long, 1996). Using quantitative neuroanatomical techniques, it has been shown that with aging there is a non-random or systematic loss of LC neurons in the human brain (Manaye et al., 1995). Cell loss in specific brain regions is associated with several age-related neurodegenerative disorders. Significant decreases in the numbers of LC neurons are found in the brains of patients with Alzheimer's disease (Manaye et al., 1995), Down's syndrome (Mann et al., 1984) and Pick's disease (Arima and Akashi, 1990). Substantial cell loss also occurs in the dopamine nigrostriatal system during normal aging with a subregional pattern that differs from that in Parkinson's disease (Kish et al., 1992).

While it has been hypothesized that stress is one of the causes of physiological decline with aging, aging has been proposed to alter the ability to respond to stress. The effects of aging on baseline hypothalamic-pituitary-adrenocortical (HPA) function of both rats and humans remains controversial (Sapolsky et al., 1986; Sonntag et al., 1987; Waltman et al., 1991; Cizza et al., 1995). Depending on the stress paradigm, different effects on plasma catecholamine levels are observed. Aged rats have enhanced plasma CA responses to acute restraint stress, cold stress, swimming in cold water and footshock, compared to young adults (Palmer et al., 1978; McCarty, 1986; Mabry et al., 1995a, b). With prolonged exposure to an intense stressor (cellular glucoprivation following administration of 2-deoxy-D-glucose) 24-month-old rats exhibited an exaggerated adrenomedullary response and a greater mortality rate compared to 6-month-old rats (McCarty, 1984, 1986). However, some studies found similar increases in plasma NE and EPI in both young adults and aged rats, in response to immobilization (IMO) stress (Milakofsky et al., 1993; McCarty et al., 1997) and to a novel stress (Mabry et al., 1995b). Others observed a diminished response to stress in old rats (McCarty, 1986; Cizza et al., 1995). The magnitude of the increase in both NE and EPI during IMO stress was significantly smaller in old than in young rats (Cizza et al., 1995). Old rats also exhibited a diminished sympathetic-adrenomedullary response to brief intermittent footshock (McCarty, 1986).

The observed changes in plasma NE and EPI levels may reflect age-related alterations in the expression pattern of the genes involved in CA biosynthesis. It has been reported that tyrosine hydroxylase (TH) activity (Kvetňanský et al., 1978; Strong et al., 1990; Tumer and LaRochelle, 1995), DBH activity (Banerji et al., 1984), PNMT activity (Kvetňanský et al., 1978) and TH mRNA levels (Strong et al., 1990; Voogt et al., 1990; Tumer and

LaRochelle, 1995) increase with age in the rat adrenal medulla. These were accompanied by elevated neuropeptide (NPY) immunoreactivity and mRNA levels (Higuchi *et al.*, 1991). In this study we examine the effects of aging on the regulation of basal levels of gene expression for the CA biosynthetic enzymes in adrenal medulla and central catecholaminergic neurons as well as on the response to stress.

Methods

Immobilization stress

All animal experiments were approved by the Animal Care and Use Committee. Adult, pathogen-free male, young (3–4 months of age) and old (24 months of age) Fischer 344/N or young (2.5–3 months) Sprague-Dawley rats were utilized.

Immobilization stress was applied for 2 hours as previously described in McMahon *et al.* (1992). All experiments were performed between 8 a.m. and 1 p.m. and there were 6 to 8 animals per group. Rats were decapitated immediately after last episode of stress. The adrenal medullae and a number of catecholaminergic brain regions were immediately dissected, separately frozen in liquid nitrogen and stored at –70°C until extraction. Dissection of the brain was performed by using a tissue slicer with digital micrometer (Stoeltin, Wood Dale, IL). Frontal sections 9.2 to 10.4 mm from the bregma or 4.8 to 5.5 mm from bregma were taken for LC, VTA and substantia nigra (SN) sampling respectively, and chilled in ice-cold saline. The regions of the LC and VTA were isolated by taking bilateral punches with a 15-gauge needle.

Isolation of RNA and Northern blots

The relative levels of mRNAs for TH, dopamine-β-hydroxylase (DBH), phenylethanolamine N-methyl-transferase (PNMT), proenkaphalin (PE) and NPY were determined by Northern blot analysis. The tissue samples from each animal were homogenized in RNAzol (Tel-Test). Total amount of RNA from each animal separately was then isolated and fractionated on 1.2% agarose gels. The RNA was transferred to Gene-Screen Plus membranes (New England Nuclear) and consecutive hybridizations were subsequently performed, as described (Nankova *et al.*, 1994; Serova *et al.*, 1998) with the rat TH, DBH, PNMT, PE, NPY cDNA probes and a DNA for 18S rRNA. The probes were labeled with [^{32}P]α -dCTP (6000 Ci/mmol, New England Nuclear) by using the random primer method (Megaprime, Amersham) and purified on Nuc Trap columns (Stratagene). Hybridization was performed at 42°C in a solution containing $5 \times$ SSPE (0.15M NaCl, 10mM NaH$_2$PO$_4$, 1mM EDTA), 50% formamide, $5 \times$ Denhardt's, and 1% SDS, and 10^6 dpm of ^{32}P-labeled probes. The hybridization with the DNA probes and washing of the filters were done as previously described (Nankova *et al.*, 1994). Following exposure to X-ray film (Kodak) within the liner range of the signal, autoradiograms were scanned and analyzed by using Image- Pro-Analysis software (Media Cybernetics).

Statistical Analysis

Data are expressed as mean $\pm$ SEM. One way ANOVA, followed by Fisher's LSD test for comparison of the means (for more than 2 experimental groups) or Student's t-test (for two experimental groups) were used to evaluate the data. A level of P < 0.05 was considered significant.

Results

Adrenal medulla

Initially we studied the effect of aging on the basal levels of expression of several neurotransmitter genes in adrenal medulla of male rats. Northern blot analysis revealed significant differences in the relative mRNA levels for most of the genes of interest with age (Figure 1). The steady-state levels of TH, DBH, PE and NPY mRNA were all significantly higher in old compared to young rats. Only PNMT mRNA levels were unchanged.

A single immobilization stress markedly increased all mRNA levels tested (Figure 2 and Table 1). Despite the enhanced basal mRNA levels for some genes in old animals, the magnitude of elevation in response to stress was not different from that of the young rats. There were about 4–6 fold inductions for TH and PE, 2-fold for DBH and PNMT and 3-fold for NPY mRNA levels.

Locus Coeruleus

In the LC, in contrast to the findings in the adrenal medulla, aging was not found to affect the basal levels of TH expression (Figure 3). Exposure to acute stress triggered small, but statistically significant increases in

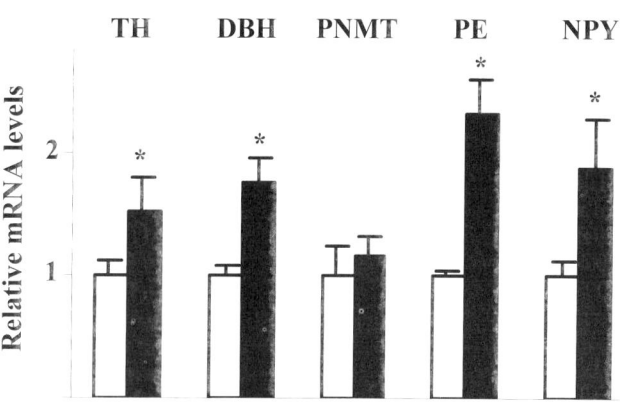

Fig. 1 Increased basal levels of neurotransmitter gene expression in adrenal medulla of old rats. Basal levels of TH, DBH, PNMT, PE and NPY mRNA, calculated relative to 18S mRNA levels, were determined in young (open bars) and old rats (close bars). The levels of mRNA for each gene in young animals are taken as 1. Data are expressed as mean $\pm$ SEM (n = 6 to 8 rats per group). * P < 0.05 vs appropriate young group.

Table 1 Effect of immobilization stress on mRNA levels for catecholamine biosynthetic enzymes and neuropeptides in adrenal medulla in young and old rats.

	4 month		24 month	
	control	1 × IMO	control	1 × IMO
TH	1 ± 0.14	6.00 ± 0.44**	1.75 ± 0.21*	7.77 ± 1.18**
DBH	1 ± 0.09	1.80 ± 0.18**	1.48 ± 0.17*	2.60 ± 0.33**
PNMT	1 ± 0.21	2.60 ± 0.43**	1.16 ± 0.16	3.20 ± 0.51**
NPY	1 ± 0.13	2.65 ± 0.31**	1.67 ± 0.15*	4.71 ± 0.85**
PE	1 ± 0.15	6.72 ± 0.43**	2.35 ± 1.12*	12.32 ± 1.75**

The levels of mRNA for each gene in young (4 months) rats were taken as 1.
*$P < 0.05$ vs 4 month controls; ** $P < 0.05$ vs control group of the same age.

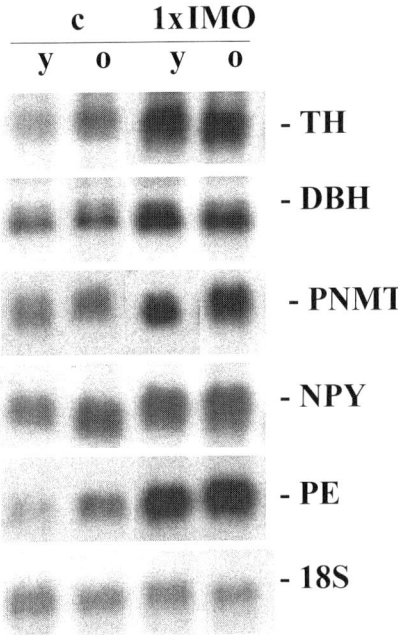

Fig. 2 Effect of immobilization stress on neurotranmitter gene expression in adrenal medulla of young and old rats. Representative Northern blot analyses show consecutive hybridizations for all genes of interest in response to immobilization stress in 4-month-old (young, y) or 24-month-old (o) rats. c – unstressed control, 1 × IMO – 2 hours immobilization stress.

the relative TH mRNA levels in young rats. Notably, the stress response was impaired in old animals. A single IMO stress did not significantly elevate TH mRNA levels in the old rats.

Substantia Nigra and Ventral Tegmental Area

To test whether immobilization stress may affect TH gene expression in these brain regions, young male Sprague-Dawley rats were exposed to stress for different amounts of time. Total RNAs isolated from SN and VTA of individual animals were used for Northern blots analysis. Acute as well as repeated

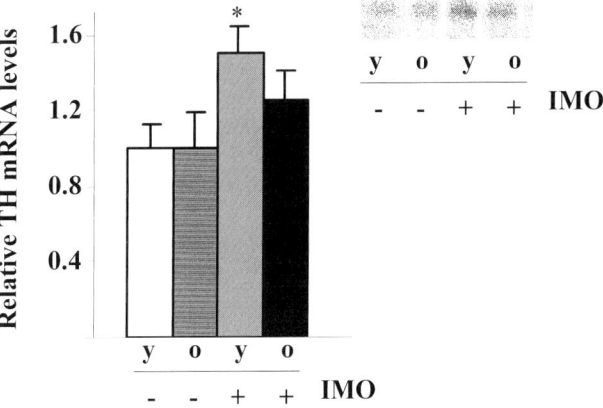

Fig. 3 Effect of immobilization stress on TH mRNA levels in locus coeruleus in old and young rats. Representative Northern blot and summary data are shown for TH mRNA levels. Total RNA from LC punches obtained from individual rats were analyzed separately by Northern blots. Data are expressed as mean ± SEM (n = 6 to 8 rats per group). * $P < 0.05$ vs young unstressed group.

immobilization stress lead to a 2-fold elevation in TH mRNA levels in SN (Figure 4). Exposure to 7 repeated episodes of immobilization stress did not cause further increases in TH mRNA levels. In the VTA, repeated stress was required for significant increases in TH mRNA levels (Figure 4).

Aging did not alter significantly the basal levels or stress-elicited induction of TH mRNA in SN in Fischer rats (Figure 5). In the VTA, significantly elevated basal levels of TH expression were observed with age. TH mRNA was about twice as high in the VTA of old rats. Exposure to a single immobilization stress did not affect TH mRNA levels in all experimental groups (Figure 5).

Discussion

In aged rats, tissue specific differences were observed in the basal levels and in the magnitude of stress-elicited

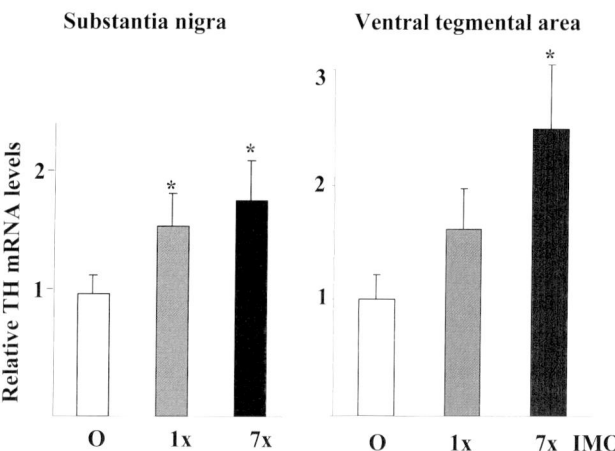

Fig. 4 Effect of stress on TH mRNA levels in dopaminergic cell bodies in VTA and SN. Sprague-Dawley rats were unstressed (0) or exposed to single (1x) and repeated (7x) immobilization stress. The results are calculated relative to 18S mRNA levels. The levels in TH mRNA in unstressed animals are taken as 1. Data are expressed as mean ± SEM (n = 6 to 8 rats per group). * P < 0.05.

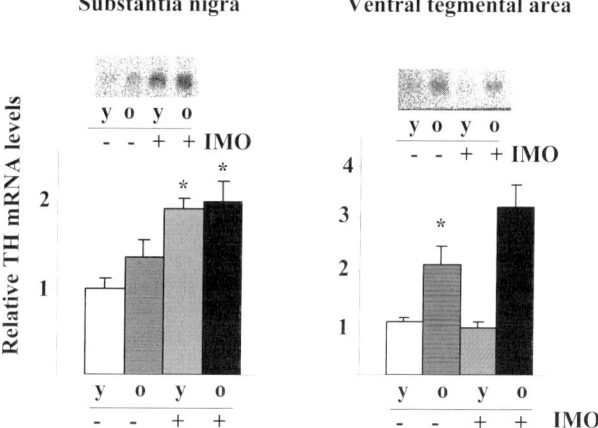

Fig. 5 Effect of immobilization stress on TH mRNA levels in substantia nigra and ventral tegmental area of young (y), and old (o) Fischer rats. The results are calculated relative to 18S mRNA levels. The levels in young unstressed animals are taken as 1. Data are expressed as mean ± SEM (n = 6 to 8 rats per group). * P < 0.05 vs appropriate unstressed control of the same age.

induction of expression for the various neurotransmitter related genes. Age-associated changes were found in the gene expression of CA biosynthetic enzymes and for PE and NPY in adrenal medulla. Steady-state mRNA levels for TH, DBH, PE and NPY were significantly higher in old versus young male rats, although animals of both ages responded similarly to a single IMO stress. In the LC, stress triggered induction of TH was blunted in senescent rats. In the major catecholaminergic cell bodies, increased basal TH mRNA levels were observed only in the VTA, but not in the LC or SN.

The elevated basal adrenomedullary mRNA levels for TH and DBH with aging are in agreement with previous findings of age related increases in adrenal catecholamine biosynthetic enzymes, TH mRNA levels and EPI and NE concentrations in Wistar and Fischer 344/N rats (Kvetňanský et al., 1978; Banerji et al., 1984; Voogt et al., 1990; Tumer and LaRochelle, 1995). The increased basal steady-state TH and DBH mRNA levels are likely responsible for the higher activity of these enzymes in the adrenal medulla of aged rats.

In the central nervous system, changes in TH mRNA levels were over twice as high in older animals in the VTA. Despite of a reported prominent age-related loss of noradrenergic and dopaminergic cell bodies from the LC and SN, basal TH mRNA levels were unchanged in those structures. Similar data were obtained by *in situ* hybridization (Voogt et al., 1990; Cizza et al., 1995). These results are somewhat surprising given the reduced CA content observed with aging in a number of brain regions (Morgan and Finch, 1988; Woods and Druse, 1996). Perhaps the reduced brain catecholamines reflect increased catabolism by the reported higher activity of monoamine oxidase (Alper et al., 1999), coupled with unchanged or higher basal TH gene expression. It is attractive to speculate that increased catecholamine synthesis and degradation may be one of the causative factors in free radical formation and cellular damage elicited by increased oxidative stress during aging.

The extent to which age may impair the ability to respond to stress is controversial. Progressive hypothalamic CRH deficiency and diminished relative elevations in peripheral CA system levels in response to acute immobilization stress were observed by some investigators (Cizza et al., 1995), while others found that a single immobilization stress elicited a similar rise in plasma epinephrine (McCarty et al., 1997).

The results of the present study reveal that despite the elevated basal levels of TH, DBH, PE and NPY mRNAs, the adrenomeduallary response to a single immobilization stress is not impaired in aged Fischer 344N rats. The magnitude of the increase in mRNA levels for these genes, as well as PNMT, was similar to that observed in young rats. In contrast, TH gene expression in senescent rats (24-month-old Fischer 344/N rats) is reported to be unresponsive to cold stress. Exposure to cold significantly increased TH mRNA, immunoreactivity and enzymatic activity in young but not old rats (Tumer and LaRochelle, 1995). The difference between these findings and the present results with immobilization stress, may be due to distinct, stressor-specific pathways utilized, since the old and young rats are reported to also respond similarly to forskolin.

The results indicate that similar to CA biosynthetic enzymes, NPY and proenkephalin mRNAs are higher with age. IMO stress elicits a marked and similar induction in PE and NPY mRNA in young and old rats. In contrast, Silverstein et al. (1998) reported stress-impaired induction of adrenal NPY.

The young Fischer 344/N rats exhibited a significant, but much smaller (about 60%) rise of TH mRNA in LC neurons. Parallel increases in TH, DBH and GTPCH mRNA levels of about 300% to 400% over control levels were observed with a single IMO in LC of adult Sprague-Dawley rats (Serova *et al.*, 1999). Thus, although the response to IMO stress in the adrenal medulla was similar in young adult Sprague-Dawley and Fischer 344/N rats, there appear to be strain-specific differences in the responsiveness of the LC to acute immobilization. Previously pronounced differences in the response of the hypothalamic-pituitary-adrenal axis to acute and prolonged stress were reported in these genetically related strains (Dhabhar *et al.*, 1997).

Our study indicated that aging impaired the stress-triggered induction of TH mRNA in LC. Similar phenomenon may underlay the limited ability of older individuals to handle challenging conditions and reflect reduced vigilance. However, because of the small response of the LC in Fischer rats to stress, the extent of the impairment is still unclear.

The molecular mechanisms mediating the differential regulation of the TH gene in CA nuclei is still not clear. We have shown that IMO stress leads to a substantial rise in the relative rate of TH gene transcription in LC (Serova *et al.*, 1999). Several conditions that activate TH gene expression in LC were found to increase phosphorylation of CREB (Nestler and Aghajanian, 1997). Direct infusion of CREB antisense oligonucleotides into the LC prevented the morphine-induced up-regulation of TH (Lane-Ladd *et al.*, 1997). Aging decreases CRE-binding activity in many brain regions (Asanuma *et al.*, 1996). An age dependent decline in CRE-binding activity may mediate impairments in the response of TH gene transcription to stress in the LC of aged rats.

In different brain regions, the mechanism of IMO elicited activation of the TH gene may be different. In the VTA, in contrast to the LC, post-transcriptional rather than transcriptional mechanisms have been shown to mediate morphine-triggered up-regulation of TH gene expression (Boundy *et al.*, 1998). The induction of TH mRNA in the VTA by repeated immobilization stress may also be a consequence of post-transcriptional mechanisms.

In conclusion, this study revealed multiple age-related changes in neurotransmitter gene expression and in the response to stress. Notably, differential patterns of age-associated alterations in their regulation are manifested in the distinct catecholaminergic nuclei.

Acknowledgments

We gratefully acknowledge support by grants NS28869 and NS32166 from National Institutes of Health, grant 0251 from Smokeless Tobacco Research Council, Fogarty International Collaboration Award TW00984 and grant 2- 610999 from Slovak Grant-Agency for Science.

References

Alper, G., Girgin, F.K., Ozgonul, M., Mentes, G., and Ersoz, B. (1999). MAO inhibitors and oxidant stress in aging brain tissue. *European Neuropsychopharmacology* **9**, 247–252.

Arima, K., and Akashi, T. (1990). Involvement of the locus coeruleus in Pick's disease with or without Pick body formation. *Acta Neuropathologia (Berlin)* **79**, 629–633.

Asanuma, M., Nishibayash, S., Iwa, E., Kondo, Y., Nakanishi, T., Vargas, M.G., and Ogawa, N. (1996). Alterations of cAMP response element-binding activity in the aged rat brain in response to administration of rolipram, a cAMP-specific phosphodiesterase inhibitor. *Molecular Brain Research* **41**, 210–215.

Banerji, T.K., Parkening, T.A., and Collins, T.J. (1984). Adrenomedullary catecholaminergic activity increases with age in male laboratory rodents. *Journal of Gerontology* **39**, 264–268.

Boundy, V.A., Gold, S.J., Mess, C.J., Chen, J., Son, J.H., Joh, T.H., and Nestler, E.J. (1998). Regulation of tyrosine hydroxylase promoter activity by chronic morphine in TH9.0-LacZ transgenic mice. *Journal of Neuroscience* **18**, 9989–9995.

Cizza, G., Gold, P.W., and Chrousos, G.P. (1995). Aging is associated in the 344/N Fischer rat with decreased stress responsivity of central and peripheral catecholaminergic systems and impairment of the hypothalamic-pituitary-adrenal axis. *Annals of the New York Academy of Sciences* **771**, 491–511.

Date, I., Felten, D.L., and Felten, S.Y. (1990). Long-term effect of MPTP in the mouse brain in relation to aging: neurochemical and immunocytochemical analysis. *Brain Research* **519**, 266–276.

Dawson, R.J., and Meldrum, M.J. (1992). Norepinephrine content in cardiovascular tissues from the aged Fischer 344 rat. *Gerontology* **38**, 185–191.

Dhabhar, F.S., McEwen, B.S., and Spencer, R.L. (1997). Adaptation to prolonged or repeated stress – comparison between rat strains showing intrinsic differences in reactivity to acute stress. *Neuroendocrinology* **65**, 360–368.

Folkow, B., and Svanborg, A. (1993). Physiology of cardiovascular aging. *Physiological Review* **73**, 725–764.

Goldstein, D., Lenders, J.W.M., Kaler, S.G., and Eisenhofer, G. (1996). Catecholamine phenotyping: Clues to the diagnosis, treatment, and pathophysiology of neurogenetic disorders. *Journal of Neurochemistry* **67**, 1781–1790.

Goudsmit, E., Feenstra, M.G., and Swaab, D.F. (1990). Central monoamine metabolism in the male Brown-Norway rat in relation to aging and testosterone. *Brain Research Bulletin* **25**, 755–763.

Gozlan, H., Daval, G., Verge, D., Spampinato, U., Fattaccini, C.M., Gallissot, M.C., and Hamon, M. (1990). Aging associated changes in serotoninergic and dopaminergic pre- and postsynaptic neurochemical markers in the rat brain. *Neurobiology of Aging* **11**, 437–449.

Hebert, M.A., Larson, G.A., Zahniser, N.R., and Gerhardt, G.A. (1999). Age-related reductions in [³H]WIN 35,428 binding to the dopamine transporter in nigrostriatal and mesolimbic brain regions of the Fischer 344 rat. *Journal of Pharmacology and Experimental Therapeutics* **288**, 1334–1339.

Higuchi, H., Yokokawa, K., Iwas, A., Yoshida, H., and Mik, N. (1991). Age-dependent increase in neuropeptide Y gene expression in rat adrenal gland and specific brain areas. *Journal of Neurochemistry* **57**, 1840–1847.

Irwin, I., DeLanney, L.E., McNeill, T., Cha, P., Forno, L.S., Murphy, G.M. Jr., Di M.D.A., Sandy, M.S., and Langston, J.W. (1994). Aging and the nigrostriatal dopamine system: a non-human primate study. *Neurodegeneration* **3**, 251–265.

Ivanisevic-Milovanovic, O.K., Demajo, M., Loncar-Stevanovic, H., Karakasevic, A., and Pantic, V. (1997-98). Basal and stress induced concentrations of adrenal gland catecholamines and plasma ACTH during aging. *Acta Physiologica Hungarica* **85**, 65–75.

Kish, S.J., Shannak, K., Rajput, A., Deck, J.H., and Hornykiewicz, O. (1992). Aging produces a specific pattern of striatal dopamine loss: implications for the etiology of idiopathic Parkinson's disease. *Journal of Neurochemistry* **58**, 642–648.

Kvetňanský, R., Jahnova, E., Torda, T., Strbak, V., Balaz, V., and Mach, L. (1978). Changes of adrenal catecholamines and their synthesizing enzymes during ontogenesis and aging in rats. *Mechanisms of Ageing and Development* **7**, 209–216.

Lane-Ladd, S.B., Pineda, J., Boundy, V.A., Pfeuffer, T., Krupinski, J., Aghajanian, G.K., and Nestler, E.J. (1997). CREB (cAMP response element-binding protein) in the locus coeruleus: biochemical, physiological, and behavioral evidence for a role in opiate dependence. *Journal of Neuroscience* **17**, 7890–7901.

Mabry, T.R., Gold, P.E., and McCarty, R. (1995a). Age-related changes in plasma catecholamine and glucose responses of F-344 rats to a single footshock as used in inhibitory avoidance training. *Neurobiology of Learning and Memory* **64**, 146–155.

Mabry, T.R., Gold, P.E., and McCarty, R. (1995b). Age-related changes in plasma catecholamine responses to chronic intermittent stress. *Physiology and Behavior* **58**, 49–56.

Manaye, K.F., McIntire, D.D., Mann, D.M., and German, D.C. (1995). Locus coeruleus cell loss in the aging human brain: a non-random process. *Journal of Comparative Neurology* 358, 79–87.

Mann, D.M., Yates, P.O., and Marcyniuk, B. (1984). Monoaminergic neurotransmitter systems in presenile Alzheimer's disease and in senile dementia of Alzheimer type. *Clinical Neuropathology* 3, 199–205.

McCarty, R. (1984). Effects of 2-deoxyglucose on plasma catecholamines in adult and aged rats. *Neurobiology of Aging* 5, 285–289.

McCarty, R. (1986). Age-related alterations in sympathetic-adrenal medullary responses to stress. *Gerontology*, **32**, 172–183.

McCarty, R., Pacak, K., Goldstein, D.S., and Eisenhofer, G. (1997). Regulation of peripheral catecholamine responses to acute stress in young adult and aged F-344 rats. *Stress* **2**, 113–122.

McMahon, A., Kvetňanský, R., Fukuhara, K., Weise, V.K., Kopin, I., and Sabban, E.L. (1992). Regulation of tyrosine hydroxylase and dopamine β-hydroxylase mRNA levels in rat adrenals by a single and repeated immobilization stress. *Journal of Neurochemistry* **5**, 2124–2130.

Milakofsky, L., Harris, N., and Vogel, W.H. (1993). Effect of repeated stress on plasma catecholamines and taurine in young and old rats. *Neurobiology of Aging* **14**, 359–366.

Morgan, D.G., and Finch, C.E. (1988). Dopaminergic changes in the basal ganglia. A generalized phenomenon of aging in mammals. *Annals of the New York Academy of Sciences* **515**, 145–160.

Morrow, L.A., Linares, O.A., Hill, T.J., Sanfield, J.A., Supiano, M.A., Rosen, S.G., and Halter, J.B. (1987). Age differences in the plasma clearance mechanisms for epinephrine and norepinephrine in humans. *Journal of Clinical Endocrinology and Metabolism* **65**, 508–511.

Nankova, B., Kvetňanský, R., McMahon, A., Viskupic, E., Hiremagalur, B., Frankle, G., Fukuhara, K., Kopin, I.J., and Sabban, E.L. (1994). Induction of tyrosine hydroxylase gene expression by a nonneuronal nonpituitary-mediated mechanism in immobilization stress. *Proceedings of the National Academy of Sciences USA* **91**, 5937–5941.

Nestler, E.J., and Aghajanian, G.K. (1997). Molecular and cellular basis of addiction. *Science* **278**, 58–63.

Palmer, G., Ziegler, M.G., and Lake, R. (1978). Response of norepinephrine and blood pressure to stress increases with age. *Journal of Gerontology* **33**, 482–487.

Sapolsky, R.M., Krey, L.C., and McEwen, B.S. (1986). The neuroendocrinology of stress and aging: the glucocorticoid cascade hypothesis. *Endocrine Reviews* **7**, 284–301.

Scarpace, N.T., and Lowenthal, D.T. (1994). Sympathetic nervous system: aging and exercise. *Southern Medical Journal* **87**, S42–S46.

Serova, L.I, Saez E., Spiegelman, B.M, and Sabban, E.L. (1998). c-fos deficiency inhibits induction of mRNA for some, but not all, neurotransmitter biosynthetic enzymes by immobilization stress. *Journal of Neurochemistry* **70**, 1935–1940.

Serova, L.I., Nankova, B.B., Feng, Z., Hong, J.S., Hutt, M., and Sabban, E.L. (1999). Heightened transcription for enzymes involved in norepinephrine biosynthesis in the rat locus coeruleus by immobilization stress. *Biological Psychiatry* **45**, 853–862.

Silverstein, J.H., Beasley, J., Mizuno, T.M., London, E., and Mobbs, C.V. (1998). Adrenal neuropeptide Y mRNA but not preproenkephalin mRNA induction by stress is impaired by aging in Fischer 344 rats. *Mechanism of Ageing and Development* **101**, 233–243.

Sonntag, W.E., Goliszek, A.G., Brodish, A., and Eldridge, J.C. (1987). Diminished diurnal secretion of adrenocorticotropin (ACTH), but not corticosterone, in old male rats: possible relation to increased adrenal sensitivity to ACTH *in vivo*. *Endocrinology* **120**, 2308–2315.

Strong, R., Moore, M.A., Hale, C., Wessels-Reiker, M., Armbrecht, H.J., and Richardson, A. (1990). Modulation of tyrosine hydroxylase gene expression in the rat adrenal gland by age and reserpine. *Brain Research* **525**, 126–132.

Tumer, N., and LaRochelle, J.S. (1995). Tyrosine hydroxylase expression in rat adrenal medulla: influence of age and cold. *Pharmacology Biochemistry and Behavior* **51**, 775–780.

Unnerstall, J.R., and Long, M.M. (1996). Differential effects of the intraventricular administration of 6-hydroxydopamine on the induction of type II beta-tubulin and tyrosine hydroxylase mRNA in the locus coeruleus of the aging Fischer 344 rat. *Journal of Comparative Neurology* **364**, 363–381.

Voogt, J., Arbogast, L.A., Quadri, K.S., and Andrews, G. (1990). Tyrosine hydroxylase messenger RNA in the hypothalamus, substantia nigra and adrenal medulla of old female rats. *Molecular Brain Research* **8**, 55–62.

Waltman, C., Blackman, M.R., Chrousos, G.P., Riemann, C., and Harman, S.M. (1991). Spontaneous and glucocorticoid-inhibited adrenocorticotropic hormone and cortisol secretion are similar in healthy young and old men. *Journal of Clinical Endocrinology and Metabolisim* **73**, 495–502.

Woods, J.M., and Druse, M.J. (1996). Effects of chronic ethanol consumption and aging on dopamine, serotonin, and metabolites. *Journal of Neurochemistry* **66**, 2168–2178.

Woods, J.M., Ricken, J.D., and Druse, M.J. (1995). Effects of chronic alcohol consumption and aging on dopamine D1 receptors in Fischer 344 rats. *Alcohol: Clinical and Experimental Research* **19**, 1331–1337.

32. Regulation of Tyrosine Hydroxylase with Age: Modulation by Cold and Exercise

N. Tümer

Introduction

A fundamental theory that underlies much current aging research states that aging systems do not respond or adapt to physiological challenges as well as younger systems. This is clearly a problem in human aging where it is known, for instance, that elderly individuals do not regulate either blood pressure (Tuck, 1989) or body temperature (Collins and Exton-Smith, 1983) as well as young people in response to environmental stimuli or disease. These kinds of observations have a dramatic impact on human mortality and morbidity and, in many ways, dictate the "quality of life" of our aging population. The regulation of catecholamine biosynthesis is a particularly good model for studying this age-related loss of plasticity. Tyrosine hydroxylase (TH) is the rate-limiting enzyme in catecholamine biosynthesis. In this chapter we describe age-related decreases in the ability to both up and down regulate TH induction in the rat adrenal medulla in response to a cold stimulus and exercise training.

Catecholamine Biosynthesis

The catecholamines, including dopamine, norepinephrine (NE), and epinephrine (EPI), are formed from their amino acid precursor tyrosine in the brain, chromaffin cells of the adrenal medulla, and sympathetic nerves. TH facilitates the hydroxylation of tyrosine, producing dopamine, whereas dopamine-beta-hydroxylase activity facilitates the conversion of dopamine to NE. TH is generally regarded as the rate-limiting step in the biosynthesis of catecholamines, and its activity is the important regulatory step in the biosynthesis pathway (Nagatsu et al., 1964). Among the mechanisms that control the rate of this reaction, two types of changes in TH activity are particularly important. First, rapid (within minutes) changes in TH activity are modulated by phosphorylation/dephosphorylation, in particular, cAMP dependent protein kinase A phosphorylation of serine 40, which increases the activity of TH (Daubner et al., 1992; Acheson and Zigmond, 1981). Second, prolonged activation of catecholaminergic cells leads to a gradual increase in TH activity, caused by the apparent formation of new TH molecules as a result of increases in TH mRNA levels (Acheson and Zigmond, 1981). The former, the minute-to-minute regulation of TH by phosphorylation, is not the focus of this chapter (for review see Kumer and Vrana, 1996).

This chapter is focused on the long-term regulation of TH enzyme levels and TH gene expression with age, and the modulation by cold exposure and exercise training.

Molecular Regulation of TH Activity

The chromaffin cells of the adrenal medulla are innervated by typical preganglionic sympathetic nerve terminals, whose neurotransmitter is acetylcholine. In addition to cholinergic agonists, TH gene expression is induced by numerous agents including Angiotensin II (Yu et al., 1996), NPY (Hong et al., 1995), and glucocorticoids, as well as agonists that increase cAMP (Yoon and Chikaraishi, 1992). Genomic regulator elements have been identified in the promoter region of the TH gene, two of the most important being the AP1 regulatory element and the cAMP response element (CRE). The AP1 regulatory element binds products of the immediate early genes, the c-Fos and c-Jun related proteins, whereas the cAMP response element binds the cAMP response element binding protein (CREB transcription factor). The sequence that is reportedly necessary to maintain basal expression of the TH gene is a 23 base pair region (−205 to −182) that includes the AP1 regulatory site (Yoon and Chikaraishi, 1992). The c-Fos and c-Jun related proteins that activate this site are the products of c-fos and c-jun, and expression of these genes precedes the induction of TH expression under certain conditions (Wessel and Joh, 1992). Another important regulatory element is the CRE site, which is strategically located in close proximity to the TATA box. The CRE sequence is also an important regulatory site for both basal and inducible TH gene transcription (Carroll et al., 1991). Both the AP1 and CRE sites have been implicated in the physiologic induction of TH (Miner et al., 1992). The actual sequence of events that leads to gene induction in TH is not completely known. However, there is evidence that TH induction is mediated by the adenylyl cyclase-cAMP signal transduction pathway (Wessels-Reiker et al., 1991) and modulated by cholinergic nicotinic receptors (Stachowiak et al., 1988). Adenylyl cyclase may be activated by either receptors coupled to the G-protein or by acetylcholine-nicotinic receptors through activation of calmodulin. Acetylcholine through the nicotinic receptor increases intracellular calcium concentration, which increases calcium bound to calmodulin. The calcium-calmodulin can then activate specific isoenzymes of adenylyl cyclase (types I, III, and VIII) that are sensitive to calcium-calmodulin. The generated cAMP binds to the regulatory subunit of protein kinase A and causes the dissociation and activation of the catalytic unit of protein kinase A. The drug forskolin can bypass the necessity for receptor participation and acts directly at the catalytic subunit of adenylyl cyclase to increase cAMP (Seaman and Daly, 1981). The cAMP in turn activates protein kinase A. The catalytic unit of protein kinase A is translocated to the nucleus where it phosphorylates the CREB transcription factor at Ser 133. Both the phosphorylated CREB (P-CREB) and unphosphorylated CREB can bind to CRE on the promoter region of the TH gene, but only P-CREB can promote the transcription of the TH gene with an increase in the production of TH mRNA (Karin and Smeal, 1992). In a similar fashion, the nicotinic-induced increase in calcium can also activate protein kinase C. Eventually, this results in phosphorylation of c-Fos-c-Jun complexes, promoting their binding to the AP1 site on the TH gene promoter (Taylor and Merritt, 1986). In addition, angiotensin II through the AT_1 receptor can induce TH expression through activation of phospholipase C (Yu et al., 1996). Phospholipase C hydrolyzes phosphatidyl inositide, resulting in the formation of the dual second messengers diacyl glycerol and polyphosphoinositides, most notably IP_3. IP_3 mobilizes Ca^{2+} and this ion, together with diacyl glycerol, activates protein kinase C, promoting the subsequent phosphorylation of c-Fos-c-Jun complexes and TH induction.

Catecholamines Increase with Age

The catecholamine content of the rat adrenal medulla increases with age. There is a progressive increase in the synthesis of both EPI and NE content with age (Roberts and Tümer, 1987; Martinez et al., 1981). Epinephrine in the circulation originates in the adrenal medulla and represents adrenal medullary activity. The adrenal medulla contains considerable amounts of NE; however, the majority of NE found in the circulation originates in the nerve terminals and represents adrenergic neural activity (see Roberts and Tümer, 1987).

In humans, circulating levels of NE increase with age, whereas epinephrine levels do not seem to change with age. In most studies with rats, investigators have reported that basal circulating levels of NE increase with age (Irwin et al., 1992; Kiritsy-Roy et al., 1992; Milakofsky et al., 1993; Chiueh et al., 1980). In contrast, in some studies, basal values of NE were found to be unchanged with age (McCarty et al., 1996; and see Roberts and Tümer, 1987). Circulating levels of EPI also increase with age in rats (Roberts and Tümer, 1987; Hassler and Christ, 1984; Ito et al., 1986). This may be due to the increase with age in adrenomedullary activity, as measured by TH and dopamine-beta-hydroxylase activity (Tümer et al., 1992; Banerji et al., 1984). Secretion rates of EPI from the adrenal gland under resting conditions are two to four times greater in old rats (Ito et al., 1986). In addition, Ito et al. (1986) found that sympathetic efferent nerves innervating the adrenal medulla showed increased neural activity with age in resting conditions. This increase in neural activity paralleled the increased catecholamine secretion from the medulla.

TH Activity and TH mRNA Increase with Age

The effect of age on the biosynthesis of catecholamines has been studied in a variety of tissues, including the adrenal medulla and sympathetic ganglia (see Roberts and Tümer, 1987). In the rat adrenal medulla, there is a 50–75% increase in catecholamine content with age. Parallel to these changes, TH activity and TH mRNA increase with age (Strong et al., 1990; Kedzierski and Porter, 1990; Voogt et al., 1990). Our data indicate that TH activity increases by 3-fold in the adrenal glands of senescent compared with young rats (Figure 1) (Tümer and LaRochelle, 1995). To determine if this increase with age was due to a different requirement for the cofactor $6-MPH_4$ in young compared with old rats, pterin dose response curves were assessed. However, we found that the optimum concentration of pterin for maximum enzyme activity is about 1 mM in the adrenal medulla of both young and old rats (Figure 2). Above this concentration there is inhibition. An Eadie Hofstee plot of these data indicated that the K_m is unchanged whereas the V_{max}, the maximum activity, is increased with age (data not shown). The elevated TH activity in the adrenal medulla with age could be due to a increase in the amount of enzyme with age. To examine this possibility, TH immunoreactivity and TH mRNA levels were determined in the adrenal gland with age. Both parameters were significantly higher in senescent rats. TH immunoreactivity demonstrated a 120% increase whereas TH mRNA levels increased 62% with age

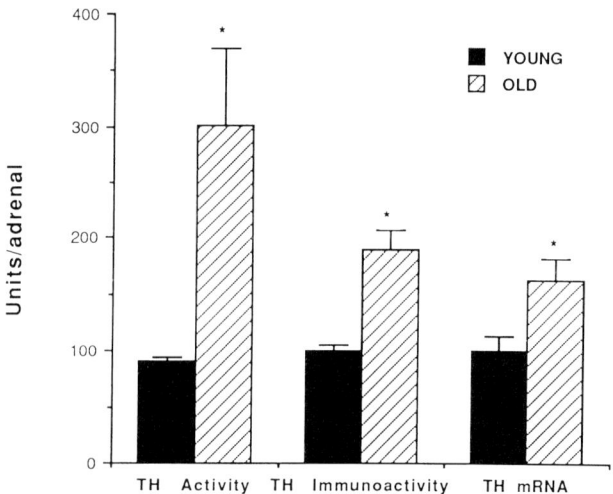

Fig. 1 TH activity, TH immunoreactivity, and TH mRNA in adrenal medulla from 3 month (young) and 24 month (old) rats. Values represent mean ± SEM of 8–10 rats. * Significantly different from corresponding young rats, $P < 0.001$.

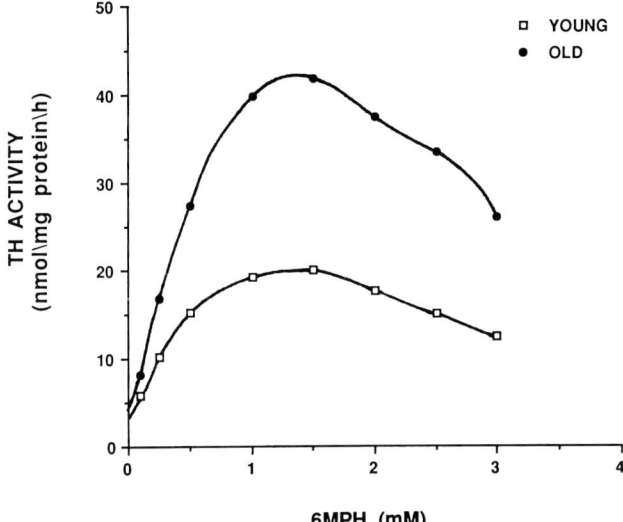

Fig. 2 TH activity dose response curves as a function of 6MPH$_4$ concentration in the adrenal medulla of young (□) and old (●) rats. Data reflect activity curves from a single pair of rats, representative of 4 experiments. Maximum enzyme activity was achieved at 1.4 mM pterin concentration in both age groups. Above this concentration there was inhibition of TH activity.

(Figure 1) (Tümer and LaRochelle, 1995). In contrast to the adrenal medulla, TH mRNA levels do not change in the superior cervical ganglion between 10 and 24 months (Kedzierski and Porter, 1990).

Cold Exposure Increases TH Activity and TH mRNA

Chronic cold exposure is associated with a gradual increase in TH activity and the relative abundance and cell-free translation activity of TH mRNA (Fluharty *et al.*, 1985; Stachowiak *et al.*, 1986; Baruchin *et al.*, 1990; Stachowiak *et al.*, 1985; Tank *et al.*, 1985). The identity of the agent or agents that mediate the cold induction of TH activity are still speculative. The induction of TH mRNA may be mediated through the transsynaptic activation of cholinergic receptors (Stachowiak *et al.*, 1986; Stachowiak *et al.*, 1988). Miner *et al.* (1992) observed an apparent down-regulation of adrenal muscarinic receptors and an up-regulation of adrenal nicotinic receptors following cold exposure (Miner *et al.*, 1989). Another study reported that pretreatment with a nicotinic antagonist prevented the cold-induced increase in TH in sympathetic ganglia (Hanbauer *et al.*, 1973). The strongest evidence to support the induction of TH following cold exposure by the cAMP and protein kinase C pathways comes from a report examining transcription factor binding following cold exposure (Miner *et al.*, 1992). The protein binding activity to DNA fragments of the proximal region of the TH gene promoter were assessed following cold exposure. Fragments containing the AP1 and CRE sites demonstrated increased protein binding activity (Miner *et al.*, 1992).

Effect of Cold Exposure on TH Activity with Age in the Adrenal Medulla

We assessed TH activity, TH immunoreactivity, and TH mRNA with age in the adrenal medullae of 3- and 24-month-old male F-344 rats following both a 1-hour and a 48-hour cold exposure at 8°C. We choose this temperature based on our preliminary studies that indicated some aged rats did not survive below this temperature. Following cold exposure, TH activity was significantly elevated by approximately 35% at 1 hr and 48 hr in young rats (Figure 3). In contrast to young rats, there was no significant increase in TH activity following cold exposure in the old rats (Figure 3). TH immunoreactivity was also assessed following both 1 and 48 hr cold exposure in young and old rats. In young rats TH immunoreactivity increased following both 1 hr and 48 hr cold exposures at 8°C; however, the increase was significant only at the 48 hr time point (Figure 4). In old rats, although TH immunoreactivity was significantly elevated with age, there was no change in TH immunoreactivity in 24 month animals following either 1 hr or 48 hr cold exposures (Figure 4). To assess whether alterations in TH activity and TH protein are a consequence of changes in gene expression, TH mRNA levels were measured. Similar to our previous studies, TH mRNA was significantly elevated by 50% in adrenals from old rats compared with young rats, and cold exposure increased TH mRNA in young rats (Figure 5). In contrast there was no increase in TH mRNA with cold exposure in the senescent rats (Figure 5).

These data, an increase in TH gene expression, TH immunoreactivity, and TH activity in the adrenal medullae of young rats but not in old rats following

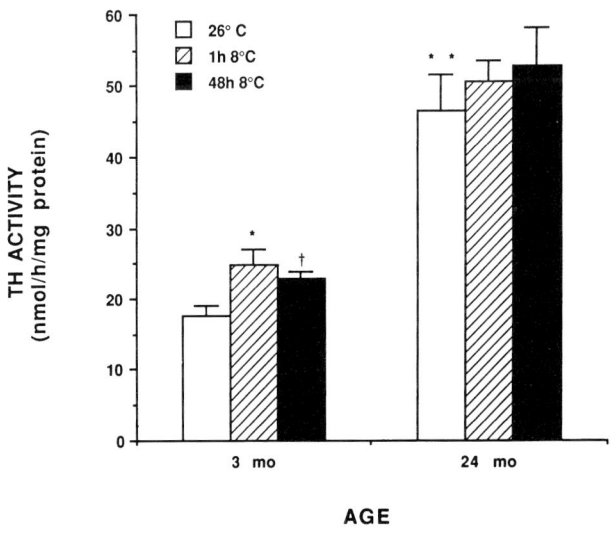

Fig. 3 TH activity with age and following either 1 hr or 48 hr cold exposure. Values represents mean ± SEM of 6–8 rats assayed in duplicate. *$P < 0.005$ for difference with cold by two-way ANOVA. *, † $P < 0.05$ for difference from young control by post hoc analysis. **$P < 0.005$ for difference with age by two-way ANOVA.

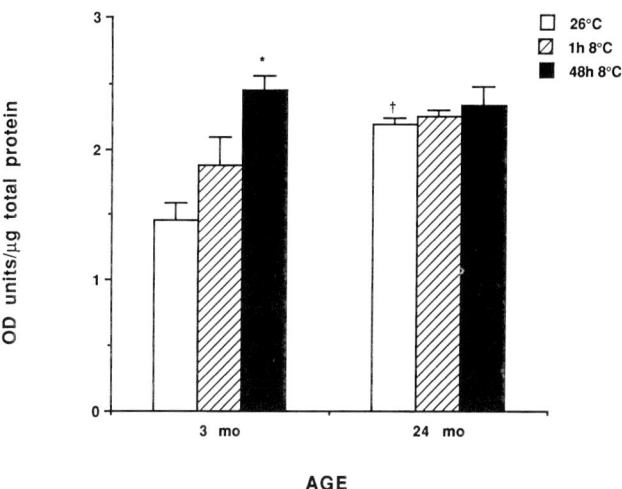

Fig. 4 TH immunoreactivity with age and following either 1 hr or 48 hr cold exposure. Values represent mean ± SEM of 6–8 rats. *P < 0.005 for difference with cold by two-way ANOVA. *, † P < 0.05 for difference from young control by post hoc analysis. †P < 0.005 for difference with age by two-way ANOVA.

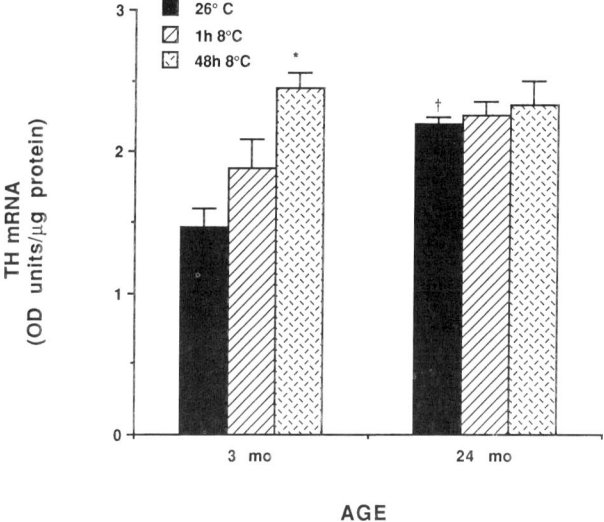

Fig. 5 TH mRNA levels with age and following either 1 hr or 48 hr cold exposure. Values represent mean ± SEM of 6–8 rats. *Significantly different from age-matched control, P < 0.005. †Significantly different from 3-month-old controls, P < 0.001.

cold exposure suggest several possibilities. First, this lack of an increase in TH synthesis is apparently not a result of an inadequate signal to the adrenals (Avakian *et al.*, 1984). We previously demonstrated that choline acetyl transferase activity is unchanged in the adrenal medulla with age (Tümer *et al.*, 1992). Furthermore, in response to acute cold, plasma EPI increases equally in young and old rats (Avakian *et al.*, 1984). This increase in EPI represents the release of stored adrenal medullary epinephrine and suggests that the adrenal gland is

receiving adequate signals to respond to the cold stress. Collectively, these data suggest that the impairment in the cold-induced TH induction is within the adrenal gland. The exact mediator of the cold-induced stimulation of TH gene expression is unknown, and this increased expression is most likely the consequence of the integration of several signals, including both the cAMP and AP-1 signal transduction pathways (Miner *et al.*, 1992; Yoon and Chikaraishi, 1992). The failure to increase TH activity in senescent rats following cold exposure may be a result of impaired signal transduction of the cAMP and AP-1 pathways. Impairment of one or more of these signals with age could result in a failure in inducible TH gene expression. Alternatively, because basal TH mRNA levels, TH immunoreactivity, and TH enzyme activity are two to three fold higher in older compared with younger rats (Tümer and La-Rochelle, 1995). Thus, TH gene expression could already be maximally activated in the basal state of the senescent rats, such that further stimulation by cold exposure is ineffectual.

Adenylyl Cyclase Mediated Signal Transduction Increases with Age

An impaired ability to activate adenylyl cyclase with age may be contributing to the impaired regulation of TH induction in senescent rats. It is well known that the capacity to stimulate adenylyl cyclase activity and cAMP production decline with age in a wide variety of tissues (Scarpace *et al.*, 1991). We have demonstrated that in the rat heart there are decreases with age in G-protein function and the number of catalytic units of adenylyl cyclase (Scarpace *et al.*, 1991). Thus an important regulatory step in TH induction, especially in senescent animals, may be signal transduction and the production of cAMP. On the other hand, there may be adequate activation of cAMP, but a failure in the activation or binding of the CREB transcription factor to the TH gene promoter. To examine whether the sensitivity or efficacy for adenylyl cyclase stimulation is altered with age, we determined the vasoactive intestinal peptide (VIP) and forskolin activation of adenylyl cyclase in adrenal medullary membranes from young and senescent male F-344 rats. Stimulation by VIP (10 μm) and GTP (10 μm) were unchanged with age. In contrast, the maximal stimulation by forskolin was increased nearly 2-fold (25.1 ± 4.6 vs 43.5 ± 4.5 pmol cAMP/min/mg protein, p < 0.001), and the sensitivity for forskolin stimulation was increased 5-fold with age (Tümer *et al.*, 1996). These data suggest that the impaired cold-induction of TH activity apparently is not a consequence of a deficient cAMP pathway with age. However, these data do suggest that the increased adenylyl cyclase stimulation with age may be involved in the elevated TH gene expression with age (Tümer *et al.*, 1996). Increase basal levels of the cAMP/CREB pathway could be contributing to the elevation of basal levels of TH mRNA and TH enzyme activity with age.

In vivo Induction of TH by Forskolin

Our data, indicating that forskolin-stimulated adenylyl cyclase is elevated with age, support a role for an over-active cAMP/CREB pathway with age. We examined the induction of TH mRNA and TH activity by the direct activation of adenylyl cyclase *in vivo* by administering forskolin to young and senescent rats. Forskolin is a direct activator of the enzyme adenylyl cyclase and results in a rise in intracellular cAMP levels (Seamon and Daly, 1986). cAMP can both activate existing TH enzyme by promoting phosphorylation of inactive enzyme and can stimulate the synthesis of new enzyme through phosphorylation of CREB and the binding of CREB to the cAMP response element on the promoter region of the TH gene (Carroll *et al.*, 1991). However, the activation of existing TH enzyme by cAMP is rapidly reversed during the tissue preparation, and our assay determined only forskolin induction of TH synthesis (Tümer *et al.*, 1997a).

TH mRNA levels were examined five hours after administration of a single dose of forskolin (1.8 mg/kg) to young and old rats (Tümer *et al.*, 1997a). Similar to our previous findings (Tümer and LaRochelle, 1995), TH mRNA levels in control senescent rats were greater than 2-fold higher than the mRNA levels in control young rats (Figure 6). Following forskolin administration, the incremental increase in TH mRNA was the same in both young and senescent rats (Figure 6); however, because of the higher baseline value in the older rats, the percent increase in these rats was less (79% vs. 178%). After forskolin stimulation, the levels of

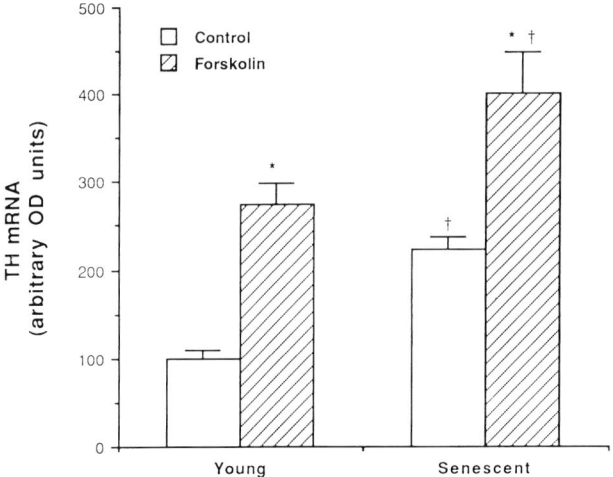

Fig. 6 TH mRNA levels in the adrenal medulla of young and old rats administered a single dose of forskolin (1.8 mg/kg). Data represent the mean ± SEM of 4–6 rats in each group. The TH mRNA level for young controls was arbitrarily set to 100. *P < 0.001 for difference with forskolin treatment by two-way ANOVA; P < 0.001 (young) and P < 0.003 (old) for difference with forskolin from corresponding controls by post-hoc analysis. †P < 0.001 for difference with age by two-way ANOVA. P < 0.001 (control) and P < 0.05 (forskolin) for difference with age by post-hoc analysis.

TH mRNA remained significantly higher in the adrenal medulla from senescent compared with young rats (Figure 6) (Tümer *et al.*, 1997a). These data indicate that activation of the cAMP pathway *in vivo* stimulates TH gene expression in both young and senescent rats. Furthermore, these data indicate that the already elevated level of TH gene expression in the adrenal medulla from senescent rats is still capable of further stimulation, and thus the elevated basal level of TH mRNA with age is not maximally activated.

AP-1 Transcription Factor Binding with Age and Cold

The AP-1 regulatory element binds a sequence-specific DNA transcription factor that is a dimer of two proteins, typically c-Fos and c-Jun, or related proteins. One measure of the level of this transcription factor can be assessed by its binding activity to the consensus sequence of the AP-1 regulatory element. This AP-1 regulatory element has been implicated in the cold-induced TH expression in the adrenal medulla. We examined the cold-stimulated AP-1 transcription factor binding to an oligonucleotide with the consensus sequence of the AP-1 response element in nuclear extracts from adrenal medulla of young and senescent rats (Tümer *et al.*, 1997b). AP-1 transcription factor binding activity diminished by 38% with age in the unstimulated adrenal medulla. Following cold stimulation, the AP-1 binding activity increased by 21 to 25% in the adrenal medulla of both young and senescent rats (Figure 7). However, the level of AP-1 binding in cold-stimulated senescent rats was still less than in cold-stimulated younger rats (Figure 7). For comparison, we assessed binding activity to an oligonucleotide with the consensus sequence of the AP-3 response element. There were no changes in AP-3 binding activity with either age or cold exposure in the adrenal medulla (Figure 7). These data indicate that there is reduced AP-1 binding activity in senescent control rats. This suggests that transcription factor binding to the AP-1 regulatory element is not involved in the increase in TH mRNA levels with age. Moreover, the demonstration that a cold stimulus evokes similar increases in AP-1 binding activity in both young and old rats suggests that the stimulation pathway that increases the AP-1 transcription factor is maintained in the senescent animals (Tümer *et al.*, 1997b).

Exercise and Ageing

Exercise can increase cardiovascular functional capacity and decrease myocardial oxygen demand in healthy people, as well as in most patients with cardiovascular disease (McHenry *et al.*, 1990). Exercise is also beneficial in controlling hypertension (McHenry *et al.*, 1990) and can increase longevity in rats (Holloszy, 1993). Two categories of exercise are generally recognized; acute exercise and exercise training. Acute exercise rapidly stimulates sympathetic outflow (Mazzeo, 1991). Mea-

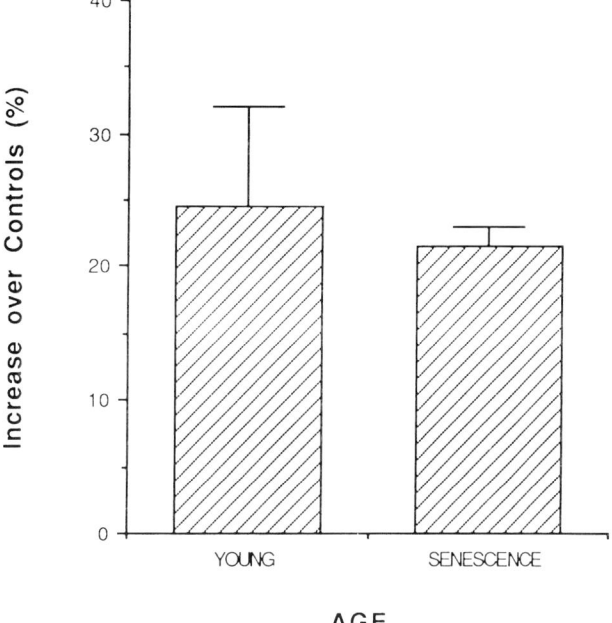

Fig. 7 Transcription factor binding to the oligonucleotide consensus sequence for the AP-1 regulatory element in adrenal medulla from young and old rats following 48 hr cold exposure. Data represent the percent increase in binding compared to thermoneutral controls. AP-1 binding activity in thermoneutral controls was determined in 5 separate experiments. AP-1 binding activity with cold exposure was determined twice.

surement of parameters usually occurs during or immediately after the exercise period. Exercise training results in long-term adaptation to the repeated effects of exercise. Measurement of parameters usually occurs 1–2 days after the last exercise period to eliminate the effects of acute exercise.

The findings concerning the effect of exercise training on circulating catecholamines in humans are controversial. In some studies, regular exercise does not appear to influence plasma NE levels in healthy, elderly subjects who do not have elevated catecholamines (Cononie et al., 1996; Kohrt et al., 1993). However, there is some evidence that regular aerobic exercise results in lower plasma NE concentrations in elderly humans with elevated baseline levels (Seals et al., 1994). In addition, exercise training reduces both systolic and diastolic blood pressure in young and elderly hypertensive subjects (Hagberg and Seals, 1986; Tipton, 1991).

Mazzeo et al. (1986) reported the effects of training on tissue catecholamine content in F-344 rats. Resting cardiac catecholamine content declined significantly with age. In response to an acute exercise bout, EPI levels in the heart were greatly reduced with age. Similarly, the NE content in the heart in response to acute exercise was also markedly reduced with advancing age. In contrast, adrenal catecholamine levels showed an increase with age. It was concluded that

when challenged with strenuous physical stress, cardiac catecholamine content is markedly diminished with age, suggesting synthesis may be impaired. In other studies, turnover and circulating levels of peripheral catecholamines measured both at rest and in response to exercise were reduced by chronic exercise in both humans and rodents (Mazzeo et al., 1986; Mazzeo, 1991).

Exercise Decreases TH Gene Expression in the Adrenal Medulla of Young but not Old Rats

We tested the hypothesis that exercise training reduces plasma catecholamines by modulating TH activity at the level of TH mRNA concentrations (Tümer et al., 1992). Female F-344 rats of 5 and 25 months of age were exercised by treadmill running on a progressive training program over 10 weeks. Initial training sessions began at 17.5 meter/min at level grade for 15 min/day, progressing to 60 min/day, by the end of the fifth week. From the sixth to tenth week the grade was gradually increased to 6% with a speed of 20 meters/min. Non-exercised animals were placed into the treadmill for equal lengths of time but the apparatus was kept off. At the end of the training period, we assessed adrenal medulla TH activity, TH mRNA, and maximum oxygen consumption. There were no significant changes in the weights of the animals. However, there was a significant elevation in maximum O_2 consumption in trained animals of both age groups as compared to their respective control groups, indicating training had occurred. Although the levels of oxygen consumption differed significantly between the two age groups (79.4 ± 1.9 ml O_2/kg/min, 5 mo; 59.4 ± 3.0, 25 mo, $p < 0.05$), both groups showed a similar magnitude of increase in oxygen consumption associated with training (94.4 ± 0.7, 5 mo; 72.8 ± 1.2, 25 mo, $p < 0.05$). These data indicate that training occurred equally in both groups (Tümer et al., 1992).

To determine the effect of training and age on the capacity for catecholamine biosynthesis, TH activity was assessed. Among control groups, TH activity was significantly greater in the 25-month-old compared with 5-month-old animals (Table 1) (Tümer et al., 1992). Among 5-month-old animals, TH activity was significantly lower (Table 1) in the trained group as compared with the young control group. However, in the older rats there was no significant difference in TH activity associated with training (Tümer et al., 1992). To determine if alternations in TH activity were a consequence of changes in gene expression, we measured TH mRNA levels. Among control animals, a comparison between age groups revealed that TH mRNA was significantly elevated by 48% in adrenals from old rats compared with young rats (Table 1). In young animals, TH mRNA in adrenals of exercised rats was significantly lower as compared with the control group. In contrast to the young rats, there was no significant effect of training on TH mRNA in 25-month-old animals (Table 1).

Table 1 Effect of aging and exercise training on tyrosine hydroxylase (TH) activity and TH mRNA

	TH activity (pmol/mg Protein/hr)			THmRNA (Optimal density units)	
	5-month-old rats	25-month-old rats		5-month-old rats	25-month-old rats
Control	3654 ± 184	$4433 \pm 253^{*}$	Control	0.912 ± 0.10	$1.24 \pm 0.20^{*}$
Trained	$2908 \pm 264^{\dagger}$	3944 ± 577	Trained	$0.66 \pm 0.04^{\dagger}$	1.14 ± 0.10

Values represent means $\pm$ SEM of 9 to 11 rats.
* Significantly different from 5-month-old controls (p < .01).
† Significantly different from age-matched controls (p < .01).

The primary finding from these studies is that chronic exercise training is associated with a decrease in TH gene expression and TH activity in the adrenal gland of young rats but not old rats. The adrenal gland is innervated by preganglionic sympathetic nerve terminals whose neurotransmitter is acetylcholine. Exercise may diminish the sympathetic stimulation of the adrenal gland and, thus, decrease TH gene expression. Therefore, we examined cholinergic nerve activity by measuring choline acetyl transferase (ChAT) activity (Tümer et al., 1992). In contrast to the findings for TH activity and TH mRNA, there was no significant effect of age on ChAT activity in control animals (9.06 ± 0.58 and 9.67 ± 0.78 nmol/mg protein/hr; 5 mo and 25 mo, respectively, $p > 0.05$). On the other hand, ChAT activity was significantly lower by 30% in the young exercised rats as compared with the control group, but ChAT activity was unchanged with exercise training in the senescent animals (6.57 ± 0.41 and 8.59 ± 0.57 nmol/mg protein/hr; 5 mo and 25 mo, respectively, $p < 0.05$). These data suggest that exercise training may reduce TH gene expression by reducing cholinergic nerve activity in young rats but that this process is impaired in senescent rats. These findings of a reduction in ChAT activity in young rats are consistent with the finding that exercise decreases ChAT activity in the brainstem, an area important in the control of autonomic function (Somani et al., 1991).

Conclusions

The incidence of cardiovascular disease increases in the elderly. In particular, the elderly do not regulate blood pressure (Tuck, 1989) or body temperature (Collins and Exton-Smith, 1983) as well as young people (Tuck, 1989). In addition, there are increased adverse reactions in elderly hypertensives to medications prescribed for their conditions (Tuck, 1989).

An elevated sympathoadrenal activity with age might play an important role in the pathogenesis of genetically determined blood pressure elevations (Sabol and Higuchi, 1990). The sustained elevation of catecholaminergic activity most likely requires increases in TH gene expression to maintain the augmented catecholamines. This is supported by our data indicating that basal levels of TH activity, TH immunoreactivity, and TH mRNA are elevated with age. However, the mechanism of this increased expression with age remains elusive. The AP-1 regulatory element has been implicated in the basal expression of TH mRNA; however, levels of transcription factor that bind to this regulatory element are unchanged with age. Forskolin stimulated adenylyl cyclase activity is elevated with age, and elevated levels of this pathway with age may account for the increased basal levels of TH mRNA with age, but direct assessment of CREB transcription factor levels with age has not been determined.

Increased autonomic dysfunction in the elderly may be due to a failure of homeostatic regulation. Autonomic dysfunction not only contributes to medical problems in the elderly, but strongly affects the quality of life in relatively healthy individuals. TH plays a central role in the neuronal transmission and hormonal action of catecholamines. Besides hypertension and cardiovascular disease, the pathogenesis of some disorders of catecholaminergic neurons, such as Parkinson's disease (Hassler and Christ, 1984), may be related to changes in TH. Catecholamines are diminished in the adrenal medulla of patients with this disease (Hassler and Christ, 1984). The impaired ability of senescent rats to upregulate TH activity following cold exposure is another example of homeostatic failure of autonomic function. This impaired upregulation of TH activity, TH immunoreactivity, and TH mRNA were not due to an already maximally stimulated induction pathway in the basal state of the senescent rats, such that further stimulation by cold exposure would be ineffectual. Forskolin was capable of inducing TH gene expression in both young and old rats, and cold exposure increased AP-1 binding activity in both young and old rats. Furthermore, neither the forskolin induction of TH gene expression nor the cold-stimulated increase in AP-1 binding activity were diminished with age. These data suggest that these pathways are not responsible for the impaired cold induction of TH with age. It is possible that other pathways are involved in the cold-induction of TH or that a coordinated signal is necessary and this coordinated signal is disrupted in the aging animal.

Exercise is known to help control hypertension, emotional stress, and to reduce risk factors for coronary artery disease and prevention of cardiovascular disease (McHenry *et al.*, 1990). Moreover, increased levels of physical activity are associated with longevity and improved performance on various cognitive tasks (Blumenthal *et al.*, 1989). Exercise may reduce blood pressure by decreasing TH activity, and our data indicate that TH activity and TH mRNA are reduced following exercise training in young but not senescent rats. The mechanism may involve an exercise training reduction in cholinergic nerve activity in young rats, but this process appears to be impaired in senescent rats. These data suggest that the beneficial effects of exercise may be blunted in the elderly.

References

Acheson, A.L., and Zigmond, M.J. (1981). Short and long term changes in tyrosine hydroxylase activity in rat brain after subtotal destruction of central noradrenergic neurons. *Journal of Neuroscience* **1**, 493–504.

Arvakian, E.V., Horvath, S.M., and Colburn, R.W. (1984). Influence of age and cold stress on plasma catecholamine levels in rats. *Journal of the Autonomic Nervous System* **10**, 127–133.

Banerji, T.K., Parkening, T.A., and Collins, T.J. (1984). Adrenomedullary catecholaminergic activity increases with age in male laboratory rodents. *Journal of Gerontology* **39**, 264–268.

Baruchin, A., Weisberg, E.P., Miner, L.L., Ennis, D., Nisenbaum, L.A., Naylor, E., Stricker, E.M., Zigmond, M.J., and Kaplan, B.B. (1990). The effects of cold exposure on rat adrenal tyrosine hydroxylase: an analysis of RNA, protein, enzyme activity and cofactor levels. *Journal of Neurochemistry* **54**, 1769–1775.

Blumenthal, J.A., Emery, C.F., Madden, D.J., George, L.K., Coleman, R.E., Riddle, M.W., McKee, D.C., Reasoner, J., and Williams, R.S. (1989). Cardiovascular and behavioral effects of aerobic exercise training in healthy older men and women. *Journal of Gerontology* **44**, M147–M157.

Carroll, J.M., Kim, K.S., Kim, K.T., Goodman, H.M., and Joh, T.H. (1991). Effects of second messenger system activation on functional expression of tyrosine hydroxylase fusion gene constructs in neuronal and nonneuronal cells. *Journal of Molecular Neuroscience* **3**, 65–74.

Chiueh, C.C., Nespor, S.M., and Rapoport, S.I. (1980). Cardiovascular, sympathetic and adrenal cortical responsiveness of aged Fischer-344 rats to stress. *Neurobiology of Aging* **1**, 157–163.

Collins, K.J., and Exton-Smith, A.N. (1983). Thermal homeostasis in old age. *Journal of The American Geriatrics Society* **31**, 519–524.

Cononie, C.C., Graves, J.E., Pollock, M.L., Phillips, M.I., Sumners, C., and Hagberg, J.M. (1996). Effect of exercise training on blood pressure in 70- to 79-yr-old men and women. *Medicine and Science in Sports and Exercise* **23**, 505–511.

Daubner, S.C., Lauriano, C., Haycock, J.W., and Fitzpatrick, P.F. (1992). Site-directed mutagenesis of serine 40 of rat tyrosine hydroxylase. *Journal of Biological Chemistry* **267**, 12639–12646.

Fluharty, S.J., Rabow, L.E., Zigmond, M.J., and Stricker, E.M. (1985). Tyrosine hydroxylase activity in the sympathoadrenal system under basal and stressful conditions: Effect of 6-hydroxydopamine. *Journal of Pharmacology and Experimental Therapeutics* **235**, 354–360.

Hagberg, J., and Seals, D. (1986). Exercise training and hypertension. *Acta Medica Scandinavica* **177**, 131–136.

Hanbauer, I., Kopin, I.J., and Costa, E. (1973). Mechanisms involved in the trans-synaptic increase of tyrosine hydroxylase and dopamine β-hydroxylase activity in sympathetic ganglia. *Naunyn-Schmiedebergs Archives of Pharmacology* **280**, 39–48.

Hassler, R.G., and Christ, J.F. (1984). Parkinson-Specific Motor and Mental Disorders. R.G. Hassler and J.F. Christ (Eds.). *Advances in Neurology*, Vol. 40. Raven Press, New York.

Holloszy, J.O. (1993). Exercise increases average longevity of female rats despite increased food intake and no growth retardation. *Journal of Gerontology* **48**, B97–B100.

Hong, M., Li, S., Fournier, A., St-Pierre, S., and Pelletier, G. (1995). Role of neuropeptide Y in the regulation of tyrosine hydroxylase gene expression in rat adrenal glands. *Molecular Neuroendocrinology* **61**, 85–88.

Irwin, M., Hauger, R., and Brown, M. (1992). Central corticotropin-releasing hormone activates the sympathetic nervous system and reduces immune function: increased responsivity of the aged rat. *Endocrinology* **131**, 1047–1053.

Ito, K., Sato, A., Sato, Y., and Suzuki, H. (1986). Increases in adrenal catecholamine secretion and adrenal sympathetic nerve unitary activities with aging in rats. *Neuroscience Letters* **69**, 263–268.

Karin, M., and Smeal, T. (1992). Control of transcription factors by signal transduction pathways: the beginning of the end. *Trends in Biochemical Sciences* **17**, 418–422.

Kedzierski, W., and Porter, J.C. (1990). Quantitative study of tyrosine hydroxylase mRNA in catecholaminergic neurons and adrenals during development and aging. *Molecular Brain Research* **7**, 45–51.

Kiritsy-Roy, J.A., Halter, J.B., Smith, M.J., and Cass-Terry, L. (1992). Selective impairment of neuroendocrine and hemodynamic responses to a mu-opioid peptide in aged rats. *Journal of Gerontology* **47**, B89–B97.

Kohrt, W.M., Spina, R.J., Ehsani, A.A., Cryer, P.E., and Holloszy, J.O. (1993). Effects of age, adiposity, and fitness level on plasma catecholamine responses to standing and exercise. *Journal of Applied Physiology* **75**, 1828–1835.

Kumer, S.C., and Vrana, K.E. (1996). Intricate regulation of tyrosine hydroxylase activity and gene expression. *Journal of Neurochemistry* **67**, 443–462.

Martinez, J.L., Vasquez, B.J., Messing, R.B., Jensen, R.A., Liamg, K.C., and McGaugh, J.L. (1981). Age-related changes in the catecholamine content of peripheral organs in male and female F-344 rats. *Journal of Gerontology* **36**, 280–284.

Mazzeo, R.S., Colburn, R.W., and Horvath, S.M. (1986). Effect of aging and endurance training on tissue catecholamine response to strenuous exercise in Fischer-344 rats. *Metabolism* **35**, 602–607.

Mazzeo, R.S. (1991). Catecholamine responses to acute and chronic exercise. *Medicine and Science in Sports and Exercise* **23**, 839–844.

McCarty, R., Mabry, T.R., Foster, T.C., and Gold, P.E. (1996). Stress, peripheral catecholamines and age-related memory deficits. In: R. McCarty, G. Aguilera, E. Sabban, and R. Kvetňanský (Eds.). *Stress: Molecular Genetic and Neurobiological Advances* pp. 967–979. New York: Gordon and Breach Science.

McHenry, P.L., Ellestad, M.H., Fletcher, G.F., Froelicher, V., Hartley, H., Mitchell, J.H., and Froelicher, E.S.S. (1990). A position for health professionals by the committee on exercise and cardiac rehabilitation of the council on clinical cardiology, American Heart Association. Special Report, *Circulation* **81**, 396–398.

Milakofsky, L., Harris, N., and Vogel, W.H. (1993). Effect of repeated stress on plasma catecholamines and taurine in young and old rats. *Neurobiology of Aging* **14**, 359–366.

Miner, L.L., Baruchin, A., and Kaplan, B.B. (1989). Effect of cold stress on cholinergic receptors in the rat adrenal gland. *Neuroscience Letters* **106**, 339–344.

Miner, L.L., Pandalai, S.P., Weisberg, E.P., Sell, S.L., Kovacs, D.M., and Kaplan, B.B. (1992). Cold-induced alterations in the binding of adrenomedullary nuclear proteins to the promoter region of the tyrosine hydroxylase gene. *Journal of Neuroscience Research* **33**, 10–18.

Nagatsu, T., Levitt, M., and Udenfriend, S. (1964). Tyrosine hydroxylase: The initial step in norepinephrine biosynthesis. *Journal of Biological Chemistry* **238**, 2910–2917.

Roberts, J., and Tümer, N. (1987). Age-related changes in autonomic function of catecholamines. In: M. Rothstein and A.R. Liss (Eds.), *Review of Biological Research in Aging*, Vol. 3, pp. 257–298. New York.

Sabol, S.L., and Higuchi, H. (1990). Transcriptional regulation of the neuropeptide Y gene by nerve growth factor: antagonism by glucocorticoids and potentiation by adenosine 3′, 5′-monophosphate and phorbol ester. *Molecular Endocrinology* **4**, 384–392.

Scarpace, P.J., Tümer, N., and Mader, S.L. (1991). β-adrenergic function in aging. Basic mechanisms and clinical implications. *Drugs & Aging* **1**, 116–129.

Seals, D.R., Taylor, J.A., Ng, A.V., and Esler, M.D. (1994). Exercise and aging: autonomic control of the circulation. *Medicine and Science in Sports and Exercise* **26**, 568–576.

Seaman, K., and Daly, J.W. (1981). Activation of adenylate cyclase by the diterpene forskolin does not require the guanine nucleotide regulatory protein. *Journal of Biological Chemistry* **256**, 9799–9801.

Seamon, K.B., and Daly, J.W. (1986). Forskolin: its biological and chemical properties. In: P. Greengard and G.A. Robinson (Eds.), *Advances in Cyclic Nucleotide and Protein Phosphorylation Research*, Raven Press, New York.

Somani, S.M., Babu, S.R., Arneric, S.P., and Dube, S.N. (1991). Effect of cholinesterase inhibitor and exercise on choline acetyltransferase and acetylcholinesterase activities in rat brain regions. *Pharmacology Biochemistry and Behavior* **39**, 337–343.

Stachowiak, M.K., Sebbane, R., Stricker, E.M., Zigmond, M.J., and Kaplan, B.B. (1985). Effect of chronic cold exposure on tyrosine hydroxylase mRNA in rat adrenal gland. *Brain Research* **359**, 356–359.

Stachowiak, M.K., Fluharty, S.J., Stricker, E.M., Zigmond, M.J., and Kaplan, B.B. (1986). Molecular adaptations in catecholamine biosynthesis induced by cold stress and sympathectomy. *Journal of Neuroscience Research* **16**, 13–24.

Stachowiak, M., Striker, E.M., Zigmond, M.J., and Kaplan, B.B. (1988). A cholinergic antagonist blocks cold stress-induced alterations in rat adrenal tyrosine hydroxylase mRNA. *Molecular Brain Research* **3**, 193–196.

Strong, R., Moore, M.A., Hale, C., Wessels-Reiker, M., Armbrecht, H.J., and Richardson, A. (1990). Modulation of tyrosine hydroxylase gene expression in the rat adrenal gland by age and reserpine. *Brain Research* **525**, 126–132.

Tank, A.W., Lewis, E.J., Chikaraishi, D.M., and Weiner, N. (1985). Elevation of RNA coding for tyrosine hydroxylase in rat adrenal gland by reserpine treatment and exposure to cold. *Journal of Neurochemistry* **45**, 1030–1033.

Taylor, C.W., and Merritt, J.E. (1986). Receptor coupling to polyphosphatidyl inositol turnover: a parallel with the adenylate cyclase system. *Trends in Pharmacological Sciences* **7**, 238–242.

Tipton, C.M. (1991). Exercise training and hypertension. *Exercise and Sports Sciences Reviews*. J. Hollowzy (Ed.) pp. 447–505. Baltimore, Williams and Wilkins Company.

Tuck, M.L. (1989). Treatment of hypertension in the elderly. In: J.A. Armbrecht, R. Coe, and N. Wongsurawat (Eds.), *Endocrine Function and Aging*, pp. 147–160. Springer-Verlag. New York.

Tümer, N., Hale, C., Lawler, J., and Strong, R. (1992). Modulation of tyrosine hydroxylase gene expression in the rat adrenal gland by exercise: Effects of age. *Molecular Brain Research* **14**, 51–56.

Tümer, N., and LaRochelle, J.S. (1995). Tyrosine hydroxylase expression in rat adrenal medulla: Influence of age and cold. *Pharmacology, Biochemistry and Behavior* **51**, 775–780.

Tümer, N., Sego, R.L., and Scarpace, P.J. (1996). Atypical pattern of adenylyl cyclase activity in the adrenal medulla with age. *Experimental Gerontology* **31**, 571–576.

Tümer, N., Bowman, C.J., LaRochelle, J.S., Kelley, A., and Scarpace, P.J. (1997a). Induction of tyrosine hydroxylase by forskolin: modulation with age. *European Journal of Pharmacology* **324**, 57–62.

Tümer, N., Scarpace, P.J., Baker, H.V., and LaRochelle, J.S. (1997b). AP-1 transcription factor binding activity in the adrenal medulla and hypothalamus with age and cold exposure. *Neuropharmacology* **36**, 1065–1069.

Voogt, J.L., Arbogast, L.A., Quadri, S.K., and Andrews, G. (1990). Tyrosine hydroxylase messenger RNA in the hypothalamus, substantia nigra and adrenal medulla of old female rats. *Molecular Brain Research* **8**, 55–62.

Wessel, T.C., and Joh, T.H. (1992). Parallel upregulation of catecholamine-synthesizing enzymes in rat brain and adrenal gland: effects of reserpine and correlation with immediate early gene expression. *Molecular Brain Research* **15**, 349–360.

Wessels-Reiker, M., Haycock, J.W., Howlett, A.C., and Strong, R. (1991). Vasoactive intestinal polypeptide induces tyrosine hydroxylase in PC12 cells. *Journal of Biological Chemistry* **266**, 9347–9350.

Yoon, S.O., and Chikaraishi, D.M. (1992). Tissue-specific transcription of the rat tyrosine hydroxylase gene requires synergy between an AP-1 motif and an overlapping E box-containing Dyad. *Neuron* **9**, 55–67.

Yu, K., Lu, D., Rowland, N.E., and Raizada, M.K. (1996). Angiotensin II regulation of tyrosine hydroxylase gene expression in the neuronal cultures of normotensive and spontaneously hypertensive rats. *Endocrinology* **137**, 3566–3576.

33. Effects of Neonatal Treatment with Antisense Oligodeoxynucleotide to Alpha2A-Adrenergic Receptor on the Noradrenergic System of Rat Brain

N.N. Dygalo, T.S. Kalinina, N. Yu. Sournina and L.B. Melnikova

Introduction

Alpha2-adrenoreceptors of A, B and C subtypes are differentially distributed in cells and tissues, clearly endowing the receptors with different physiological functions. Immunohistochemical localization of alpha2A-adrenergic receptors (alpha2A-ARs) in catecholaminergic neurons of brain stem (Aoki et al., 1994; Lee et al., 1998) suggest that alpha2A-ARs are the autoreceptors of these neurons and regulate their activity. Alpha2A-AR mRNA is detected in noradrenergic neurons of rat fetal brain beginning at embryonic day 14. High levels of perinatal expression of this adrenoreceptor type may indicate a specific functional role for the alpha2A-ARs in the developing noradrenergic system of the brain (Winzer-Sherhan et al., 1997). Available ligands for alpha2-ARs have only marginal subtype selectivity and this has limited insight into the functional role of each subtype. Antisense oligodeoxynucleotides offer the potential to block the expression of specific genes within cells (Wagner, 1994). In this study, direct injection of antisense oligonucleotide to alpha2A-AR mRNA into the locus coeruleus region of rat pups was used to investigate the effects of the receptor subtype on norepinephrine levels and AR densities in the developing rat brain.

Materials and Methods

Two five-day-old Wistar rat pups remaining with their original dams were used in experiments. Dams had unlimited access to food and water. On days 2, 3 and 4 of life, animals in the experimental group received three intrabrain injections of a 18-mer antisense phosphorotioate oligodeoxynucleotide to alpha2A-AR mRNA (A-ODN) into the locus coeruleus region. The antisense was targeted to the area of the alpha2A-AR mRNA sequence that bridges the initiation codon (from $-11°$ to $+7°$, 5'-agcccatgggcgcaaagc-3'). As a control, a phosphorotioate – modified oligodeoxynucleotide with the same proportion of bases as the alpha2A-AR antisense, but having a random sequence (R-ODN; 5'-gacgacccagtgagcacg-3'), was similarly prepared. The sequences of A-ODN and R-ODN have relatively low homology with any of the other known mammalian sequences found in the Gene Bank database. The oligodeoxynucleotides were dissolved in sterile saline.

Injections were performed on cold-anesthetized pups and aimed at the locus coeruleus region using a modified rat stereotaxis. The dose of A-ODN or R-ODN per one injection was 0.12 μg DNA dissolved in 5 μl of saline. In addition to A-ODN and R-ODN treated groups a group of animals which received 3 intrabrain injections of 5 μl saline was also used in the study. The pups were killed on day 5, 24 hours after the last intrabrain injection. The frontal cortex and the brain stem region which included pons and medulla oblongata were dissected.

Total cellular RNA from frontal cortex and brain stem was isolated by a single step acid guanidinium-phenol-chloroform extraction method (Chomczynski and Sacchi, 1987). Two micrograms of total mRNA was used as template for first strand cDNA synthesis in 20 μl reaction volume containing 67 mM Tris-HCl, pH 8.8, 17mM KCl, 1mM $MnCl_2$, 1mM each of dNTP, 100 ng oligo-dT-primer, 2.5% glycerol and 5U TET-z DNA Polymerase, which has reverse transcriptase activity in the presence of Mn^{2+}. Reaction temperatures were 25°C for 10 min for annealing of primer, 42°C for 1 hour for extension of primer. After that 5 μl reaction mixture was increased to 50 μl with 45 μl of PCR mix (67mM Tris-HCl, pH 8.0, 2.5 mM $MgCl^2$, 0.01M 2-mercaptoethanol, 0.01% Tween-20, 0.2mM of forward and reverse for alpha2A-AR or β-actin PCR primers and 2.5U Tag Polymerase). Oligonucleotide primers were designed from published sequences of cytoplasmic β-actin (Nudel et al., 1983) and alpha2A-AR (Kobilka et al., 1987). Oligonucleotide for alpha2A-AR forward primer sequence was 5'-tgcgagatcaacgaccagaa-3', and reverse was 5'-cacgaacgtgaagcgcttct-3'. The expected size of the amplified fragment was 560 base pairs. For β-actin, primers were 5'-tccctcatgccatcctgcgt-3' and 5'-ggaacctctcattgccgata-3' to produce a 255 bp fragment. Amplifications were performed in a programmable thermal cycler with an initial template denaturation at 95°C for 3 min, annealing at 52°C for 1.5 min and extension of primers for 2 min. at 70°C. This first cycle was followed by 40 amplification cycles of denaturation at 95°C for 0.8 min, annealing at 52°C for 0.8 min and extension at 70°C for 1 min. The final cycle lasted 10 min at 70°C. In all experiments, the presence of possible contaminants was checked by a control reaction in which amplification was carried out on samples in which reverse transcriptase was omitted from the reverse transcription reaction mixture. The amplification products were separated on ethidium bromide stained 1.5% agarose gel. Gene expression level of alpha2A-AR was quantified relative to β-actin by scanning densitometry (Biodoc II Video Documentation System, Biometra GmbH, Germany).

For radioligand binding assays, brain tissue was homogenized in 20 vol of ice-cold 50 mM Tris-HCl buffer (pH 7.7) and centrifuged at 20,000 g for 15 min. The pellets were rehomogenized in another portion of the buffer, then centrifuged again. The final pellets were resuspended in 140 vol of the same buffer. 3.2 nM ^{3}H-Clonidine (23.2 Ci/mmol, Amersham), imidazoline with alpha2-adrenoreceptor agonist specificity, or 1.6 nM ^{3}H-dihydroalprenolol (DHA, 75 Ci/mmol, Amersham), the beta-adrenoreceptor antagonist, were added to 0.8 ml of membrane suspensions and the samples in final volumes of 1 ml were incubated: for ^{3}H-clonidine binding for 30 min at 25°C in the absence (total binding) or presence (nonspecific binding) of 100 μM norepinephrine (Koch-Light); and for ^{3}H-DHA binding for 20 min at 23°C with or without 10 μM propranolol (Sigma). The reactions were terminated by rapid filtration under a vacuum through glass fiber filters (GF/B, Whatman). Filters were washed three times with 5 ml of ice-cold buffer and radioactivity was measured in a Delta-300 liquid scintillation counter with an efficiency of 35%. Specific binding was expressed as fmol radioligand bound per mg protein which was determined by the method of Lowry *et al.* (1951).

Norepinephrine and dopamine levels were determined fluorimetrically (Jacobowitz and Richardson, 1978). Data were analyzed by one-way ANOVA.

Results

To determine the effects of treatment with A-ODN on the mRNA for the alpha2A-ARs, the proportions of this mRNA to the mRNA for beta-actin were determined in the brain stem and frontal cortex of 5-day-old pups by RT-PCR. Figure 1 shows the results of a representative example of such an experiment. The ratio of the optical density of the alpha2A-AR band to the optical density of the beta-actin band of the same cDNA sample were determined and presented in Figure 2 as means from 5 to 6 rat pups per treatment group for each brain region. Treatment with R-ODN caused a two-fold decrease of alpha2A-AR mRNA concentrations in the brain stem ($F_{2,14}$ = 20.744; p < 0.0001). A-ODN treatment decreased alpha2A-AR mRNA concentrations in the brain stem below the threshold of sensitivity of the method. The levels of alpha2A-AR mRNA in the cortex were significantly increased in A-ODN injected rat pups as compared with saline or R-ODN treated animals ($F_{2,14}$ = 13.021; p < 0.0006).

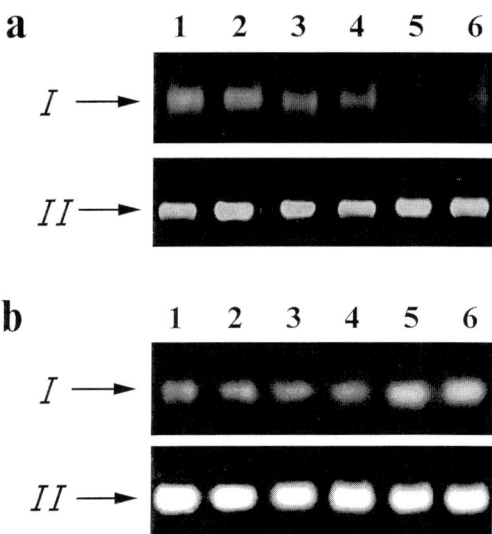

Fig. 1 RT-PCR analysis of alpha2A-AR mRNA expression in the brain of 5-day-old rat pups after intrabrain injections of antisense oligonucleotide to this mRNA on days 2, 3 and 4 of life. Agarose gel electrophoresis of alpha2A-AR and beta-actin PCR products amplified from the same cDNAs that were reverse transcribed from total RNA samples of 5-day-old rat brain stem and cortex. **a** – brain stem; **b** – cortex. *I* – alpha2A-AR; *II* – beta-actin. Lane 1 and 2 – intrabrain injection of saline; 3 and 4 – oligonucleotide of random sequence (R-ODN); 5 and 6 – antisense oligonucleotide to alpha2A-AR mRNA (A-ODN).

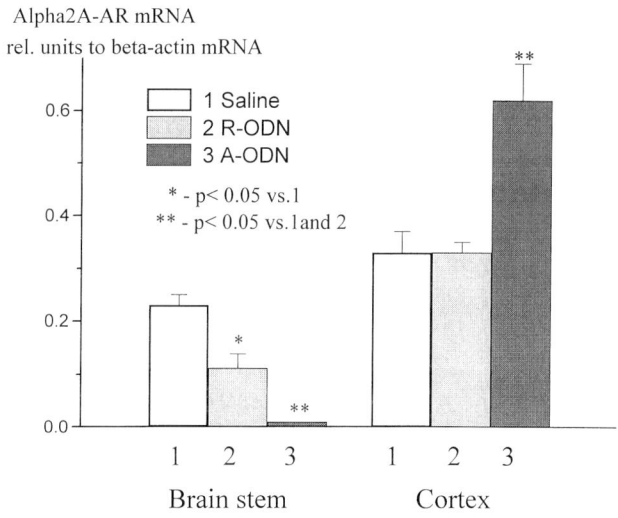

Fig. 2 Relative concentrations of alpha2A-AR mRNA to beta-actin mRNA in the brain stem and cortex of 5-day-old rat pups after intrabrain injections of antisense oligonucleotide to alpha2A-AR mRNA on days 2, 3 and 4 of life. 1 – saline; 2 – oligonucleotide of random sequence (R-ODN); 3 – antisense oligonucleotide to alpha2A-AR mRNA (A-ODN).

Treatment with both ODNs had no appreciable effects on [^{3}H]clonidine and [^{3}H]DHA binding in the brain stem (Figure 3). The density of [^{3}H]clonidine binding sites was increased in the cortex of A-ODN

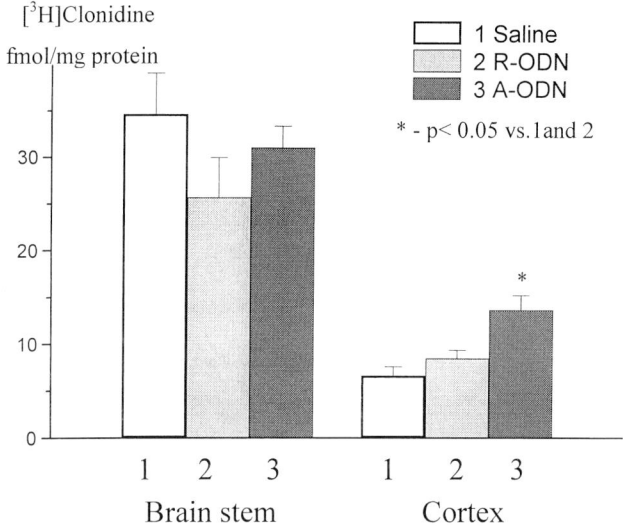

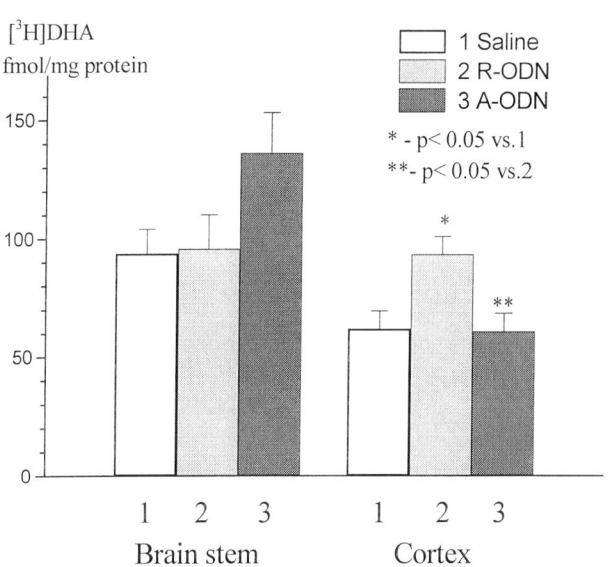

Fig. 3 Densities of alpha2A-ARs ([³H]clonidine binding sites) and beta-ARs ([³H]DHA binding sites) in the brain stem and cortex of 5-day-old rat pups after intrabrain injections of antisense oligonucleotide to alpha2A-AR mRNA on days 2, 3 and 4 of life. 1 – saline; 2 – oligonucleotide of random sequence (R-ODN); 3 – antisense oligonucleotide to alpha2A-AR mRNA (A-ODN).

treated pups as compared with saline and R-ODN injected animals ($F_{2,33} = 10.268$; $p < 0.0003$). R-ODN increased [³H]DHA binding in the frontal cortex and A-ODN treatment returned it back to the level of saline injected controls ($F_{2,34} = 5.341$; $p < 0.01$).

A-ODN treatment increased norepinephrine levels in the brain stem compared to saline or R-ODN injected animals (Figure 4; $F_{2,12} = 3.992$; $p < 0.05$). Dopamine levels in the brain stem did not differ significantly between the animals of all studied groups. Treatment with both ODNs decreased norepinephrine and dopamine levels in the cortex in comparison with saline

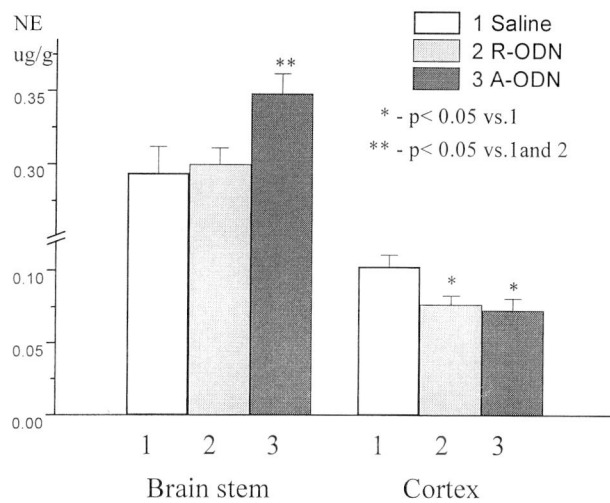

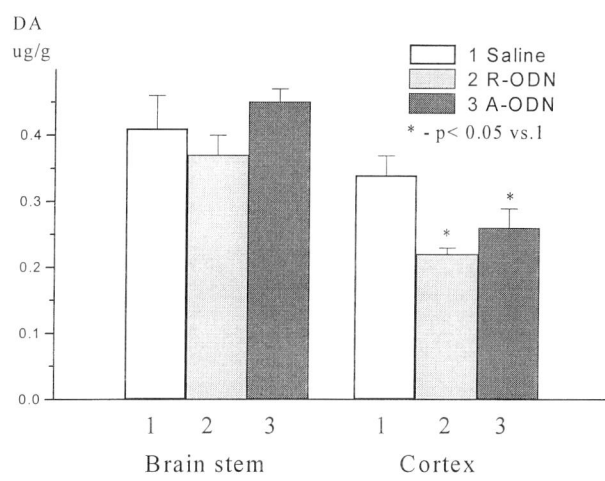

Fig. 4 Norepinephrine and dopamine levels in the brain stem and cortex of 5-day-old rat pups after intrabrain injections of antisense oligonucleotide to alpha2A-AR mRNA on days 2, 3 and 4 of life. 1 – saline; 2 – oligonucleotide of random sequence (R-ODN); 3 – antisense oligonucleotide to alpha2A-AR mRNA (A-ODN).

treated animals ($F_{2,15} = 5.290$; $p < 0.018$ and $F_{2,15} = 5.195$; $p < 0.019$ respectively).

Discussion

Antisense ODN to alpha2A-AR injected into the neonatal rat brain has specific effects on some parameters of brain noradrenergic system function. These effects could be inferred from the biological effects of antisense as compared to control ODN. Antisense ODN decreased alpha2A-AR mRNA levels in the brain stem. However, the total amplitude of this effect probably includes both specific effects on target RNA and nonspecific effects of phosphorotioated ODNs and their breakdown products on the function of brain stem neurons. These nonspecific effects are evident from a comparison of R-ODN treated pups and saline injected animals. R-ODN decreased alpha2A-AR mRNA levels in the brain stem, norepinephrine and dopamine levels in the cortex and increased [³H]DHA binding site densities in the cortex. All these changes seem to be related to the toxic effects of phosphorotioated ODNs and their breakdown products on cell proliferation, differentiation and function (Wagner, 1994). Toxic inhibition of axonal outgrowth into the cortex will result in a decrease in norepinephrine and dopamine concentrations in this brain region. The decrease in norepinephrine concentrations can lead to an increase in [³H]DHA binding site densities in the cortex of rat pups (Dygalo et al., 1993).

While antisense ODN significantly decreased alpha2A-AR mRNA levels in the brain stem, it had no appreciable effect on [³H]clonidine binding site densities in this brain region. Clonidine is an alpha2-ARs agonist with low subtype specificity and it can also label imidazoline binding sites (MacDonald et al., 1997) which are abundant in the brain stem (Bricca et al., 1989). It should be mentioned that injections of antisense ODNs for the alpha2A-AR subtype into the locus coeruleus which decreased hypnotic responses to dexmedetomidine (an alpha2-AR – agonist) of adult rats had no effect on [³H]atipamezole (an alpha2-AR ligand) binding site densities in their brain stem (Mizobe et al., 1996). The lack of subtype specificity of the ligands used for measurements of alpha2-AR binding site densities can be one of the reasons for the discrepancies between radioligand binding and functional data or mRNA levels.

The presence of high amounts of alpha2A-AR mRNA in the noradrenergic cell body regions (Sheinine et al., 1994; Dygalo et al., 1999), and immunohistochemical localization of alpha2A-ARs in noradrenergic neurons (Aoki et al., 1994; Lee et al., 1998) suggest that some receptors of this subtype are presynaptic autoreceptors in the rat brain and regulate the activity of noradrenergic neurons. Application of alpha2-AR agonists inhibits firing of locus coeruleus neurons (Williams et al., 1991) as well as synthesis and release of norepinephrine in these neurons and in their terminals (Callado and Stamford, 1999). The expected result of the blockade of the expression of alpha2A-AR will be an increase in neurotransmitter synthesis and release by noradrenergic neurons and their terminals. Increases in norepinephrine concentrations in the brain stem and decreases in [³H]DHA binding site densities in the cortex of A-ODN pups in comparison to R-ODN animals may be evidences for an increase in neurotransmitter synthesis and releases respectively. Density of beta-ARs ([³H]DHA binding sites) is down regulated by norepinephrine in the brain cortex of neonatal rats (Dygalo et al., 1993).

Antisense ODN specifically increased alpha2A-AR mRNA levels and [³H]clonidine binding site densities in the cortex. If A-ODN indeed increases norepinephrine release in the cortex, then the above data suggest up-regulation of alpha2A-AR subtype expression in neurons of the brain cortex by neurotransmitter. Such possibility could not be ruled out. This receptor subtype has very low sensitivity to down regulation by norepinephrine (Heck, and Bylund, 1998), and incubation of rat cerebral cortical slices with an adrenergic agonist causes an increase in alpha2-AR binding in addition to a decrease in beta-AR binding (Maggi et al., 1980).

In conclusion, inhibition of the expression of alpha2A-ARs in the brain stem has diverse effects on the function of the brain noradrenergic system. The results support autoreceptor function for some adrenergic receptors of the alpha2A subtype in the developing rat brain.

Acknowledgment

This work was supported by Grant N 98–04–49651 of the Russian Fund for Basic Research.

References

Aoki, C., Go, C.-G., Venkatesan, C., and Kurose, H. (1994). Perikaryal and synaptic localization of alpha2A-adrenergic receptor-like immunoreactivity. Brain Research **650**, 181–204.

Bricca, G., Dontenwill, M., Molines, A., Feldman, J., Belcourt, A., and Bousquet, P. (1989). The imidazaline preferring receptor: Binding studies in bovine, rat and human brainstem. European Journal of Pharmacology **162**, 1–9.

Callado, L.F., and Stamford, J.A. (1999). Alpha2A-but not alpha2B/C-adrenoceptors modulate noradrenaline release in locus coeruleus: voltametric data. European Journal of Pharmacology **366**, 35–39.

Chomczynski, P., and Sacchi, N. (1987). Single step method of RNA isolation by acid guanidinium thiocyanate-phenol-chloroform extraction. Analytical Biochemistry **162**, 156–159.

Dygalo, N.N., Kalinina, T.S., Sournina, N. Yu., Nosova, A.V., Shishkina, G.T. (1999). Alpha2A-adrenergic receptor mRNA concentrations and the densities of the binding sites for their agonist in the brain regions. Dokl. Akad. Nauk **364**, 417–419.

Dygalo, N.N., Shishkina, G.T., and Milova, A.A. (1993). Beta-adrenoceptors and norepinephrine in the brain cortex of the rat pups after treatments during early ontogenesis. Ontogenez **24**, 93–97.

Fornai, C., Blandizzi, C., and Tacca, M. (1990) Central alpha2 adrenoceptors regulate central and peripheral functions. Pharmacology Research **22**, 541–554.

Heck, D.A., and Bylund, D.B. (1998). Differential down-regulation of alpha2-adrenergic receptor subtypes. Life Sciences **62**, 1467–1472.

Jacobowitz, D.M., and Richardson, J.S. (1978). Method for the rapid determination of norepinephrine, dopamine and serotonin in the same brain region. Pharmacology Biochemistry and Behavior **8**, 515–519.

Kobilka, B., Matsui, H., Kobilka, T., Yang-Feng, T.L., Francke, U., Caron, M.G., Lefkowitz, R.J., and Regan, J.W. (1987). Cloning, sequencing and expression of the gene coding for the human platelet alpha2-adrenergic receptor. Science **238**, 650–656.

Lee, A., Rosin, D.L., and Van Bockstaele, E.J. (1998). Alpha2A-adrenergic receptors in the rat nucleus locus coeruleus: subcellular localization in catecholaminergic dendrites, astrocytes, and presynaptic axon terminals. Brain Research **795**, 157–169.

Lowry, O.H., Rosenbrough, N.J., Farr, A.L., and Randall, R.J. (1951). Protein measurement with the Folin phenol reagent. Journal of Biological Chemistry **193**, 265–275.

MacDonald, E., Kobilka, B.K., and Scheinin, M. (1997). Gene targeting – homing in on α_2-adrenoceptor subtype function. TiPS **18**, 211–219.

Maggi, A., U'Prichard, D.C., and Enna, S.J. (1980). Beta-adrenergic regulation of alpha2-adrenergic receptors in the central nervous system. Science **207**, 645–647.

Mizobe, T., Maghsoudi, K., Tianzhi, G., Ou, J., and Maze, M. (1996). Antisense technology reveals the alpha2A adrenoceptor to be the subtype mediating the hypnotic response to the highly selective agonist, dexmedetomidine, in the locus coeruleus of the rat. Journal of Clinical Investigation **98**, 1076–1080.

Nudel, U., Zacut, M., Neuman, S., Levy, Z., and Yaffe, D. (1983). The nucleotide sequence of the rat cytoplasmic β-actin gene. Nucleic Acids Research **11**, 1759–1771.

Ordway, G.A. (1995). Effect of noradrenergic lesions on subtypes of alpha2-adrenoceptors in rat brain. *Journal of Neurochemistry* **64**, 1118–1126.

Scheinin, M., Lomasney, J.W., Hayden-Hixson, D.M., Schambra, U.B., Caron, M.G., Lefkowitz, R.J., and Fremeau, Jr. R.T. (1994). Distribution of alpha2-adrenergic receptor subtype gene expression in rat brain. *Molecular Brain Research* **21**, 133–149.

Wagner, R.W. (1994). Gene inhibition using antisense oligodeoxynucleotides. *Nature* **372**, 333–335.

Winser-Serhan, U.H., Raymon, H.K., Broide, R.S., Chen, Y., and Leslie, F.M. (1997). Expression of alpha 2 adrenoceptors during rat brain development–1. Alpha 2A messenger RNA expression. *Neuroscience* **76**, 241–260.

34. Prenatal Stress Alters Endocrine and Immune Reactions in Neonatal Pigs

W. Otten, E. Kanitz and M. Tuchscherer

Introduction

Prenatal stress in humans or experimental animals yields transient and long-term effects on structure and function of the hypothalamic-pituitary-adrenocortical (HPA) and the sympatho-adrenomedullary (SAM) systems resulting in an altered coping during stressful situations (Weinstock, 1997). Because the release of stress hormones can affect the activation or suppression of immune reactions, an alteration of neuroendocrine stress systems in neonates by prenatal stress may also modify their immune function. To elucidate prenatal stress effects in domestic pigs, it was the aim of this study to investigate the effects of a repeated prenatal restraint stress in sows during the last third of gestation (1) on endocrine reactions of neonates to an acute stressor, (2) on the immune response of the neonates and (3) on changes of thymus, adrenal gland and glucocorticoid receptors in brain.

Methods

Animals and Prenatal Stress Treatment

Thirty-nine primiparous sows (Landrace × Duroc) and their litters were used in this experiment. During the last five weeks of gestation, twenty sows were restrained with a nose sling for five minutes per day.

Endocrine Reactions of the Neonates

After farrowing, three male and three female piglets with similar birth weights were selected from each litter. For the investigation of endocrine reactions to an acute stressor and of adrenocortical capacity, piglets were subjected to an immobilization and ACTH test during the suckling period at the age of 3, 7, 21 and 35 days. Each piglet was put on its back and restrained by the legs for 2 min. Immediately afterwards, a blood sample was taken by jugular venipuncture (epinephrine, norepinephrine, cortisol, CBG). Synthetic $ACTH_{1-24}$ (Synacthen, Ciba-Geigy, Basel, Switzerland) was injected intramuscularly at a dose of 0.5 IU per kg body weight (5 μg/kg body weight). Piglets were returned to their pens and 60 min later a second blood sample was taken (cortisol).

Hormone Analyses

Catecholamines were extracted from the plasma samples by absorption on aluminium oxide and the analysis of epinephrine and norepinephrine was performed using high pressure liquid chromatography with electrochemical detection (Otten et al., 1997). Plasma concentrations of cortisol were measured in ethanol-extracted samples using a single-antibody radioimmunoassay technique as described previously (Kanitz et al., 1999). Plasma CBG was measured by a modification of the method described by Lundström et al. (1983).

Immune Response of the Neonates

For the investigation of immune competence, blood samples from the neonates were taken at the age of 1, 7, 21 and 35 days. Concanavalin A (ConA), a T cell specific mitogen, was chosen for testing the capacity of cell mediated immunity, using a lymphocyte proliferation assay as described previously (Tuchscherer et al., 1998).

Tissue Collection and Brain Glucocorticoid Receptor Analysis

Two neonatal pigs from each litter were euthanized by barbiturate overdose at the age of 1 day and another two piglets at the age of 35 days. Thymus and adrenal glands were removed, weighed and the areas of the adrenal cortex and medulla were measured in median adrenal slices. Glucocorticoid receptor binding was analyzed in brain areas (hippocampus, amygdala, hypothalamus) using a cytosol binding technique (Kanitz et al., 1998).

Statistical Analysis

Data were analysed by a repeated measures analysis of variance using the GLM procedure of SAS®.

Results

Prenatal stress, as daily restraint of sows for 5 min with a nose sling during the last 5 weeks of gestation, had neither an effect on the duration of gestation, nor on the number of piglets born or on the body weight of the neonates. However, the number of piglets per sow which perished during the first 10 days of life was significantly higher in the prenatal stressed litters than in the control litters (1.3 ± 0.2 vs. 0.5 ± 0.3, P < 0.05).

Endocrine Reactions of the Neonates

Plasma epinephrine concentrations after the immobilization test decreased from day 3 to day 35 in both prenatally stressed and control piglets (Figure 1).

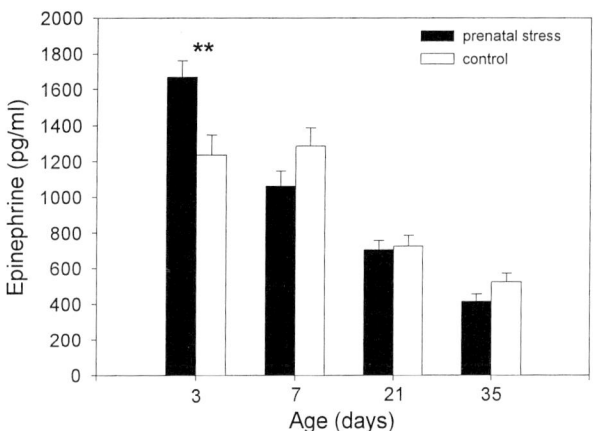

Fig. 1 Effects of prenatal stress on plasma epinephrine concentrations in neonatal pigs after an immobilization for 2 min.

Prenatally stressed piglets showed significantly higher plasma epinephrine concentrations after immobilization at the age of 3 days. Plasma norepinephrine concentrations also decreased with age, but at no age did norepinephrine concentrations differ significantly between piglets from stressed and control sows.

A significant decrease with age was also found for plasma concentrations of cortisol after immobilization and after ACTH stimulation (Figure 2). Prenatally stressed piglets showed significantly lower cortisol levels after ACTH stimulation at day 3 and after immobilization at day 3 and day 35. Plasma concentrations of CBG were also significantly influenced by prenatal stress. However, in contrast to cortisol levels, piglets from stressed sows showed increased plasma CBG concentrations at day 3 (Figure 3).

Immune Response of the Neonates

During the first week of life, prenatal stress caused a significant decrease in the lymphocyte response to the T cell mitogen ConA (Figure 4).

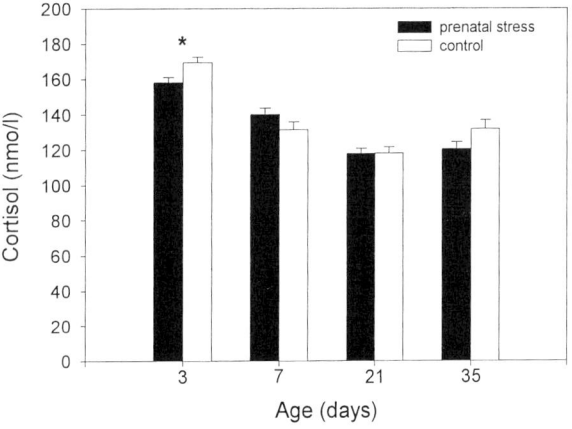

Fig. 2 Effects of prenatal stress on plasma cortisol concentrations in neonatal pigs 60 min after ACTH stimulation.

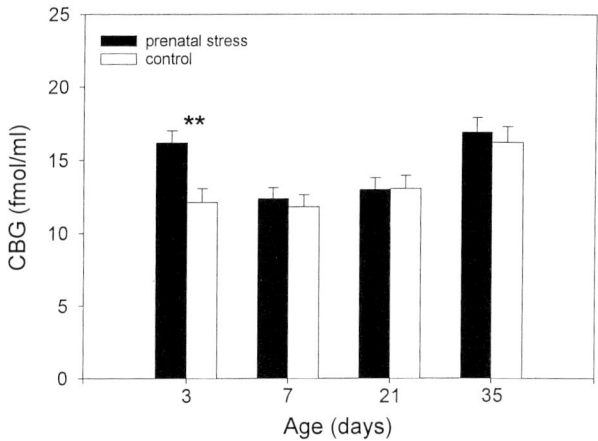

Fig. 3 Effects of prenatal stress on plasma CBG concentrations in neonatal pigs after an immobilization for 2 min.

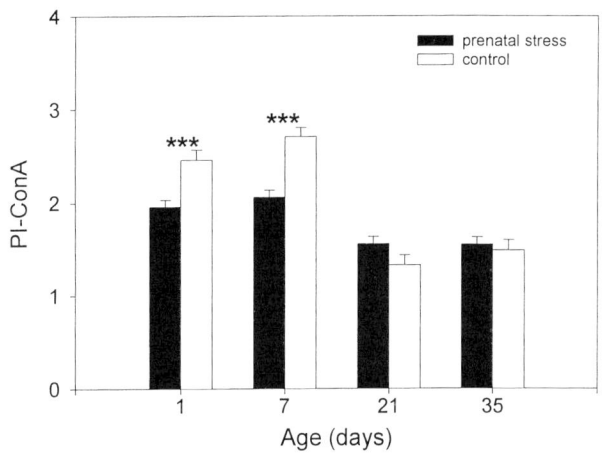

Fig. 4 Effects of prenatal stress on the lymphocyte proliferation index in response to ConA (T cell mitogen) in neonatal pigs.

Thymus, Adrenal Gland and Glucocorticoid Receptors in Brain

The thymus weight of prenatally stressed piglets sacrificed at days 1 and 35 was significantly reduced compared with the control piglets. As a result of the prenatal stress treatment the area of the adrenal cortex was significantly increased in neonates at day 1, and there was also a tendency for an increased area of the adrenal medulla at this age. In brain, prenatal stress caused a significant decrease in glucocorticoid receptor density in the hypothalamus on day 1.

Discussion

It has previously been shown that prenatal stress causes increased plasma corticosterone levels in neonatal as well as in adult rats under basal and stressful conditions (McCormick *et al.*, 1995). Our results with neonatal pigs, however, revealed lower plasma cortisol levels after immobilization and after ACTH stimulation in prenatally stressed animals than in control animals at day 3. In contrast to cortisol, plasma CBG concentrations at day 3 were increased by the prenatal stress treatment of the sows. The increased CBG concentrations indicate that cortisol levels in the stressed fetus may have been elevated before birth, possibly by higher circulating levels of maternal glucocorticoids. The reduced cortisol response to stress after birth indicates an impaired responsiveness of the HPA axis in prenatally stressed piglets, which is supported by the significant decrease in glucocorticoid receptors in the hypothalamus. Higher amounts of circulating cortisol, e.g. after repeated exposure to stressful stimuli or gestational stress, can decrease the number of glucocorticoid receptors in various brain regions (Henry *et al.*, 1994; Kanitz *et al.*, 1998). The transient decrease of glucocorticoid receptor binding in the hypothalamus in our study may be caused by the action of increased maternal cortisol on the developing fetal brain.

On day 3 plasma epinephrine concentrations after immobilization were significantly higher in prenatally stressed than in control piglets, whereas norepinephrine levels were not different between the two groups. It was previously shown that prenatal stress induces enhanced activation of the sympathetic nervous system in response to footshock stress in rats, probably mediated by an increased activation of tyrosine hydroxylase (Weinstock *et al.*, 1998). Our findings also indicate a greater reactivity of the adrenomedullary system, because only epinephrine but not norepinephrine levels were increased in prenatally stressed animals after immobilization.

Our results show that prenatal stress in pigs has an immunosuppressive effect on the neonates as shown by the decreased lymphocyte proliferation to the T cell mitogen ConA during the first days of life. In addition, at days 1 and 35 after birth the thymus weight of prenatally stressed piglets was significantly reduced. It is assumed that a lower lymphocyte proliferation *in vitro* indicates less effective cell function resulting in a higher susceptibility to pathogens. A suppressed immune function caused by prenatal stress may be the reason for the higher mortality found in this group.

In general, morphological, endocrine and immune effects of prenatal stress treatment were only observed in piglets during the first days after birth. We suppose that prenatal stress during late gestation in pigs affects the ontogeny of the fetal neuroendocrine system mediated by repeatedly increased maternal stress hormone concentrations. These effects however, disappear during the early postnatal ontogeny.

References

Henry, C., Kabbaj, M., Simon, H., Le Moal, M., and Maccari, S. (1994). Prenatal stress increases the hypothalamic-pituitary-adrenal axis response to stress in young and adult rats. *Journal of Neuroendocrinology* **6**, 341–345.

Kanitz, E., Manteuffel, G., and Otten, W. (1998). Effects of weaning and restraint stress on glucocorticoid receptor binding capacity in limbic areas of domestic pigs. *Brain Research* **804**, 311–315.

Kanitz, E., Otten, W., Nürnberg, G., and Brüssow, K.P. (1999). Effects of age and maternal reactivity on the stress response of the pituitary-adrenocortical axis and the sympathetic nervous system in neonatal pigs. *Animal Science* **68**, 519–526.

Lundström, K., Dahlberg, E., Nyberg, L., Snochowski, M., Standal, N., and Edquist, L.E. (1983). Glucocorticoid and androgen characteristics in two lines of pigs selected for rate of gain and thickness of backfat. *Journal of Animal Science* **56**, 401–409.

McCormick, C.M., Smythe, J.W., Sharma, S., and Meaney, M.J. (1995). Sex-specific effects of prenatal stress on hypothalamic-pituitary-adrenal responses to stress and brain glucocorticoid receptor density in adult rats. *Developmental Brain Research* **84**, 55–61.

Otten, W., Puppe, B., Stabenow, B., Kanitz, E., Schön, P.C., Brüssow, K.P., and Nürnberg, G. (1997). Agonistic interactions and physiological reactions of top and bottom ranking pigs confronted with a familiar and unfamiliar group: preliminary results. *Applied Animal Behaviour Science* **55**, 79–90.

Tuchscherer, M., Puppe, B., Tuchscherer, A., and Kanitz, E. (1998). Effects of social status after mixing on immune, metabolic, and endocrine responses in pigs. *Physiology and Behavior* **64**, 353–360.

Weinstock, M. (1997). Does prenatal stress impair coping and regulation of hypothalamic-pituitary-adrenal axis? *Neuroscience and Biobehavioral Reviews* **21**, 1–10.

Weinstock, M., Poltyrev, T., Schorer-Apelbaum, D., Men, D., and McCarty, R. (1998). Effect of prenatal stress on plasma corticosterone and catecholamines in response to footshock in rats. *Physiology and Behavior* **64**, 439–444.

35. Human Fetal and Maternal Stress Responses

R. Gitau, N. Fisk and V. Glover

Introduction

Recent studies have suggested that the human fetus can mount hormonal stress responses to invasive stimuli, with rises in β-endorphin, cortisol (Giannakoulopoulos et al., 1994), and norepinephrine (Giannakoulopoulos et al., 1999). The aim of this study was first to confirm that the human fetus has a cortisol and β-endorphin response to direct intrauterine needling and further characterize the response. Maternal responses were measured in parallel. We also looked at the relationship between baseline fetal and maternal levels of these two hormones.

Fetal responses to transfusion in which the fetal intrahepatic vein (IHV) was needled, were compared with those obtained by needling the placental cord insertion (PCI). The PCI is not innervated and transfusion via this site would not be expected to be stressful to the fetus. Needling via the IHV is considered safer but involves transgression of the fetal trunk and is potentially stressful. Maternal cortisol and β-endorphin responses were analyzed in order to characterize the maternal hypothalamo-pituitary-axis (HPA) responses to the stress of intrauterine needling, and basal concentrations were compared to fetal levels to determine whether any increase due to maternal stress may have a secondary effect on the fetus.

Methods

Women with singleton pregnancies undergoing clinically indicated fetal blood sampling or transfusion were eligible to participate in the study. All fetuses included in the analyses were structurally normal, appropriately grown for gestational age, with no evidence of hydrops and end-diastolic frequencies present in the umbilical artery. The site of sampling was chosen by the operator on the basis of technical factors and ease of approach. As anatomical factors often dictate one approach in favor of the other, randomization of site sampling was not appropriate. The indications for intrauterine transfusion were fetal anemia or thrombocytopenia in alloimmunized pregnancies. 27 fetuses satisfied the criteria for the transfusion group. Paired maternal and fetal samples were available from 47 patients, including both pre transfusions and from fetal blood samplings. Indications for fetal blood sampling were rapid karyotyping or for evaluation of fetal anaemia. Maternal blood samples were collected before, and within 20 minutes after the transfusion. 2 ml of fetal blood were withdrawn following the collection of clinical samples, immediately on accessing the fetal vein, and on completion of the transfusion.

Cortisol levels were assayed using a standard solid phase radioimmunoassay (by DPC, Los Angeles, USA). Plasma β-endorphin levels were determined using a solid phase two-site immunoradiometric assay (by Nichols Institute Diagnostics, CA), (cross reactivity 14% with β-lipotropin).

Results

Fetal Responses

Figure 1 shows that transfusion carried out via the IHV resulted in a significant rise in fetal plasma cortisol levels (rise 50.7 nmol/l; confidence interval [CI] [22 to 80]; p = 0.002), and β-endorphin levels (rise

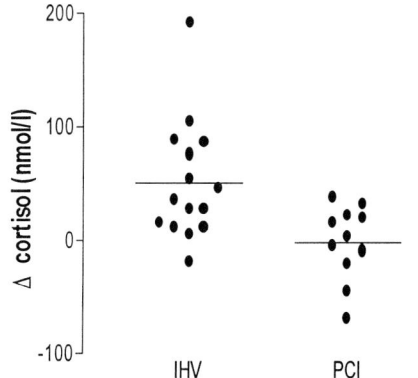
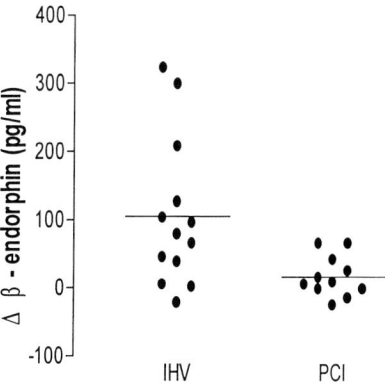

Fig. 1 Changes in cortisol and β-endorphin (delta = post-pre) with transfusion at either the placental cord insertion (PCI) or the intrahepatic vein (IHV).

105 pg/ml; [39 to 171]; p = 0.005). Such responses were absent when the same procedure was done via the PCI, cortisol (–1.9 nmol/l; [–24 to 20]; p = 0.9), β-endorphin (16.6 pg/ml; [–4 to 37]; p = 0.1). The magnitude of the response (post-pre value) was not related to the age of the fetus (cortisol r = 0.3 ns, β-endorphin r = –0.09 ns) over the age range studied of 20–35 weeks.

The time taken to administer the transfusion correlated with the rise of both hormones in the IHV group, (cortisol r = 0.7 [0.22 to 0.91], p = 0.01; β-endorphin r = 0.8 [0.36 to 0.94], p = 0.004).

Maternal Responses

Maternal cortisol and β-endorphin levels did not change with transfusion via either site, (cortisol; IHV, p = 0.6, PCI, p = 0.9. β-endorphin; IHV. p = 0.9, PCI, p = 0.5) (Table 1).

Relationship Between Basal Maternal and Fetal Levels

Basal fetal cortisol levels were linearly related to their paired maternal basal cortisol levels. (r = 0.53, p = 0.0001). There was no correlation with paired β-endorphin levels (r = –0.08 ns). (Table 2)

There was no significant correlation with gestational age with either fetal or maternal basal hormone levels over the age range studied of 17–35 weeks.

Discussion

These results confirm and extend the findings of Gianakoulopoulos et al. (1994) that invasive procedures cause the fetus to mount a significant HPA response. This study included younger fetuses than previously studied. It is of interest that the two 20-week fetuses in the cohort showed typically large responses, and that over the age range 20–35 weeks the magnitude of cortisol and β-endorphin responses were independent of gestational age. As the hormone levels did not change significantly with transfusion at the PCI, the increase seen in the IHV group appears not to be due to any effects of the transfused blood or the transfusion procedure.

There was no evidence of a relationship between fetal basal hormone levels and gestational age. This is similar to the findings of Economides et al. (1988) for cortisol. Donaldson et al., did find an increase with gestation (Donaldson et al., 1991). However, the rise in cortisol

Table 1 Maternal responses to transfusion via the intrahepatic vein (IHV) or the placental cord insertion (PCI).

	Maternal cortisol (nmol/l)		Maternal β-endorphin (pg/ml)		Maternal norepinephrine (nmol/l)*	
	IHV	PCI	IHV	PCI	IHV	PCI
n	7	6	6	6	6	9
Pre (mean)	493	696	31.8	36	3.86	3.72
Post (mean)	521	677	31.7	36.5	8.67	11.5
Mean Δ	28.1	–18.3	–0.17	–5.5	6.47	9.49
Paired t-test (p)	ns	ns	ns	ns	0.01	< 0.005

*From (Giannakoulopoulos et al., 1999).

Table 2 Paired basal maternal and fetal hormone levels.

	Fetal mean	Fetal range	Maternal mean	Maternal range	Correlation
Cortisol (nmol/l)	50	13–184	590	88–1344	0.53 p < 0.001
β-endorphin (pg/ml)	90	18–409	43	14–119	–0.08 ns
Norepinephrine (nmol/l)*	0.4	0.1–0.8	4.2	1.9–8	0.08 ns

*From (Giannakoulopoulos et al., 1999).

they found was predominantly at the very end of gestation, a period not included in this study.

Maternal levels of either hormone did not change with transfusion to the fetus, indicating no HPA activation. This contrasts with the response seen for norepinephrine which increased substantially (Giannakoulopoulos *et al.*, 1999). However others have observed a partial desensitization of the maternal HPA response during pregnancy, probably occurring as a result of the release of corticotrophin releasing hormone (CRH) from the placenta (Schulte *et al.*, 1990). Our results may well be a reflection of this desensitization. The lack of any significant maternal response confirms that the fetal increases are fetal in origin and not a result of increased placental passage of maternal hormones.

As was the case with norepinephrine (Giannakoulopoulos *et al.*, 1999), maternal β-endorphin levels were not directly linked with fetal levels. The significant correlation between maternal and fetal cortisol levels is suggestive of there being some placental passage from the maternal circulation into the fetus of this hormone. It has been suggested that high levels of placental 11β-hydroxysteroid-dehydrogenase type 2 activity excludes maternal cortisol from the fetus (Pearson Murphy *et al.*, 1974; Lopez-Bernal *et al.*, 1980; Seckl, 1993). However, a study by Pearson Murphy *et al.* (1974) showed 15% of ^{3}H-cortisol crossed the placenta unmetabolized in fetal-placental units before abortion. Our finding is compatible with substantial metabolism of maternal cortisol during passage through the placenta, as fetal concentrations are about 11-fold lower, a contribution of as little as 10–20% of maternal levels could still double fetal levels.

Maternal stress in pregnancy is associated with low birth weight and impaired brain development (Lou *et al.*, 1994), and in animal studies, with altered long-term HPA responses in the resultant offspring (Clarke *et al.*, 1993). Fetal stress can thus have long-term implications (Sapolsky, 1996). The developing nervous system appears to be at a very plastic stage and is vulnerable to insult. The passage of cortisol to the fetus may help to explain how maternal stress in pregnancy has a negative effect on fetal development.

References

Clarke, A.S., and Schneider, M.L. (1993). Prenatal stress has long-term effects on behavioural responses to stress in juvenile rhesus monkeys. *Developmental Psychobiology* **26**, 293–304.

Donaldson, A., Nicolini, U., Symes, E.K., Rodeck, C.H., and Tannirandorn, Y. (1991). Changes in concentrations of cortisol, dehydroepiandrosterone sulphate and progesterone in fetal and maternal serum during pregnancy. *Clinical Endocrinology* **35**, 447–51.

Economides, D.L., Nicolaides, K.H., Linton, E.A., Perry, L.A., and Chard, T. (1988). Plasma cortisol and adrenocorticotropin in appropriate and small for gestational age fetuses. *Fetal Therapy* **3**, 158–64.

Giannakoulopoulos, X., Sepulveda, W., Kourtis, P., Glover, V., and Fisk, N.M. (1994). Fetal plasma cortisol and beta-endorphin response to intrauterine needling. *Lancet* **344**, 77–81.

Giannakoulopoulos, X., Teixeira, J., Glover, V., and Fisk, N. (1999). Human fetal and maternal noradrenaline responses with intrauterine needling. *Pediatric Research* **45**, 494–499.

Lou, H., Hansen, D., Nordentoft, M., Pyrds, O., Jensenn, F., Nim, J., and Hemminsen, R. (1994). Prenatal stressors of human life affect fetal brain development. *Developmental Medicine and Child Neurology* **36**, 826–832.

Pearson Murphy, B.E., Clark, S.J., Donald, I.R., Pinsky, M., and Vedady, D. (1974). Conversion of maternal cortisol to cortisone during placental transfer to the human fetus. *American Journal of Obstetrics and Gynecology* **118**, 538–41.

Sapolsky, R.M. (1996). Why stress is bad for your brain. *Science* **273**, 749–50.

Schulte, H.M., Weisner D., and Allolio B. (1990). The corticotrophin releasing hormone test in late pregnancy: lack of adrenocorticotrophin and cortisol response. *Clinical Endocrinology* **33**, 99–106.

Seckl, J. (1993). 11β-Hydroxysteroid dehydrogenase isoforms and their implications for blood pressure regulation. *European Journal of Clinical Investigation* **23**, 589–601.

36. Overview of Corticotropin-Releasing Factor Receptors and Binding Protein

E.B. De Souza and D.E. Grigoriadis

Introduction

Corticotropin-releasing factor (CRF), a 41-residue peptide, plays a crucial role in integrating the body's overall response to stress (for reviews see De Souza and Grigoriadis, 1994; De Souza and Nemeroff, 1990; Dunn and Berridge, 1990; Owens and Nemeroff, 1991). The CRF-adrenocorticotropin (ACTH)-glucocorticoid axis is central to the endocrine response to stress. In addition to its endocrine effects, immunohistochemical localization of CRF has demonstrated that the peptide has a broad extrahypothalamic distribution in the central nervous system (CNS). CRF produces a wide spectrum of autonomic, electrophysiological and behavioral effects consistent with a neurotransmitter or neuromodulator role in the brain. Intracerebroventricular administration of CRF provokes stress-like responses including activation of the sympathetic nervous system, and inhibition of the parasympathetic nervous system with consequential increases in plasma concentrations of epinephrine, norepinephrine and glucose; increases in heart rate and mean arterial blood pressure; and inhibition of gastrointestinal functions, including inhibition of gastric acid secretion. The behavioral profile following central administration of CRF is also characteristic of a compound that increases arousal and emotional reactivity to the environment. These effects of CRF include general arousal, as exhibited by increased locomotion, sniffing, grooming, and rearing in familiar surroundings and increased agitation in unfamiliar surroundings. In contrast, sexual receptivity and feeding are decreased.

The recent cloning of multiple CRF receptor subtypes and binding protein (see Chalmers et al., 1996; De Souza et al., 1995) along with the recent identification of urocortin, a novel peptide with sequence identity to urotensin I (63%) and CRF (45%) (Donaldson et al., 1996; Vaughan et al., 1995), has precipitated a new era in CRF research. Here, we describe some of our recent studies on a variety of CRF related targets including three functionally distinct receptor subtypes and the CRF-binding protein (CRF-BP). The characteristics of CRF receptors including their sequence homologies, pharmacological profiles and second messenger activities will be described. In addition, the differential distribution of mRNA for CRF receptors and CRF-BP will be reviewed.

Molecular Biology/Receptor Structure of CRF Receptors and Binding Protein

The CRF receptors belong to the growing family of "gut-brain" neuropeptide receptors, which includes receptors for calcitonin, vasoactive intestinal peptide (VIP), parathyroid hormone, secretin, pituitary adenylate cyclase-activating peptide, glucagon, and growth hormone-releasing hormone (GHRH) (Segre and Goldring, 1993). These receptors all have seven putative transmembrane domains, share considerable sequence homology, and all stimulate adenylate cyclase in response to agonist activation.

CRF$_1$ Receptors

The CRF$_1$ receptor was cloned from several species including human (Chen et al., 1993; Vita et al., 1993), mouse (Vita et al., 1993) and rat (Chang et al., 1993; Perrin et al., 1993). All three species of CRF$_1$ receptor mRNAs encode proteins of 415 amino acids that are 98% identical to one another. In general, the CRF$_1$ receptor is approximately 30% identical to all other members of the gut-brain neuropeptide receptor family. Characteristic of most G-protein coupled receptors, the CRF$_1$ receptor has putative N-linked glycosylation sites on the N-terminal extracellular domain (Figure 1). There are five predicted sites on CRF$_1$, substantiating the glycosylation profiles determined in chemical affinity cross-linking studies (Grigoriadis and De Souza, 1989). In addition, there are potential protein kinase C phosphorylation sites in the first and second intracellular loops and in the C-terminal tail, as well as casein kinase II and protein kinase A phosphorylation sites in the third intracellular loop (Chen et al., 1993). These sites, although not fully characterized, may serve as regulatory elements in the control of receptor expression and/or function.

The human CRF$_1$ receptor gene contains at least two introns and is found in at least one splice variant form that produces a 29 amino acid insert in the first cytoplasmic loop (Chen et al., 1993). This is in contrast to the mouse where the gene has been identified to contain at least 12 introns (Chen et al., 1994). The physiological significance or relevance of this splice variant is currently undefined (Chen et al., 1993). The human gene has been localized on chromosome 17 at position 17q12–22 (Polymeropoulos et al., 1995).

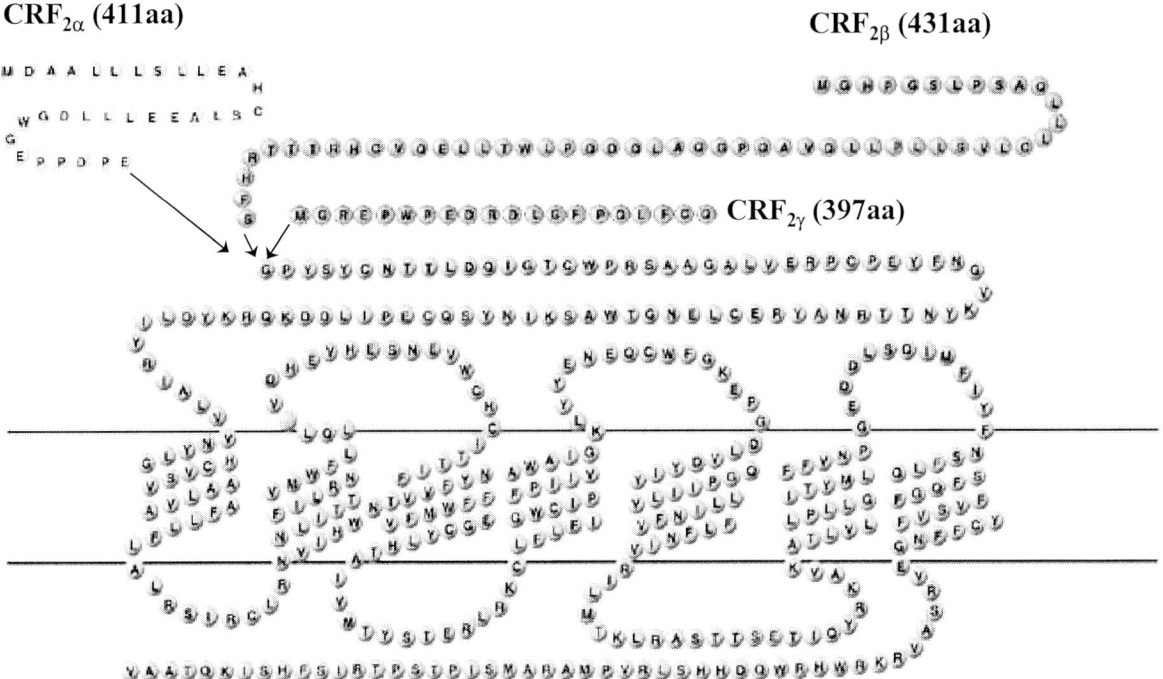

Fig. 1 Amino acid sequences of CRF₁, CRF₂(a) CRF₂(b) and CRF₂(c) receptors: The CRF₁, CRF₂(a), CRF₂(b) and CRF₂(c) receptor forms are shown with the predicted seven transmembrane spanning domains. The arrows indicate where CRF₂(a), CRF₂(b) and CRF₂(c) diverge at the N-terminus. Amino acids shaded in gray are those that are different between CRF₁ and CRF₂ receptors. Potential N-glycosylation sites are marked by and potential sites for phosphorylation of protein kinase C are denoted by H. (*From Chalmers et al., 1996*).

CRF₂ Receptors

There are currently three known forms of the CRF₂ receptor, CRF₂(a), CRF₂(b) and CRF₂(c). The CRF₂(a) receptor, which was originally described by Lovenberg *et al.* (Lovenberg *et al.*, 1995b), is a 411 amino acid protein with approximately 71% identity to the CRF₁ receptor (Figure 1). The CRF₂(b) receptor, which has been cloned from rat (Lovenberg *et al.*, 1995b), mouse (Kishimoto *et al.*, 1995; Perrin *et al.*, 1995) and human (Kostich *et al.*, 1996), is 431 amino acids and differs from CRF₂(a) in that the first 34 amino acids in the N-terminal extracellular domain are replaced by a unique sequence, 54 amino acids in length. The CRF₂(c) receptor has most recently been identified in human brain (Sperle *et al.*, 1997). This splice variant uses yet a different 5′ alternative exon for its amino terminus and replaces the first 34 amino acid sequence of the CRF₂(a) receptor with a unique 20 amino acid sequence. Figure 1 illustrates the exact difference between CRF₂(a), CRF₂(b) and CRF₂(c) in the N-terminal extracellular domain. Amino acids which are shaded highlight the difference between CRF₂ and CRF₁ receptors. It is interesting to note that there are very large regions of amino acid identity between CRF₁ and CRF₂ receptors, particularly between transmembrane domain five and transmembrane domain six. This similarity argues strongly for conservation of biochemical function since it is this region which is thought to be the primary site of G-protein coupling and signal transduction. All three CRF₂ receptor subtypes contain five potential N-

glycosylation sites, which are analogous to those found in CRF₁. The full characterization of the CRF₂(c) subtype has not yet been elucidated in terms of native characteristics and function.

The genomic structure of the human CRF₂ receptor gene is similar to that of the mouse CRF₁ receptor described above and has 12 introns, the last ten of which interrupt the coding region in identical positions. These gene sequences however diverge significantly at the 5′ end. The chromosomal mapping of the human CRF₂ gene has been localized to chromosome 7 p21-p15 (Liaw *et al.*, 1996).

CRF Receptor Splice Variants

The first group to report the cloning of the CRF receptor from human pituitary (Chen *et al.*, 1993) identified a clone which contained an additional 87 nucleotides in-frame with the coding sequence, which resulted in the addition of 29 amino acids in the first intracellular loop. This intriguing finding was followed up by Xiong *et al.* (Xiong *et al.*, 1995) who showed that this alternatively-spliced form of the receptor was able to bind CRF, albeit with a 2-fold lower affinity, whereas the ability of CRF to stimulate adenylate cyclase was reduced 100-fold. Since no species homologues have been identified and no tissue expression analyses of this receptor splice variant have been performed, it is too early to hypothesize whether the reduced efficacy of CRF has a functional significance. In rat brain, Chang *et al.*, (Chang *et al.*, 1993) found that an alternative splicing event with the CRF_1 mRNA generated a second, relatively abundant, gene product. This alternate gene product encoded only a truncated 224 amino acid form of the receptor that was presumed to be non-functional in that it lacked the third and half of the fourth transmembrane domains. This missing region turns out to be the entire sixth exon of the CRF_1 receptor gene (unpublished observations). Through RNase protection assays, they showed that this truncated form of the receptor was expressed in parallel to the full length mRNA, albeit in approximately a 1:2 ratio. There is as of yet no predicted function for this truncated transcript and it may simply represent a sequence-dependent splicing mistake by the CNS processing machinery. As described above, three splice forms of the CRF_2 receptor ($CRF_{2(a)}$, $CRF_{2(b)}$ and $CRF_{2(c)}$) have been identified. Thus far, only the $CRF_{2(a)}$ and $CRF_{2(b)}$ isoforms have been detected in rat (Lovenberg et al., 1995b) while all three splice variants have now been identified in human (Kostich et al., 1996; Liaw et al., 1996; Sperle et al., 1997). We have subsequently demonstrated by RNase protection assays and in situ hybridization that $CRF_{2(a)}$ and $CRF_{2(b)}$ are differentially expressed in the CNS and periphery and possess unique pharmacological profiles (Lovenberg et al., 1995a). RT-PCR analysis of human brain mRNA demonstrated expression of the $CRF_{2(c)}$ receptor subtype in amygdala and hippocampus while Southern analysis of rat genomic DNA yielded negative results, suggesting that this subtype does not exist in rat. In addition to the $CRF_{2(a)}$,$CRF_{2(b)}$ and $CRF_{2(c)}$ forms, we have identified many additional splice forms of the rat and human CRF_2 receptor mRNAs with many different variations of intron-containing mRNAs as well as mRNAs with deletions (unpublished observations).

CRF-Binding Protein

Plasma CRF, which under normal circumstances is undetectable, has been shown to be dramatically elevated during the third trimester of human pregnancy (Linton et al., 1988), possibly relating to events leading to parturition. However, despite the increased levels of plasma CRF, there is no marked elevation in either ACTH secretion or circulating glucocorticoids. The lack of CRF-induced ACTH secretion in this situation was postulated to be a consequence of CRF inactivation in plasma by a circulating binding protein. This hypothesis was validated by the isolation of a CRF-binding protein (CRF-BP) from human plasma and its subsequent cloning and expression (Behan et al., 1993; Potter et al., 1991). Screening of a human liver cDNA library using oligonucleotide probes generated from the original amino acid sequence of the CRF-BP revealed a full length cDNA encoding a predicted protein of 322 amino acids (Potter et al., 1991). Subsequent screening of a rat brain cDNA library resulted in the isolation of a rat CRF-BP cDNA encoding a 322 amino acid protein which was 85% identical to the human CRF-BP (Potter et al., 1991). Both proteins possess one putative N-linked glycosylation site and 11

conserved cysteine residues. The mouse (Cortright et al., 1995) and sheep (Behan et al., 1996) CRF-BPs have also been cloned. The sheep binding protein has 85% and 87% amino acid homology to the rat and human CRF-BP, respectively. Similarly, the mouse binding protein is 94% and 86% homologous to the rat and human CRF-BP, respectively. The CRF-BP appears to be at least partially membrane associated; however, the mechanism for this association remains undefined as no obvious transmembrane domains are evident in the CRF-BP amino acid sequence (Potter et al., 1991).

The human CRF-BP gene has been cloned and was found to be 18kb, consisting of seven exons and six introns (Behan et al., 1993), which is in agreement with the exon/intron organization of the mouse gene (Cortright et al., 1995). The gene was mapped to the distal region of chromosome 13 and loci 5q in the mouse and human genomes, respectively.

Pharmacological Characteristics of CRF Receptors and Binding Protein

We have stably transfected both the human and rat cloned CRF_1 receptor genes into COS-7 (monkey kidney) and Ltk^- (mouse fibroblast) cells (neither of which normally express CRF receptors) and characterized their pharmacological profiles (Grigoriadis et al., 1996b). The receptors were subcloned in the mammalian expression plasmid pCDM-7amp containing the human cytomegalovirus (CMV) promoter. Cells were cotransfected with the CRF_1 or $CRF_{2(a)}$ receptor plasmids and pSV-2neo as a selectable marker and stable clones were identified by either radioligand binding of $[^{125}I]Tyr^\circ$-ovine CRF to CRF_1 receptors (De Souza, 1987) or $[^{125}I]Tyr^\circ$-sauvagine binding to CRF_1 or CRF_2 receptors (Grigoriadis et al., 1996a).

Ligand Binding Profile of CRF_1 Receptors

CRF_1 receptors expressed in mammalian cell lines described above demonstrated reversible, saturable, high-affinity binding to CRF and its related peptides with the pharmacological and functional characteristics comparable to those found in the variety of animal or human tissues described above. Human and rat CRF_1 receptors in stably transfected Ltk^- cells demonstrated binding to a single homogeneous population of receptors with apparent affinities (K_D) of 130 and 168 pM and receptor densities (B_{max}) of 97 and 588 fmol/mg protein, respectively (Grigoriadis et al., 1994). The pharmacological rank orders of potency in the stably transfected cell lines with either the human or rat CRF_1 receptors were identical to the established profile for the CRF receptor in the rat frontal cortex where urocortin = urotensin I = sauvagine > ovine CRF (oCRF) = rat/human CRF (r/h CRF) = bovine CRF > D-Phe r/hCRF(12–41) > α-helical ovine CRF(9–41) >> r/hCRF(6–33), r/hCRF(9–33), r/hCRF(1–41)OH, VIP, arginine vasopressin (AVP).

Ligand Binding Profile of CRF$_2$ Receptors

The CRF$_2$ receptors have been expressed in stable mammalian cell lines and have been shown to bind CRF-related peptides and function through stimulation of cAMP production. In general, the CRF$_2$ splice variants have a similar pharmacological profile to each other but one that is quite distinct from either the CRF$_1$ receptor subtype or the CRF binding protein. When expressed in stable cell lines, these receptors exhibit the following pharmacological rank order profile: urocortin = sauvagine = urotensin I > astressin > r/hCRF > D-Phe CRF(12–41) = α-helical CRF(9–41) >> oCRF >> r/hCRF(6–33), oCRF(6–33), r/hCRF(1–41)OH, GHRH, AVP and VIP. The limited data thus far on the CRF$_{2(c)}$ receptor suggests a rank order of potencies similar to the other two splice variants with sauvagine > urotensin I > r/hCRF (Sperle et al., 1997).

Ligand Binding Profile of CRF-Binding Protein

The CRF-BP (purified following expression in COS cells) exhibits a pharmacological profile that is different from that observed with either the CRF$_1$ or CRF$_2$ receptor subtypes. In both human and rat, CRF-BP exhibits high affinity for r/hCRF ($K_D \sim 200$ pM), urocortin ($K_i \sim 2$ nM) and very low affinity for oCRF ($K_i \sim 250$–400 nM). Furthermore, r/hCRF fragments that are inactive at CRF receptors such as CRF(1–41)OH, CRF(6–33) and CRF(9–33) have high affinity for the CRF-BP ($K_i \sim 0.2$–5 nM). The rank order ligand binding profile of purified human CRF-BP was as follows: urotensin I > r/hCRF(1–41)OH = r/hCRF = α-helical CRF(9–41) > r/hCRF(6–33) > urocortin > sauvagine > astressin >> D-Phe CRF(12–41) > oCRF. Thus, although there may be some similarities in the binding domains of the CRF-BP and the CRF receptors discussed above, these are distinct proteins each with unique and discernible pharmacological characteristics. The ability of the CRF-BP to bind and functionally inactivate CRF and/or urocortin is somewhat analogous to the role of high-affinity monoamine transporter molecules and represents an additional CRF/urocortin "modulatory" molecule.

Guanine Nucleotide Characteristics of CRF Receptors

Receptor systems that are coupled to an adenylate cyclase second messenger system are almost exclusively linked to guanine nucleotide binding proteins. Guanine nucleotides have been shown to decrease the specific binding of agonists for a number of seven transmembrane neurohormone or neurotransmitter receptors coupled to G-proteins. CRF$_1$ receptors transfected in stable mammalian cell lines were used to determine the guanine nucleotide sensitivity of the cloned receptors. The effects of guanine nucleotides on the binding of [^{125}I]oCRF were examined in order to determine whether the expressed human and rat CRF receptors are G-protein-coupled. All guanine nucleotides examined inhibited binding by 50–70% (where 100% inhibition was defined by 1 μM unlabeled r/hCRF) with the non-hydrolyzable forms of guanine nucleo-

tides (Gpp(NH)p and GTP-γ-S) being more potent than GTP itself. The order of potencies was GTP-γ-S > Gpp(NH)p > GTP with ED$_{50}$ values of ~ 20–45, 200–800 and 2,500–3,000 nM, respectively. These effects on the binding of [^{125}I]oCRF in both human and rat CRF$_1$-receptor stable cell lines were specific for the guanine nucleotides since the adenosine nucleotide ATP was ineffective in altering the binding at equimolar concentrations (ED$_{50}$: > 10,000 nM).

For the CRF$_2$ receptor splice variants, studies using the stable cell line expressed receptors demonstrated an equally efficient coupling to a putative Gs G-protein. In radioligand binding assays using [^{125}I]sauvagine, the binding was inhibited dose-dependently with the guanine nucleotides with a rank order of potencies being GTP-γ-S > Gpp(NH)p > GTP (Grigoriadis et al., 1996b). This profile was consistent for both the CRF$_{2(a)}$ and CRF$_{2(b)}$ receptors. No data are yet available for the CRF$_{2(c)}$ receptor on the effects of guanine nucleotides per se; however, it has been demonstrated that these receptors function through stimulation of cAMP, suggesting interaction with a Gs protein (Sperle et al., 1997).

Second Messenger Characteristics of CRF Receptors

Previous studies have demonstrated that the second messenger involved in transducing the signal of CRF receptors involves stimulation of cAMP production in the pituitary (Aguilera et al., 1988) and in brain (Battaglia et al., 1987). The stimulation of cAMP production and POMC-secretion in anterior pituitary was dose-related and exhibited the appropriate pharmacology (Aguilera et al., 1988; Bilezikjian and Vale, 1983; Giguere et al., 1982). To characterize further the stable CRF receptor transfectants, the ability of CRF-related and unrelated peptides to stimulate CRF$_1$ receptor-mediated cAMP production from the cells was examined. The peptides oCRF, r/hCRF, bCRF and urocortin all had apparent EC$_{50}$ values of approximately 3–5 nM as did the related peptides sauvagine and urotensin I. Peptide fragments and the deamidated form of bovine or human CRF were all without activity as were the unrelated peptides VIP, AVP and GHRH. The rank order of the peptides in stimulating cAMP production in these cells was in keeping with their rank order in inhibiting [^{125}I]oCRF or [^{125}I]sauvagine binding to the CRF$_1$ receptor, and this rank order was virtually identical to the second messenger pharmacological profile previously reported in brain (Battaglia et al., 1987).

When expressed in mouse Ltk$^-$ cells, the CRF$_{2(a)}$ receptor also stimulated cAMP production in a dose-dependent manner in response to CRF, urocortin and the known non-mammalian CRF-related peptides sauvagine and urotensin I. The rank order of potency however was different from that of the CRF$_1$ receptor as urocortin, sauvagine and urotensin I demonstrated higher affinity for this receptor subtype (EC$_{50}$: 0.5–2 nM). The rank order profile of the CRF-like peptides for CRF$_{2(a)}$ receptors was: sauvagine = urocortin > urotensin I > r/hCRF > oCRF.

In addition to the stimulation studies, CRF-stimulated cAMP production could be competitively inhibited by the putative peptide antagonists D-Phe CRF(12–41) and α-helical CRF(9–41) in both CRF$_1$ and CRF$_{2(a)}$ expressing cell lines, demonstrating that the expressed CRF receptors in these cell lines were functional. Interestingly, both D-Phe CRF(12–41) and α-helical CRF(9–41) inhibited the stimulated cAMP response with the same EC$_{50}$ in both receptor subtype preparations. This suggests that although there is some fundamental difference in the pharmacological specificity of these two receptor subtypes when examining agonists, they must still possess some similarities in their structure of the binding site at least as far as antagonists are concerned. These results clearly suggest that not only subtype-specific and subtype-selective compounds can be identified for these two receptor subtypes but that mixed antagonists are possible and could bring to light possible subtle functions of the two receptor subtypes.

Localization of Ligand-Binding Domains of CRF Receptors: Chimera and Mutational Studies

In an attempt to understand the structural requirements for ligand binding to the CRF$_1$ and CRF$_2$ receptor subtypes, chimeric and point mutation studies were performed initially looking at only the differences between the two receptor subtypes. Both ligand binding studies and stimulation of cAMP-dependent β-galactosidase activity were used to determine altera-

tions in ligand/receptor interactions (Liaw *et al.*, 1997a; Liaw *et al.*, 1997b). Mutant receptors (substituting amino acid residues in the CRF$_1$ receptor sequence for the corresponding CRF$_2$ amino acids) demonstrated that three distinct regions were critical for the binding of r/hCRF and/or receptor activation (Figure 2). For example, one region identified, encompassed the junction of the third extracellular domain and the fifth transmembrane domain. Substitution of Val^{266}, Tyr^{267} and Thr^{268} for the corresponding CRF$_2$ sequence (Asp^{266}, Leu^{267} and Val^{268}) decreased the affinity and activity of r/hCRF approximately 10-fold (Liaw *et al.*, 1997b). Interestingly, differences were observed in the binding sensitivity of the peptides sauvagine and urocortin to these same amino acid substitutions. Thus, while a 10-fold difference in affinity was observed for r/hCRF, the affinity for sauvagine was decreased by almost 40-fold while the affinity of urocortin remained relatively unchanged (Liaw *et al.*, 1997a). The affinities were altered from 0.2 nM to 1.5 nM; 0.15 nM to 7.1 nM and 0.2 nM to 0.3 nM for r/hCRF, sauvagine and urocortin, respectively (Liaw *et al.*, 1997a). These data suggested that although all three of these CRF-related peptides have the ability to bind and function through the CRF$_1$ and CRF$_2$ receptor subtypes, their structural requirements for binding and/or activation are not entirely overlapping. This possibly provides some insight as to the differences in binding affinity and activity of these CRF-related peptides at these two receptor subtypes.

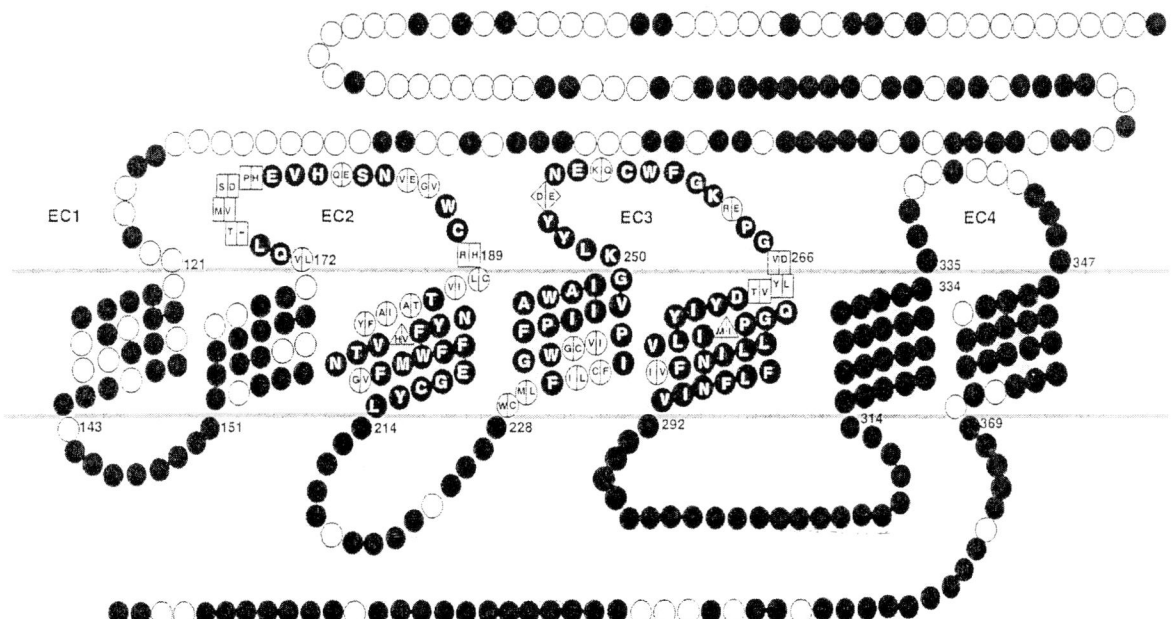

Fig. 2 Schematic model of the human CRF receptor. Sequence alignment of CRF$_1$ and CRF$_2$ receptors are shown between extracellular domain 2 (EC2) and transmembrane domain 5 (TM5) with the amino acid of CRF$_1$ receptor shown on the *left* and those of CRF$_2$ receptor shown on the *right*. The conserved amino acids are shown in *solid circles* and divergent amino acids are shown in *open symbols*. The regions that are important for binding all three peptide ligands r/hCRF, urocortin and sauvagine, are shown in *squares*; the region that is important for binding sauvagine only is shown as a *diamond*; the two amino acids that are important for binding of a small molecule selective CRF$_1$ antagonist (NBI 27914) are shown in *triangles*. The numbers indicate the amino acid positions of those residues flanking the postulated TMs. All numberings are based on the sequence of the CRF$_1$ receptor. (*From Liaw et al., 1997a*).

Distribution of CRF Receptors and Binding Protein

CRF Receptors

In situ hybridization histochemistry studies indicate heterogeneous anatomical distribution patterns for CRF_1 and CRF_2 receptor subtypes in brain (Chalmers *et al.*, 1995; Chalmers *et al.*, 1996; Potter *et al.*, 1994). While CRF_1 receptor mRNA expression is most abundant in neocortical, cerebellar and sensory relay structures, CRF_2 receptor expression is generally localized to specific sub-cortical structures, most notably lateral septal nuclei and various hypothalamic areas (Figure 3). This heterogeneous distribution of CRF_1 and CRF_2 receptor mRNA suggests distinctive functional roles for each receptor in CRF-related systems. Thus, the selective expression of CRF_2 receptor mRNA within hypothalamic nuclei such as the ventromedial nucleus (VMH) and the paraventricular nucleus (PVH) indicates that the anorexic actions of CRF in these nuclei are likely to be CRF_2 receptor-mediated. On the other hand, within the pituitary, CRF_1 receptor expression predominates over CRF_2 expression in both the intermediate and anterior (AP) lobes, indicating that CRF_1 receptors are primarily responsible for CRF-induced changes in ACTH release.

Examination of CRF_2 receptor splice variants, $CRF_{2(a)}$ and $CRF_{2(b)}$, indicates distinctive anatomical distributions: The $CRF_{2(a)}$ form being primarily expressed within the brain and the $CRF_{2(b)}$ variant being found in both the CNS and periphery (Lovenberg *et al.*, 1995a). Within brain, it appears that the CRF_2 form represents the predominant neuronal CRF_2 variant while the $CRF_{2(b)}$ splice form is localized on non-neuronal elements, the choroid plexus and cerebral blood vessels. In all of these brain areas, $CRF_{2(a)}$ mRNA is present only in low densities. In peripheral organs, $CRF_{2(b)}$ mRNA is expressed at high levels in both cardiac and skeletal muscle with lower levels evident in both lung and intestine.

CRF-Binding Protein

Although the human and rat forms of the CRF-BP are homologous, there is a somewhat different anatomical distribution pattern in the two species. The human form of the binding protein has been found abundantly in tissues such as liver, placenta and brain while in the rat, levels of mRNA for the binding protein have only been localized in brain and pituitary (Potter *et al.*, 1991). Peripheral expression of the binding protein may have its greatest utility in the modulation and control of the elevated circulating levels of CRF induced by various normal physiological conditions (see above). In addition, expression of this binding protein in the brain and pituitary offers additional mechanisms by which CRF-related neuronal or neuroendocrine actions may be modulated.

CRF-BP has been localized to a variety of brain regions including neocortex, hippocampus (primarily in the dentate gyrus) and olfactory bulb. In the basal forebrain, mRNA is localized to the amygdaloid complex with a distinct lack of immunostained cells in the medial nucleus. CRF-BP immunoreactivity is also present in the brain stem particularly in the auditory, vestibular and trigeminal systems, raphe nuclei of the midbrain and pons and the reticular formation (Potter *et al.*, 1992). In addition, high expression levels of binding protein mRNA are seen in the anterior pituitary, predominantly restricted to the corticotrope cells. Expression of this protein in the corticotropes strongly suggests that the CRF-BP is involved in the regulation of neuroendocrine functions of CRF by limiting and/or affecting the interactions of CRF with its receptor, which is also known to reside on corticotropes. The detailed role of the binding protein in regulating pituitary-adrenal function remains to be elucidated.

CRF_1 Receptor mRNA

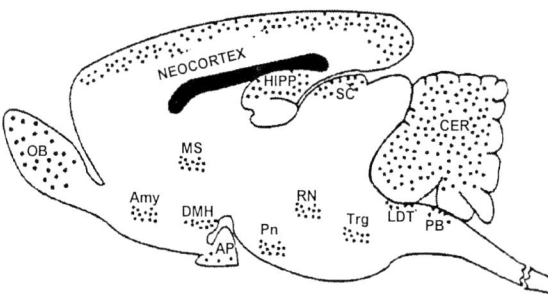

CRF_2 Receptor mRNA

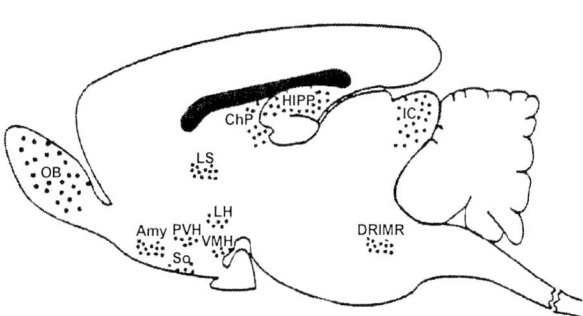

Fig. 3 Schematic illustration of the distribution of CRF_1 and CRF_2 receptor mRNAs in rat brain saggital section. Abbreviations: Amy, amygdala; AP, anterior pituitary; CER, cerebellum; ChP, choroid plexus; DMH, dorsomedial hypothalamic nucleus; DR/MR, dorsal, medial raphe; HIPP, hippocampus; IC, inferior colliculus; LDT, lateral dorsal tegmental nuclei; LH, lateral hypothalamus; LS, lateral septum; MS, medial septum; OB, olfactory bulb; PB, parabrachial nuclei; Pn, pontine nuclei; PVH, paraventricular nuclei; RN, red nuclei; SC, superior colliculus; So, supraoptic nuclei; Trg, trigeminal nuclei; VMH, ventromedial hypothalamic nucleus; (*From De Souza et al.*, 1995).

Summary and Conclusions

Corticotropin-releasing factor is a multifunctional neurohormone and neurotransmitter that is directly involved in coordinating the endocrine, autonomic, and behavioral responses to stress via mechanisms in both the brain and the periphery. CRF mediates its actions through cell surface receptors coupled to the activation of adenylyl cyclase. As described in this chapter, CRF receptors can be classified into four functionally distinct subtypes, CRF_1, $CRF_{2(a)}$, $CRF_{2(b)}$ and $CRF_{2(c)}$ which encode 411, 415, 431 and 397 amino acid proteins, respectively. These receptors differ with respect to their structure, tissue distribution, and pharmacological specificity for CRF-related ligands. Urocortin, a mammalian CRF-related peptide with close sequence homology to fish urotensin, interacts with high affinity with all three CRF receptors and may represent the naturally occurring ligand for $CRF_{2(a)}$ receptors. In addition to the receptors, the actions of CRF and urocortin in brain can also be modulated by a binding protein of 322 amino acids. The ligand requirements of CRF receptors and the CRF-BP can be distinguished in that mid-portion fragments of human CRF (human CRF_{6-33} and human CRF_{9-33}) have high affinity for CRF-BP but low affinity for the receptor. In total, the areas of distribution of these CRF receptors and CRF-BP in brain are correlated well with the immunocytochemical distribution of CRF and urocortin pathways and pharmacological sites of action of CRF and urocortin. It is quite conceivable that the multiple receptor subtypes and CRF-BP described here may mediate the different functional and pathophysiological actions of this family of neuropeptides.

Acknowledgements

The work described has resulted from the collaborative efforts of many individuals at Neurocrine Biosciences, Inc. We especially thank Drs. Dominic Behan, Derek Chalmers, Stephen Heinrichs, Chen Liaw, Nicholas Ling, Timothy Lovenberg and Tilman Oltersdorf for their scientific contributions.

References

Aguilera, G., Abou-Samra, A.B., Harwood, J.P., and Catt, K.J. (1988). Corticotropin releasing factor receptors: Characterization and actions in the anterior pituitary gland. *Advances in Experimental and Medical Biology* **245**, 83–106.

Battaglia, G., Webster, E.L., and De Souza, E.B. (1987). Characterization of corticotropin-releasing factor receptor-mediated adenylate cyclase activity in the rat central nervous system. *Synapse* **1**, 572–581.

Behan, D.P., Cepoi, D., Fisher, W.H., Park, M., Lowry, P.J., Sutton, S., and Vale, W.W. (1996). Characterization of a sheep brain corticotropin releasing factor binding protein. *Brain Research* **709**, 265–274.

Behan, D.P., Potter, E., Lewis, K.A., Jenkins, N.A., Copeland, N., Lowry, P.J., and Vale, W. (1993). Cloning and structure of the human corticotrophin releasing factor-binding protein gene (CRHBP). *Genomics* **16**, 63–68.

Bilezikjian, L.M., and Vale, W.W. (1983). Glucocorticoids inhibit corticotropin-releasing factor-induced production of adenosine 3',5'-monophosphate in cultured anterior pituitary cells. *Endocrinology* **113**, 657–662.

Chalmers, D.T., Lovenberg, T.W., and De Souza, E.B. (1995). Localization of novel corticotropin-releasing factor receptor (CRF2) mRNA to specific sub-cortical nuclei in rat brain: Comparison with CRF1 receptor mRNA expression. *Journal of Neuroscience* **15**, 6340–6350.

Chalmers, D.T., Lovenberg, T.W., Grigoriadis, D.E., Behan, D.P., and De Souza, E.B. (1996). Corticotropin-releasing factor receptors: From molecular biology to drug design. *Trends in Pharmacological Sciences* **17**, 166–172.

Chang, C.P., Pearse, R.I., O'Connell, S., and Rosenfeld, M.G. (1993). Identification of a seven transmembrane helix receptor for corticotropin-releasing factor and sauvagine in mammalian brain. *Neuron* **11**, 1187–1195.

Chen, R., Lewis, K.A., Perrin, M.H., and Vale, W.W. (1993). Expression cloning of a human corticotropin-releasing-factor receptor. *Proceedings of the National Academy of Sciences (USA)* **90**, 8967–8971.

Chen, R., Lewis, K.A., Perrin, M.H., and Vale, W.W. Characterization of the mouse CRF receptor gene., in *76th Annual Meeting of the Endocrine Society*. Endocrine Society, Anaheim, CA, 217.

Cortright, D.N., Nicoletti, A., and Seasholtz, A.F. (1995). Molecular and biochemical characterization of the mouse brain corticotropin-releasing hormone-binding protein. *Molecular and Cellular Endocrinology* **111**, 147–157.

De Souza, E.B. (1987). Corticotropin-releasing factor receptors in the rat central nervous system: Characterization and regional distribution. *Journal of Neuroscience* **7**, 88–100.

De Souza, E.B., and Grigoriadis, D.E. (1994). Corticotropin-releasing factor: Physiology, pharmacology and role in central nervous system and immune disorders. In F.E. Bloom, D.J. Kupfer (eds.), *Psychopharmacology: The Fourth Generation of Progress*, pp. 505–517. New York: Raven Press.

De Souza, E.B., Lovenberg, T.W., Chalmers, D.T., Grigoriadis, D.E., Liaw, C.W., Behan, D.P., and McCarthy, J.R. (1995). Heterogeneity of corticotropin-releasing factor receptors: Multiple targets for the treatment of CNS and inflammatory disorders. *Annual Reports in Medicinal Chemistry* **30**, 21–30.

De Souza, E.B., and Nemeroff, C.B. (1990) *Corticotropin-releasing factor: Basic and clinical studies of a neuropeptide*. CRC Press, Inc., Boca Raton, Florida, 365.

Donaldson, C.J., Sutton, S.W., Perrin, M.H., Corrigan, A.Z., Lewis, K.A., Rivier, J.E., Vaughan, J.M., and Vale, W.W. (1996). Cloning and characterization of human urocortin. *Endocrinology* **137**, 2167–2170.

Dunn, A.J., and Berridge, C.W. (1990). Physiological and behavioral responses to corticotropin-releasing factor administration: is CRF a mediator of anxiety and stress responses? *Brain Research Reviews* **15**, 71–100.

Giguere, V., Labrie, F., Cote, J., Coy, D.H., Sueiras-Diaz, J., and Schally, A.V. (1982). Stimulation of cyclic AMP accumulation and corticotropin release by synthetic ovine corticotropin-releasing factor in rat anterior pituitary cells: Site of glucocorticoid action. *Proceedings of the National Academy of Sciences (USA)* **79**, 3466–3469.

Grigoriadis, D.E., and De Souza, E.B. (1989). Heterogeneity between brain and pituitary corticotropin-releasing factor receptors is due to differential glycosylation. *Endocrinology* **125**, 1877–1888.

Grigoriadis, D.E., Liaw, C.W., Oltersdorf, T., and De Souza, E.B. (1994). Characterization of stable expression of cloned corticotropin-releasing factor receptors. *Neuroscience Abstracts* **20**, 1345.

Grigoriadis, D.E., Liu, X.J., Vaughn, J., Palmer, S.F., True, C.D., Vale, W.W., Ling, N., and De Souza, E.B. (1996a). ^{125}I-Tyr^0-Sauvagine: A novel high affinity radioligand for the pharmacological and biochemical study of human corticotropin-releasing factor$_{2\alpha}$ receptors. *Molecular Pharmacology* **50**, 679–686.

Grigoriadis, D.E., Lovenberg, T.W., Chalmers, D.T., Liaw, C., and De Souza, E.B. (1996b). Characterization of corticotropin-releasing factor receptor subtypes. In J. N. Crawley, S. McLean, (eds.), *Neuropeptides: Basic and Clinical Advances*, pp. 60–80. New York: The New York Academy of Sciences.

Kishimoto, T., Pearse II, R.V., Lin, C.R., and Rosenfeld, M.G. (1995). A sauvagine/corticotropin-releasing factor receptor expressed in heart and skeletal muscle. *Proceedings of the National Academy of Sciences (USA)* **92**, 1108–1112.

Kostich, W., Chen, A., Sperle, K., Horlick, R.A., Patterson, J., and Largent, B.L. (1996). Molecular cloning and expression analysis of human CRH receptor type 2 α and β isoforms. *Neuroscience Abstracts* **22**, 1545.

Liaw, C.W., Grigoriadis, D.E., Lorang, M.T., De Souza, E.B., and Maki, R.A. (1997a). Localization of agonist- and antagonist-binding domains of human corticotropin-releasing factor receptors. *Molecular Endocrinology* **11**, 2048–2053.

Liaw, C.W., Grigoriadis, D.E., Lovenberg, T.W., De Souza, E.B., and Maki, R.A. (1997b). Localization of ligand-binding domains of human corticotropin-releasing factor receptor: a chimeric receptor approach. *Molecular Endocrinology* **11**, 980–985.

Liaw, C.W., Lovenberg, T.W., Barry, G., Oltersdorf, T., Grigoriadis, D.E., and De Souza, E.B. (1996). Cloning and characterization of the human CRF2 receptor gene and cDNA. *Endocrinology* **137**, 72–77.

Linton, E.A., Wolfe, C.D.A., Behan, D.P., and Lowry, P.J. (1988). A specific carrier substance for human corticotropin-releasing factor in late gestational maternal plasma which could mask the ACTH-releasing activity. *Clinical Endocrinology* **28**, 315–324.

Lovenberg, T.W., Chalmers, D.T., Liu, C., and De Souza, E.B. (1995a). $CRF_{2\alpha}$ and $CRF_{2\beta}$ receptor mRNAs are differentially distributed between the rat central nervous system and peripheral tissues. *Endocrinology* **136**, 4139–4142.

Lovenberg, T.W., Liaw, C.W., Grigoriadis, D.E., Clevenger, W., Chalmers, D.T., De Souza, E.B., and Oltersdorf, T. (1995b). Cloning and characterization of a functionally distinct corticotropin-releasing factor receptor subtype from rat brain. *Proceedings of the National Academy of Sciences (USA)* **92**, 836–840.

Owens, M.J., and Nemeroff, C.B. (1991). Physiology and pharmacology of corticotropin-releasing factor. *Pharmacological Reviews* **43**, 425–473.

Perrin, M., Donaldson, C., Chen, R., Blount, A., Berggren, T., Bilezikjian, L., Sawchenko, P., and Vale, W. (1995). Identification of a second corticotropin-releasing factor receptor gene and characterization of a cDNA expressed in heart. *Proceedings of the National Academy of Sciences (USA)* **92**, 2969–2973.

Perrin, M.H., Donaldson, C.J., Chen, R., Lewis, K.A., and Vale, W.W. (1993). Cloning and functional expression of a rat brain corticotropin releasing factor (CRF) receptor. *Endocrinology* **133**, 3058–3061.

Polymeropoulos, M.H., Torres, R., Yanovsky, J.A., Chandrasekharrappa, S.C., and Ledbetter, D.H. (1995). The human corticotropin-releasing hormone receptor (CRHR) gene maps to chromosome 17 q12–22. *Genomics* **28**, 123–124.

Potter, E., Behan, D.P., Fischer, W.H., Linton, E.A., Lowry, P.J., and Vale, W.W. (1991). Cloning and characterization of the cDNAs for human and rat corticotropin-releasing factor-binding proteins. *Nature* **349**, 423–426.

Potter, E., Behan, D.P., Linton, E.A., Lowry, P.J., Sawchenko, P.E., and Vale, W.W. (1992). The central distribution of a corticotropin-releasing factor (CRF)-binding protein predicts multiple sites and modes of interaction with CRF. *Proceedings of the National Academy of Sciences (USA)* **89**, 4192–4196.

Potter, E., Sutton, S., Donaldson, C., Chen, R., Perrin, M., Lewis, K., Sawchenko, P.E., and Vale, W.W. (1994). The distribution of corticotropin releasing factor receptor mRNA expression in the rat brain and pituitary. *Proceedings of the National Academy of Sciences (USA)* **91**, 8777–8781.

Segre, G.V., and Goldring, S.R. (1993). Receptors for secretin, calcitonin, parathyroid hormone (PTH)/PTH-related peptide, vasoactive intestinal peptide, glucagon-like peptide 1, growth hormone-releasing hormone, and glucagon belong to a newly discovered G-protein-linked receptor family. *Trends in Endocrinology and Metabolism* **4**, 309–314.

Sperle, K., Chen, A., Kostich, W., and Largent, B.L. (1997). $CRH_{2\gamma}$: A novel CRH2 receptor isoform found in human brain. *Neuroscience Abstracts* **23**, 689.614.

Vaughan, J., Donaldson, C., Bittencourt, J., Perrin, M.H., Lewis, K., Sutton, S., Chan, R., Turnbull, A.V., Lovejoy, D., Rivier, C., Rivier, J., Sawchenko, P.E., and Vale, W. (1995). Urocortin, a mammalian neuropeptide related to fish urotensin I and to corticotropin-releasing factor. *Nature* **378**, 287–292.

Vita, N., Laurent, P., Lefort, S., Chalon, P., Lelias, J.M., Kaghad, M., Le, F.G., Caput, D., and Ferrara, P. (1993). Primary structure and functional expression of mouse pituitary and human brain corticotrophin releasing factor receptors. *FEBS Letters* **335**, 1–5.

Xiong, Y., Xie, L.Y., and Abou-Samra, A.-B. (1995). Signaling properties of mouse and human corticotropin-releasing factor (CRF) receptors: Decreased coupling efficiency of human type II CRF receptor. *Endocrinology* **136**, 1828–1834.

37. Pituitary CRH and VP Receptors and their Role in Corticotroph Responsiveness During Stress

C. Rabadan-Diehl and G. Aguilera

Introduction

Activation of the hypothalamic-pituitary-adrenal (HPA) axis, with increases in circulating ACTH and corticosterone, is the major hormonal response to a stressor. While responses to an acute somatosensory or metabolic stressor are marked and self-limited, responses to a repeated stressor of the same type vary depending on the stress paradigm. On the other hand, even when there is desensitization to the repeated stressor, there is facilitation of the responses to a novel stimulus (Dallman, 1992; Aguilera, 1994). The main regulators of ACTH secretion in the pituitary corticotroph are the 41 amino acid peptide, corticotropin releasing hormone (CRH), and the nonapeptide, vasopressin (VP). These neuropeptides are synthesized by parvocellular neurons in the hypothalamic paraventricular nucleus (PVN), and secreted into the pituitary portal circulation through nerve terminals in the external zone of the median eminence (Vale et al., 1983; Antoni, 1993). Although VP is a weak ACTH stimulus on its own, it acts synergistically with CRH, and plays an important role in sustaining pituitary responsiveness during chronic stress (Gilles et al., 1982; Antoni, 1993, Aguilera, 1994). Evidence suggests that there is differential regulation of CRH and VP in parvocellular hypothalamic neurons, with predominant secretion of VP during chronic stress (deGoeij et al., 1991, 1992; Aguilera, 1994; Ma et al., 1999). In addition to the levels of CRH and VP changes in the number of receptors for these hormones, or alterations in the coupling of the receptor to signal transduction systems are likely to play an important role in modulating ACTH responses.

The actions of CRH and VP in the pituitary are mediated by plasma membrane receptors coupled to their respective effector systems through guanyl nucleotide binding proteins (G-proteins). Molecular cloning techniques have identified two major types of CRH receptors, of which the type 1 is the main subtype in the pituitary (Potter et al., 1994). The CRH receptor is coupled to adenyl cyclase/protein kinase A and depends on signaling through the G protein, Gs. On the other hand, the pituitary vasopressin receptor, the V1b subtype, is coupled to phospholipase C leading to inositol phospate and diacylglycerol formation, probably through the G protein, Gq. Since stress affects CRH and VP receptor levels (Wynn, 1985; Hauger, 1988; Aguilera et al., 1994), regulation of these receptors is

likely to play a role in the adaptation of the HPA axis to the stress response. This paper presents information on the contribution of CRH and VP receptor regulation determining corticotroph responsiveness during stress, and the mechanisms responsible for the regulation of CRH and VP receptor content and CRH and VP receptor mRNA levels.

Materials and Methods

Animal Procedures

Male Sprague-Dawley rats (Zivic-Miller, Zelienople, PA) weighing 320–350 g were subjected to the following two stress paradigms. Rats were restrained for one hour, by placing them into a 2.5 × 6 inch plastic restrainer, and sacrificed four hours later, or subjected to repeated restraint for one hour daily for 8 to 14 days. Alternatively, rats received an i.p. hypertonic saline injection (1.5 ml of 5M NaCl), a painful stressor with a transient osmotic component, and were sacrificed 4 hours later. Groups subjected to repeated stress were killed 24 h after the last stress exposure. To investigate the role of glucocorticoids on stress-induced changes in CRH receptor mRNA, the effect of acute stress was studied in 6 day adrenalectomized animals with and without glucocorticoid replacement.

Analysis of Receptor mRNA Levels

CRH receptor and V1b receptor mRNA levels in the pituitary were measured by Northern Blot analysis or by in situ hybridization techniques. Briefly, for Northern blot analysis poly (A) + RNA from pools of four to five pituitaries were isolated and hybridized with random primed ^{32}P labeled DNA probes against the coding region of the receptors. Hybridization with a probe for cyclophilin was also used for correction of RNA loading. In situ hybridization was performed on slide-mounted pituitary sections and hybridized with ^{35}S labelled riboprobes directed against the coding region of CRH and V1b receptors.

Inositol Phosphate Determination

Quartered hemipituitaries were incubated with 30 μCi of myo-[^{3}H] inositol (100 Ci/ mmol, Amersham, Arlington Heights, IL) in inositol free medium 199 and incubated with and without 100 nM arginine vasopressin for 15 min. Total IP was extracted and separated by anion exchange chromatography.

Western Blot Analysis

Proteins (50 μg) from single pituitaries from control and dexamethasone-treated rats, or cultured anterior pituitary cells treated with dexamethasone, were separated by electrophoresis on a 12% polyacrylamide gel. Gq content was determined using a polyclonal antibody directed to the carboxy terminus of Gαq/α11 (Calbiochem, San Diego, CA) and an enhanced chemiluminescence detection system (Amesham, Arlington Heights, IL).

Results and Discussion

Regulation of Pituitary CRH Receptors

Effect of Stress on CRH Receptor mRNA and CRH Binding

Alterations of the HPA axis following adrenalectomy, glucocorticoid administration or stress are all associated with downregulation of the number of CRH receptors in the pituitary. However, changes in CRH binding do not correlate with the responsiveness of the axis to a novel super-imposed stressor. For example, during long-term adrenalectomy or chronic somatosensory stress, there are increases in pituitary POMC mRNA levels and ACTH responses to a novel stressor in spite of marked downregulation and desensitization of CRH receptors. While studies using CRH antagonists and CRH knock out mice show that CRH is indispensable for the stress response, the lack of correlation between CRH receptor number and pituitary responsiveness suggests that a small number of CRH receptors is sufficient to obtain a full ACTH response.

To examine the effect of stress on pituitary CRH receptor mRNA, *in situ* hybridization and Northern blot analysis were performed on pituitaries from animals subjected to different stress paradigms. Acute stress causes biphasic changes in pituitary CRH-R mRNA, with a decrease by 2 h followed by recovery, or an increase by 4 h after initiation of the stressor. In some acute stress models, e.g. immobilization, increases in CRH-R mRNA are accompanied by increases in CRH binding, while in others (ip hypertonic saline or lipopolysaccharide injection) there is a more prolonged decrease of CRH-R mRNA and CRH binding does not increase (Rabadan-Diehl *et al.*, 1996, Aubry *et al.*, 1997). Northern Blot analysis of pituitary RNA from rats acutely stressed showed downregulation of CRH receptor mRNA levels after a single i.p. hypertonic saline injection (levels decreased to 41 ± 6.5% of the control values) while a single immobilization markedly increased receptor mRNA (135 ± 9.1%) (Figure 1). Rats subjected to a single injection of CRH and VP presented an increase in CRH receptor mRNA similar to the one obtained after acute immobilization, supporting a role for these hypothalamic regulators on CRH receptor regulation (132.8 ± 3.9%).

During repeated stress, CRH-R mRNA levels decrease transiently after each stress exposure (Makino *et al.*, 1995) but they were elevated 24 h after the last stress exposure, in spite of CRH receptor down-

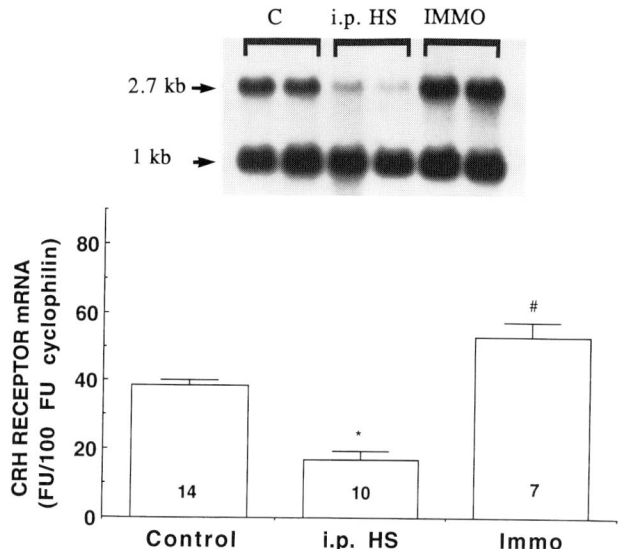

Fig. 1 Northern blot analysis of pituitary receptor mRNA 4 h after i.p. hypertonic saline injection (i.p. HS) or 1 h immobilization (Immo). Autoradiogram shows results of one representative experiment in pools of five pituitaries per experimental group. Arrows indicate a 2.7-Kb band corresponding to the CRH receptor mRNA and a 1-Kb band corresponding to cyclophilin mRNA. Bars represent mean and SE of values in number of experiments indicated in bars. *, $P < 0.01$ lower than controls; #, $P < 0.01$ higher than controls.

regulation (Rabadan-Diehl *et al.*, 1996). *In situ* hybridization studies on mRNA in pituitaries of chronically stressed rats, 8 and 14 days immobilization or i.p. hypertonic saline injection, showed a 30 to 60% increase in CRH receptor mRNA 24 h after exposure to the last stressor. Consistent with previous reports, this effect was accompanied by downregulation of CRH receptor binding. (Figure 2). Daily injections of CRH (1 μg) and VP (1 μg) subcutaneously for 14 days yielded a similar effect to chronic stress with an increase in CRH receptor mRNA (142.0 ± 6.3%) and a decrease in receptor binding. These results suggest that increased release of VP and CRH from the hypothalamus during stress exposure contributes to the changes in pituitary CRH receptor expression during chronic stress. The lack of correlation between CRH receptor mRNA levels and CRH receptor binding suggests that the downregulation observed after stress is due to higher receptor utilization and internalization, or to inhibition of receptor synthesis at the translational level. The observations *in vivo* are at variance with *in vitro* studies in primary pituitary cell cultures that have shown a negative effect of hypothalamic factors on CRH receptor mRNA levels (Pozzoli, 1996), indicating that interaction between the regulators and the pattern of exposure of the corticotroph to the regulators are critical during physiological regulation of pituitary CRH receptors.

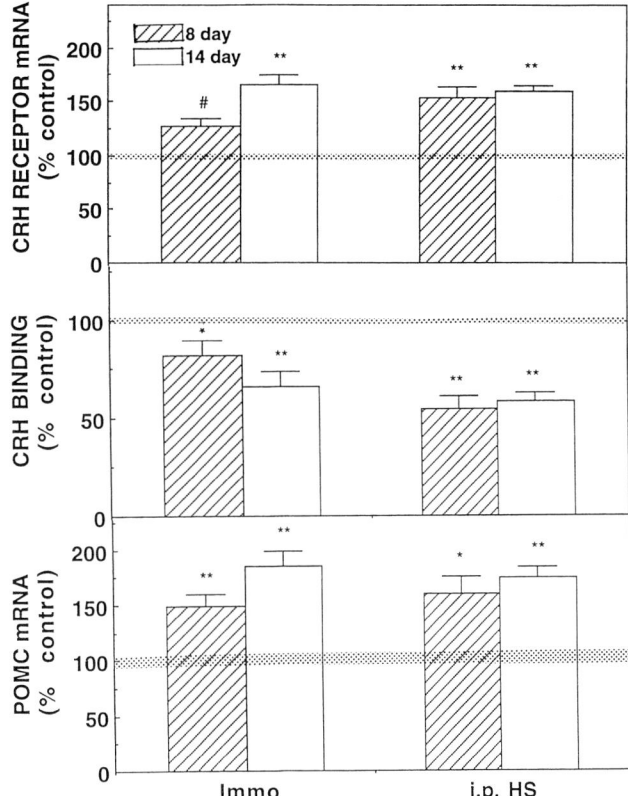

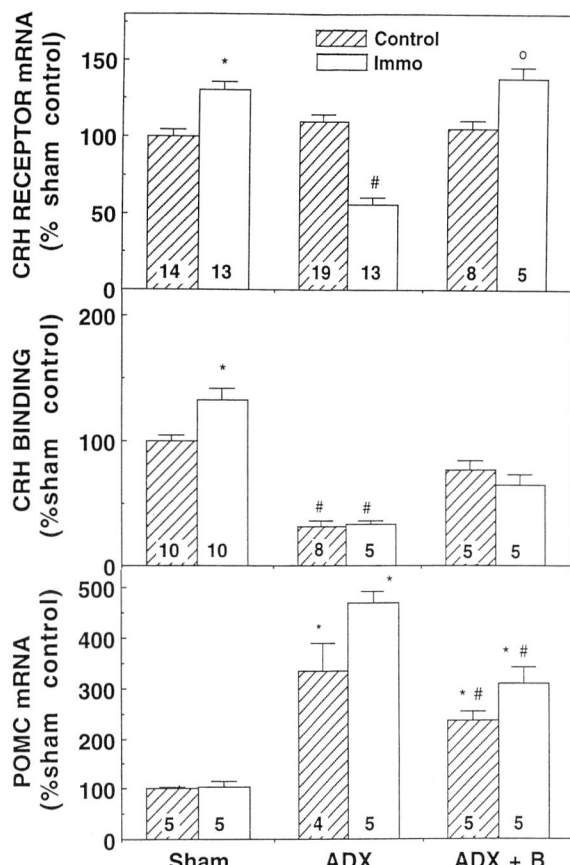

Fig. 2 Effect of repeated immobilization (Immo) or i.p. hypertonic saline injection (i.p. HS) for 8 or 14 days on CRH receptor mRNA (upper), CRH binding (middle) and POMC mRNA (lower). Bars are the mean and SE of data obtained by *in situ* hybridization or binding autoradiography in 9–15 rats per experimental group. Shaded band indicates two SE of controls. #, $P < 0.05$ vs. controls (Fisher's test only); *, $P < 0.01$ vs. controls; **, $P < 0.01$ *vs.* controls.

Fig. 3 Effect of adrenalectomy without (ADX) and with corticosterone replacement in drinking water (ADX + B) on changes in CRH receptor mRNA (upper), CRH binding (middle), and POMC mRNA (lower) after acute immobilization stress (Immo). Six days after sham surgery or adrenalectomy, rats were immobilized for 1 h and sacrificed 3 h later. Bars represent mean and SE of values obtained by *in situ* hybridization or binding autoradiography and number of rats is indicated in columns. *, $P < 0.05$ *vs.* sham controls; #, $P < 0.05$ *vs.* ADX control; o, $P < 0.05$ *vs.* ADX + B controls.

Glucocorticoid Dependence of the Effect of Stress on Pituitary CRH Receptor Expression

To determine whether the elevations in circulating glucocorticids can influence CRH receptor expression during stress, rats were subjected to acute stress in the absence or presence of glucocorticoids after surgical adrenalectomy or glucocorticoid administration. It has been shown in our laboratory that CRH receptor mRNA levels transiently decrease 18 hours after adrenalectomy, recovering to basal levels 4 to 6 days after surgery (Luo *et al.*, 1995). Based on that observation, 6 day-adrenalectomized animals with and without glucocorticoid replacement were subjected to a single 1 h immobilization and sacrificed four hours later. *In situ* hybridization studies showed that in adrenalectomized animals, stress induced a downregulation of CRH receptor mRNA (57 ± 4.5% of the control values), an effect that was prevented by replacement of glucocorticoids in the drinking water (Figure 3). The data show that while CRH and VP are likely to mediate pituitary CRH receptor regulation, glucocorticoids are necessary and modulate the stimulatory effect of the hypothalamic factors during stress.

Regulation of Pituitary V1b Receptors

Unlike CRH receptor regulation, there is a direct correlation between V1b receptor expression in the pituitary and responsiveness of the HPA axis to stress. For example, stress paradigms that cause ACTH hypersecretion as a response to a novel stimulus, such as immobilization and i.p. hypertonic injection, are accompanied by an increase in V1b receptor number (Figure 4). Therefore, it is reasonable to speculate that regulation of V1b receptors might play an important role in corticotroph adaptation and in the response mechanisms of the HPA axis to stress.

Regulation of V1b Receptors by Stress

Since the number of VP receptors during stress changes in parallel with the changes in corticotroph responsiveness, studies were performed to examine steady state levels of V1b receptor mRNA in rats subjected to stress paradigms associated with a hyperresponsiveness of

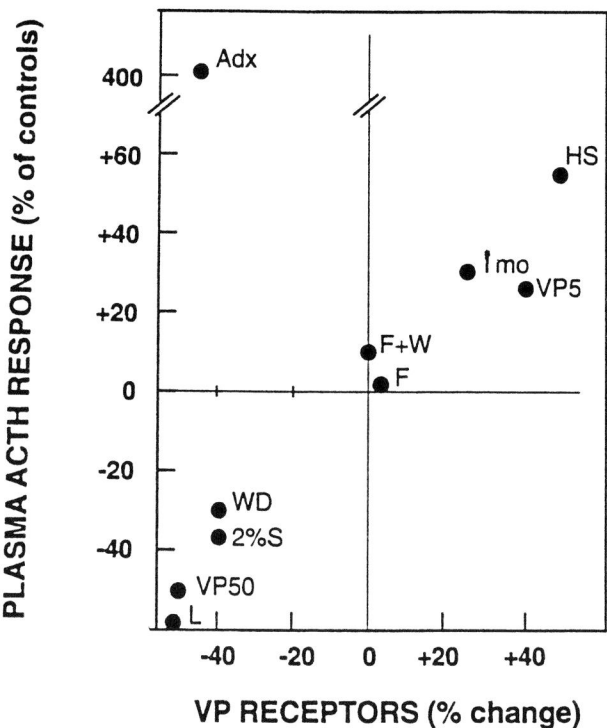

Fig. 4 Pituitary vasopressin receptors and plasma ACTH responses in animals subjected to different stress paradigms.

the HPA axis. *In situ* hybridization analysis showed increases of 63 and 84% in V1b mRNA after daily immobilization for 2 h for 8 or 14 days, respectively. Similar increases of 100 and 72% were observed after 8 or 14 days, respectively of repeated i.p. hypertonic saline injection. The parallel increases in V1b receptor mRNA levels and VP binding suggest that receptor upregulation might be determined by an increase in gene transcription, although other factors, such as increased mRNA stability, are possible. While 4h after immobilization for 1 h, V1b receptor mRNA levels were $145.0 \pm 11.4\%$ of the control values, a marked decrease $(54.0 \pm 16\%)$ was observed 4 h after a single i.p. hypertonic saline injection. The mechanism of this decrease is likely to involve increases in VP secretion from magnocellular neurons, due to the osmotic component of the stressor, since the effect was prevented by a V1 antagonist and was mimicked by injection of VP (Rabadan-Diehl *et al.*, 1995).

Glucocorticoids increase V1b receptor coupling to phospholipase C

Glucocorticoids regulate vasopressin receptors in a complex manner. Removal of endogenous glucocorticoids by surgical adrenalectomy as well as glucocorticoid administration decrease vasopressin binding to rat anterior pituitary membranes (Antoni *et al.*, 1985; Lutz-Bucher *et al.*, 1986). To investigate whether the decrease in VP binding after glucocorticoid treatment was also associated with changes in biological responses to VP, the ability of VP to stimulate inositol phosphate (IP)

formation was measured in rats that had received chronic administration of dexamethasone for 7 days. The data showed (Figure 5) that in spite of receptor downregulation, inositol phosphate formation after 15 minutes of VP stimulation was potentiated by 33% after glucocorticoid treatment, suggesting an increase in the coupling properties of the V1b receptor. Inositol phosphate formation measured in primary pituitary cell cultures preincubated with dexamethasone was also potentiated, indicating that there is a direct effect of glucocorticoids on the V1b receptor that

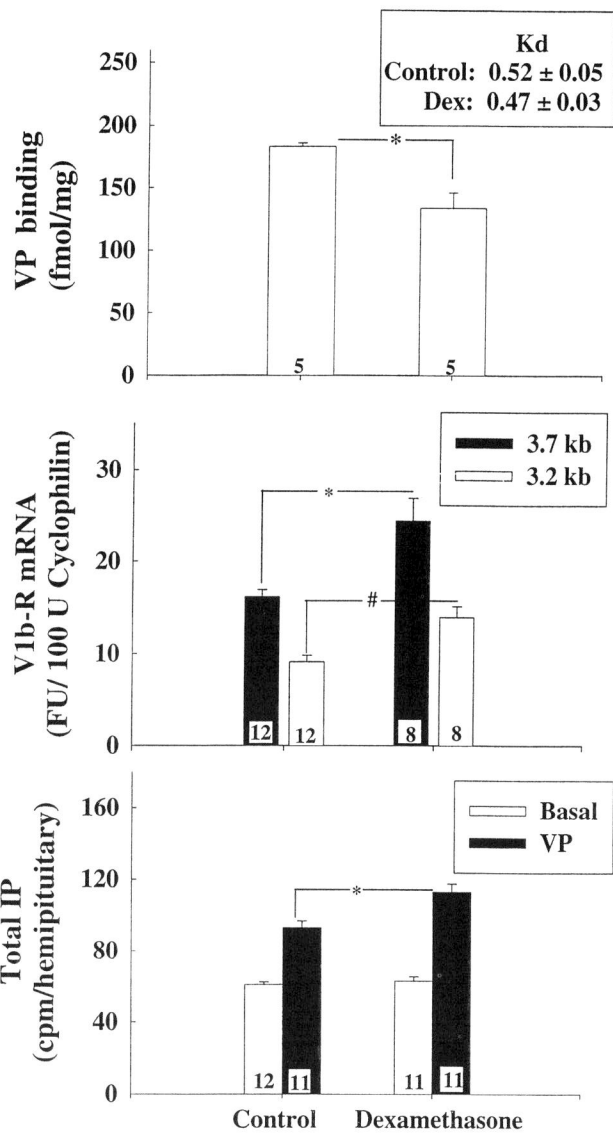

Fig. 5 Effect of injection of dexamethasone, 100 μg, sc, on pituitary VP binding (upper), V1b receptor (V1b-R) mRNA (middle) and VP-stimulated total inositol phosphate (IP) formation (lower). Bars are the mean and the SE of the values obtained in the number of experiments shown inside the bars. VP receptor number and affinity were determined by Scatchard analysis, V1b receptor mRNA by Northern blot, and total IP formation by anion exchange chromatography. *, $P < 0.01$ *vs.* control; #, $P < 0.05$ *vs.* controls.

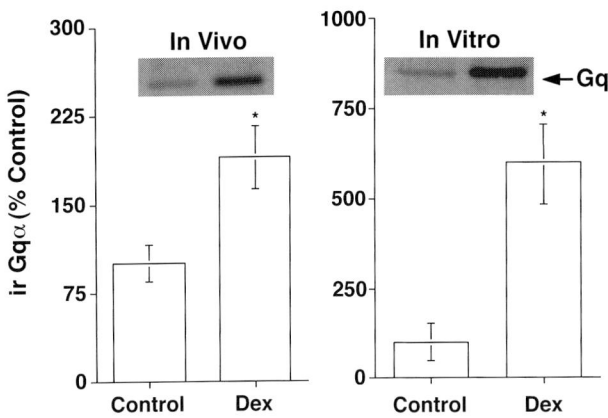

Fig. 6 Changes in immunoreactive Gq in membranes from single pituitaries of rats receiving injections of dexamethasone (100 μg) or vehicle for 7 days (*In vivo*) and from anterior pituitary calls cultured for 7 days with or without 10 nM dexamethasone (*In vitro*). The Bars are the mean and SE of the values obtained in 6 rats per group (*In vivo*) and the pool of the values obtained in three different experiments (*In vitro*). *, $P < 0.01$ *vs.* controls.

is not mediated by hypothalamic factors (data not shown).

To study the possibility that glucocorticoids potentiate ligand-stimulated IP formation at the level of the G-protein, the content of Gq was measured by Western Blot analysis in rat pituitaries and primary cells after dexamethasone pre-incubation. The results show (Figure 6) that in animals treated with dexamethasone, the levels of immunoreactive Gq increased by 2-fold when compared with untreated animals. A similar increase (119%) was observed in dexamethasone treated cells. Upregulation of the coupling protein, Gq, provides a possible mechanism by which the V1b receptor is highly efficient in spite of receptor downregulation. Also, the increase in coupling provides a mechanism in which VP facilitates corticotroph responsiveness in spite of elevated levels of circulating glucocorticoids during stress.

References

Aguilera, G. (1994). Regulation of pituitary ACTH secretion during chronic stress. *Frontiers in Neuroendocrinology* **15**, 321–350.

Aguilera, G., Pham, Q., and Rabadan-Diehl, C. (1994). Regulation of pituitary vasopressin receptors during chronic stress: relationship with corticotroph responsiveness. *Journal of Neuroendocrinology* **6**, 299–304.

Antoni, F.A. (1986). Hypothalamic control of adrenocorticotropin secretion: advances since the discovery of 41-residue CRF. *Endocrine Reviews* **7**, 351–358

Antoni, F.A. (1993). Vasopressinergic control of pituitary adrenocorticotropin secretion comes with age. *Frontiers in Neuroendocrinology* **14**, 76–122.

Antoni, F.A., Holmes, M.C., and Kiss J.Z. (1985). Pituitary binding of vasopressin is altered by experimental manipulations of the hypothalamo-pituitary-adrenocortical axis in normal as well as homozygous (di/di) Brattleboro rats. *Endocrinology* **117**, 1293–1299.

Aubry, J.-M., Turnbull, A.V., Pozzoli, G., Rivier, C., and Vale, W. (1997). Endotoxin decreases corticotropin-releasing factor receptor 1 messenger ribonucleic acid levels in the rat pituitary. *Endocrinology* **138**, 1621–1626.

Dallman, M.F., Akana, S.F., Scibner, K.A., Bradbury, M.J., Walker, C.-D., Strack, A.M., and Casio, C.S. (1992). Stress, feedback and facilitation in the hypothalamus pituitary adrenal axis. *Journal of Neuroendocrinology* **4**, 517–526.

deGoeij, D.C.E., Jezova, D., and Tilders, F.J.H. (1992). Repeated stress enhances vasopressin synthesis in corticotropin releasing factor neurons in the paraventricular nucleus. *Brain Research* **577**, 165–168.

deGoeij, D.C.E., Kvetňanský, R., Whitnall, M.H., Jezova, D., Berkenbosh, T., and Tilders, F.J.H. (1991). Repeated stress-induced activation of corticotropin-releasing factor neurons enhances vasopressin stores and colocalization with corticotropin releasing factor in the median eminence of rats. *Neuroendocrinology* **53**, 150–159.

Hauger, R.L., Millan, M.A., Lorang, M., Harwood, J.P., and Aguilera, G. (1988). Corticotropin-releasing factor receptors and pituitary adrenal responses during immobilization stress. *Endocrinology* **123**, 396–405.

Levin, N., and Roberts, J.L. (1991). Positive regulation of proopiomelanocortin gene expression in corticotropes and melanotropes. *Frontiers in Neuroendocrinology* **12**, 1–22.

Luo, X., Kiss, A., Rabadan-Diehl, C., and Aguilera G. (1995). Regulation of hypothalamic and pituitary corticotropin-releasing hormone receptor messenger ribonucleic acid by adrenalectomy and glucocorticoids. *Endocrinology* **136**, 3877–3883.

Lutz-Bucher, B., Kovacs, K., Makara, G., Stark, E., and Koch, B. (1986). Central nervous system control of pituitary vasopressin receptors: evidence for involvement of multiple factors. *Neuroendocrinology* **43**, 618–624.

Ma, X.-M., Lightman, S.L., and Aguilera, G. (1999). Vasopressin and corticotropin releasing hormone gene responses to novel stress in rats adapted to repeated restraint. *Endocrinology* **140**, 3623–3632.

Makino, S., Schulkin, J., Smith, M. A., Pacack, K., Palkovits, M., and Gold, P.W. (1995). Regulation of corticotropin releasing hormone receptor messenger ribonucleic acid in the rat brain and pituitary by glucocorticoids and stress. *Endocrinology* **136**, 4517–4525.

Potter, E., Sutton, S., Donaldson, C., Chen, R., Perrin, M., Lewis, K., Sawchenko, P., and Vale, W. (1994). The distribution of CRF receptor mRNA expression in the rat brain and the pituitary. *Proceedings of the National Academy of Sciences USA* **91**, 8777–8781.

Pozzoli, G., Bilezikjian, L.M., Perrin, M.H., Blount, A.L., and Vale, W.W. (1996). Corticotropin-releasing factor (CRF) and glucocorticoids modulate the expression of type 1 CRF receptor messenger ribonucleic acid in rat anterior pituitary cell cultures. *Endocrinology* **137**, 65–71.

Rabadan-Diehl, C., Kiss A., Camacho, C., and Aguilera, G. (1996). Regulation of messenger ribonucleic acid for corticotropin releasing hormone receptor in the pituitary during stress. *Endocrinology* **137**, 3808–3814.

Rivier, C., Smith, M., and Vale, W. (1990). Regulation of adrenocorticotropic hormone (ACTH) secretion by croticotropin releasing factor (CRF). In: E. DeSouza, and C. Nemeroff (Eds.), *Corticotropin releasing-factor: basic and clinical studies of a neuropeptide* pp. 175–189. Florida, Boca Raton: CRC Press.

Vale, W., Spiess, J., and Rivier, C.J.R. (1981). Characterization of a 41-residue ovine hypothalamic peptide that stimulates secretion of corticotropin and β-endorphin. *Science* **213**, 1394–1395.

Vale, W.W., Rivier, C., and Brown, M.R. (1983). Chemical and biological characterization of corticotropin releasing factor. *Recent Progress in Hormone Research* **39**, 245–270.

Wynn, P.C., Harwood, J.P., Catt, K.J., and Aguilera, G. (1985). Regulation of corticotropin releasing factor (CRF) receptors in the rat pituitary gland: effects of adrenalectomy on CRF receptors and corticotroph responses. *Endocrinology* **116**, 1653–1659.

38. Effect of 2-Deoxy-D-Glucose – Induced Inhibition of Glucose Utilization on Nuclear All-Trans Retinoic Acid Receptor Status in Rat Liver

J. Brtko, D. Macejova, J. Knopp and R. Kvetňanský

Introduction

Retinoic acids (all-trans, 9-cis-RA), the biologically active vitamin A metabolites, exert a substantial regulatory role in cell growth and differentiation and many other actions (Pfahl and Chytil, 1996), e.g. they can inhibit or reverse the process of malignant transformation in some cell types (Simon et al., 1998). They are also involved in the regulation of dopaminergic receptor pathways (Wolf, 1998). Retinoic acids act via their cognate nuclear receptors – ligand-inducible transcription factors, members of the steroid/thyroid/retinoid receptor family. As true retinoic acid-inducible transcription factors, they are capable of increasing the transcription of direct target genes by binding to cis-acting retinoic acid-responsive elements on the DNA. These effects are primarily mediated by three distinct alpha, beta, and gamma forms of nuclear all-trans retinoic acid receptors (RAR) or three distinct alpha, beta, and gamma forms of nuclear 9-cis retinoic acid receptors (RXR) (Evans, 1988).

Our previous data demonstrated that immobilization stress markedly decreased RAR concentration in the rat liver while its affinity remained unchanged, but in the second stress model, in which a 70% hepatectomy was performed, no significant changes in samples of liver nuclei for RAR concentration or affinity were observed (Brtko et al., 1996). Since all-trans retinoic acid (RA) has been found to stimulate glucose transport and expression of the gene for phosphoenolpyruvate carboxykinase via its cognate nuclear receptors (Pan et al., 1990), the RA-RAR complex thus may play a role in regulation of gluconeogenesis.

The aim of the present work was to investigate the effect of 2-deoxy-D-glucose (2DG), a metabolic stressor causing intracellular glucopenia, on RAR binding parameters in the rat liver in order to examine the possible modulation of RAR concentration in rat liver by 2DG.

Materials and Methods

Male Wistar rats weighing 250–340 g, maintained on a 12:12 light-dark cycle and fed standard pelleted diet, were used in the experiments. Adult rats were injected either with a single dose of 2DG (500 mg/kg body wt, ip) or saline, and were decapitated after different time intervals (1 h, 5 h, 10 h or 24 h). In the second set of experiments, a 2DG adapted group of rats (6 doses of 2 DG during 6 days, 500 mg/kg body wt, ip) was decapitated 24 h after the last 2DG dose or the above 2DG adapted group of rats received the next dose of 2DG, and the animals were decapitated 5 h after the seventh dose of 2 DG.

Determination of Rat Liver Nuclear All-Trans Retinoic Acid Receptors

Rat liver nuclei were prepared according to the procedure of DeGroot and Torresani (1975). Assays of labeled [11,12-^{3}H(N)] all-trans retinoic acid binding to nuclear receptors were performed at $20°C$ in the dark in a high ionic strength buffer (0.3 M KCl, 1 mM $MgCl_2$, 10 mM Tris-HCl, pH 7.0) according to a previously described method (Brtko, 1994).

Results

In comparison with the control group of animals, Scatchard analysis showed a single dose effect of 2DG (500 mg/kg) in liver which resulted in a significant ($p \leq 0.0005$) approximately 200% increase of the RAR maximal binding capacity (B_{max}) 10 h after 2DG administration (Figure 1), while the equilibrium association constant (K_a) remained unchanged (Figure 2). Shorter time intervals, 1 or 5 h after 2DG administration were ineffective on either the RAR B_{max} and K_a. In the 2DG adapted rats (6 doses of 2 DG, 500 mg/kg; 1 dose/day) decapitated 24 h after the last 2DG dose, the RAR B_{max} was found to be approximately 100% higher ($p < 0.05$) when compared to the control group of animals, but approximately 100% lower ($p < 0.01$) in comparison to animals which received 1 dose of 2DG, and were decapitated 10 h after 2DG administration (Figure 1). No further effect of the seventh dose of 2DG to repeatedly injected rats on either the RAR B_{max} or K_a was observed when animals were decapitated 5 h after the seventh dose of 2DG (Figures 3 and 4).

These data suggest that intracellular glucopenia up-regulates RARs in liver, and in this manner the effect of

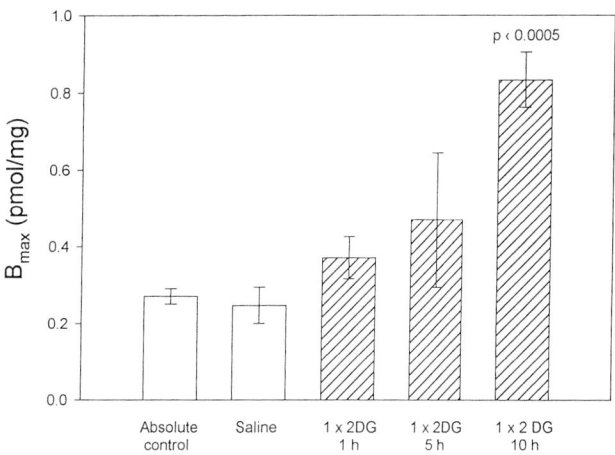

Fig. 1 Time course of the effect of 2-deoxy-D-glucose (2DG, 500 mg/kg) on maximal binding capacity (B_{max}) of nuclear all-trans retinoic acid receptors in rat liver.

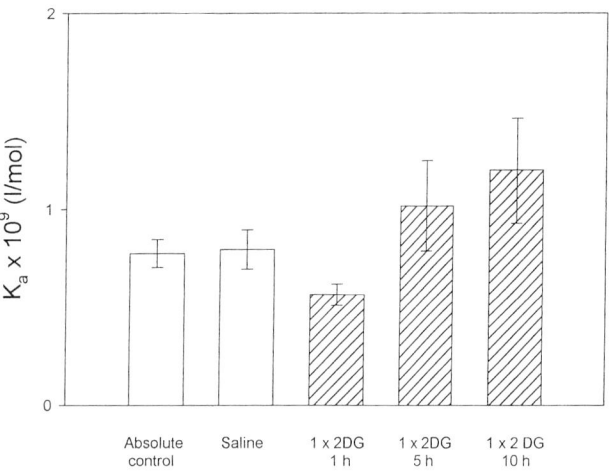

Fig. 2 Time course of the effect of 2-deoxy-D-glucose (2DG, 500 mg/kg) equilibrium association constant (K_a) of nuclear all-trans retinoic acid receptors in rat liver.

2DG specifically differs from that of immobilization stress.

Discussion

In recent years, great attention has been focused on the role of nuclear retinoic acid receptors – regulators of cell signaling pathways. RARs are capable of functioning as transcriptional repressors in the absence of RA and as potent activators upon binding of RA. To date, a number of retinoic acid target genes have been identified which include genes encoding rat growth hormone (Bedo *et al.*, 1989), osteocalcin (Schule *et al.*, 1990), oxytocin (Lipkin *et al.*, 1992) and many other genes encoding cell surface receptors, transcription factors, structural proteins or enzymes (Glass *et al.*, 1991). Our recent data have shown that immobilization stress down-regulates RARs in liver (Brtko *et al.*, 1996).

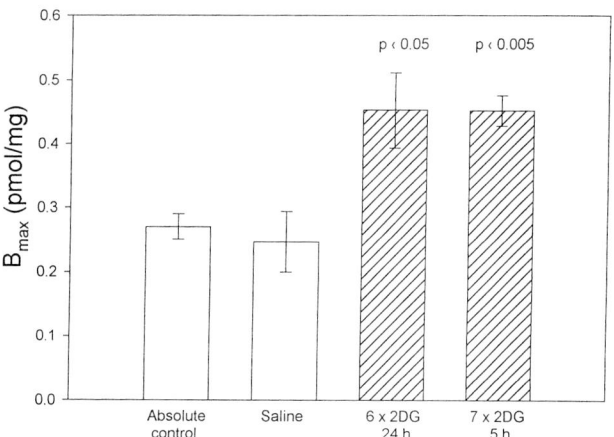

Fig. 3 Nuclear all-trans retinoic acid receptor maximal binding capacity (B_{max}) in rat liver of 2-deoxy-D-glucose (2DG) adapted animals (6 doses of 2DG, 500 mg/kg; 1 dose/ day) decapitated 24 h after the last 2DG dose; $7 \times 2DG$; 2DG adapted rats receiving the seventh dose of 2DG (500 mg/kg) and decapitated 5 h after the seventh dose of 2DG.

Since retinoic acid and its derivatives bound to their nuclear receptors have been found to induce stimulation of 2DG uptake due to the increase in the synthesis and recycling of hexose transporters (Hasegawa and Nishino, 1984; Sleeman *et al.*, 1995), we examined the effect of 2DG on nuclear all-trans retinoic acid status in rat liver. Metabolic stress *in vivo* using 2DG relies upon the assumption that the phosphorylated form, 2-deoxy-D-glucose 6-phospate can not be further metabolized, which results in intracellular glucopenia (Colwell *et al.*, 1996). Contrary to the data showing down-regulation of RARs after immobilization (Brtko *et al.*, 1996), intracellular glucopenia was found to up-regulate RARs in liver. In the 2DG adapted rats, no further effect of the next dose of 2DG was shown when animals were decapitated after the seventh dose of 2DG. Retinoids are thought to stimulate 2DG uptake via their cognate receptors. On the other hand, 2DG seems to enhance a population of retinoic acid receptors, ready for RA binding, probably in order to increase synthesis of hexose transporters. This still unknown regulatory mechanism of retinoid receptors in metabolic stress needs to be evaluated further.

Acknowledgements

This work has been supported in part by VEGA grant No. 2/6085/99. Authors wish to acknowledge the technical assistance of Mária Danihelová.

References

Bedo, G., Santisteban, P., and Aranda, A. (1989). Retinoic acid regulates growth hormone expression. *Nature* **339**, 231–234.

Brtko, J. (1994). Accurate determination and physicochemical properties of rat liver nuclear retinoic acid (RA) receptors. *Biochemical and Biophysical Research Communications* **204**, 439–445.

Brtko, J., Knopp, J., Hudecova, S., and Kvetňanský, R. (1996). Nuclear retinoic acid receptor binding and gene expression in rat liver: Effects of immobilization stress and partial hepatectomy. In: R. McCarty,

G. Aguilera, E. Sabban, and R. Kvetňanský (Eds.), *Stress: Molecular Genetic and Neurobiological Advances*, pp. 591–598. New York: Gordon and Breach.

Colwell, D.R., Higgings, J.A., and Denyer, G.S. (1996). Incorporation of 2-deoxy-D-glucose into glycogen. Implications for measurement of tissue-specific glucose uptake and utilization. *International Journal of Biochemistry and Cellular Biology* 28, 115–121.

DeGroot, L.J., and Torresani, J. (1975). Triiodothyronine binding to isolated liver cell nuclei. *Endocrinology* 96, 357–369.

Evans, R.M. (1988). The steroid and thyroid hormone receptor superfamily. *Science* 240, 889–895.

Glass, C.K., DiRenzo, J., Kurokawa, R., and Han, Y. (1991). Regulation of gene expression by retinoic acid receptors. *DNA Cell Biology* 10, 623–638.

Hasegawa, T., and Nishino H. (1984). Stimulatory effect of polyprenoic acid E5166 on 2-deoxy-D-glucose uptake by Swiss 3T3 cells. *Acta Vitaminologica et Enzymologica* 6, 251–257.

Lipkin, S.M., Nelson, C.A., Glass, C.K., and Rosenfeld, M.G. (1992). A negative retinoic acid response element in the rat oxytocin promoter restricts transcriptional stimulation by heterologous transactivation domains. *Proceedings of the National Academy of Sciences, USA* 89, 1209–1213.

Pan, C.J., Hoeppner, W., and Chou, J.Y. (1990). Induction of phosphoenolpyruvate carboxykinase gene expression by retinoic acid in an adult rat hepatocyte line. *Biochemistry* 29, 10883–10888.

Pfahl, M. and Chytil, E. (1996). Regulation and metabolism by retinoic acid and its nuclear receptors. *Annual Review of Nutrition* 16, 257–283.

Schule, R., Umesono, K., Mangelsdorf, D.J., Bolado, I., Pike, J.W., and Evans, R. M. (1990). Jun-Fos and receptors for vitamin A and D recognize a common response element in the human osteocalcin gene. *Cell* 61, 497–504.

Simon, D., Köhrle, J., Reiners, C., Boerner, A., Schmutzler, C., Mainz, K., Goretzki, P.E., and Röher, H.D. (1998). Redifferentiation therapy with retinoids: Therapeutic option for advanced follicular and papillary thyroid carcinoma. *World Journal of Surgery* 22, 569–574.

Sleeman, M.W., Zhou, H., Rogers, S., Ng, K.W., and Best, J.D. (1995). Retinoic acid stimulates glucose transporter expression in L6 muscle cells. *Molecular and Cellular Endocrinology* 108, 161–167.

Wolf, G. (1998). Vitamin A functions in the regulation of the dopaminergic system in the brain and pituitary gland. *Nutrition Reviews* 56, 354–355.

39. Mineralocorticoid Receptor-Mediated Effects on Hippocampal Synaptic Plasticity are Modulated by the β-adrenergic System

H. Saito and M. Smriga

Introduction

Several reports showed inhibitory effects of adrenal steroids on the sympatho-adrenal system (Munck et al., 1984; Brown and Fischer, 1986; Munck and Naray-Fejes-Toth, 1992; Kvetňanský et al., 1995). On the other hand, the central catecholaminergic neurons appear to play a stimulatory role upon the hypothalamo-pituitary-adrenal (HPA) axis activity. According to several research groups (Plotsky, 1987; Szafarczyk et al., 1987; Pacak et al., 1993; Kvetňanský et al., 1995) these effects are brain-mediated. Catecholamines trigger release of both corticotropin-releasing factor and corticotropin (Plotsky, 1987; Szafarczyk et al., 1987), corticotropin-releasing factor regulates sympatho-adrenal outflow and sympathetic nerve activity (Kvetňanský et al., 1995), glucocorticoids inhibit the central noradrenergic system (Kvetňanský et al., 1995), adrenalectomy augments norepinephrine (NE) release in the hypothalamus (Pacak et al., 1993) and so on. In addition, cognition and emotion are under direct regulation from the hormones of both axes (Joels and deKloet, 1989; Joels et al., 1991; Conrad and Roy, 1995; McGaugh et al., 1996; Gould and Cameron, 1996; Hu et al., 1997).

The dentate gyrus (DG) hippocampal area was directly implicated in associative memory acquisition (Bliss and Collingridge, 1993) and consolidation (Feldman et al., 1995; McGaugh et al., 1996). Both adrenal steroids (Joels et al., 1991; Conrad and Roy, 1995; Hu et al., 1997) and norepinephrine (Stanton and Sarvey, 1985; Joels and deKloet, 1989) affect the DG.

Stress-exposed rodents are characterized by impairments of synaptic plasticity in the DG (Foy et al., 1987). Adrenal steroids were thought to be responsible for this effect (Filipini et al., 1991; Pavlides et al., 1993; Smriga et al., 1996). On the other hand, long-lasting removal of the circulating steroids has negative effects on the neuronal morphology and synaptic plasticity in the DG (Gould et al., 1990; Diamond et al., 1992; Conrad and Roy, 1995; Hu et al., 1997). According to electrophysiological studies (Joels and deKloet, 1992), concentration-dependent biphasic actions of adrenal steroids were due to opposing roles for each steroid receptor type, with mineralocorticoid receptors (MR) facilitating and glucocorticoid receptors (GR) suppressing neuronal excitability. MR agonists prevented adrenalectomy-induced apoptosis, while GR agonists did not (Woolley et al., 1991). Moreover, mineralocorticoids prolonged the duration of hippocampal long-term potentiation (Pavlides et al., 1994).

EPI, released from the adrenal medulla, also has dramatic impact upon DG neuron properties (Stanton and Sarvey, 1985), as well as general learning performance (McGaugh et al., 1996). EPI was assumed to act indirectly via NE release in several brain regions (McGaugh et al., 1996).

In spite of the above-findings, the central interplay between the two hormonal systems is largely unknown. According to McGaugh (McGaugh et al., 1996), amygdaloid nuclei serve as an interactive site for the post-training, memory-stabilizing influences of adrenal hormones. However, post-training release of a hormone excludes its effect on the memory acquisition (Sandi et al., 1997). Thus, hormonal influences on memory acquisition (pre-training effects) may be conveyed via distinct central areas and different cellular mechanisms (McGaugh et al., 1996).

The present study investigated electrophysiological interactions between the EPI and mineralocorticoids in the brain, especially with regards to the hippocampal neurons. To free the central MRs from their endogenous ligands and to avoid the isolation of hippocampal areas from peripheral and central inputs, in vivo recordings from adrenalectomized (ADX) rats were used.

Materials and Methods

Electrophysiological Recordings

Male Wistar rats (200 g) were anesthetized with a combination of urethane and α-chloralose. They were placed in a stereotaxic frame and prepared unilaterally with 250 μm stainless recording electrodes (insulated except at the 500 μm cut tip) in the hilus of the DG, and an identical bipolar stimulating electrode in the angular bundle to activate medial perforant path fibers (Smriga et al., 1996). The depths of the electrodes were adjusted to produce maximal field excitatory postsynaptic potentials (positive-going) and population spikes (PS). The test stimulation (diphasic; 80 μsec, 450–600 μA) was adjusted in amplitude to produce a 3–5 mV PS. Rats in which the above described stimulation pulse failed to induce the required PS during the stabilization period were not included in further experiments. Although the PS does not provide full information embraced in the field potential (only its linear approximation), it correlates highly with the responses of DG granule cells (Sclabassi *et al.*, 1988). Additionally, it reflects changes in the humoral environment (Pavlides *et al.*, 1993). Test pulses were applied at 30 sec intervals. After at least 60 min of basal recording one set of high-frequency tetanic stimulation composed 20 pulses (60 Hz) was delivered through the stimulating electrodes and responses were followed for 60 min. PS amplitudes and latencies were expressed as percent change from the pre-tetanic baseline values and averaged across 2 min epochs. In order to readably evaluate changes in the magnitude of the tetanically-evoked potentiation among several experimental groups, we have established the area under curve (Figure 1B) as the potentiation measure. Since the drugs used did not significantly influence kinetic profiles of potentiation (i.e., decay rate), the magnitude (area under curve) was a justified approximation. On completion of the electrophysiological experiments, lesions were made at each electrode tip (400 μA, 20 s) in order to verify their placement. Brains were sliced and examined. The subjects with improperly placed electrodes were excluded from analysis. Experiments were done between 8:00 a.m. and 2:00 p.m. in order to avoid impacts of the circadian rhythm. Rectal temperature was monitored by a probe. Data were statistically evaluated by one way ANOVA (post hoc Duncan's multiple range test). ADX and sham-operation, were done 24 h before the electrophysiological experiments. Procedures (including the CORT measurement) were described previously (Smriga *et al.*, 1996).

Drugs and Treatments

Aldosterone (ALDO) (Sigma) was applied subcutaneously (sc) 60 min prior to tetanic stimulation. The time-span was chosen to allow proper passage of the studied hormone through the blood-brain barrier and its binding to allow for intracellular MR. MR antagonist spironolactone (Sigma) was given sc 10 min before ALDO (70 min before tetanus). EPI (Sigma) was administered (ip, 100 μg/kg) 20 min prior to tetanus. Adrenergic antagonists yohimbine HCl (3 mg/kg), prazosine HCl (3 mg/kg), dibenamine (4 mg/kg) and d,l-propranolol (2 mg/kg) (all from Sigma), were injected ip immediately before EPI (approximately 21 min before tetanus), or 20 min before ALDO. The doses of agonists and antagonists were based on previously reported behavioral studies (Introini-Collison *et al.*, 1995). In one experimental set, propranolol (9 nmol, 3 mM $\times$ 3 μl) was infused directly into the recording area or into the central nucleus of amygdala (CEA) (injection time; 3 min).

Results

A single 20 pulses (60 Hz) tetanic stimulation of the medial perforant pathway produced a marked increase in PS amplitude in vehicle treated rats. This increase reached approximately 135% of baseline levels 5 min

after the tetanus and remained stable above the 118–120% level until the end of experimental session (Figure 1A, white circles). The potentiation in the sham and ADX rats showed comparable magnitude and time-course (Figure 1A and B, white circles). Thus, the mild level of synaptic potentiation evoked by the tetanus remained unaltered by prior removal of adrenal glands (24 h removal).

ALDO applied (sc, 100 μg/kg) 1 h before tetanus to the shams did not affect either the magnitude, or the time-course of the post-tetanic potentiation (Figure 1A, black circles). However, the mineralocorticoid significantly enhanced the extent of PS potentiation in the ADX rats (Figure 1B, black circles). The shaded portion (Figure 1B) illustrates area under curve. Area under curve served as a measure of tetanically-evoked potentiation magnitude in all of the following experiments. Since the effect of ALDO was observed only in the ADX animals, subsequent experiments were undertaken exclusively in animals that were adrenalectomized.

The level of plasma corticosterone served as the analytical measure of adrenalectomy. The mean corticosterone level in sham-operated animals was 19.2 $\pm$ 0.8 μg/dl. Removal of adrenal glands 24 h before the electrophysiological recordings resulted in a significant decrease of the plasma corticosterone (5.2 $\pm$ 1.4 μg/dl, $p < 0.01$ from sham-operated rats (t-test)). Pharmacological treatments did not cause significant variations in the plasma corticosterone levels in ADX subjects. Figure 1C illustrates that the effect of ALDO in ADX subjects (shown also in panel B) was significantly canceled by pretreatment with the MR antagonist spironolactone (sc, 20 mg/kg) administered 10 min before ALDO (70 min before tetanus). The effects of ALDO and spironolactone on basal PS latency, which was evaluated immediately before tetanus (inset to Figure 1A), are documented in Figure 1D. Adrenalectomy resulted in significant latency increases, as compared to sham animals. ALDO caused a marked decrease in the mean latency value in ADX, but not sham, rats. The effect was restored by spironolactone. Figure 2 depicts the effect of EPI (ip, 100 μg/kg) applied 20 min before tetanus in ADX rats. EPI significantly increased the magnitude of post-tetanic potentiation (Figure 2A) without any effect upon basal PS latency (Figure 2B). The effect of EPI was markedly blocked by the β-adrenergic antagonist propranolol (ip, 2 mg/kg) given immediately before EPI (Figure 2A). The alpha$_1$-adrenergic antagonist prazosin (ip, 3 mg/kg), or the α_2-adrenergic antagonist yohimbine (ip, 3 mg/kg) were ineffective (data not shown). The MR antagonist spironolactone (sc, 20 mg/kg) given 50 min before EPI (70 min before tetanus) also failed to affect potentiation levels in the EPI group (Figure 2A). The adrenoceptor antagonists and spironolactone administered alone did not significantly modify the various measures.

Interactions between ALDO and adrenergic system are documented in Figures 3A and B. The effect of ALDO in ADX rats was not affected by the pretreat-

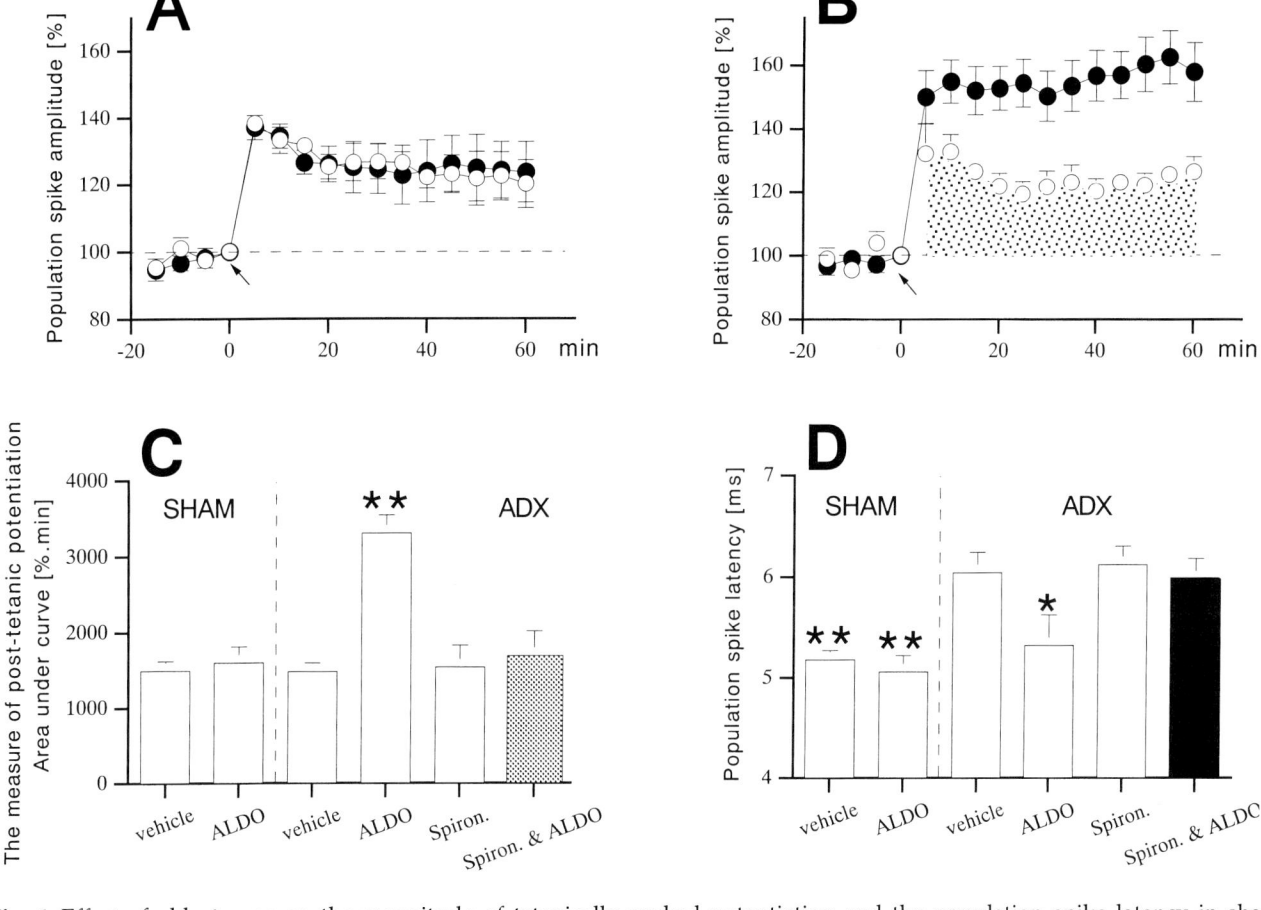

Fig. 1 Effect of aldosterone on the magnitude of tetanically-evoked potentiation and the population spike latency in sham-operated and adrenalectomized rats *in vivo.*

A) Aldosterone (sc, 100 μg/kg) (●) applied 60 min prior to tetanus (20 pulses, 60 Hz, arrow) did not affect the post-tetanic population spike in sham-operated rats. Time-course of changes is compared to vehicle treated rats (○). Results are expressed as percent differences between basal (100%) and stimulated levels. Means of 4 rats ± SEM.

B) Aldosterone (sc, 100 μg/kg) (●) applied 60 min prior to tetanus (20 pulses, 60 Hz, arrow) markedly enhanced the post-tetanic population spike in adrenalectomized rats. Time-course of changes is compared to vehicle (sc) treated rats (○). Means of 5–6 rats ± SEM. The shaded portion represents area under curve.

C) Effect of aldosterone (sc, 100 μg/kg) on the magnitude of the tetanically-evoked population spike in ADX rats was significantly canceled by spironolactone (sc, 20 mg/kg) (**$p < 0.01$, Duncan's multiple range test from all other groups). Area under curve from 5 to 60 min after tetanus (see panel B) served as an overall measure of potentiation. Means of 4–6 rats ± SEM.

D) Adrenalectomy resulted in a significant increase in population spike latency. Aldosterone significantly restored the latency levels in ADX rats and its effect was canceled by spironolactone (sc, 20 mg/kg) (*$p < 0.05$, **$p < 0.01$, Duncan's multiple range test from all other groups). Means of 4–6 rats ± SEM.

ment with high dose (4 mg/kg) of non-selective adrenoceptor blocker dibenamine (ip), but it was partly decreased by propranolol (ip, 2 mg/kg) (Figure 3A). Additionally, propranolol significantly increased the basal PS latency in the ALDO group (Figure 3B). The effects of propranolol were not observed when the antagonist was administered 40 min after ALDO (20 min before tetanus) (data not shown).

The final experiment (Figure 4A, B) examined the involvement of the central β-adrenergic system in ALDO-enhanced synaptic plasticity and excitability. Figure 4A documents that propranolol infused 10 min before ALDO directly into the DG granule cell layer (9 nmol, 3 mM × 3 μl), but not into the CEA, partially canceled the effect of the mineralocorticoid.

The effect of centrally-infused propranolol, although stronger than the one observed after its peripheral administration (Figure 3A), failed to reach significant levels. Since propranolol alone could directly influence the level of synaptic responses in DG neurons (Stanton and Sarvey, 1985), higher doses of the antagonist were not evaluated. At the present dose (9 nmol) propranolol did not affect the measured characteristics (Figure 4A, B). Figure 4B illustrates that β-adrenergic antagonist pre-treatment into the DG neurons significantly restored PS latency in the ALDO group. Amygdaloid infusion failed to mimic the effectiveness of direct DG application.

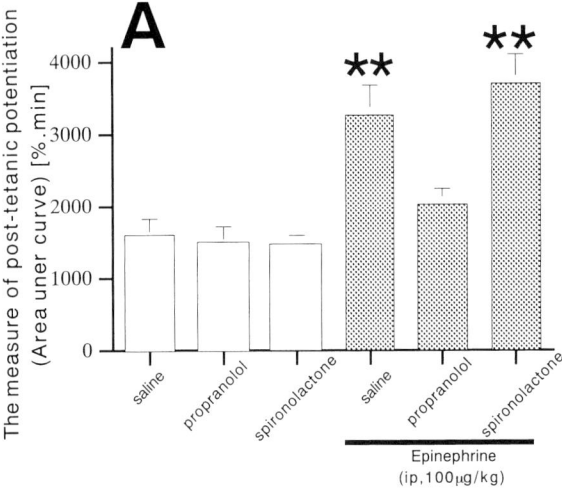

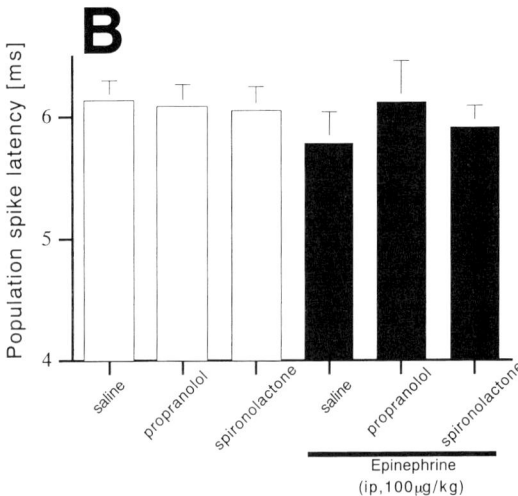

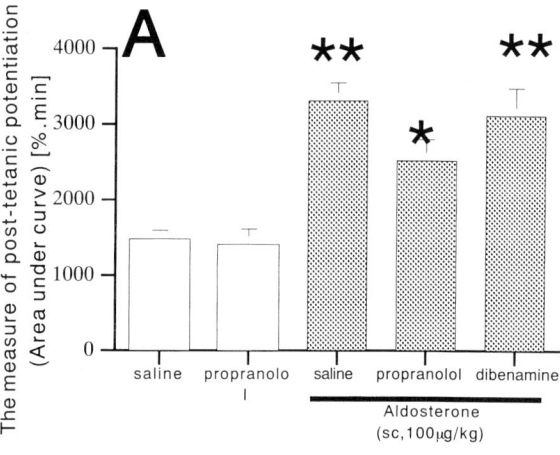

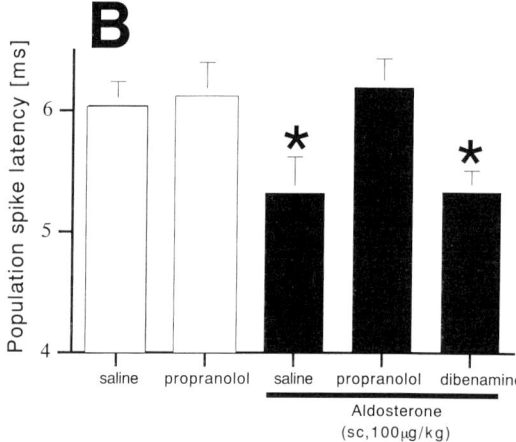

Fig. 2 Effect of epinephrine on the magnitude of tetanically-evoked potentiation and the population spike latency in adrenalectomized rats *in vivo*.

A) Effect of epinephrine applied 20 min prior to tetanus (20 pulses, 60 Hz) on the magnitude of tetanically-evoked population spike was significantly canceled by propranolol (ip, 2 mg/kg), but not spironolactone (sc, 20 mg/kg) (**$p < 0.01$, Duncan's multiple range test from all other groups). Means of 4–6 rats ± SEM.

B) Epinephrine did not significantly affect the basal population spike latency. Means of 4–6 rats ± SEM.

Fig. 3 Aldosterone-induced electrophysiological effects in the dentate gyrus of adrenalectomized rats are dependent on the propranolol-sensitive pathways.

A) Propranolol (ip, 2 mg/kg), but not dibenamine (ip, 4 mg/kg), partly canceled the aldosterone-induced enhancement of the tetanically-evoked population spike amplitude. Results are expressed as described in Figure 1C (*$p < 0.05$, **$p < 0.01$, Duncan's multiple range test from saline group). Means of 5–6 rats ± SEM.

B) Propranolol (ip, 2 mg/kg), but not dibenamine (ip, 4 mg/kg), significantly canceled the aldosterone-induced decrease of the basal population spike latency. Results are expressed as described in the fig. 1D (*$p < 0.05$, Duncan's multiple range test from all groups). Means of 5–6 rats ± SEM.

Discussion

Adrenal steroids affect specific neuronal populations via centrally distributed MR and GR (Sloviter *et al.*, 1989; McEwen 1994). In parallel with the previous studies (Pavlides *et al.*, 1994), we have found enhancement of the activity-dependent synaptic responses after a single ALDO administration in the DG of ADX rats. This effect was preceded by a marked decrease of the basal PS latency, indicating a drug-induced elevation of neuronal excitability within DG granule cells. No effect of the mineralocorticoid was recorded in sham-operated animals. This could be a consequence of MR blockade in shams (Joels and deKloet, 1992). Moreover,

ALDO-induced effects in ADX rats were blocked by the centrally-acting MR antagonist (spironolactone), confirming the MR-mediated effects.

The short time-span of adrenalectomy (24 h) was introduced to avoid DG granule cell apoptosis that results from the prolonged removal of adrenal glands (Gould *et al.*, 1990; Hu *et al.*, 1997). Indeed, adrenalectomy alone did not result in changes of synaptic plasticity. However, the operation caused a marked increase of the population spike latency, suggesting impaired levels of neuronal excitability. Thus, although short-term adrenalectomy does not trigger morpholo-

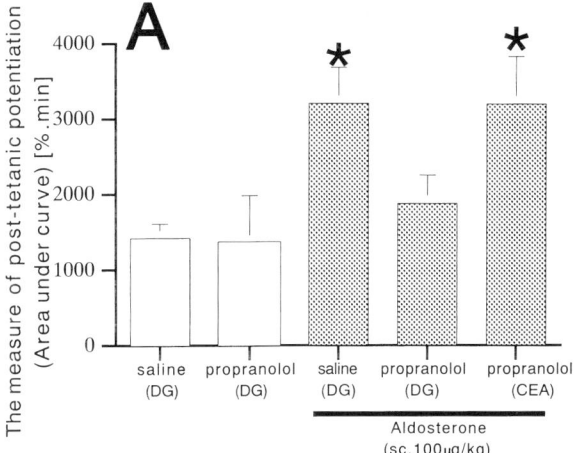

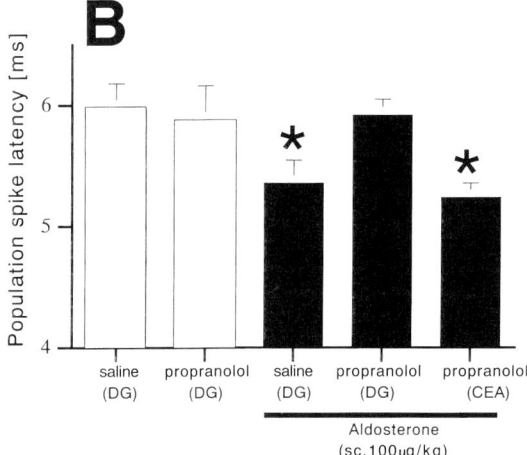

Figure 4 Aldosterone-induced electrophysiological effects in the dentate gyrus of adrenalectomized rats are dependent on the local β-adrenergic system.
A) Propranolol (9nmol), directly infused into the dentate gyrus granule cells (DG), but not into the central nucleus of amygdala (CEA) (injection time, 3 min), partly canceled the aldosterone-induced enhancement of the tetanically-evoked population spike amplitude. Results are expressed as described in the fig. 1C (*p < 0.05, Duncan's multiple range test from saline group). Means of 4–6 rats ± SEM.
B) Propranolol (9nmol), directly infused into the DG, but not into CEA (injection time, 3 min), significantly canceled the aldosterone-induced decrease of the basal population spike latency. Results are expressed as described in the fig. 1D (*p < 0.05, Duncan multiple range test from all groups). Means of 4–6 rats ± SEM.

gical degeneration in the DG, it changes the rate of neuronal excitability. One of the reasons could be a drop in MR occupancy in ADX rats, since the mineralocorticoid ALDO significantly increased excitability in ADX subjects, in other words, restored it back to the levels found in shams. However, the problem seems to be more complex, because MR antagonist alone did not affect the excitability. Hence, the changes induced by adrenalectomy involve other mechanisms.

EPI represents the second class of adrenal hormones closely involved in the regulation of DG activity (Weil-Malherbe *et al.*, 1959; Gold and van Buskirk, 1978; Stanton and Sarvey, 1985; Dahl and Sarvey, 1989; Introini-Collison *et al.*, 1992). The enhancement of tetanically-evoked synaptic plasticity following the peripheral EPI administration in ADX rats agrees with the previous observations in intact rats (Dahl and Sarvey, 1989). Since the HPA axis is assumed to have inhibitory effects on the sympatho-adrenal system (Munck *et al.*, 1984; Kvetňanský *et al.*, 1995), there could be partial differences in the EPI effects in ADX and control rats. However, by extracellular recording, we did not detect significant differences between the effect of EPI in sham and ADX rats (unpublished results). As β-adrenoceptor antagonist, but not α_1- and α_2-adrenoceptor antagonists, canceled EPI-induced enhancement in ADX rats, the involvement of the β-adrenergic pathway was suggested. The enhancing effect of EPI on synaptic plasticity was not accompanied by significant changes in the PS latency, suggesting an absence of non-selective changes in local neuronal excitability.

The results showed that both ALDO and EPI significantly enhanced tetanically-evoked synaptic plasticity in ADX animals. The effect of ALDO was mediated via MR, while the effect of EPI via β-adrenergic receptors. (The use of propranolol as a non-selective blocker did not enable pharmacological distinction among β-adrenoceptor subtypes.) These findings are in parallel with the previous indications of a positive involvement of MR (Joels and deKloet, 1992; Pavlides *et al.*, 1994) and β-adrenoceptors (Inroini-Collison *et al.*, 1992; McGaugh *et al.*, 1996) in memory-related processes.

The main part of the present study was concerned with the inter-relationship between the MR- and adrenoceptor-mediated effects in the DG. We found that peripheral pre-treatment with propranolol partly canceled the effect of ALDO on the magnitude of tetanically-evoked potentiation and significantly eliminated ALDO-induced changes in the basal PS latency. Propranolol did not act when it was given 40 min after ALDO (20 min before tetanus). Interactions between the two pathways might result in a modulation of the basal membrane potential, which could be the trigger of detected latency variations, as suggested by the previous data (Joels and deKloet, 1989, 1992). Though tetanically-evoked synaptic plasticity is also dependent on membrane potential (Gustafsson *et al.*, 1987), propranolol reduced only partially the ALDO-induced enhancement of synaptic plasticity. Thus, while the MR-mediated pathway seems to converge with the β-adrenergic system in its influences on the membrane potential properties in DG neurons, the interplay is not fully translated into plastic synaptic changes. Although one can argue that higher doses of propranolol might have been more effective, we have avoided their use, because an extensive blockade of the β-adrenergic system can trigger irreversible central changes. Since *in vivo* electrophysiological recording did not provide a tool for the evaluation of nonspecific effects of the antagonist, we have evaluated only a relatively moderate dose, which did not affect overall neurotransmission (Stanton and Sarvey, 1985).

To determine whether the above-described results were consequences of direct interactions within DG neurons, we injected propranolol into the DG and CEA. The CEA has rich noradrenergic connections with hippocampal area, which could be involved in several physiologically significant functions (McGaugh et al., 1996). However, while the DG infusion of propranolol mimicked the influence of its peripheral treatment, infusion into the CEA did not. These results pointed out the interactions of β-adrenergic- and MR-mediated pathways at the level of DG neurons. The assumption agrees with the distribution pattern of MR that are abundant in the hippocampus, but virtually absent in the rest of the brain (Chao et al., 1989).

Our results partially contradict the observations that the amygdaloid nuclei are specifically involved in the interactions between EPI and adrenal steroids released during training (McGaugh et al., 1996). Post-training (in the case of electrophysiological models, post-tetanic) administration of the adrenal steroids seems to produce qualitatively different effects than those produced by their pre-treatment (compare Filipini et al., 1991; Pavlides et al., 1993; Smriga et al., 1996; Seidenbecher et al., 1997). Post-tetanic drug administration excludes any effect on memory acquisition (Seidenbecher et al., 1997). Based on our data, the acquisition-modulating hormonal interactions that are operative before tetanus appear to be conveyed by hippocampal neurons, rather than amygdaloid ones. This distinction could be one of the sources for the above-mentioned differences.

Taken together, we found that the functional β-adrenergic system in the DG was necessary for the MR-mediated increase of neuronal excitability and, partially, also for the enhancement of the tetanically-evoked synaptic plasticity in vivo. On the other hand, the functional mineralocorticoid pathway was not necessary for the EPI-induced effects.

Iintracellular data are required to explore the above-outlined interactions. Presently, it is possible only to say that the hippocampal β-adrenergic system may participate in steroid-modulated synaptic processes, at least during periods when these effects are mediated predominantly by MR. As the plasma levels of adrenal steroids undergo marked oscillations (Dana and Martinez, 1984; Sapolsky and Meaney, 1986; Gould and Cameron, 1996), this participation could have important physiological implications.

References

Bliss, T.V.P., and Collingridge, G.L. (1993). A synaptic model of memory: long-term potentiation in the hippocampus. Nature 361, 31–39.

Brown, M.R., and Fisher, L.A. (1986). Glucocorticoid suppression of the sympathetic nervous system and adrenal medulla. Life Sciences 39, 1003–1012.

Chao, H.M., Choo, P.H., and McEwen, B.S. (1989). Glucocorticoid and mineralocorticoid receptors mRNA expression in rat brain. Neuroendocrinology 50, 365–371.

Conrad, C.D., and Roy, E.J. (1995). Dentate gyrus destruction and spatial learning impairment after corticosteroid removal in young and middle-aged rats. Hippocampus 5, 1–15.

Dahl, D., and Sarvey, J.M. (1989). Norepinephrine induces pathway-specific long-term potentiation and depression in the hippocampal dentate gyrus. Proceedings of the National Academy of Sciences USA 6, 4776–4780.

Dana, R.C., and Martinez, J.L. (1984). Effect of adrenalectomy on the circadian rhythm of LTP. Brain Research 308, 392–395.

Diamond, D.M., Bennett, M.C., Fleshner, M., and Rose, G.M. (1992). Inverted-U relationship between the level of peripheral corticosterone and the magnitude of hippocampal primed burst potentiation. Hippocampus 2, 421–430.

Feldman, S., Conforti, N., and Weidenfeld, J. (1995). Limbic pathways and hypothalamic neurotransmitters mediating adrenocortical responses to neural stimuli. Neuroscience and Behavioral Reviews 19, 235–240.

Filipini, D., Gijsberg, K., Birmingham, M.K., and Dubrovsky, B. (1991). Effects of adrenal steroids and their reduced metabolites on hippocampal long-term potentiation. Journal of Steroid Biochemistry and Molecular Biology 40, 87–92.

Foy, M.R., Stanton, M.E., Levine, S., and Thompson, R.F. (1987). Behavioral stress impairs long-term potentiation in rodent hippocampus. Behavioral and Neural Biology 48, 138–149.

Gold, P.E., and van Buskirk, R. (1977). Post-training brain norepinephrine concentrations: correlation with retention performance of avoidance training and peripheral epinephrine modulation of memory processing. Behavioral Biology 24, 168–184.

Gould, E., and Cameron, H.A. (1996). Regulation of neuronal birth, migration and death in the rat dentate gyrus. Developmental Neuroscience 18, 22–35.

Gould, E., Woolley, C.S., and McEwen, B.S. (1990). Short-term glucocorticoid manipulations affect neuronal morphology and survival in the adult dentate gyrus. Neuroscience 37, 367–375.

Gustafsson, B., Wigstrom, H., and Abraham, W.C. (1987). Long-term potentiation in the hippocampus using depolarizing current pulses as the conditioning stimulus to single volley synaptic potentials. Journal of Neuroscience 5, 774–780.

Hu, Z., Yuri, K., Ozawa, H., Lu, H., and Kawata, M. (1997). The in vivo time course for elimination of adrenalectomy-induced apoptotic profiles from the granule cell layer of the rat hippocampus. Journal of Neuroscience 17, 3981–3989.

Introini-Collison, I., Saghafi, D., Novack, G.D., and McGaugh, J.L. (1992). Memory-enhancing effects of post-training dipivefrin and epinephrine: involvement of peripheral and central adrenergic receptors. Brain Research 572, 81–86.

Joels, M., and deKloet, E.R. (1989), Effects of glucocorticoids and norepinephrine on the excitability in the hippocampus. Science 245, 1502–1505.

Joels, M., Hesen, W., and deKloet, E.R. (1991). Mineralocorticoid hormones suppress serotonin-induced hyperpolarization of rat hippocampal CA1 neurons. Journal of Neuroscience 11, 2288–2294.

Joels, M., and deKloet, E.R. (1992). Control of neuronal excitability by corticosteroid hormones. Trends in Neuroscience 15, 25–30.

Kvetňanský, R., Pacak, K., Fukuhara, K., Viskupic, E., Hiremagalur, B., Nankova, B., Goldstein, D., Sabban, E.L., and Kopin, I.J. (1995). Sympathoadrenal system in stress. Annals of the New York Academy of Sciences 771, 131–158.

McEwen, B.S. (1994). Corticosteroids and hippocampal plasticity. Annals of the New York Academy of Sciences 746, 134–142.

McGaugh, J.L., Cahill, L., and Roozendaal, B. (1996). Involvement of the amygdala in memory storage: Interaction with other brain systems. Proceedings of the National Academy of Sciences USA 93, 13508–13514.

Munck, A., Guyre, P.M., and Holbrook, N.J. (1984). Physiological function of glucocorticoids in stress and their relation to pharmacological actions. Endocrine Reviews 5, 25–44.

Munck, A., and Naray-Fejes-Toth, A. (1992). The ups and downs of glucocorticoid physiology. Permissive and suppressive effects revisited. Molecular and Cellular Endorinology 90, C1–C4.

Pacak, K., Kvetňanský, R., Palkovits, M., Fukuhara, K., Yadid, G., Kopin, I.J., and Goldstein, D. (1993). Adrenalectomy augments in vivo release of norepinephrine in the paraventricular nucleus during immobilization stress. Endocrinology 133, 1404–1410.

Pavlides, C., Watanabe, Y., and McEwen, B.S. (1993). Effects of glucocorticoids on hippocampal long-term potentiation. Hippocampus 3, 183–192.

Pavlides, C., Kimura, A., Magarinos, A.M., and McEwen, B.S. (1994). Type I adrenal steroid receptor prolongs hippocampal long-term potentiation. *Neuroreport* **5**, 2673–2677.

Plotsky, P.M. (1987). Facilitation of immunoreactive corticotropin-releasing factor secretion into the hypophysial-portal circulation after activation of catecholaminergic pathways or central norepinephrine injection. *Endocrinology* **121**, 924–930.

Sandi, C., Loscertales, C., and Guaza, C. (1997). Experience-dependent facilitating effect of corticosterone on spatial memory formation in the water maze. *European Journal of Neuroscience* **9**, 637–642.

Sapolsky, R.M., and Meaney, M.J. (1986). Maturation of the adrenocortical stress response: Neuroendocrine control mechanisms and the stress hyperresponsive period. *Brain Research Review* **11**, 65–67.

Sclabassi R.J., Eriksson, J.L., Port, R.L., Robinson, G.B., and Berger, T.W. (1988). Nonlinear systems analysis of the hippocampal perforant path-dentate projections (theoretical and interpretational considerations). *Journal of Neurophysiology* **60**, 1066–1076.

Seidenbecher, T., Reymann, K.G., and Balschun, D. (1997). A post-tetanic time window for the reinforcement of long-term potentiation by appetitive and aversive stimuli. *Proceedings of the National Academy of Sciences USA* **94**, 1494–1499.

Sloviter, R.S., Valiquette, E., Abrams, G.M., Ronk, E.C., Sollas, A.L., Paul, L.A., and Neubert, S. (1989). Selective loss of hippocampal granule cells in the mature rat brain after adrenalectomy. *Science* **243**, 535–538.

Smriga, M., Saito, H., and Nishiyama, N. (1996). Hippocampal long- and short-term potentiation is modulated by adrenalectomy and corticosterone. *Neuroendocrinology* **64**, 35–41.

Stanton, P.K., and Sarvey, J.M. (1985). Depletion of norepinephrine, but not serotonin, reduces long-term potentiation in the dentate gyrus of rat hippocampal slices. *Journal of Neuroscience* **5**, 2169–2176.

Szafarczyk, A., Malaval, F., Laurent, A., Gibaud, R., and Assenmacher, I. (1987). Further evidence for a central stimulatory action of catecholamines on adrenocorticotropin release in the rat. *Endocrinology* **121**, 883–892.

Weil-Malherbe, H., Axelrod, J., and Tomchick, R. (1957). Blood-brain barrier for adrenaline. *Science* **129**, 1226–1228.

Woolley, C.S., Gould, E., Sakai, R.R., Spencer, R.L., and McEwen, B.S. (1991). Effects of aldosterone or RU28362 treatment on adrenalectomy-induced cell death in the dentate gyrus of the adult rat. *Brain Research* **541**, 312–315.

40. Alpha and Beta Adrenoceptor-Mediated Molecular and Electrocardiographic Responses to Emotional Stress

T. Ueyama, K. Yoshida and E. Senba

Introduction

Emotional stress and the associated activation of the sympatho-adrenergic system underlie cardiac incidents. Immobilization (IMO) of rats is a useful model of emotional stress, since it activates the sympatho-adrenergic system and hypothalamo-pituitary-adrenocortical axis (Kvetňanský et al., 1979).

When cells are confronted with changes in their environment, a set of genes and their proteins are rapidly induced. Immediate-early genes (IEGs)/proto-oncogenes such as c-fos and c-jun (Senba and Ueyama, 1997) and heat shock proteins (HSP) (Itoh and Tashima, 1991) are the typical stress-induced substances. Fos family and Jun family proteins form hetero-(Fos-Jun) or homo-(Jun-Jun) dimers (AP-1), which may elicit expression of target genes and contribute to the cellular response to the primary stimuli. HSPs are also important since they act as molecular chaperones to prevent the misfolding of cellular proteins. In situ hybridization histochemistry (ISH) is a very effective method to detect discrete cells activated by stress. Monitoring mRNAs for IEGs or HSPs has demonstrated the spatial and temporal spread of excitation from the whole body to the intracellular molecules (Senba and Ueyama, 1997).

By use of ISH and IMO model, we demonstrated the molecular and electrocardiographic changes, which are mediated by activation of alpha and beta adrenoceptors in the heart.

Methods

Male Wistar rats, 6 weeks old, were restrained by securing them on their back to a board using adhesive tape. The rats were immediately decapitated at 15, 30, 45, 60, and 120 min from the start of IMO. Other rats were also decapitated at 1 h, 4 h after the end of 2 h IMO. The heart was rapidly removed and immediately frozen. ECG under IMO was also monitored. After administration of prazosin (1 mg/kg, po, 45 min before), metoprolol (10 mg/kg, ip, 10 min before), amosulalol (50 mg/kg, po, 60 min before), the rats were exposed to IMO for 20 min and the hearts were removed.

The hearts of rats were perfused initially for 10–20 min with a modified Krebs-Henseleit solution at a constant pressure of 80 cm H_2O as described previously (Mizukami et al., 1997). An alpha 1-agonist, phenylephrine (10 μM), a beta-agonist, isoproterenol (0.1 μM), or the combination of these drugs were added to the perfusion medium. After drug administration for 15 min, the hearts were immediately frozen.

Frozen sections of 10 μm in thickness were cut in a cryostat and thaw-mounted onto silane-coated slides. The probe sequence for c-fos (45-mer) was complementary to the nucleotides spanning amino acids 1–15. The probes for HSP70 (30-mer), Atrial natriuretic peptide (ANP; 44-mer and 45-mer) and Brain natriuretic peptide (BNP; 42-mer and 36-mer) were complementary to nucleotides 863–892, 472–515 and 231–275, 214–255 and 358–393. ISH was preformed as described previously (Ueyama et al., 1996).

Results

The elevation of the ST segment at II, III, aVF and precordial leads occurred soon after the onset of IMO. ECG returned to the pre-IMO basal level soon after the end of IMO, suggesting that IMO induces a transient change just like vasospastic angina. Pretreatment with a combination of prazosin and metoprolol or with amosulalol completely abolished the stress-induced tachycardia and the elevation of ST segment (Figure 1). Vasodilators such as nitroglycerine or diltiazem were ineffective (data not shown).

Signals for c-fos mRNA were observed from 15 min to 45 min and they returned to control levels at 60 min under IMO. They were strongly expressed in the myocardium in the area surrounding the left ventricular cavities and in the smooth muscle cells of

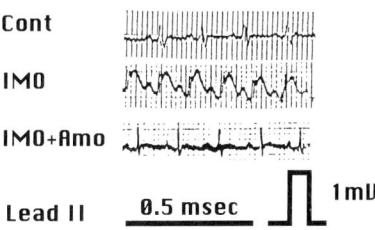

Fig. 1 ECG recordings of control (Cont), stressed (IMO; immobilization) and stressed + amosulalol pretreated rats (Amo).

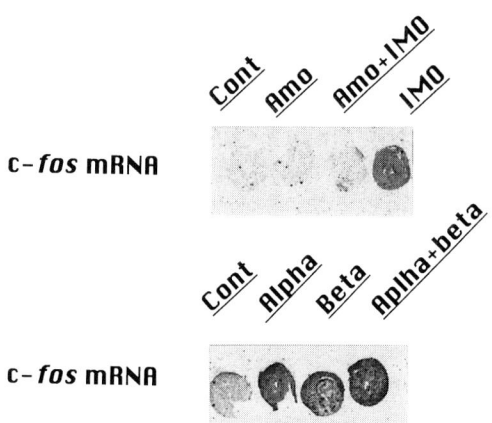

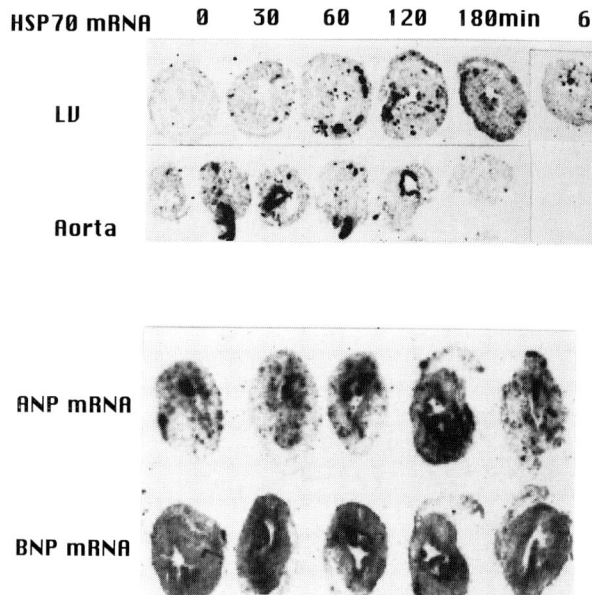

Fig. 2 The upper column demonstrates film autoradiography showing the expression of *c-fos* mRNA in response to immobilization stress with or without pretreatment of amosulalol. The lower column indicates film autoradiography showing the expression of *c-fos* mRNA in the heart perfused with buffer alone (Cont), or with buffer containing phenylephrine (Alpha), isoproterenol (Beta), or phenylephrine plus isoproterenol (Alpha/Beta).

Fig. 3 Film autoradiography showing the time-course for expression of HSP, ANP and BNP mRNAs in the heart. Axial sections of the heart were taken from an unstressed animal (0) and animals subjected to immobilization stress for 30, 60, 120 min, and those sacrificed at 1 h (180 min), 4 h (6 h) after the end of 2 h stress.

coronary arteries. Pretreatment with prazosin and metoprolol, or with amosulalol, abolished the stress-induced expression of *c-fos* mRNA (Figure 2). Treatment of perfused heart with phenylephrine, isoproterenol, or their combination, elicited a strong expression of *c-fos* mRNA (Figure 2).

HSP70 mRNA was observed from 30 min to 6 h, and the maximum expression was at 180 min (Figure 3). Strong signals were detected in the smooth muscle cells of aorta and coronary arteries. They were scattered in the myocardium of left ventricle. Signals for ANP and BNP mRNAs were observed in the control heart, while they were upregulated in response to IMO and reached a maximum at 3h from the onset of IMO (Figure 3).

Discussion

IMO of rats induced a transient change in ECG just like a vasospastic angina and the upregulation of *c-fos* mRNA in the discrete types of cardiac cells. As the induction of *c-fos* mRNA indicates cellular activation, it is suggested that this stress activates individual types of cardiac cells.

Failure of the potent vasodilator to inhibit these ECG changes and expression of *c-fos* mRNA suggests that ischemia is not responsible for them. Complete inhibition of the stress-induced changes by combined blockade of alpha- and beta-adrenoceptors indicates that blocking both alpha 1- and beta 1-adrenoceptors is essential. Perfusion studies indicate that activation of either alpha- or beta-adrenoceptors is sufficient to mimic the response evoked by IMO in the rat.

IMO induced the upregulation of ANP and BNP mRNAs in the left ventricle, where induction of *c-fos* mRNA was also observed. In addition, colocalization of

ANP mRNA and c-Fos protein was observed (data not shown), suggesting c-Fos is involved in the upregulation of these natriuretic peptides.

In conclusion, activation of alpha or beta adrenoceptors is the primary trigger of stress-induced molecular and electrophysiological changes in the heart (Figure 4). Sequential upregulation of genes can be considered as an adaptive response.

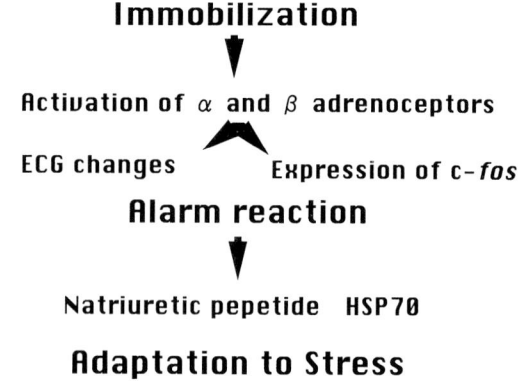

Fig. 4 A simplified cascade of events that take place in the heart of emotionally stressed rats.

References

Itoh, H., and Tashima, Y. (1991). The stress (heat shock) proteins. *International Journal of Biochemistry* 23, 1185–1191.

Kvetňanský, R., McCarty, R., Thoa, N.B., Lake, C.R., and Kopin, I.J. (1979). Sympatho-adrenal responses of spontaneously hypertensive rats to immobilization stress. *American Journal of Physiology* **236**, H457–H462.

Mizukami, Y., Yoshioka, K., Morimoto, S., and Yoshida, K. (1997). A novel mechanism of JNK1 activation. Nuclear translocation and activation of JNK1 during ischemia and reperfusion. *Journal of Biological Chemistry* **272**, 16657–16662.

Senba, E., and Ueyama, T. (1997). Stress-induced expression of immediate early genes in the brain and peripheral organs of the rat. *Neuroscience Research* **29**, 183–207.

Ueyama, T., Umemoto, S., and Senba, E. (1996). Immobilization stress induced *c-fos* and *c-jun* immediate early genes expression in the heart. *Life Sciences* **59**, 339–347.

41. Stressor-Specific Activation of Catecholaminergic Systems: Clinical Demonstrations

D.S. Goldstein, I. Elman, S.M. Frank and J.W.M. Lenders

Introduction

Current concepts in neuroendocrinology consider stress to be characterized by a unitary "stress syndrome," elicited when the intensity of any stressor exceeds a threshold. This proposition continues Selye's "doctrine of nonspecificity," which defines stress as the nonspecific response of the body to any demand. A recent study of rats (Pacak et al., 1998), however, showed that different stressors evoke markedly different neuroendocrine and central neurochemical response patterns, in a manner inconsistent with both Selye's doctrine of nonspecificity and with the notion of a unitary "stress syndrome" and supporting an alternative theory, which posits that each stressor has a neurochemical "signature," with quantitatively if not qualitatively distinct central and peripheral mechanisms. This presentation applies the concept of "primitive specificity" to humans exposed to acute glucoprivation induced by 2-deoxyglucose, simulated orthostasis induced by exposure to lower body negative pressure, or environmental cold. Glucoprivation evokes profound and selective adrenomedullary activation, as indicated by plasma epinephrine levels, with little or no sympathoneural activation, as indicated by plasma norepinephrine levels. In marked contrast, both lower body negative pressure and exposure to cold evoke mainly sympathoneural activation. The findings fit with the notion that humans maintain "thermostasis" and "barostasis" at least partly by sympathetically-mediated redistribution of blood volume, and humans maintain "glucostasis" at least partly by epinephrine-induced release of glucose by the liver. Whereas even mild hypoglycemia elicits mainly adrenomedullary activation, cold exposure probably elicits adrenomedullary activation only when the blood temperature falls, and orthostasis elicits adrenomedullary activation mainly when the blood pressure falls.

The Doctrine of Nonspecificity

In 1936, Hans Selye reported that exposure to any of several noxious agents produced the same pathological changes – adrenal enlargement, gastrointestinal ulceration, and thymicolymphatic involution (Selye, 1936). From this triad he developed a theory of stress that attained wide popularity and aroused intense research interest but also incited controversy that persists to this day. He defined stress as the non-specific response of the body to any demand (Selye, 1974), emphasizing that the same pathologic triad would result from exposure to any stressor. We call this the doctrine of nonspecificity.

Numerous studies have demonstrated different neuroendocrine responses during exposure to different stressors. This patterning does not of itself refute the doctrine of nonspecificity, because the different patterns could result from superimposition of different stressor-specific homeostatic responses on the same nonspecific stress response. According to Selye's theory, stress refers only to the nonspecific component revealed after subtraction of the specific components from the total response.

A recent study of laboratory animals, based on several experiments involving various intensities of different stressors and multiple peripheral and central neurochemical dependent measures, confirmed the marked heterogeneity of neuroendocrine responses (Pacak et al., 1998). Glucoprivation, regardless of severity, markedly stimulated adrenomedullary secretion, as reflected by plasma epinephrine levels, with much less intense sympathoneural activation, as reflected by plasma norepinephrine levels. In contrast, cold exposure produced much larger proportionate increments in plasma norepinephrine levels than in plasma epinephrine or ACTH levels.

As noted above, this heterogeneity does not of itself support or refute Selye's stress theory. To explain the heterogeneity of neuroendocrine responses to stressors, Selye hypothesized the existence of specific reactions. Only after subtraction of these from consideration could one identify the core, non-specific, shared element – the stress – and its stereotypical consequence, the stress response.

It can be shown mathematically that, without simplifying assumptions, the doctrine of nonspecificity cannot be disproved (Pacak et al., 1998). Even with simplifying assumptions, appropriate testing of the doctrine of nonspecificity would require a very complex study, involving multiple stressors at different intensities and multiple simultaneously assessed dependent measures. Although an enormous literature described effects of graded intensities of stressors on neuroendocrine dependent measures, none of this literature applied to the issue of the validity of the doctrine of nonspecificity, until a study by Pacak *et al.* (Pacak *et al.*, 1998) The doctrine of nonspecificity could be tested, given one of two assumptions about the existence of a threshold stressor intensity for the nonspecific response or about the magnitude of the specific response above the threshold stressor intensity. Data about ACTH and epinephrine responses to hemorrhage and to formalin subcutaneous injection fit the assumptions required to test the doctrine of nonspecificity. The doctrine of nonspecificity failed to predict the experimental results.

This study led to the following conclusions. Without simplifying assumptions, which may or may not be acceptable for particular stressors, the doctrine of nonspecificity is impossible to disprove and is therefore of little scientific value. Yet with acceptable simplifying assumptions, the doctrine of nonspecificity fails to predict the obtained experimental results. Given the simplifying assumptions, then, the data were inconsistent with Selye's stress theory and refuted the existence of a unitary "stress syndrome."

Primitive Specificity of Stress Responses

The results of the study of Pacak *et al.* (1998) are more consistent with an alternative proposition – that each stressor has a neurochemical "signature," with quantitatively if not qualitatively distinct central and peripheral mechanisms. These neurochemical changes would occur not in isolation but in concert with physiological, behavioral, and even experiential changes. In evolutionary terms, natural selection would have favored this patterning of stress responses, by enhancing the protection and propagation of genes (Dawkins, 1989). We call this the doctrine of "primitive specificity."

The main thesis of this presentation is that activities of stress effector systems are coordinated in relatively specific patterns, including neuroendocrine patterns. These patterns, produced by the actions of different "homeostats," serve different homeostatic needs. In stress, the organism senses a disruption or a threat of disruption of homeostasis. This sensation requires a comparative process, where the brain compares available information with setpoints for responding. Consider the analogy of a thermostat. Feedback about temperature reaches a thermostat set for a certain temperature. A sufficiently large discrepancy between the measured temperature and the set temperature turns on the furnace, and sufficient reduction of the discrepancy shuts down the furnace, keeping room temperature within a certain range, and with the average temperature corresponding to the thermostat setting.

The body has many such homeostatic comparators; they can be called "homeostats." Each homeostat compares information with a setpoint for responding, determined by a regulator (Figure 1). Homeostats typically use one or more effectors to change values for a monitored variable.

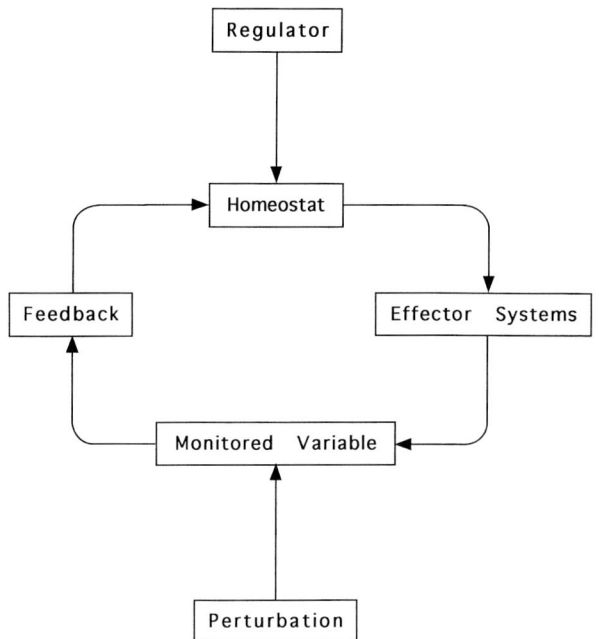

Fig. 1 A homeostatic system.

The notion of stressor-specific response patterns disagrees with the theories of both Cannon and Selye. They viewed stress responses in terms of only one effector system – in Cannon's case the "sympathico-adrenal" system and in Selye's the pituitary-adrenocortical system. Predictably, both concluded that acute exposure to any of several different stressors would produce essentially the same neuroendocrine response. Cannon, largely ignoring other systems, asserted that "sympathico-adrenal" activation meets most or all important threats to the internal environment (Cannon, 1939). According to Cannon, the neuronal and hormonal components of the sympathico-adrenal system function as a unit.

Selye overemphasized responses of the pituitary-adrenocortical system. This activation probably produces the triad of adrenal hypertrophy, gastrointestinal ulceration, and thymicolympathic degeneration but only provides a glimpse at the spectrum of systemic responses to stress. Developing a concept based on heterogeneous neuroendocrine responses during stress has required measures of several endogenous substances indicating activities of different stress systems, and descriptions of assays for these measurements appeared only long after Cannon and Selye published their unitary theories.

According to the present conception, the body's homeostats interact continuously to maintain the internal environment, via primitively specific patterns of response. Multiple effectors tend to evolve over time. Effector redundancy enhances survival, such as by compensatory activation of alternative effectors when one effector fails; and this redundancy comes at the relatively small cost of ever-increasing complexity of regulation. Effector redundancy, in turn, could have provided a basis for the evolution of adaptively advantageous, stressor-specific patterns.

Here we present patterns of effector system responses upon exposure of humans to three different stressors – glucoprivation, orthostasis, and cold. As will be seen, differences in relative extents of activation of the adrenomedullary hormonal system and sympathetic nervous system during different stress responses ironically provide some of the most convincing evidence for distinctive neuroendocrine stress response patterns. This is ironic because it was Cannon, so instrumental in describing and studying these components, who most forcefully propounded the combined and undifferentiated activation of the "sympathico-adrenal" system in response to stress (Cannon, 1929, 1939).

Three effector systems dominate in determining the body's ability to regulate blood glucose levels – insulin, glucagon, and epinephrine (Cryer, 1993). As indicated in Figure 2, the effectors are hormonal. This makes sense teleologically, in that glucoprivation threatens all the cells of the body, and local changes, such as sympathetically-induced alterations in regional distribution of blood flows, do not suffice. Predictably, then, administration of 2-deoxyglucose to normal volunteers elicits large increases in circulating levels of epinephrine, exceeding by far those in levels of norepinephrine, the sympathetic neurotransmitter (Figure 3). Regardless of the intensity of glucoprivation, the adrenomedullary response exceeds the sympathoneural response.

Exposure of humans to lower body negative pressure (LBNP) temporarily decreases venous return to the heart. This decreases cardiac stroke volume, and, probably via both cardiopulmonary and arterial baroreceptors, sympathetic neuronal outflows increase (Figure 4). Perhaps in contrast with simply standing up, where plasma norepinephrine levels and rates of peroneal sympathetic nerve traffic approximately double within minutes, LBNP elicits relatively small increases in total body norepinephrine spillover and small but statistically significant increases in total body epinephrine spillover

(Figure 5). This might reflect a degree of anxiety among healthy people exposed to this novel stressor. Of note, limb norepinephrine spillover increases during LBNP, consistent with increased sympathetic neuronal outflows to skeletal muscle.

Without an intact sympathetic nervous system, mammals cannot survive environmental cold. Stimulation of cutaneous temperature sensors and decreased temperature of the blood both act to increase sympathoneural outflows (Frank *et al.*, 1997), eliciting well-known and obvious effects such as shivering, piloerection, cutaneous vasoconstriction, and, in some species, thermogenesis. These responses all counter the initial perturbation and help maintain homeostasis of body temperature (Figure 6).

The plasma catecholamine responses of humans to cold depend importantly on the core temperature (Figure 7). When exposed to a cold environment, vasoconstriction and other sympathetically-mediated

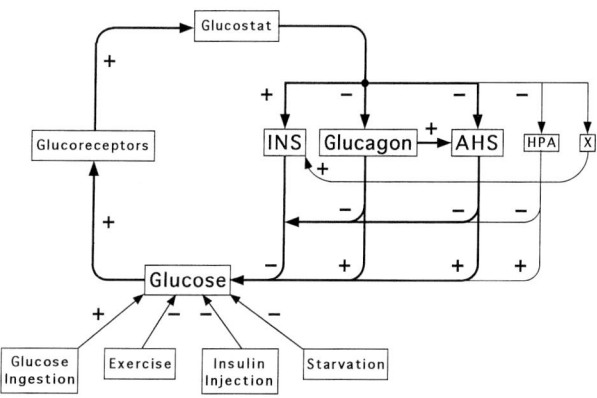

Fig. 2 Homeostatic regulation of glucose. Abbreviations: INS = insulin; AHS = adrenomedullary hormonal system; HPA = hypothalamo-pituitary-adrenocortical axis; X = vagus nerve.

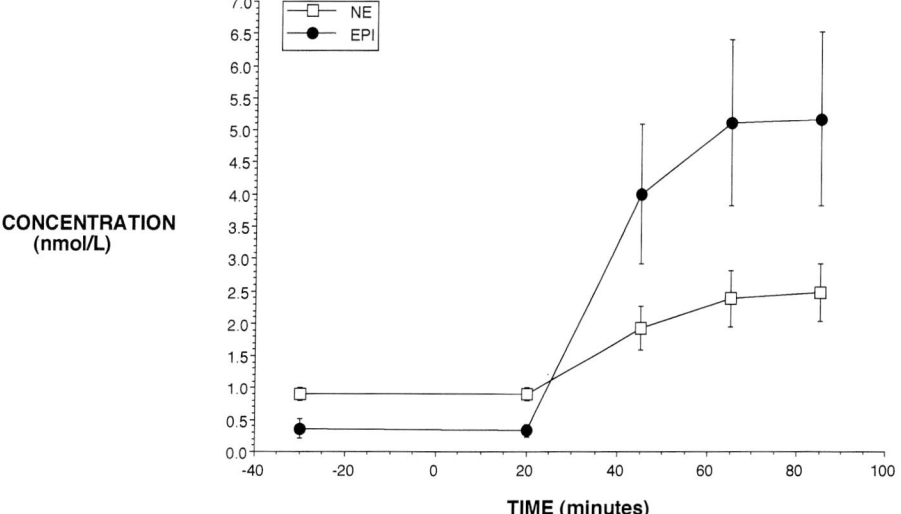

Fig. 3 Effects of 2-deoxyglucose administration on plasma levels of epinephrine (EPI) and norepinephrine (NE) in humans.

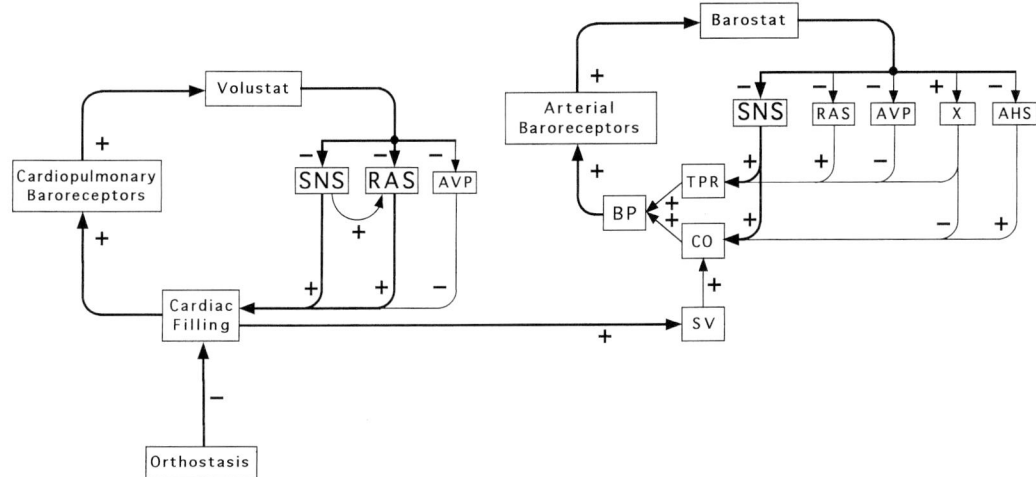

Fig. 4 Homeostatic regulation of cardiac filling and blood pressure during orthostasis. Additional abbreviations: SNS = sympathetic nervous system; RAS = renin-angiotensin-aldosterone system; AVP = arginine vasopressin; BP = blood pressure; TPR = total peripheral resistance; CO = cardiac output; SV = stroke volume.

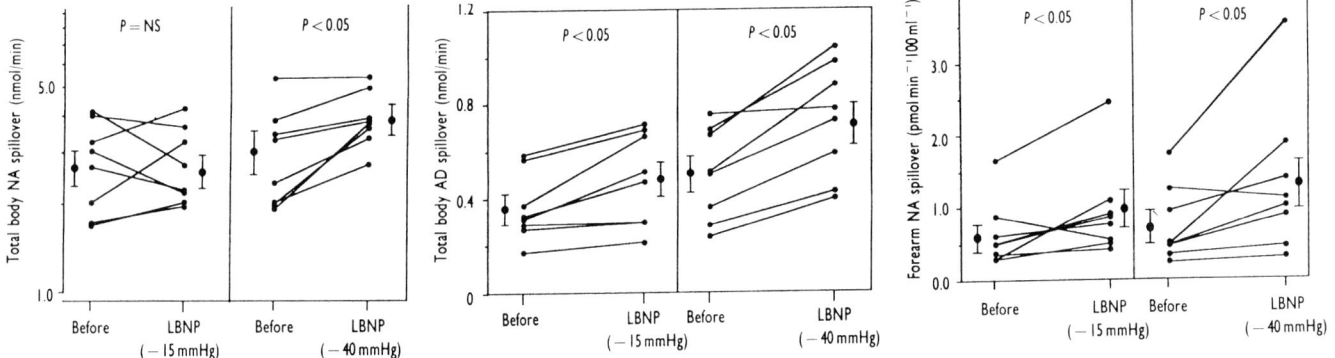

Fig. 5 Plasma catecholamine responses to lower body negative pressure. NA: norepinephrine, AD: epinephrine.

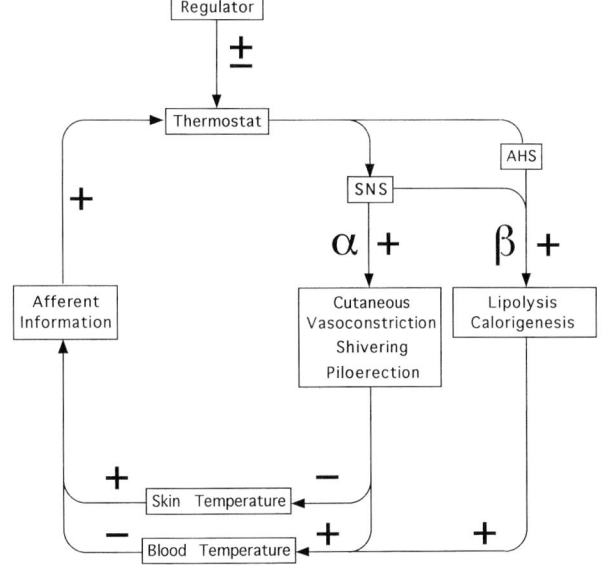

Fig. 6 Homeostatic regulation of body temperature.

responses can suffice to maintain core temperature. When they do, the plasma norepinephrine response exceeds by far the plasma epinephrine response. During a decrease in core temperature, achieved by i.v. infusion of chilled saline, then the plasma norepinephrine level increases further, but the plasma epinephrine level increases as well. These results suggest that the thermostat regulating autonomic responses to decreased skin temperature may differ from that regulating autonomic responses to decreased core temperature.

Perspective

The present findings confirm in humans the stressor-specificity of neuroendocrine patterns of response found in numerous studies of laboratory animals. The evolution of multiple, shared effectors has allowed the expression of different central neural patterns to different stressors in humans as well as in laboratory animals. Applying the concept of "primitive specificity" in studies relating neurochemical alterations in

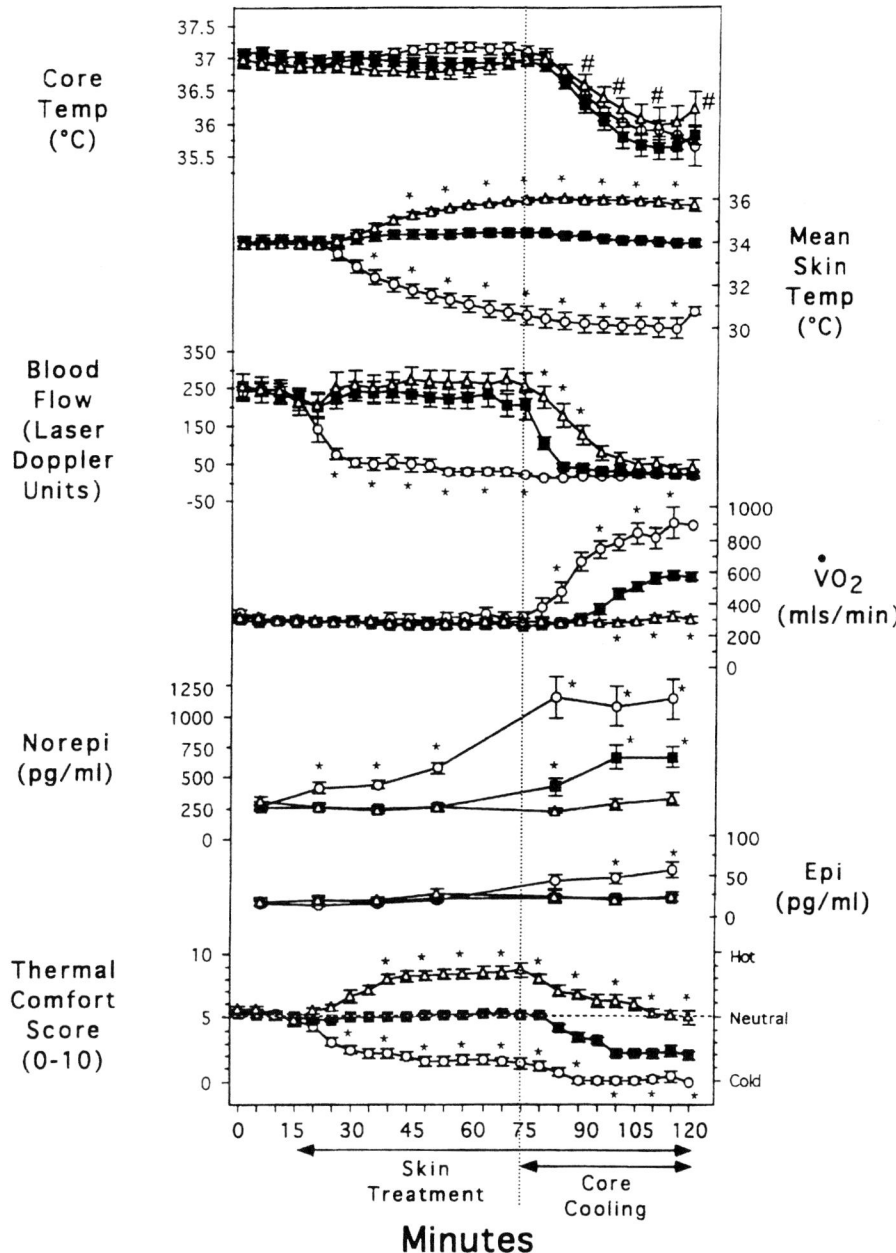

Fig. 7 Plasma catecholamine responses to external cooling and i.v. infusion of chilled saline.

particular brain pathways with simultaneously monitored dependent neuroendocrine, physiological, and behavioral variables, may provide a means to identify the components and elucidate the regulation of these patterns in stress and disease.

References

Cannon, W.B. (1929) *Bodily Changes in Pain, Hunger, Fear and Rage*. New York: D. Appleton & Co.

Cannon, W.B. (1939) *The Wisdom of the Body*. New York: W.W. Norton.

Cryer, P.E. (1993) Glucose counterregulation: prevention and correction of hypoglycemia in humans. *American Journal of Physiology* **264**, E149-E155.

Dawkins, R. (1989) *The Selfish Gene*. New York: Oxford University Press.

Frank, S.M., Higgins, M.S., Fleisher, L.A., Sitzmann, J.F., Raff, H., and Breslow M.J. (1997) Adrenergic, respiratory, and cardiovascular effects of core cooling in humans. *American Journal of Physiology* **272**, R557–R562.

Pacak, K., Palkovits, M., Yadid, G., Kvetňanský, R., Kopin, I.J., and Goldstein D.S. (1998) Heterogeneous neurochemical responses to different stressors: a test of Selye's doctrine of nonspecificity. *American Journal of Physiology* **275**, R1247–R1255.

Selye, H. (1936) A syndrome produced by diverse nocuous agents. *Nature* **138**, 32.

Selye, H. (1974) *Stress without Distress*. New York: New American Library.

42. Hypocortisolism and the Triad of Fatigue, Pain, and Stress

C. Heim and D.H. Hellhammer

Introduction

Increasing evidence from neuroendocrinological studies suggests that the adrenal gland is hypoactive in some stress-related states. This hypocortisolism represents a paradox in the context of current concepts of stress research. Stress has long been associated with activation of the hypothalamic-pituitary-adrenal (HPA) axis, resulting in increased secretion of cortisol from the adrenal glands. Early studies from the 1960s and 1970s first revealed hypocortisolism in healthy individuals who lived under conditions of ongoing stress. The phenomenon has regained considerable attention with accumulating evidence for hypocortisolism in individuals who had been exposed to a traumatic event and subsequently developed post-traumatic stress disorder (PTSD; DSM-IV 309.81). At the same time, similar findings have been reported for patients with bodily disorders, namely chronic fatigue syndrome (CFS) and fibromyalgia syndrome (FMS) among others, and these disorders have arguably been related to stress experiences (see Heim et al., 2000, for review).

Although hypocortisolism appears to be a widespread phenomenon, its precise definition and epidemiology as well as the underlying mechanisms and physiological significance remain speculative. Hypocortisolism may reflect many facets of neuroendocrine dysregulation, which may differ within and across populations, ultimately causing a lack of cortisol effects on the organism. Thus, hypocortisolism may reflect: (1) reduced synthesis or release of hypothalamic CRF or AVP; (2) reduced basal or stimulated synthesis or release of ACTH; (3) reduced basal or stimulated synthesis or release of cortisol; (4) enhanced negative feedback; (5) reduced availability of free cortisol; and (6) reduced effects of cortisol on target cells. We briefly summarize findings of hypocortisolism in populations with stress-related disorders and subsequently present evidence that hypocortisolism is not a specific correlate of these disorders, but can also be observed in subclinical states of stress, exhaustion and pain. We further propose that hypocortisolism is the neurobiological substrate of this triad and that the manifestation of clinical syndromes represents the end of the continuum. A pre-existing disposition for hypocortisolism may become manifest over time during states of prolonged stress and may be causally involved in the induction of heightened stress vulnerability and pain sensitivity. This pathophysiological model may have important implications for the prevention, diagnosis and treatment of stress-related bodily disorders.

Hypocortisolism in Stress-Related Disorders

In a remarkable series of studies, Yehuda and her colleagues have most prominently described the phenomenon of hypocortisolism for patients suffering from PTSD (for review see Yehuda, 1997). Several studies revealed decreased 24 h urinary cortisol excretion in Vietnam veterans and Holocaust survivors suffering from PTSD (Mason et al., 1986; Yehuda et al., 1990; Yehuda et al., 1993; Yehuda et al., 1995). Moreover, decreased cortisol concentrations have been measured in single plasma or saliva samples obtained from patients with PTSD (Boscarino, 1996; Goenjian et al., 1996; Stein et al., 1997). In a chronobiological study, Vietnam veterans with PTSD demonstrated a decreased nadir, but increased peak of cortisol release throughout the diurnal cycle, possibly suggesting tight regulation of the HPA axis in PTSD (Yehuda et al., 1996a). Consistent with exaggerated control of the HPA axis, Vietnam veterans with PTSD were shown to demonstrate increased numbers of glucocorticoid receptors in lymphocytes as compared to controls and other diagnostic groups, lending support for a hypothesis of increased feedback sensitivity of the HPA axis in this disorder (Yehuda et al., 1991). Thus, when given a reduced dose of dexamethasone, Vietnam veterans with PTSD demonstrated enhanced suppression of cortisol relative to controls (Yehuda et al., 1993b). The observation of exaggerated ACTH responses to the β-hydroxylase inhibitor, metyrapone, which produces a state of pharmacological "adrenalectomy" in these patients, was furthermore interpreted as reflecting increased feedback inhibition of the HPA axis in PTSD (Yehuda et al., 1996b). The latter result also shows that there is a pronounced central corticotropin-releasing activity in patients with PTSD despite their hypocortisolism. Consistently, blunted ACTH and normal cortisol responses have been reported for Vietnam veterans with PTSD and for sexually abused girls (Smith et al., 1989; DeBellis et al., 1994), and Vietnam war veterans with PTSD further show increased cerebrospinal fluid (CSF) CRF immunoreactivity (Bremner et al., 1997). Thus, in the case of PTSD, hypocortisolism may reflect a peripheral adaptation to central hyperactivity.

The phenomenon of hypocortisolism was also observed in patients with several bodily disorders, namely CFS and FMS among others. Demitrack et al., (1991) identified reduced 24 h urinary cortisol excretion, low basal plasma cortisol concentrations, blunted ACTH responses to CRF stimulation and decreased cortisol responses to maximal doses of $ACTH_{1-24}$ (250 μg) as correlates of CFS. Recently, Scott et al. (1998a) replicated the finding of blunted ACTH responses to CRF. These authors also observed blunted cortisol responses to a low dose of $ACTH_{1-24}$ (1 μg) in CFS patients (Scott et al., 1998b). In contrast, the adrenal cortex seems to be sensitized to minimal doses of $ACTH_{1-24}$ (0.01 μg/kg), possibly reflecting up-regulation of adrenal ACTH receptors as a consequence of pituitary hypoactivity or alterations at higher levels of the HPA system (Demitrack et al., 1991). Since CRF is involved in the regulation of arousal and vigilance, it has been suggested that a CRF deficiency may cause symptoms of fatigue and exhaustion (Sternberg, 1993), although, in a single study, normal CSF CRF levels were measured in patients with CFS (Demitrack et al., 1991).

Similar findings have been reported for patients with FMS and other chronic pain syndromes. Three studies found reduced 24 h urinary cortisol excretion in patients with FMS relative to healthy controls (Crofford et al., 1994; Griep et al., 1998; McCain and Tilbe, 1989). Furthermore, low morning cortisol levels have been measured in patients with idiopathic pain syndromes of diverse location (Alfvén et al., 1994; Valdés et al., 1989; von Knorring and Almay, 1989). Neuroendocrine challenge studies provide evidence for adrenocortical impairment in FMS. Patients with FMS demonstrate reduced adrenocortical reactivity in the CRF stimulation test, with ACTH responses being either normal (Crofford et al., 1994) or increased (Griep et al., 1993, 1998). Several investigators have applied a standard dexamethasone suppression test to patients with FMS. Although two studies report increased rates of non-suppressors among FMS patients (Ferraccioli et al., 1990; McCain and Tilbe, 1989), several other studies revealed strikingly low non-suppressors in this patient population (Hudson et al., 1984; Griep et al., 1993, 1998), which are even lower than non-suppressor rates in

healthy individuals (Stokes et al., 1984). Non-suppression only seems to occur in patients with chronic pain and comorbid major depression (Hudson et al., 1984). Based on these findings, one may expect that, similar to findings in PTSD, FMS and other idiopathic pain syndromes may be related to increased negative feedback sensitivity. Despite pronounced suppression of cortisol by dexamethasone, reduced glucocorticoid receptor affinity has recently been measured in lymphocytes of patients with FMS, suggesting reduced effects of already low cortisol levels in peripheral target cells (Lentjes et al., 1998). Other bodily disorders, for which hypocortisolism has been described, include rheumatoid arthritis (Chikanza et al., 1992; Hedman et al., 1992; Cash et al., 1992), asthma (Kruger and Spieker, 1994), and atypical depression (Vanderpool et al., 1991).

Taken together, different facets of hypocortisolism have been identified in diverse stress-related bodily disorders (see Table 1), although some of the facets have not been sufficiently explored in some of the disorders, i.e. adrenal reactivity in PTSD and feedback sensitivity in FMS (see Heim et al., in press). Given the considerable evidence for hypocortisolism in patients suffering from diverse stress-related disorders, one may speculate that these disorders represent a spectrum of related disorders with similar psychological and neuroendocrine correlates. We believe that all of these disorders are interconnected syndromes, which all are characterized by stress sensitivity, pain and fatigue. Thus, PTSD patients should be at higher risk for bodily disorders and bodily disorders should be related to stress experiences and enhanced risk for the development of PTSD. Indeed, there is considerable comorbidity between PTSD and the above bodily disorders (Davidson et al., 1991; Culclasure et al., 1993; Amir et al., 1997). This increased stress vulnerability may be caused by a lack of cortisol effects in peripheral target cells and at the central nervous system level, resulting in disinhibition of immune parameters and central stress reactions (see Heim et al., 2000). However, to date, there are virtually no integrative studies assessing stress history (i.e. abuse) and psychopathology (i.e. PTSD) together with HPA axis function (i.e. hypocortisolism) in patients with bodily disorders (i.e. CFS and FMS).

Table 1 Possible manifestations of hypocortisolism and their occurrence in stress-related disorders. Abbreviations: ACTH = adrenocorticotropin, AVP = arginine vasopressin, CRF = corticotropin-releasing factor, CSF = chronic fatigue syndrome, FMS = fibromyalgia syndrome, PTSD = posttraumatic stress disorder.

Manifestations of hypocortisolism	Stress-related disorders
1: reduced synthesis or release of hypothalamic CRF or AVP	CFS (?)
2: reduced (stimulation of) synthesis or release of ACTH	CFS, PTSD
3: reduced (stimulation of) synthesis or release of cortisol	CSF, FMS, PTSD
4: enhanced negative feedback	PTSD, FMS (?)
5: reduced availability of free cortisol	CFS, FMS, PTSD
6: reduced effects of cortisol on target cells	FMS

The Triad of Fatigue, Pain, and Stress

Stress and Fatigue

A role for stress in the development of CFS has long been postulated and has now been supported by evidence from empirical studies. Psychometric testing using the Minnesota Multiphasic Personality Inventory (MMPI) revealed high levels of emotional stress along with bodily complaints in patients with CFS. Actually, the authors point out that MMPI profiles in patients with CFS resemble those of patients with chronic pain. In another study, perceived emotional trauma was significantly related to symptoms of fatigue in a sample of patients with FMS (Aaron et al., 1997). Interestingly, a combat-related fatigue syndrome has recently been identified in military personnel (Takla et al., 1994) and gulf war veterans have been shown to demonstrate a higher prevalence of chronic fatigue syndrome than non-battlefield military personnel (Iowa Persian Gulf Study Group, 1997). Moreover, it has been reported that hurricane Andrew induced relapses of CFS and exacerbations of CFS symptoms in a civilian sample of patients with CFS. Most interestingly, the patients' behavioral stress response to the hurricane, i.e. the extent of shock, fear, anxiety, grief, and despair, was the single and strongest predictor of the likelihood and severity of the relapse and functional impairment (Lutgendorf et al., 1995). There also seems to be a link between sexual abuse experiences and symptoms of fatigue in female psychiatric patients and in patients with FMS (Craine et al., 1988; Alexander et al., 1998). In addition, burnout, a syndrome sharing many features with CFS, often occurs in caregivers, social service employees, hospital staff and teachers, in response to chronic stress (see Pruessner et al., 1999). Thus, taken together, there is considerable evidence for a relationship between stress experience and fatigue symptoms. In a recent study, our group has evaluated the relationship between perceived stress, burnout and HPA axis activity in a sample of 66 school teachers (Pruessner et al., 1999). As an index of HPA axis function, salivary cortisol responses to awakening were determined on three consecutive days. In the evening before the third day, 0.5 mg of dexamethasone was administered to test feedback sensitivity of the HPA axis. Cortisol responses to awakening were previously identified as a reliable indicator of adrenal reactivity with high intraindividual stability (Pruessner et al., 1997). Psychometric tests were used to assess perceived stress, burnout, self competence and physical symptoms. Results revealed that teachers scoring high on burnout showed lower overall cortisol secretion on all sampling days as well as higher suppression of cortisol by dexamethasone. Perceived stress correlated with higher increases of cortisol in the first hour after awakening after dexamethasone intake. Thus, teachers with high levels of perceived stress and high levels of burnout demonstrated lower cortisol secretion on the first two days with stronger increases during the first hour after awakening after dexamethasone. This subgroup of teachers also showed low levels of self-esteem,

external control attribution and high levels of physical complaints, i.e. pain. The latter finding is consistent with an earlier study by our group, in which nurses with burnout and high levels of bodily complaints demonstrated decreased salivary cortisol secretion in the morning (Hellhammer, 1990).

Stress and Pain

A potent role of stressful experiences has also been documented for the development of chronic pain syndromes. Thus, patients with FMS reported higher levels of major life events as well as more daily hassles than controls (Ahles et al., 1984; Dailey et al., 1990; Uveges et al., 1990). Increased levels of major life events have also been reported for women suffering from chronic pelvic pain (CPP; Harrop-Griffiths et al., 1988). Interestingly, more recently, strong associations between sexual or physical abuse and chronic pain syndromes have been identified. Patients with FMS report more often sexual abuse experiences than controls and symptom severity is related to the severity of the abuse (Boisset-Pioro et al., 1995; Taylor et al., 1995; Walker et al., 1997). Moreover, there seems to be an association between emotional trauma as well as lack of parental care and FMS (Aaron et al., 1997; Zant et al., 1997). Sexual and physical abuse experiences are also over-represented among women suffering from CPP (Walker et al., 1992; Walling et al., 1994). In addition, chronic pain syndromes of other localization have been linked to sexual or physical abuse (Rapkin et al., 1990; Goldberg et al., 1999). Also supporting a link between stress sensitivity and pain are findings of high comorbidity between PTSD and FMS and other states of chronic pain. For example, 80% of combat veterans with PTSD report chronic pain (Beckham et al., 1997). In a prospective study, history of PTSD significantly increased the risk to develop pain in the course of the study (Andreski et al., 1998). Consistently, Culclasure et al. (1993) presented a case report in which a PTSD patient turned into a FMS patient over time. Furthermore, there are numerous reports on comorbidity between PTSD and FMS in individuals exposed to trauma (Baker et al., 1982; Amir et al., 1997; Iowa Persian Gulf Study Group, 1997). Thus, there appears to be a rather strong relationship between adverse experiences and the development of pain. To explore further the relationship between stress, psychopathology, HPA axis dysfunction and pain symptoms, we performed a series of studies in women suffering from CPP. Women with CPP with no identified organic correlate (as verified by diagnostic laparoscopy) reported increased prevalence rates of sexual and physical abuse experiences and suffered more often from PTSD as compared to controls. Moreover, these women reported higher numbers of major life events than did controls. With respect to HPA axis measures, we found normal to low diurnal salivary cortisol in women with CPP. In the CRF stimulation test (100 μg hCRF), women with CPP exhibited normal plasma ACTH, but reduced salivary cortisol concentrations as compared to controls. After a low dose of dexamethasone (0.5 mg), women with

CPP demonstrated enhanced suppression of cortisol secretion over the diurnal cycle (Heim et al., 1998). Interestingly, we also measured normal to decreased GR binding in peripheral lymphocytes in these women despite low adrenal activity, suggesting an amplification of a lack of cortisol on peripheral target cells (Heim et al., 1997a). Stress and neuroendocrine abnormalities may also contribute to the development of chronic pain in the face of an organic correlate, since we recently observed increased rates of abuse along with hypocortisolism in a sample of women with CPP and verified pelvic adhesions (Heim et al., 1999a). In a subsequent study, we determined cortisol levels after awakening as well as chronic stress in the past year in women with CPP (Heim et al., 1999b). As compared to controls, women with CPP demonstrated markedly decreased salivary cortisol levels after awakening along with increased ratings of chronic stress. These findings are consistent with the above findings in burned out teachers, who showed low cortisol levels after awakening, supersuppression of cortisol and high levels of physical complaints, i.e. pain (Pruessner et al., 1999).

Pain and Fatigue

It has been suggested in the literature that FMS and CFS represent related syndromes with overlapping features. Thus, patients with FMS frequently report marked symptoms of fatigue and sleep disturbance and often meet full diagnostic criteria for CFS (Hudson and Pope, 1994; Yunus, 1994; Clauw and Chrousos, 1997). Conversely, patients with CFS exhibit marked symptoms of chronic pain, such as myalgia, arthralgia and headache, and these pain symptoms have recently been included into the definition and diagnostic criteria of CFS (Fukuda et al., 1994). In a large study comprising 554 individuals with FMS, 72% of these patients suffered from chronic fatigue as compared to 17.8% of controls. Interestingly, 66.1% of patients with FMS also suffered from CPP as compared to 23.1% of controls (Waylonis and Heck, 1992). As presented above, both pain and fatigue are frequently associated with stress experiences and PTSD. Thus, Persian Gulf war (PGW) veterans more often suffer from PTSD, FMS and chronic fatigue as compared to non-PGW military personnel (Iowa Persian Gulf Study Group, 1997). In women with FMS, sexual abuse experiences predict the severity of fatigue symptomatology (Alexander et al., 1998). Thus, it appears that several stress-related clinical syndromes represent one spectrum of disorders with the common feature of pain and fatigue and potentially similar mechanisms, one of which may be a lack of cortisol. To explore further the relationship between stress, pain, fatigue and hypocortisolism, our group evaluated stress, adrenal activity, immunological features and symptoms of fatigue in patients who developed persisting sciatic pain after discectomy (Geiss et al., 1997). Patients who demonstrated persisting sciatic pain eight weeks after surgery exhibited markedly decreased salivary cortisol levels after awakening as compared to patients with low

complaints after surgery and healthy, pain-free controls. Moreover, serum concentrations of interleukin (IL)-6 were measured in 45 minute-intervals during the day and evening. Patients with persisting pain showed significantly increased IL-6 concentrations in the evening as compared to patients with low complaints and controls and IL-6 levels were correlated with pain intensity. Importantly, none of these patients showed signs of inflammation as evidenced by normal C-reactive protein levels and body temperature. Interestingly, patients with persisting pain after discectomy reported more chronic stress related to work as well as maladaptive stress coping strategies in the year preceding the study. These patients also suffered more frequently from symptoms of fatigue and depression as compared to patients with low complaints and pain-free controls (Geiss et al., 1997).

Hypotheses on Links between Stress, Fatigue, and Pain

In general, multiple bodily complaints without an organic correlate often occurring in relation to stress experiences have been categorized as somatization disorders in the Diagnostic and Statistical Manual for Mental Disorders Fourth Edition (DSM-IV; American Psychiatric Association, 1994). While the DSM-IV does not specify a theoretical framework of this symptom constellation, several authors have formulated theoretical concepts or hypotheses on the link between stress, pain, and fatigue. For example, in a review paper, Hudson and Pope (1994) point out that FMS shares frequent comorbidity not only with CFS, but also with irritable bowel syndrome and chronic headache as well as depression and anxiety disorders, including PTSD. Since patients with all of these disorders typically respond to antidepressant treatment, the authors suggest that these disorders represent a family of "affective spectrum disorders" with a common pathophysiology. However, the authors do not specify potential biological mechanisms nor do they elaborate on the role of stress experience in the development of the spectrum of disorders.

A similar concept has been formulated by Yunus (1994), who suggests the existence of a "dysfunctional spectrum syndrome", including FMS, CFS, irritable bowel syndrome, headache, dysmenorrhea, and restless leg syndrome among others. The author develops a theoretical model, in which stress, trauma, genetic disposition, infection, and inflammation promote a heterogeneous neuroendocrine-immune dysfunction, which ultimately alters central mechanisms involved in the mediation of pain. This central dysfunction is supposed to cause fatigue and pain amplification. The author emphasizes that the spectrum of disorders is different from depression in terms of neurobiological changes.

In a more recent publication, Clauw and Chrousos (1997) elegantly delineate the overlapping features of

FMS and CFS and relate these disorders with the concept of somatization disorder and multiple chemical sensitivity and other allied symptoms. The authors emphasize that a dualism of psychiatric versus physical causes may be inappropriate in the definition of these disorders. Similar to the above concept, they propose that genetic disposition along with environmental triggers and psychological co-factors may result in a sustained dysregulation of the human stress response. The dysregulation of the human stress response is thought to be characterized by (a) loss of descending pain inhibitory function, (b) low autonomic basal activity and reactivity to stress, and (c) blunted neuroendocrine responsiveness of the hypothalamic-pituitary axis to stress. In particular, lower than normal central CRH release along with low sympathetic ouput is causally implicated in the development of fatigue symptoms and increased nociception. The authors suggest that differential emphasis of alterations in these neurobiological systems may cause the tremendous heterogeneity in symptoms seen in the spectrum of fatigue and pain syndromes.

We have recently suggested a hypothesis relating to the potential role of hypocortisolism in the development of stress-related bodily disorders (Heim et al., in press). We assume that chronic or traumatic stress may elicit sustained hypocortisolism in individuals predisposed due to genetic vulnerability or early adverse experiences. Such individuals may then be at greater risk to develop stress-related bodily disoders due to a lack of the protective effects of cortisol on target cells. Cortisol is known to possess pronounced immunosuppressive properties and also is involved in the regulation of the synthesis of pain substances, i.e. prostaglandins. Thus, reduced effects of the protective stress hormone, cortisol, in targets cells may promote the manifestation of autoimmune disorders, inflammation, chronic pain syndromes and other immune-related disorders. Cortisol is also known to exert effects on central structures, such as the locus coeruleus, which is the major source of noradrenergic neurons in the brain and which has been implicated in the mediation of anxiety (see Heim et al., 1997b). A lack of inhibitory control of these noradrenergic neurons may induce increased central stress sensitivity and may promote the development of stress-related psychiatric disorders, namely PTSD. Thus, the frequent co-occurrence of pain, fatigue and stress-sensitivity may be the result of the common feature of hypocortisolism.

Taken together, several authors have pointed out the relationship between pain and fatigue symptoms as well as other allied conditions on a descriptive level (Hudson and Pope, 1994; Yunus, 1994). In addition, specific hypotheses on the underlying pathophysiology have been developed recently, one focusing on central alterations (Clauw and Chrousos, 1997) and one focusing on the effects of hypocortisolism in target cells (Heim et al., 2000).

Is there a Time Course in the Development of Hypocortisolism?

It has been suggested in the literature that trauma and chronic stress may not be considered as distinct events, since a single traumatic event may induce a prolonged stress experience due to recurrent memories and continuous appraisals of situations as being threatening (see Baum et al., 1993). One may assume that hypocortisolism is a correlate of prolonged stress experiences, which develops over a time course similar to an exhaustion. We believe that a subset of individuals already exhibit low cortisol levels before the onset of a stressor and that such predisposed individuals are then at higher risk to develop stress-related pathology. Predisposing factors for hypocortisolism may be genetic determination and stress in early stages of development, both of which have been implicated in the development of PTSD and stress-related bodily disorders and both of which have been shown to set the level of HPA axis activity (see Heim et al., 2000). Predisposed individuals with hypocortisolism may activate the HPA axis in response to the onset of the stressor, but the stress response may be lower than normal. Over a time course of continuous stress and adaptation, the HPA axis may down-regulate its receptors at different levels to counterbalance the activation, ultimately resulting in even lower cortisol levels. At this point, symptoms of fatigue, pain and stress may occur. Interestingly, it has been shown that individuals who demonstrate low cortisol responses immediately after a traumatic experience are at greater risk to develop subsequent PTSD as compared to individuals who show higher cortisol responses after trauma (see Yehuda et al., 1998).

Future Directions

We have summarized the available evidence for the existence of the phenomenon of hypocortisolism in several disorders, including PTSD, CFS, and FMS. Symptoms of fatigue and pain both have been related to stressful experiences and there are significant rates of comorbidity between CFS, FMS, and PTSD. We furthermore presented evidence that hypocortisolism is not a specific correlate of these syndromal disorders, but may also be observed in sub-clinical states of stress, exhaustion and pain. We propose that hypocortisolism is the neurobiological substrate of this triad and that the manifestation of clinical syndromes represents the end of the continuum. As such, we suggest that hypocortisolism may be causally relevant for the pathophysiology of bodily disorders, i.e. pain and fatigue, due to a lack of protective effects of cortisol on target cells. The neuroendocrine abnormality may develop over a time course and may be influenced by predisposing factors, such as genetic vulnerability and stress early in life.

Future studies should be directed towards the elucidation of the mechanisms of hypocortisolism. Alterations at several levels of the stress hormone axis may contribute to the development of reduced adrenal activity. Moreover, reduced availability of free cortisol

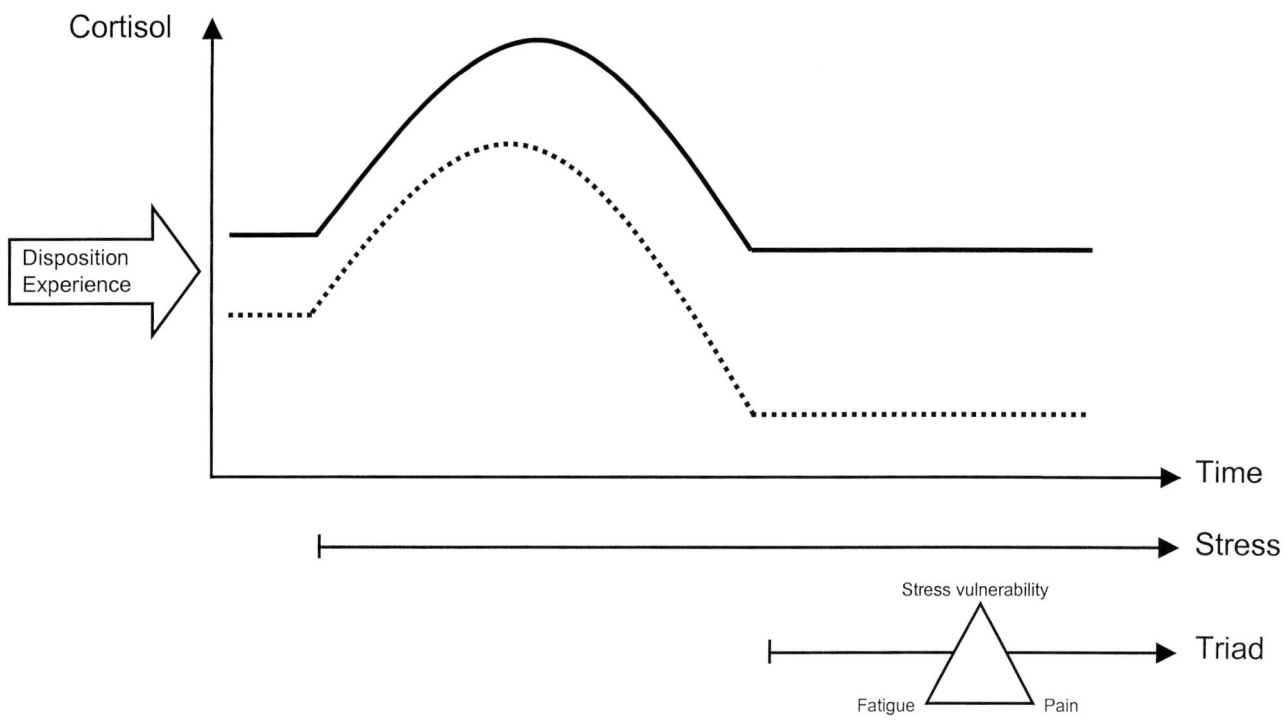

Fig. 1 Hypothetical time course of cortisol secretion during chronic stress resulting in hypocortisolism and an enhanced vulnerability for stress, pain, and fatigue in a subset of individuals with predisposition.

and impaired sensitivity of target cells (potentially due to GR dysregulation) may contribute to a hypocortisolism-like state. Mechanisms of hypocortisolism may be complex and heterogeneous between and within clinical and sub-clinical populations. Moreover, mechanisms may be differentially expressed over a time course of development. For example, enhanced feedback inhibition of the HPA axis may over time result in adrenal atrophy and insufficiency. Another goal for future research is to prove a causal role of hypocortisolism, if present, in the pathophysiology of stress-related disorders. Thus, the effects of hypocortisolism on target tissues need to be investigated in individuals with PTSD, CFS, and FMS and related sub-clinical states. Moreover, the consequences of hypocortisolism on central structures involved in the mediation of stress responses and anxiety should be further considered. Finally, efficiency of glucocorticoid replacement in the treatment of both bodily symptoms and symptoms of stress and anxiety should be investigated. Such studies may have important implications for our understanding of this complex triad of symptoms and for the implementation of therapeutic approaches to alleviate such symptoms.

References

Aaron, L.A., Bradley, L.A., Alarcon, G.S., Triana-Alexander, M., Alexander, R.W., Martin, M.Y., and Alberts, K.R. (1997). Perceived physical and emotional trauma as precipitating events in fibromyalgia. Associations with health care seeking and disability status but not pain severity. *Arthritis and Rheumatism* **40**, 453–460.

Ahles, T.A., Yunus, M.B., Riley, S.D., Bradley, J.M., and Masi, A.T. (1984). Psychological factors associated with primary fibromyalgia syndrome. *Arthritis and Rheumatism* **24**, 1102–1106.

Alexander, R.W., Bradley, L.A., Alarcon, G.S., Triana-Alexander, M., Aaron, L.A., Alberts, K.R., Martin, M.Y., and Stewart, K.E. (1998). Sexual and physical abuse in women with fibromyalgia: association with outpatient health care utilization and pain medication usage. *Arthritis Care Research* **11**, 102–115.

Alfvén, G., de la Torre, B., and Uvnäs-Moberg, K. (1994). Depressed concentrations of oxytocin and cortisol in children with recurrent abdominal pain of non-organic origin. *Acta Paediatrica* **83**, 1076–1080.

Amir, M., Kaplan, Z., Neumann, L., Sharabani, R., Shani, N., and Buskila, D. (1997). Posttraumatic Stress Disorder, tenderness, and fibromyalgia. *Journal of Psychosomatic Research* **42**, 607–613.

Andreski, P., Chilcoat, H., and Breslau, N. (1998). Post-traumatic stress disorder and somatization symptoms: a prospective study. *Psychiatry Research* **15**, 131–138.

Baker, D., Mendenhall, C.L., Simbratl, L.A., Magan, L.K., and Steinberg, J.L. (1982). Relationship between posttraumatic stress disorder and self-reported physical symptoms in Persian gulf war veterans. *Archives of Internal Medicine* **157**, 2076–2078.

Baum, A., Cohen, L., and Hall, M. (1993). Control and intrusive memories as possible determinants of chronic stress. *Psychosomatic Medicine* **55**, 274–286.

Beckham, J.C., Crawford, A.L., Feldman, M.E., Kirby, A.C., Hertzberg, M.A., Davidson, J.R., and Moore, S.D. (1997). Chronic posttraumatic stress disorder and chronic pain in Vietnam combat veterans. *Journal of Psychosomatic Research* **43**, 379–389.

Boisset-Pioro, M.H., Esdaile, J.M., and Fitzcharles, M.A. (1995). Sexual and physical abuse in women with fibromyalgia syndrome. *Arthritis and Rheumatism* **38**, 235–241.

Boscarino, J.A. (1996). Posttraumatic stress disorder, exposure to combat, and lower plasma cortisol among Vietnam veterans: Findings and clinical implications. *Journal of Clinical and Consulting Psychology* **64**, 191–201.

Bremner, J.D., Licino, J., Darnell, A., Krystal, J.H., Owens, M.J., Southwick, S.M., Nemeroff, C.B., and Charney, D.S. (1997). Elevated CSF corticotropin-releasing factor concentrations in PTSD. *American Journal of Psychiatry* **154**, 624–629.

Cash, J.M., Crofford, L.J., Galucci, W.T., Sternberg, E.M., Gold, P.W., Chrousos, G.P., and Wilder, R.L. (1992). Pituitary-adrenal axis responsiveness to ovine corticotropin releasing hormone in patients with rheumatoid arthritis treated with low dose prednisone. *Journal of Rheumatology* **19**, 1692–1696.

Chikanza, I.C., Petrou, P., Chrousos, G.P., Kingsley, G., and Panayi, G.P. (1992). Defective hypothalamic response to immune and inflammatory stimuli in patients with rheumatoid arthritis. *Arthritis and Rheumatism* **35**, 1281–1288.

Clauw, D.J., and Chrousos, G.P. (1997). Chronic pain and fatigue syndromes: overlapping clinical and neuroendocrine features and potential pathogenic mechanisms. *Neuroimmunomodulation* **4**, 134–153.

Craine, L.S., Henson, C.E., Colliver, J.A., and MacLean, D.G. (1988). Prevalence of a history of sexual abuse among female psychiatric patients in a state hospital system. *Hospital and Community Psychiatry* **39**, 300–304.

Crofford, L.J., Pillemer, S.R., Kalogeras, K.T., Cash, J.M., Michelson, D., Kling, M.A., Sternberg, E.M., Gold, P.W., Chrousos, G.P., and Wilder, R.L. (1994). Hypothalamic-pituitary-adrenal axis perturbations in patients with fibromyalgia. *Arthritis and Rheumatism* **37**, 1583–1592.

Culclasure, T.F., Enzenauer, R.J., and West, S.G. (1993). Posttraumatic stress disorder presenting as fibromyalgia. *American Journal of Medicine* **94**, 548.

Dailey, P.A., Bishop, G.D., Russell, I.J., and Fletcher, E.M. (1990). Psychological stress and the fibrositis/fibromyalgia syndrome. *Journal of Rheumatology* **17**, 1380–1385.

Davidson, J.R.T., Hughes, D., Blazer, D.G., and George, L.K. (1991). Posttraumatic stress disorder in the community: an epidemiological study. *Psychological Medicine* **21**, 713–721.

DeBellis, M.D., Chrousos, G.P., Dorn, L.D., Burke, L., Helmers, K., Kling, M.A., Trickett, P.K., and Putnam, F.W. (1994). Hypothalamic pituitary adrenal axis dysregulation in sexually abused girls. *Journal of Clinical Endocrinology and Metabolism* **78**, 249–255.

Demitrack, M.A., Dale, J.K., Straus, S.E., Laue, L., Listwak, S.J., Kruesi, J.P., Chrousos, G.P., and Gold, P.W. (1991). Evidence for impaired activation of the hypothalamic-pituitary-adrenal axis in patients with chronic fatigue syndrome. *Journal of Clinical Endocrinology and Metabolism* **73**, 1224–1234.

Ferraccioli, G., Cavalieri, F., Salaffi, F., Fonatana, S., Scita, F., Nolli, M., and Maestri, D. (1990). Neuroendocrinologic findings in primary fibromylagia and in other chronic rheumatic conditions. *Journal of Rheumatology* **17**, 869–873.

Fukuda, K., Straus, S.E., Hickie, I., Sharpe, M.C., Dobbins, J.G., and Komaroff, A. (1994). The chronic fatigue syndrome: a comprehensive approach to its definition and study. International Chronic Fatigue Syndrome Study Group. *Annals of Internal Medicine* **121**, 953–959.

Geiss, A., Varadi, E., Steinbach, K., Bauer, H.W., and Anton, F. (1997). Psychoneuroimmunological correlates of persisting sciatic pain in patients who underwent discectomy. *Neuroscience Letters* **237**, 65–68.

Goenjian, A.K., Yehuda, R., Pynoos, R.S. *et al.* (1996). Basal cortisol and dexamethasone suppression of cortisol among adolescents after the 1988 earthquake in Armenia. *American Journal of Psychiatry* **153**, 929–934.

Goldberg, R.T., Pachas, W.N., and Keith, D. (1999). Relationship between traumatic events in childhood and chronic pain. *Disability and Rehabilitation* **21**, 23–30.

Griep, E.N., Boersma, J.W., and deKloet, E.R. (1993). Altered reactivity of the hypothalamic-pituitary-adrenal axis in the primary fibromyalgia syndrome. *Journal of Rheumatology* **20**, 469–474.

Griep, E.N., Boersma, J.W., Lentjes, E.G., Prins, P.A., van der Korst, K.K., and deKloet, E.R. (1998). Function of hypothalamic-pituitary-adrenal axis in patients with fibromyalgia and low back pain. *Journal of Rheumatology* **25**, 1374–1381.

Harrop-Griffiths, J., Katon, W., Walker, E., Holm, L., Russo, J., and Hickok, L. (1988). The association between chronic pelvic pain, psychiatric diagnoses, and childhood sexual abuse. *Obstetrics and Gynecology* **71**, 589–594.

Hedman, M., Nilsson, E., and de la Torre, B. (1992). Low blood and synovial fluid levels of sulpho-conjugated steroids in rheumatoid arthritis. *Clinical and Experimental Rheumatology* **10**, 25–30.

Heim, C., Ehlert, U., Rexhausen, J., Hanker, J.P., and Hellhammer, D.H. (1997a). Psychoendocrinological observations in women with chronic pelvic pain. *Annals of the New York Academy of Sciences* **821**, 456–458.

Heim, C., Owens, M.J., Plotsky, P.M., and Nemeroff, C.B. (1997b). The role of early adverse life events in the etiology of major depression and posttraumatic stress disorder: focus on corticotropin-releasing factor. *Annals of the New York Academy of Sciences* **821**, 194–207.

Heim, C., Ehlert, U., Hanker, J.P., and Hellhammer, D.H. (1998). Abuse-related posttraumatic stress disorder and alterations of the hypothalamic- pituitary-adrenal axis in women with chronic pelvic pain. *Psychosomatic Medicine* **60**, 309–318.

Heim, C., Ehlert, U., Hanker, J.P., and Hellhammer, D.H. (1999a). Psychological and endocrine correlates of chronic pelvic pain associated with adhesions. *Journal of Psychosomatic Obstetrics and Gynecology* **20**, 11–20.

Heim, C., Stender, S., Ehlert, U., and Hellhammer D.H. (1999b). Hypocortisolism in chronic pelvic pain: Effects of glucocorticoid replacement therapy on pain symptomatology. International Society for Psychoneuroendocrinology XXX. *Conference Abstracts*, Orlando, Florida.

Heim, C., Ehlert , U., and Hellhammer, D.H. (2000). The potential role of hypocortisolism in the pathophysiology of stress-related bosily disorders. *Psychoneuroendocrinology* **25**, 1–35.

Hellhammer, J. (1990). Burnout bei Pflegepersonal – eine psychoendokrinologische Untersuchung. Diploma Thesis, University of Trier, Germany.

Hudson, J.I., and Pope, H.G. (1994). The concept of affective spectrum disorder: relationship of fibromyalgia and other syndromes of chronic fatigue and chronic muscle pain. *Bailliere's Clinical Rheumatology* **4**, 839–856.

Hudson, J.I., Pliner, L.F., Hudson, M.S., Goldenberg, D.L., and Melby, J.C. (1984). The dexamethasone suppression test in fibrositis. *Biological Psychiatry* **19**, 1489–1493.

Iowa Persian Gulf Study Group (1997). Self reported illness and health status among Gulf War veterans. A population-based study. *Journal of the American Medical Association* **277**, 238–245.

Kruger, U., and Spiecker, H. (1994). Die Diagnostik der Nebennierenrindeninsuffizienz bei steroidpflichtigem Asthma bronchiale: Der CRH-Test im Vergleich zu Kortisol Tagesprofil im Serum und Kortisol im 24h-Urin. *Pneumologie* **48**, 793–789.

Lentjes, E.G., Griep, E.N., Boersma, J.W., Romijn, F.P., and de Kloet, E.R. (1998). Glucocorticoid receptors, fibromyalgia and low back pain. *Psychoneuroendocrinology* **22**, 603–614.

Lutgendorf, S.K., Antoni, M.H., Ironson, G., Fletcher, M.A., Penedo, F., Baum, A., Schneiderman, N., and Klimas N. (1995). Physical symptoms of chronic fatigue syndrome are exacerbated by the stress of Hurricane Andrew. *Psychosomatic Medicine* **57**, 310–323.

Mason, J.W., Giller, E.L., Kosten, T.R., Ostroff, R.B., and Podd, L. (1986). Urinary free-cortisol levels in post-traumatic stress disorder patients. *Journal of Nervous and Mental Disease* **174**, 145–159.

McCain, G.A., and Tilbe, K.S. (1989). Diurnal hormone variation in fibromyalgia syndrome: A comparison with rheumatoid arthritis. *Journal of Rheumatology* **16**, 154–157.

Pruessner, J.C., Wolf, O.T., Hellhammer, D.H., Buske-Kirschbaum, A., von Auer, K., Jobst, S., Kaspers, F., and Kirschbaum, C. (1997). Free cortisol levels after awakening: a reliable biological marker for the assessment of adrenocortical activity. *Life Sciences* **61**, 2539–2549.

Pruessner, J., Hellhammer D.H., and Kirschbaum, C. (1999). Burnout, perceived stress and salivary cortisol upon awakening. *Psychosomatic Medicine* **61**, 197–204.

Rapkin, A.J., Kames, L.D., Darke, L.L., Stampler, F.M., and Naliboff, B.D. (1990). History of physical and sexual abuse in women with chronic pelvic pain. *Journal of Obstetrics and Gynecology* **76**, 92–96.

Scott, L.V., Medbak, S., and Dinan, T.G. (1998a). Blunted adrenocorticotropin and cortisol responses to corticotropin-releasing hormone stimulation in chronic fatigue syndrome. *Acta Psychiatrica Scandinavica* **97**, 450–457.

Scott, L.V., Medbak, S., and Dinan, T.G. (1998b). The low dose ACTH test in chronic fatigue syndrome and in health. *Clinical Endocrinology* **48**, 733–737.

Smith, M.A., Davidson, J., Ritchie, J.C., Kudler, H., Lipper, S., Chappell, P., and Nemeroff, C.B. (1989). The corticotropin-releasing hormone test in patients with posttraumatic stress disorder. *Biological Psychiatry* **26**, 349–355.

Stein, M.B., Yehuda, R., Koverola, C., and Hanna, C. (1997). Enhanced dexamethasone suppression of plasma cortisol in adult women traumatized by childhood sexual abuse. *Biological Psychiatry* **42**, 680–686.

Sternberg, E.M. (1993). Hypoimmune fatigue syndromes: diseases of the stress response? *Journal of Rheumatology* **20**, 418–421.

Stokes, P.E., Stoll, P.M., Koslow, S.H., Maas, J.W., Davis, J.M., Swann, A.C., and Robins, E. (1984). Pretreatment DST and hypothalamic pituitary adrenocortical function in depressed patients and comparison groups. A multicenter study. *Archives of General Psychiatry* **41**, 257–267.

Takla, N.K., Koffman, R., and Bailey, D.A. (1994). Combat stress, combat fatigue, and psychiatric disability in aircrew. *Aviation Space and Environmental Medicine* **65**, 858–865.

Taylor, M.L., Trotter, D.R., and Csuka, M.E. (1995). The prevalence of sexual abuse in women with fibromyalgia. *Arthritis and Rheumatism* **38**, 229–234.

Uveges, J.M., Parker, J.C., Smarr, K.L., McGowan, J.F., Lyon, M.G., Irvin, W.S., Meyer, A.A., Buckelew, S.P., Morgan, R.K., and Delmonico, R.L. (1990). Psychological symptoms in primary fibromyalgia syndrome: relationship to pain, life stress, and sleep disturbance. *Arthritis and Rheumatism* **33**, 1279–1283.

Valdés, M., Garcia, L., Treserra, J., de Pablo, J and de Flores, T. (1989). Psychogenic pain and depressive disorders: an empirical study. *Journal of Affective Disorders* **16**, 21–25.

Vanderpool, J., Rosenthal, N., Chrousos, G.P., Wher, T., and Gold, P.W. (1991). Evidence for hypothalamic CRH deficiency in patients with seasonal affective disorder. *Journal of Clinical Endocrinology and Metabolism* **72**, 1382–1387.

von Knorring, L., and Almay, B.G.L. (1989). Neuroendocrine responses to fenfluramine in patients with idiopathic pain syndromes. *Nordisk Psykiatrisk Tidsskrift* **43**, 61–65.

Walker, E.A., Katon, W.J., Hansom, J., Harrop-Griffiths, J., Holm, L., Jones, M.L., Hickok, L., and Jemelka, R.P. (1992). Medical and psychiatric symptoms in women with childhood sexual abuse. *Psychosomatic Medicine* **54**, 658–664.

Walker, E.A., Keegan, D., Gardner, G., Sullivan, M., Bernstein, D., and Katon, W.J. Psychosocial factors in fibromyalgia compared with rheumatoid arthritis: II. Sexual, physical, and emotional abuse and neglect. *Psychosomatic Medicine* **59**, 572–577.

Walling, M.K., Reiter, R.C., O'Hara, M.W., Milburn, A.K., Lilly, G., and Vincent, S.D. (1994). Abuse history and chronic pain in women: I. Prevalences of sexual abuse and physical abuse. *Journal of Obstetrics and Gynecology* **84**, 193–199.

Waylonis, G.W., and Heck, W. (1992). Fibromyalgia Syndrome. New Associations. *American Journal of Physical Medicine and Rehabilitation* **71**, 343–348.

Yehuda, R., Southwick, S.M., Nussbaum, G., Wahby, V., Giller, E.L., and Mason, J.W. (1990). Low urinary cortisol excretion in patients with posttraumatic stress disorder. *Journal of Nervous and Mental Disease* **178**, 366–369.

Yehuda, R., Lowy, M.T., Southwick, S.M., Shaffer D., and Giller, E.L. (1991b). Increased number of glucocorticoid receptor in posttraumatic stress disorder. *American Journal of Psychiatry* **148**, 499–504.

Yehuda, R., Boisoneau, D., Mason, J., and Giller, E.L. (1993a). Glucocorticoid receptor number and cortisol excretion in mood, anxiety, and psychotic disorders. *Biological Psychiatry* **34**, 18–25.

Yehuda, R., Southwick, S.M., Krystal, J.H., Bremner, D., Charney, D.S. and Mason, J.W. (1993b). Enhanced suppression of cortisol following dexamethasone administration in posttraumatic stress disorder. *American Journal of Psychiatry* **150**, 83–86.

Yehuda, R., Kahana, B., Binder-Brynes, K., Southwick, S.M., Mason, J.W. and Giller, E.L. (1995). Low urinary cortisol excretion in Holocaust survivors with PTSD. *American Journal of Psychiatry* **152**, 245–247.

Yehuda, R., Teicher, R.L., Trestman, R., Siever, L.J. (1996). Cortisol regulation in posttraumatic stress disorder and major depression: a chronobiological analysis. *Biological Psychiatry* **40**, 79–88.

Yehuda, R., Levengood, J., Schmeidler, J., Wilson, S., Guo, L.S., and Gerber, D. (1996b).. Increased pituitary activation following metyrapone administration in PTSD. *Psychoneuroendcrinology* **21**, 1–16.

Yehuda, R. (1997). Sensitization of the hypothalamic-pituitary-adrenal axis in posttraumatic stress disorder. *Annals of the New York Academy of Sciences* **821**, 57–75.

Yehuda, R., McFarlane, A.C., Shalev, A.Y. (1998). Predicting the development of posttraumatic stress disorder from the acute response to a traumatic event. *Biological Psychiatry* **44**, 1305–1313.

Yunus, M.B. (1994). Psychological aspects of fibromyalgia syndrome: a component of the dysfunctional spectrum syndrome. *Baillieres Clinical Rheumatology* **8**, 811–837.

Zant, J.L., Mooij, A., Griep, E.N., and de Kloet, E.R. (1997). Fibromyalgia in relation to process of early attachment. *Nederlands Tijdschrift voo Pijn en Pijnbestrijding* **17e**, 25–27.

43. Plasma Catecholamines, Oxidative Stress and Antioxidant Response During Different Stress Stimuli in Established Essential Hypertension

J. Koška, D. Syrová, P. Blažíček, J.D. Grňa, M. Marko, R. Kvetňanský and M. Vigaš

Introduction

Oxidative stress is considered to be an important etiopathogenetic factor in essential hypertension (Sagar et al., 1992; Parik et al., 1996). Impairment of glucose metabolism and elevation of circulating catecholamines, abnormalities frequently found in hypertensive patients (DeFronzo et al., 1991), have also been reported among the states characterized by increased production of reactive oxygen species (Singal et al., 1983; Paolisso, 1996). Physical exercise represents a physiological model of augmented ROS production, in which increased demands of energy supply for working muscles induces release of catecholamines into the circulation, mobilization of fuel sources, and increased uptake and consumption of oxygen (Li, 1995). On the other side, despite pronounced adrenomedullary responses during another stress stimulus, insulin-induced hypoglycemia, activity of antioxidant enzymes as well as levels of oxidative stress markers were decreased (Koōka et al., 1997). The present study was designed to determine the relationship between catecholamine elevations during two different stress stimuli, physical exercise and hypoglycemia, and the intensity of oxidative damage and the activity of the antioxidant enzymes, superoxide dismutase (SOD) and glutathione peroxidase (GSH-Px), in subjects with established essential hypertension.

Subjects and Methods

Subjects

Eight males with essential hypertension WHO class II (43 ± 9 years, BMI 26.9 ± 0.8 kg/m^2, SBP 149 ± 9, DBP 95 ± 7 mmHg) and nine control subjects (44 ± 4 years, BMI 25.6 ± 2.0 kg/m^2, SBP 119 ± 9, DBP 79 ± 6 mmHg) were studied. Blood pressure lowering drugs were withdrawn at least five days before testing. All subjects gave their informed written consent to participate in the study, which was approved by the ethics committee. Control blood samples and blood pressure measurements were taken at least 30 minutes after an intravenous cannula was placed into the antecubital vein.

Methods

Physical exercise of graded intensity: 1.0, 1.5 and 2.0 W/kg of b.w. was performed on a bicycle ergometer. Each period lasted six minutes followed by a one-minute rest. Blood withdrawal and blood pressure measurements were performed immediately after finishing the exercise, and then after 10, 30 and 60 minutes of rest. Hypoglycemia was induced by intravenous administration of 0.1 IU/kg b.w. of short acting insulin (Actrapid HM, NOVO-Nordisk). Blood samples were taken 30, 45, 60, 90 and 120 minutes after insulin injection. Activities of the antioxidant enzymes studied were measured photometrically using commercial kits (Randox, UK): activity of superoxide dismutase (SOD) in erythrocytes and activity of glutathione peroxidase (GSH-Px) in whole blood. Plasma malondialdehyde (MDA) was determined by HPLC according to the method of Wong et al. (1987). Plasma lipofuscin was measured by the method of Tsuchida et al. (1985). Plasma glucose was analyzed by the oxidase method (Boehringer Mannheim, Ge). Catecholamines were determined by the radioenzymatic method described by Peuler and Johnson (1977). Statistical analysis was performed with the SigmaStat program (Jandel Scientific, Ca, U.S.). All data are expressed as means ± SE and two-tailed values of $p < 0.05$ were considered to be significant.

Results

Increments of blood pressure as well as of heart rate after exercise were similar in normotensives (NT; SBP 62 ± 16, DBP 7 ± 10 mmHg, HR 88 ± 7 bpm) and hypertensives (HT; SBP 74 ± 25, DBP 9 ± 11 mmHg, HR 85 ± 14 bpm). Blood pressure and heart were not changed during hypoglycemia. Glycemia before insulin administration (HT; 5.49 ± 0.56, NT; 5.25 ± 0.42 mmol/l) as well as a decrease of glycemia 30 minutes after insulin injection (HT; 2.64 ± 0.50, NT; 2.73 ± 0.55 mmol/l) were not different between the two groups. Mean steady state values of both catecholamines studied were lower; however, only the difference in norepinephrine concentration before the exercise was statistically significant (Figure 1). A decrease of glycemia induced remarkable increases in epinephrine levels and only a slight increase

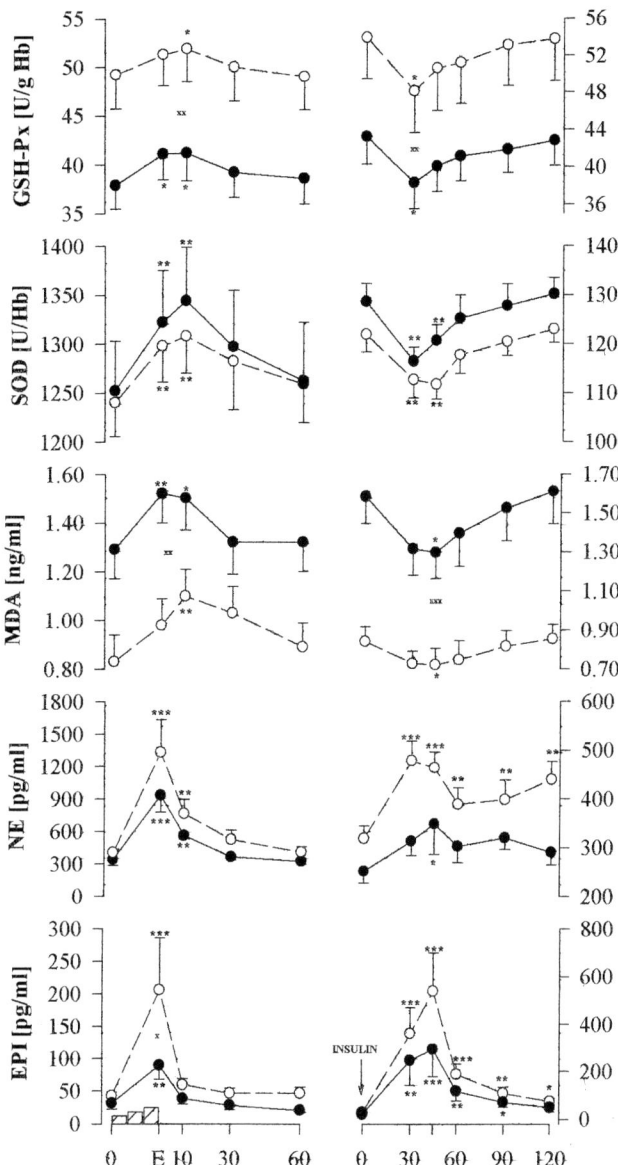

cin (exercise; 16.0 ± 1.8 v. 11.6 ± 2.1 AU, hypoglycemia 16.3 ± 2.3 v. 12.0 ± 2.3 AU, p < 0.001) while their GSH-Px activities were lower (Figure 1). The activity of SOD was similar in the two groups (Figure 1). The concentration of lipofuscin was not significantly changed during the treatments in either group. MDA concentration and the activity of the antioxidant enzymes increased during exercise and decreased during hypoglycemia (Figure 1). The changes in enzymatic activities and MDA levels during exercise were not different between the two groups studied (Figure 1). Decreases in MDA concentrations during hypoglycemia were more pronounced in patients with hypertension (Figure 1).

Discussion

Elevated concentrations of lipoperoxidation products indicate increased oxidative damage in patients with essential hypertension. Determinations of MDA and lipofuscin are used for indirect evaluation of oxidative damage (Tsuchida et al., 1985; Paolisso et al., 1996; Russo et al., 1998). Oxidative stress occurs when the dynamic balance between pro-oxidant and anti-oxidant mechanisms is impaired. In the present study, strenuous physical exercise was associated with increased oxygen utilization, producing several inter-mediates ($\bullet O^-_2$, H_2O_2 and $\bullet OH$) in mitochondria, representing the major source of free radicals (Li, 1995). A number of alternative pathways have also been proposed: (1) xanthine oxidase (XO) catalyzed reaction and (2) neutrophil (PMNL) activation as a result of muscle cell damage. In spite of pronounced sympathoadrenomedullary responses observed during physical exercise, oxidation of catecholamines was proposed as an additional source of ROS (Singal, 1983). However, elevated concentrations of catecho-lamines in essential hypertension were found in studies evaluating a series of earlier publications (Goldstein, 1983). In our investigation evaluating a group of middle-aged patients with established hypertension, concentrations of both catecholamines as well as their increments during exercise and during hypoglycemia were lower in hypertensive patients. Despite the strong adrenomedullary response during insulin-induced hypoglycemia, the concentration of MDA was significantly decreased. The close depend-ence of MDA concentrations in changes in blood sugar was already reported in our previous study, in which an oral glucose load was followed by sig-nificant elevation of MDA levels and antioxidant enzymes activity (Koōka et al., 1997). In the present study, fasting blood glucose levels and insulin-induced decreases were similar in hyper- and normo-tensives. A more pronounced MDA fall in patients with hypertension might have resulted from insulin influences on free fatty acid metabolism (Paolisso et al., 1996).

In both groups, exercise induced similar increases in MDA concentrations as well as of SOD and GSH-Px activities. Russo et al. (1998) found decreased activity of

Fig. 1 Concentrations of epinephrine, norepinephrine, mal-ondialdehyde and activities of antioxidant enzymes (super-oxide dismutase, SOD and glutathione peroxidase, GSH-Px) in 8 patients with essential hypertension and 9 normotensive control subjects during graded exercise on a bicycle erg-ometer (1.0, 1.5 and 2.0 W/kg; each step lasted 6 minutes and was followed by a one-minute rest) and after intravenous insulin administration (0.1 IU/kg). ● – hypertensive subjects; O – normotensive subjects. Means ± SE. *p < 0.05; **p < 0.01; ***p < 0.001 versus baseline (0) in each group. ×p < 0.05; ××p < 0.01, ×××p < 0.01 hypertensive versus normotensive subjects.

in norepinephrine concentrations (Figure 1). Increments of both parameters during the treatments tended to be higher in control subjects, but only the difference in epinephrine values during exercise was statistically significant (Figure 1). Before either treatment, patients with hypertension were found to have significantly elevated concentrations of MDA (Figure 1) and lipofus-

SOD in erythrocytes from hypertensive subjects. In our investigation, after exercise and during hypoglycemia, both steady state values and changes in SOD activity were comparable in the two groups. Hydrogen peroxide produced in the reaction catalyzed by SOD is removed within human cells by the action of two enzymes, catalase and GSH-Px. In our investigation, GSH-Px activity was measured in whole blood. Patients with hypertension were found to have constantly diminished GSH-Px activity. Decreased activity of GSH-Px and increased levels of lipid peroxides were also found in pregnancy complicated by hypertension (Lovero *et al.*, 1996). Normal function of GSH-Px depends on selenium. In patients with essential hypertension either diminished or normal selenium levels (Mihailovic *et al.*, Russo *et al.*, 1998) were accompanied with decreased and elevated activity of GSH-Px, respectively. The former paper reported the most pronounced decrease of selenium concentrations and GSH-Px activity in patients with established hypertension suffering from CHD, while normal selenium levels with increased GSH-Px activity were reported in patients with early hypertension without organ damage (Russo *et al.*, 1998).

In patients with established hypertension, increased oxidative damage might be due to decreases in GSH-Px activity. The ability of antioxidant enzymes to respond to increased production of ROS during short lasting physical exercise was not impaired and responses of the antioxidant enzymes to hypoglycemic stress were not altered.

References

DeFronzo, R., and Ferrannini, E. (1991). Insulin resistance. A multifaceted syndrome responsible for NIDDM, Obesity, Hypertension, and Atherosclerotic cardiovascular disease. *Diabetes Care* **14**, 173–194.

Goldstein D.S. (1983) Plasma catecholamines and essential hypertension. An analytical review. *Hypertension* **5**, 86–99.

Koška, J., Syrová, D., Blažíček, P., Marko, M., Grňa, J.D., Kvetňanský, R., and Vigaō, M. (1997). Antioxidant enzymes activity during hyper- and hypoglycemia in healthy subjects. *Annals of the New York Academy of Science* **827**, 575–579.

Li, L.J. (1995). Oxidative stress during exercise: Implication of antioxidant nutrients. *Free Radicals in Biology and Medicine* **18**, 1079–1086.

Loverro, G., Greco, P., Capuano, F., Carone, D., Cormio, G., and Selvaggi, L. (1996) Lipoperoxidation and antioxidant enzymes activity in pregnancy complicated with hypertension. *European Journal of Obstetrics and Gynecology and Reproductive Biology* **70**, 123–127.

Mihailovic, M.B., Avramovic, D.M., Jovanovic, I.B., Pesut, O.J., Matic, D.P., and Stojanov, V.J. (1998). Blood and plasma selenium levels and GSH-Px activities in patients with arterial hypertension and chronic heart disease. *Journal of Environmental Pathology Toxicology and Oncology* **17**, 285–289.

Paolisso, G., and Giugliano, D. (1996). Oxidative stress and insulin action: is there a relationship? *Diabetologia* **39**, 357–363.

Parik, T., Allikmets, K., Teesalu, R., and Zilmer, M. (1996). Oxidative stress and hyperinsulinaemia in essential hypertension: different facets of increased risk. *Journal of Hypertension* **14**, 407–410.

Peuler, J., and Johnson, G.A. (1977). Simultaneous single isotope radio-enzymatic assay of plasma norepinephrine, epinephrine and dopamine. *Life Sciences* **21**, 625–638.

Russo, C., Olivieri, O., Girelli, D., Faccini, G., Zenari, M.L., Lombardi, S., and Corrocher, R. (1998). Anti-oxidant status and lipid peroxidation in patients with essential hypertension. *Journal of Hypertension* **16**, 1267–1271.

Sagar, S., Kallo, I.J., Kaul, N., Ganguly, N.K., and Sharma, B.K. (1992). Oxygen free radicals in essential hypertension. *Molecular and Cellular Biochemistry* **111**, 103–108.

Singal, P.K., Beamish, R.E., and Dhalla, N.S. (1983). Potential oxidative pathways of catecholamines in the formation of lipid peroxides and genesis of heart disease. *Advances in Experimental Medicine and Biology* **161**, 391–401.

Tsuchida, M., Miura, T., Mizutani, K., and Aibara, K. (1985). Fluorescent substances in mouse and human sera as a parameter of *in vivo* lipidperoxidation. *Biochimica et Biophysica Acta* **834**, 196–204.

Wong, S.H., Knight, J.A., Hopfer, S.M., Zaharia, O., Leach, C.H., Jr., and Sunderman, F.W., Jr. (1987). Lipoperoxides in plasma as measured by liquid-chromatographic separation of malondialdehyde-thiobarbituric acid adduct. *Clinical Chemistry* **33**, 214–220.

44. Neuroendocrine Responses to Passive and Active Hyperthermia in Healthy Men

L. Kšinantová, J. Koška, J. Čelko, Doležal, J. Rovenský, R. Kvetňanský, D. Ježová and M. Vigaš

Introduction

Hyperthermia seems to play a significant role in the generation of signal(s) for hormone release during exercise. It is generally accepted that impulses for neuroendocrine activation during dynamic exercise are generated in motor centers in the brain (forward regulation) and in receptors (thermo-, osmo-, baro- and other) located in the working muscles or in other peripheral tissues (feedback regulation). Central regulation operates mainly during short-term exercise; however, there are some differences regarding specific hormonal responses (Kjaer et al., 1989). Exercise-induced hyperthermia may be considered an important regulatory factor and the existence of a thermal threshold for stimulation of hormone release during exercise has been suggested (Radomski et al., 1998). Most authors assume that the main regulatory input is an increase in core temperature (Weiss et al., 1988; Christensen et al., 1984; Dore, 1991).

A rise in core temperature may be achieved not only by increased heat production but also by passive heat exposure. Exogenously induced hyperthermia is known to be followed by neuroendocrine activation (Moller et al., 1989; Ježová et al., 1994). Radomski et al., (1998) have suggested that thermal responses to exercise may be causally related to the induction of hormone release. We have hypothesized that if active hyperthermia induced by working muscles was the triggering factor for neuroendocrine activation, active and passive hyperthermia of similar intensity would provoke similar changes in hormone release. The present investigation was designed to verify this hypothesis by measuring changes in hormone levels in response to hyperthermia induced by passive head-out immersion or active swimming in healthy men.

Subjects and Methods

Subjects

Eighteen healthy male volunteers (age 22.8 ± 3.5 years, BMI $22.9 \pm 2.3 \, \text{kg/m}^2$, height $180 \pm 6 \, \text{cm}$, weight $74.2 \pm 9.6 \, \text{kg}$) gave their informed written consent to participate in the study, which was approved by the Ethical Committee of the Research Institute of Rheumatic Diseases, Piestany, Slovak Republic. Starting the evening before the day of investigation, subjects were asked to fast, to refrain from alcohol and tobacco and to keep their physical activity at a minimum level. All investigations were performed in the morning.

Passive Hyperthermia

After inserting an indwelling catheter into a cubital vein, termistors of an electronic thermometer (Miniterm, Chirana, Czechoslovakia) were placed into the auditory canal and on the dorsal antebrachial skin of both arms. Thereafter the subjects were allowed to rest in a horizontal position covered with a sheet in a room with an ambient temperature of $26°C$ and humidity of 50% for at least 30 minutes prior to control blood sample collection. This trial was performed with nine volunteers.

Each subject underwent head-out immersion in water at $39–40°C$ in a semisitting position for 25 minutes, with the exception of the arm in which the catheter was placed. Temperature was measured every 5 minutes, starting immediately before entering the bath. In addition, sublingual temperature was measured using a standard mercury thermometer. Blood was taken before and at the end of the bath, as well as after 15 min of rest in a horizontal position. At the same time intervals, heart rate was measured using an auscultatory method.

Active Hyperthermia

Nine physically fit male athletes (wrestlers) without special training in swimming participated in the investigation. After 30 min of rest in a comfortable sitting position, heart rate and sublingual temperature was measured and blood samples were withdrawn for determination of control values of hormones. The subjects participated in two swimming exercise tests in the pool with water of 29 or $36°C$ in a randomized order separated by 8–11 days. They swam 22 min free style at a self-selected comfortable pace, followed by 3 min free style at maximal pace under verbal encouragement by the trainer. At the end, subjects remained sitting in the pool until measurements were performed and blood samples were withdrawn. Then the volunteers rested for 15 min in a horizontal position covered with a blanket and finally the last measurements were performed.

Hormone Analyses and Statistical Evaluation

Blood was collected into cooled polyethylene tubes using heparin as an anticoagulant. The samples were centrifuged at $4°C$, and after separation the aliquots of plasma were stored frozen (–70°C) until analyzed. All samples of one experiment were run in the same assay. Epinephrine and norepinephrine were analyzed by a radioenzymatic method (Peuler and Johnson, 1977), and GH and PRL by RIA (Immunotech Prague). Two-way ANOVA for repeated measures with consecutive post-hoc tests was used to determine the differences from baseline within each group and the differences among the treatments.

Results

Passive head-out immersion in water at 39°C led to an increment of body temperature (1.2°C) which was similar to that after swimming at 36°C (+1.3°C). Significantly lower increase of body temperature was found after swimming in water at 29°C (+0.8°C). After vigorous swimming, the rise in heart rate was the same in water of both temperatures. Passive immersion in water at 39°C led to a significantly lower increase of heart rate (Figure 1).

Exercise-induced norepinephrine and epinephrine responses were similar – higher increases were observed after swimming in warmer water, but the differences failed to be significant. Head-out immersion was followed by only a mild rise of plasma

catecholamines and the levels were significantly lower compared to high hormone concentrations induced by swimming. Growth hormone release was significantly increased in all situations studied, with highest levels after swimming in warmer water and lowest ones after head-out immersion. The difference between the levels after swimming in water of 36°C and those after immersion was statistically significant, in spite of the same increments in body temperature. Plasma prolactin was unchanged after swimming in 29°C water. The highest concentrations were found after swimming in 36°C water and the values were significantly higher compared to those after head-out immersion (Figure 1).

Discussion

These studies show that hyperthermia induced by passive immersion and by active short-term exercise is associated with different changes in hormone release. These results suggest that mechanisms other than increases in body temperature play a dominant role in exercise-induced neuroendocrine activation.

Muscular activity during workload of sufficient intensity increases the metabolic rate to produce heat and an elevation in core temperature. The body maintains thermoregulatory homeostasis by increased heat dissipation, the effectiveness of which depends on internal (circulation, hydration) and external factors (environmental humidity and temperature). Intensive hyperthermia induces secretion of several hormones (Moller et al., 1989; Ježová et al., 1994) and the rise in body temperature during exercise is considered to be one of the factors responsible for triggering hormone release (Melin et al., 1988; Brisson et al., 1987). Exercise-induced hyperthermia was reported to be a regulatory factor also for some other physiological systems, e.g. the responses induced by hyperthermia alone were very similar to exercise-induced changes in immune function (Brenner et al., 1995). Radomski et al. (1998) have suggested that exercise-induced hyperthermia actually initiates catecholamine and pituitary hormone secretion, providing a certain threshold in the core temperature is exceeded. The thermal threshold was estimated to be approximately +0.6°C. The results of the present study failed to confirm this suggestion. Increases in body temperature after immersion in hot water and after swimming in 36°C water were very similar (+1.2 and +1.3°C, respectively), but the responses of norepinephrine, epinephrine, growth hormone and prolactin were significantly higher after swimming. Similarly, a different pattern of hormone release was observed after swimming in water at 29°C, even though the rise in core temperature (+0.8°C) exceeded the estimated thermal threshold also in this case. These results do not exclude the possibility that exercise-induced hyperthermia is a stimulus for hormone release, but they do not support a dominant role for

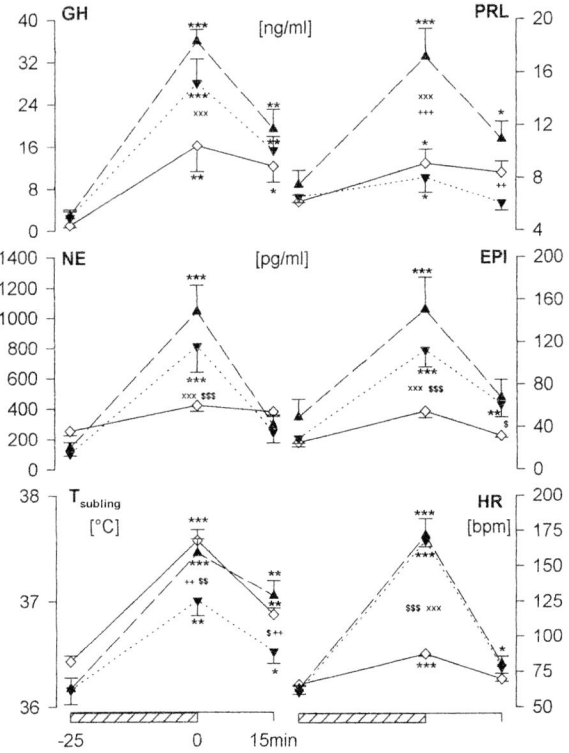

Fig. 1 Body temperature (T), heart rate (HR), norepinephrine (NE), epineprine (EPI), growth ormone (GH) and prolactin (PRL) during 25 min immersion in water of 39°C (—◇—) or during swimming in water of 29°C (---▼---) or 36°C (– –▲– –). Mean ± S.E. Significance: *v 0 min within treatment; ×◇ v ▼; $◇ v ▲; +▲ v ▼.

increases in body temperature in the control of neuroendocrine activation during short-term exercise.

In physically fit subjects, short term swimming with maximal effort induced the same increase in heart rate in water of both temperatures, indicating a similar intensity of the exercise. However, exercise-induced release of prolactin observed after swimming in the warmer water was absent when the test was performed in the cooler water (29°C). The responses of norepinephrine, epinephrine and growth hormone to swimming in the cooler water were attenuated in comparison with those in water at 36°C. These results suggest that increased body temperature could be an impulse enhancing the neuroendocrine response to other stimuli as has been demonstrated previously in studies using insulin-induced hypoglycemia (Ježová et al., 1998).

In conclusion, passive and active hyperthermia do not induce the same changes in hormone release, suggesting the involvement of different mechanisms. Nonetheless, increases in body temperature may be an important factor modulating neuroendocrine responses to exercise.

References

Brenner, I.K.M., Shek, P.N., and Shephard, R.J. (1995). Heat exposure and immune function: potential contributions to the exercise response. *Exercise Immunological Review* **1**, 49–80.

Brisson, G.R., Bouchard, J., Peronnet, F., Boisvert, P., and Garceau, F. (1987). Evidence for an interference of selective face ventilation on hyperprolactinemia induced by hyperthermic treadmill running. *International Journal of Sports Medicine* **8**, 387–391.

Christensen, S.E., Jorgensen, O.L., Moler, N., and Orskow, H. (1984). Characterization of growth hormone release in response to external heating. Comparison to exercise induced release. *Acta Endocrinologica* **107**, 295–301.

Dore, S., Brisson, G.R., Founier, A., Montpetit, R., Perrault, H., and Boisvert, D. (1991). Contribution of hGH20K variant to blood hGH response in sauna and exercise. *European Journal of Applied Physiology* **62**, 130–134.

Ježová, D., Kvetňanský, R., and Vigas, M. (1994). Sex differences in endocrine response to hyperthermia in sauna. *Acta Physiologica Scandinavica* **150**, 293–298.

Ježová, D., Kvetňanský, R., Nazar, K., and Vigaō, M. (1998). Enhanced neuroendocrine response to insulin tolerance test performed under increased ambient temperature. *Journal of Endocrinological Investigation* **21**, 412–417.

Kjaer, M., Secher, N.H., Bach, F.W., Sheikh, S., and Galbo, H. (1989). Hormonal and metabolic responses to exercise in humans: the effect of sensory nervous blockade. *American Journal of Physiology* **257**, 95–101.

Melin, B., Cure, M., Pequignot, J.M., and Bittel, J. (1988). Body temperature and plasma prolactin and noeepinephrine relationships during exercise in a warm environment: effect of dehydration. *European Journal of Applied Physiology* **58**, 146–151.

Moller, N., Beckwith, R., Butler, P.C., Christensen, N.J., Orskov, H., and Alberti, K.G.M.M. (1989). Metabolic and hormonal responses to exogenous hyperthermia in man. *Clinical Endocrinology* **30**, 651–660.

Peuler, J., and Johnson, G.A. (1977). Simultaneous single isotope radio-enzymatic assay of plasma norepinephrine, epinephrine and dopamine. *Life Sciences* **21**, 625–636.

Radomski, M.W., Cross, M., and Buguet, A. (1998). Exercise-induced hyperthermia and hormonal responses to exercise. *Canadian Journal of Physiology and Pharmacology* **76**, 547–552.

Weiss, M.F., Hack, F., Stehle, R., Pollert, R., and Weicker, H. (1988). Effect of temperature and water immersion on plasma catecholamines and circulation. *International Journal of Sports Medicine* **9**, 113–117.

45. Seventh Symposium on Catecholamines and other Neurotransmitters in Stress Smolenice Castle, Slovak Republic June 28–July 3, 1999

Symposium Summary – Richard McCarty

The 7th Symposium at Bratislava Hrad begins,
Kvetňanský has organized another big win!
Palkovits and Vivaldi – the head and the heart,
What better combination of science and art.

On day 1 at Smolenice, all are awake.
In saying this I hope there is no mistake.
Our first speaker, Chris Lowry, searched for 5-HT cells
That would respond to CRH by ringing their bells.
Jim Herman came to Smolenice without his bags,
But he brought PVN slides and a bit of jetlag.
Shame on Air France for not bringing Jim's pants!
Pain stress has never been such a delight,
Until you hear Palkovits present his insights.
Pacak has switched from immobilization stress,
Now he studies Cushing's but he remains always a Czech.
Claire Stanford studies stress and brain NA,
With lots of swords and snowflakes along the way.
Daniela described her triangular view,
With catecholamines, glutamate and stress, proving 3 is better than 2.
Almeida introduced the neurosteroid, THP,
Effects on males and females and the HPA we did see.
Following lunch we heard from the Bohus group,
Food intake and stress studies, Bela kept us in the loop.
Marta Weinstock laid out the importance of gender,
Of stress and catecholamines her results she did render.
When Zofia gave her talk on, of course, NPY,
It was easy to see why this peptide is the apple of her eye.
Dallman's Law was revealed after the break,
Prior stress equals controls – now that's a mistake!
Greti's discussion of transcription and translation
Gave new meaning to stress effects on CRH and vasopressin.
Her talk on message stability was ever so bright,
We all felt like we had seen the Light-man.
Did I say Lightman – he of pulses of B,
A more elegant technique, we will never see!
Last but not least, we heard of ulcers and stress.
After Gabor finished, we all needed a rest.

We begin day 2 after an evening of Slovak barbeque.
Esther Sabban studied adrenal as well as LC,
And looked at CREB, fos and jun with a great deal of glee.
Of super shifts and transgenes, we heard from Bill Tank.
For clarifying TH regulation, we owe Bill our thanks.

Professor Nagatsu studied GTP cyclohydrolase-one,
With PCR and clones, his lab has lots of fun.
Novel stressors are important in Kvetňanský's view.
Wonderful for him – but the rats all turn blue.
Dona Wong is the Queen of PNMT.
Her regal approach is really something to see.
Steve Thomas has developed DBH knockout mice,
But many of them die, paying the ultimate price.
Karalis has worked hard with CRH knockouts,
They have blunted adrenal responses, explaining her shouts.
CRH binding protein – now that's a lot to say,
But Seasholtz presented results that lead the way.
Krizanova has gone down a different road,
Her work on renin and heart was a most impressive load.
Stress hyporesponses in the rat neonate,
Who else but Gig Levine to bring us up to date?
And there was Plotsky with his pups that were deprived,
His model is excellent – certainly not contrived.
From young rats to old ones was Tumer's refrain,
Her message was clear – being an old rat is a pain!
We ended the day in a wonderful way,
Serova had her say, catecholamines, stress and aging she did portray.

After fireworks, birds of prey and a garden party,
We begin with stress and the immune system on meeting day 3.
Fred Tilders begins with HPA priming,
Of AVP and norepi, his results point to timing.
Milk intake in mice, IL-1 knocks it down,
Dr. Dunn tried to block it with antagonists but had a big frown.
Shawn studied sensitization by a cytokine.
How can just one dose change things for such a long time?
Dr. Kubo restrained mice as his model to test,
And examined lymphocyte subpopulations as his measure of stress.
David Jessop has been busy in Bristol, U.K.,
Studying neuropeptides and immune system by night and by day.
Hellhammer discussed his studies on human stress and coping,
For burnout and trauma, we can at least keep on hoping.

After grilled chicken and fish and way too much wine,
Would anyone hear DeSouza and arrive on time?
Day 4 started off with a large audience,
And Errol was at his best with CRH receptors and splice variants.
Of stress and AVP receptors, we heard an appeal,
And learned quite a lot from Dr. Rabadan-Diehl.
Steroid receptors, stress and anxiety,
Bohus explained the unholy 3!
Dave Goldstein convinced us that all stress is not the same,
Clinical neurocardiology is his special domain.
Jacque Lenders explained about metanephrines and stress,
His studies in humans are among the very best.
The final talk of the symposium was vintage Phil Gold,
He talked about CRH and depression and then all had been told.

As we near the end of this symposium and we head home for a rest,
I leave you with one thought – Smolenice is the best!

Keyword index

Author Index